COMPUTER PROGRAMS FOR CALCULUS TECHNIQUES AND APPLICATIONS

Program Name	Purpose	Chapter Reference
TABLE	To construct a table of function values	1
GRAPH*	To graph a function	1
LIMITS	To approximate limits of functions	2, supplement 2
DERIV	To approximate the derivative at a point	2
NEWTON	To approximate the roots of functions	3
GOLDEN	To approximate maximum/minimum points	3
INTEGRAT	To evaluate integrals by rectangular approximation	5
TRAPEZD	To evaluate integrals by trapezoidal approximation	5
DIFFEQ	To approximate solutions to differential equations	7
MULTIVAR	To approximate maximum/minimum points of multivariable functions	8
PROB	To compute expected value and variance of a discrete random variable	11

* This program requires color/graphics capability.

AN INTRODUCTION TO CALCULUS

METHODS AND APPLICATIONS

AN INTRODUCTION TO CALCULUS

METHODS AND APPLICATIONS

JAMES R. EVANS
University of Cincinnati

CHARLES W. GROETSCH
University of Cincinnati

MARYANNE WALKER

WEST PUBLISHING COMPANY
St. Paul   New York   Los Angeles   San Francisco

Copyediting: Phyllis Niklas
Design: John Edeen
Artwork: Vantage Art
Composition: Progressive Typographers
Cover art: "Italian Town" by Paul Klee. Copyright © COSMOPRESS, Geneva/
     ADAGP, Paris/VAGA, New York, 1985.
Cover design: The Quarasan Group

COPYRIGHT © 1986 By WEST PUBLISHING COMPANY
50 West Kellogg Boulevard
P.O. Box 64526
St. Paul, MN 55164-1003

Printed in the United States of America

Library of Congress Cataloging-in-Publication Data

Evans, James R. (James Robert), 1950–
   An introduction to calculus.

   Includes index.
   1. Calculus. I. Groetsch, C. W. II. Walker,
MaryAnne. III. Title.
QA303.E86   1986   515   85-22804
ISBN 0-314-93176-7

TO TONY AND RITA
TO THE MEMORY OF KENNY AND RAY
TO DUANE

CONTENTS

The purpose of this book is to provide to students of business administration, economics, the social and life sciences, and other relevant disciplines (such as computer science and information systems) an introduction to calculus from a practical, problem solving point of view. Methods and concepts of calculus have many applications in these disciplines as well as in the world around us. With this in mind, we emphasize applications drawn from all these areas and, wherever possible, from everyday situations to which undergraduate students from different disciplines can easily relate. Each chapter begins with an opening problem solving scenario that cannot be adequately solved until one of the major techniques of that chapter has been developed; in the conclusion of each chapter we present the solution. This feature is used to provide motivation and to emphasize the application of otherwise abstract concepts. While the use of calculus, and mathematics generally, in problem solving is an important theme throughout this text, this is not a book on mathematical modeling. Our primary objective is to provide a correct, yet intuitive introduction to calculus.

IMPORTANT FEATURES

Several other features of this text should be pointed out:

1. In teaching this material to students with nonmathematical majors, we have found that a major source of frustration to these students is the difficulty in mathematically formulating word problems. These arise principally in optimization applications of the derivative. We recognize that setting up these problems is a skill acquired through practice and experience. To provide the practice needed, we have included an introductory chapter focusing on modeling, and we have also provided detailed explanations throughout other parts of the text. This material extends and reinforces previous mathematics courses, and provides a transition to future quantitative methods courses which these students will likely encounter.

2. A second source of frustration to nonmathematically oriented students relates to basic algebraic skills. Students need review and reinforcement of such skills to solve problems in calculus. Throughout the text we have strategically placed "Algebra Reviews," which serve to reinforce these skills *as needed* for the text material. Brief review exercises (with answers given in the back of the book) are also included.

3. We have used chapter supplements to provide a significant amount of flexibility to the needs of individual instructors. Many advanced topics such as proofs of differentiation formulas, additional integration techniques, and peripheral topics such as multiple integration have been put in supplements for use at the instructor's discretion.

Any or all chapter supplements can be skipped without loss of continuity of the main text material.

4. Over 350 worked-out examples are presented in detail in the text, and a total of over 2200 problems are given at the end of each section to assist students in learning new concepts and techniques.

5. A chapter summary appears at the end of each chapter.

6. A glossary of new terms (boldface type in the text) can be found at the end of the book.

7. Because of the growing importance of computers and the introduction of microcomputers into curricula, we have included an optional final chapter on computers and calculus. In this chapter we present numerical methods for illustrating concepts of calculus and for problem solving. These are keyed to specific sections in the text and can be used as supplements throughout the text if desired. Diskettes with all programs for the IBM PC, compatibles, and Apple computers are available to adopters.

8. Answers to odd-numbered problems are found at the end of the book. These have been independently checked for accuracy to minimize errors.

COMMENTS ON THE USE OF THIS TEXT

The organization of this text is shown in the chart on the facing page.

Chapter 1 is a "warm-up" chapter that covers the development of mathematical models from word statements, and reviews basic precalculus material such as functions, graphs, and applications of linear and nonlinear functions. For students with appropriate backgrounds, this can easily be skipped or briefly reviewed. The material on setting up word problems, however, is recommended prior to covering optimization applications of the derivative.

Chapter 2 and its supplement introduce the notion of limits and the derivative. For a basic introduction to differentiation without an in-depth treatment of limits, Chapter 2 can be used alone. For additional treatment of limits and their applications, the supplement can be covered after the basic ideas in Chapter 2 have been presented, or after Section 2-1.

Chapter 3 introduces optimization applications of the derivative and curve-sketching techniques. Advanced methods of differentiation (the product rule, quotient rule, etc.) are not covered until Chapter 4. While this limits the mathematical scope of functions that can be treated in this chapter, our goal is to introduce *fundamental concepts* without complicating them by computational details. In Chapter 4, however, we *reinforce* these concepts through additional examples of optimization and curve sketching which use the more advanced rules of differentiation. A special section devoted to applications to business and economics is included in Chapter 3.

In Chapter 5 and its supplement we introduce concepts and techniques of integration. We place special emphasis on the interpretation of the integral in practical problems and on the use of the fundamental theorem of calculus. A separate section on business and economics applications is included.

Chapter 6, "Exponential and Logarithmic Functions," is placed after the integration chapter. Again, our goal is to introduce the fundamental concepts of integration without having to consider numerous special cases as would be

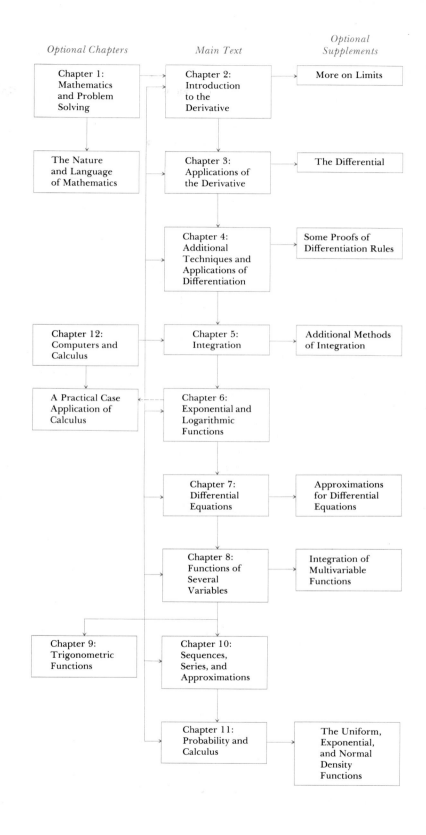

Optional Chapters   *Main Text*   *Optional Supplements*

necessary if this material were introduced earlier. However, in Chapter 6, the student has a chance to see both differentiation and integration together for specific classes of functions and can thereby appreciate the relationships between them.

Chapter 7, "Differential Equations," is concerned with the single most important tool for modeling processes that change continuously with time. In this chapter we treat only two broad classes of differential equations, but we emphasize the model building process and the extraordinary applicability of differential equations. The reader is constantly reminded to seek out analogies between models in such diverse fields as biology, physics, and economics.

Chapter 8 treats functions of several variables. A wide variety of real-life examples are used to illustrate the importance of such functions in applications. We also provide a basic treatment of the geometric interpretation of three-dimensional functions. In the final section on Lagrange multipliers, we also place emphasis on the interpretation of the Lagrange multiplier in problem solving. Multiple integration is discussed in the supplement.

Chapter 9 deals with the calculus of trigonometric functions. These are used to model cyclical events and processes that involve rotation. Since most of the applications in this chapter involve engineering and physical sciences, many instructors will treat this chapter as optional.

In Chapter 10, sequences and series, including Taylor series, are introduced. This chapter also includes two methods for generating sequences that approximate the root of an equation.

Probability and calculus is the topic of Chapter 11. Some of the material on sequences and series is put to use in studying discrete random variables. Integration is applied to investigate continuous random variables, and the fundamental theorem of calculus is interpreted in terms of density functions and cumulative distribution functions.

Chapter 12, "Computers and Calculus," is purely optional. This chapter can be used by instructors with access to microlabs or electronic classrooms to illustrate numerical methods and discuss implementation details about problem solving, or to further illustrate important concepts. Programs are included which can be used to demonstrate calculus concepts as well as numerical methods to solve problems that cannot be solved analytically or may be very tedious to solve without a computer. Each section of this chapter is keyed to a corresponding section of the text; this correspondence is summarized in an opening table. Throughout the text, a margin logo is used to designate the program in Chapter 12 that can be used to supplement the material. For example:

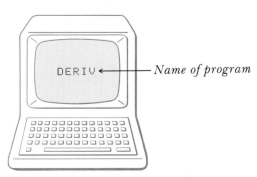

Name of program

This text can be used for a three-quarter or two-semester course sequence. For shorter sequences covering only basic topics in calculus, the authors' text *Fundamentals of Calculus: Applications to Managerial, Social, and Life Sciences,* is recommended. Possible course outlines are given below. (Instructors who wish to discuss computer-related subjects may use selected sections of Chapter 12 as appropriate.)

Two-Semester Course
First Semester

Chapter 1:   Mathematics and Problem Solving

Supplement: The Nature and Language of Mathematics

Chapter 2:   Introduction to the Derivative

Supplement: More on Limits

Chapter 3:   Applications of the Derivative

Chapter 4:   Additional Techniques and Applications of Differentiation

Chapter 5:   Integration

Supplement: Additional Methods of Integration

Chapter 6:   Exponential and Logarithmic Functions

Second Semester

Chapter 7:   Differential Equations

Supplement: Approximations for Differential Equations

Chapter 8:   Functions of Several Variables

Supplement: Integration of Multivariable Functions

Chapter 10:   Sequences, Series, and Approximations

Chapter 11:   Probability and Calculus

Supplement: The Uniform, Exponential, and Normal Density Functions

Some instructors may choose to cover Chapter 9, "Trigonometric Functions," in place of Chapter 11 and its supplement.

Three-Quarter Course
First Quarter

Chapter 1:   Mathematics and Problem Solving

Supplement: The Nature and Language of Mathematics

Chapter 2:   Introduction to the Derivative

Supplement: More on Limits

Chapter 3:   Applications of the Derivative

Supplement: The Differential

Chapter 4:   Additional Techniques and Applications of Differentiation

Supplement: Some Proofs of Differentiation Rules

Second Quarter	Chapter 5: Integration
	Supplement: Additional Methods of Integration
	Chapter 6: Exponential and Logarithmic Functions
	Chapter 7: Differential Equations
	Supplement: Approximations for Differential Equations
Third Quarter	Chapter 8: Functions of Several Variables
	Supplement: Integration of Multivariable Functions
	Chapter 10: Sequences, Series, and Approximations
	Chapter 11: Probability and Calculus
	Supplement: The Uniform, Exponential, and Normal Density Functions

Some instructors may choose to cover Chapter 9, "Trigonometric Functions," in place of Chapter 11 and its supplement.

ANCILLARIES

1. A *Student Solutions Manual* with detailed solutions for all odd-numbered problems is available from West Publishing Company.
2. An *Instructor's Manual* with detailed solutions to even-numbered problems and additional instructional notes is available to instructors.
3. A diskette with all programs contained in Chapter 12 is available to adopters for the IBM PC and its compatibles or Apple computers.
4. A computerized test-generation system is available to adopters for the above-listed microcomputers.

For the program diskette and a demonstration diskette of the test-generation system, please contact your local sales representative or contact West Publishing Company in St. Paul, Minnesota.

ACKNOWLEDGMENTS

We would like to express our sincere appreciation to the following reviewers, who read, critiqued, and provided many wonderful and invaluable suggestions for this text:

Fred Brauer
University of Wisconsin, Madison

Richard Davitt
University of Louisville

John Busovicki
Indiana University of Pennsylvania

Russ Diprizio
Oakton Community College

Samuel Councilman
California State University at Long Beach

Chaitan P. Gupta
Northern Illinois University

Kevin Hastings
University of Delaware

Robert Hunter
The Pennsylvania State University

James Jamison
Memphis State University

Anne Landry
Dutchess Community College

Deborah Frank Lockhart
Michigan Technological University

Stanley Lukawecki
Clemson University

Richard Marshall
Eastern Michigan University

Philip Montgomery
University of Kansas

Robert Moore
University of Wisconsin, Milwaukee

K. L. Nielsen
Butler University

Wayne Powell
Oklahoma State University

Thomas Roe
South Dakota State University

Robert Russell
West Valley College

Donald Shriner
Frostburg State College

Clifford Sloyer
University of Delaware

Malcolm Soule
California State University,
Northridge

John Spellman
Southwest Texas State University

Thomas Wilcox
University of Wisconsin, Whitewater

Ray Wilson
Central Piedmont Community
College

Thomas Woods
Central Connecticut State
University

Throughout our writing and production efforts we were blessed with truly remarkable editors. Jerry Westby's relentless dedication to this project and his creative insight were a source of constant encouragement and continued amazement. We thank Barb Fuller for guiding production activities and enabling the book to be completed on schedule. A very special note of appreciation goes to Phyllis Niklas for her meticulous attention to detail, and an absolutely, positively outstanding job in editing. Finally, we thank our families for their understanding during the countless hours we spent on this project.

TO THE STUDENT

If you are a student of business, economics, the social sciences, or the life sciences, you will probably not use calculus as part of your daily routine. However, this does not mean that the study of calculus and other mathematical subjects will not be useful in your future profession. Throughout this book we will see numerous examples of realistic situations where concepts of calculus are necessary for problem solving. On a more general level, everyone uses mathematics to some extent — for balancing a checkbook, presenting a graph of sales estimates, analyzing psychological or biological data, and so on. Mathematics is the understanding of structure and relationships, the development of theories and their application to real problems. Every discipline has its own structure, relationships, theories, and applications. To be successful in any profession, you have to be able to understand these structures and relationships, perhaps develop your own theories, and then apply them to situations that arise. Mathematics provides a framework for developing these thought processes that are so crucial to career success.

Many students view mathematics as a difficult subject. Everyone has different study habits, and often, this perceived difficulty is a result of poor study habits. Below, we present a few guidelines for studying the material in this book. We hope these guidelines will help you to learn mathematics more easily.

First and foremost, read the text! We have attempted to provide complete and detailed narrative explanations of the theorems and statements in each chapter. If you read only the examples, you will be missing a large part of this book.

Second, read the examples carefully. Work through each step with a pencil and paper and try to understand why each step is performed.

Third, do some problems at the end of each section. Mathematics can be learned only by working problems — lots of them. For many of the problems, an inexpensive calculator will be very useful. The problems in this book are generally graduated in difficulty. Check your answers in the back of the book. If you are having trouble, reread the text section and examples pertaining to those problems you are having difficulty with and *try again!* Don't give up too soon!

Fourth, after a chapter is completed, read the chapter review. That section highlights the important ideas that you should have learned in the chapter and provides a good outline of topics to know for exams. Work as many of the review problems at the end of the chapter as you can without referring to the text. If you feel comfortable doing this, then you have probably mastered the material quite well.

1

MATHEMATICS AND PROBLEM SOLVING

INTRODUCTION

Mathematics plays an important role in many disciplines. Problems faced by managers of business organizations, by physical and social scientists, and by all of us in our daily lives can often be effectively analyzed using mathematics. Consider the following situations:

A. The Forest Hills Township Fire Department is considering the purchase of a new rescue vehicle. The cost of a fully equipped vehicle is estimated to be $24,000. The average operating cost is determined to be $0.60 per mile. What are the total projected costs if the vehicle is driven 70,000, 80,000, or 90,000 miles?

B. A biologist has experimental data indicating that the population of a bacteria culture after t hours of growth is given by the equation $y = 2t^3 + 100$, where y is the number of bacteria (in thousands). How fast has the population grown during the third hour?

C. A manufacturer wishes to design an open cardboard box by cutting small squares out of the corners of a 28×28-inch piece of cardboard and folding up the sides. What should the dimensions of the box be in order to have the largest capacity?

D. An analyst at a state health agency knows that the spread of a flu virus occurs at the rate of

$$y = \frac{1000}{1 + 0.5e^{-0.6t}}$$

where t is in hours, y is the number of people per hour affected, and e is a constant that approximately equals 2.71828. After 10 hours, how many people are affected?

These four problems each require the use of mathematics to obtain an answer. Though they are similar in this respect, they are remarkably different. In the conclusion to this chapter (Section 1-6) we shall see that the first two problems can be answered by applying some techniques of algebra that you have previously learned. The third and fourth problems, however, cannot be solved by traditional algebraic methods. In fact, it was not until the end of the 17th century that Sir Isaac Newton (1642–1727) and Gottfried Leibniz (1646–1716) independently developed the branch of mathematics that we call *calculus*. Their motivation for this development had its roots in astronomy and other physical sciences. The study of the motion of the planets and other celestial bodies by such scientists as Copernicus (1473–1543), Kepler (1571–1630), and Galileo (1564–1642) led to many problems that could not be solved by the

mathematical methods available at that time. The work of Newton and Leibniz paved the way for the solution of many important problems. Without calculus, the technical achievements of the 20th century —the space program, modern communications, computers, and many other scientific and engineering advances—would not exist as we know them.

Although mathematics has always assumed a central role in the physical sciences, it is of greater importance than ever today in the "softer" sciences such as psychology, sociology, economics, and business. The practical uses of mathematics in business, for example, have resulted from the need to make more accurate and timely decisions involving large amounts of complex information. The impact of the computer has been unparalleled in furthering the application of mathematics in these areas.

Before starting the actual study of calculus, we will review some of the important tools and concepts that are used extensively throughout this book. Of particular importance is the process of setting up applied word problems for mathematical solution, the notion of a function, and the process of graphing equations and functions. These techniques will enable us to develop and use calculus more readily. Also, we will have the opportunity to see many of the diverse applications of mathematics in today's world.

1-1 THE PROBLEM SOLVING PROCESS

Problems that require a mathematical solution often begin with rather vague verbal statements of real situations. Problem C listed in the Introduction is an example. The manufacturer is faced with the problem of designing a box having the largest capacity. Of course, one approach to solving this problem is trial and error. The manufacturer could experiment by cutting out different sized corners from the cardboard and comparing the volumes of the resulting boxes. This approach would be quite tedious (and would waste a lot of cardboard!); also, there would be no assurance that the best solution had been found. By representing the problem in a mathematical fashion, we would reduce the amount of effort spent in solving the problem and also be able to obtain the best solution.

In order to analyze a problem using techniques of mathematics, we need to express the problem in mathematical form.

> **Mathematical modeling** is the process of translating a problem statement into a mathematical representation.

After a mathematical model is set up, various techniques are used to analyze and/or solve the model. Finally, the results obtained must be examined in light of the real problem to determine how they can be used. This process is illustrated in Figure 1-1. It is important to realize that in all situations, the model we develop is not an exact representation of the real problem but only an abstraction. The value of the mathematical model is related to how well it describes the real problem; the closer the model is to reality, the more useful the results will be. The study of the relationships between models and reality is a topic for a course in modeling, not calculus. However, to understand and use calculus effectively, it is important to be able to create models for simple situations and understand the assumptions underlying them. In this section we address these issues.

Figure 1-1
The problem solving process

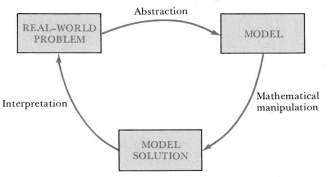

We have all seen models in one form or another. A model is simply a representation of reality. For instance, model airplanes and dolls are physical replicas of their real-life counterparts. Physical models are used extensively in engineering. To test a new airplane wing design, aeronautical engineers build a small-scale model of the wing and subject it to various tests in a wind tunnel. A contour map showing the geographical characteristics of a region is another type of model. We will not be concerned with such physical models in this book, although they are important in many problem solving situations. In this text we shall be interested in mathematical models.

SETTING UP
MATHEMATICAL
MODELS

> A **mathematical model** is a representation of a problem using mathematical symbols and relations.

Mathematical models form the basis for problem solving with mathematics. Setting up a model — that is, translating a word problem into mathematics — is a very important, though often frustrating, skill to be learned. The keys to success in doing this are:

Steps in Constructing a Mathematical Model of a Problem

> 1. Read the problem carefully.
> 2. Understand what the problem is seeking.
> 3. Properly organize the information given in the problem.

Setting up a model has little to do with mathematical skills, but much to do with problem solving in general. Whether your major interest is in business, psychology, biology, or any other field, you will always have the need to analyze and solve problems. Solving mathematical problems will help you to learn to organize your thoughts, and this will be useful to you throughout your career.

Let us consider the Forest Hills Fire Department problem given in the Introduction. We would like to be able to compute the total cost of a new rescue vehicle for several different values of miles driven. The costs that will accrue are *(1)* the cost of the vehicle, $24,000; and *(2)* the operating cost. Since it costs $0.60 per mile to operate the vehicle, the total operating cost in dollars will be 0.60 times the number of miles driven. For example, if the number of miles driven is 70,000, then the operating cost will be $(0.60)(70,000) = \$42,000$, and the total cost will be $\$24,000 + \$42,000 = \$66,000$. For 80,000 miles, the operating cost will be $(0.60)(80,000) = \$48,000$, and the total cost will be $\$24,000 + \$48,000 = \$72,000$. Finally, for 90,000 miles of use, the operating cost will be $(0.60)(90,000) = \$54,000$, and the total cost will be $\$24,000 + \$54,000 = \$78,000$. Note that the purchase cost and the cost per mile remain constant while the number of miles varies in these calculations. The number of miles is therefore called a *variable*. (At this point you may wish to

ALGEBRA REVIEW 1: VARIABLES AND CONSTANTS

In any problem, a **variable** is a quantity that is unknown and may assume a variety of values. A **constant** is a quantity that has a fixed value. For example, consider the problem "John has $3 in dimes and quarters. If he has 18 coins, how many of each does he have?" The variables are the number of dimes and the number of quarters, while the constants are $3 and 18 coins. Usually, variables are denoted by letters toward the end of the alphabet such as x, y, w, or z. Constants that are not specified are often symbolized by letters near the beginning of the alphabet such as a, b, h, or k. However, many times it is useful to use symbols

Quantity	Symbol
Nickels	n
Distance	d
Time	t
Profit	P

that relate to the name of the variable in the specific problem, as indicated in the table.

As long as you define your variables and constants properly by a statement such as "let n = the number of nickels," then any symbol will serve.

study Algebra Review 1.) If we let $x =$ number of miles driven, then the operating cost is written as $0.60x$. The total cost (TC) can then be stated as

$$TC = 24,000 + 0.60x$$

PRINCIPLES OF MATHEMATICAL MODELING

There are three major components to every mathematical model. Being able to categorize each piece of information in a problem statement as one of these components brings us much closer to creating a correct model and solving the problem at hand. These components are:

a. Unknown quantities, or variables
b. Known quantities, or constants
c. Relationships among the variables and constants

Unknowns are the quantities that must be found in order to arrive at a numerical answer. Since we do not know these in advance, they are the variables in the problem. The *knowns* are simply any constants that are specified in the problem. Finally, *relationships* are those aspects of the problem that tie the variables and constants together. They enable us to put the pieces together into a complete mathematical model. In many problems, relationships may be stated explicitly in the problem statement; in others, we may have to rely on experience and general knowledge to discover them. In the fire department example, we had to realize that total cost is the sum of the purchase cost plus the cost per mile times the number of miles driven. In other problems, specific facts and assumptions might be given. Thus, the first step in modeling is to be able to extract the important constants, define the variables, and specify the relationships from a possibly ambiguous problem statement.

EXERCISES*

Define the variables and constants in each of the following:

a. How long will it take to travel 300 miles at a fixed speed of m miles per hour?

b. A man is 10 times as old as his daughter. The sum of their ages is 33.

c. How many gallons of gasoline will an automobile use if it gets 22 miles per gallon and is driven 500 miles per month?

d. What is the total value of timber worth $3000 per square mile lost in a forest fire destroying x square miles?

e. What is the cost of a cubic box with side x if the material costs $0.30 per square foot?

---

* Answers to algebra review exercises can be found at the end of the answer section in the back of the book.

Once a problem has been formulated in this manner and symbols have been defined for the variables, we must translate the relations into mathematical expressions. These will most often be algebraic expressions. (For further discussion, see Algebra Review 2.)

Example 1   Define the unknowns and express the following relations as algebraic equations.

    a. Bill is 8 years older than Lorrie.
       Let $x =$ Bill's age and $y =$ Lorrie's age. Then $x = y + 8$.
    b. Find the time to travel x miles at a speed of 55 miles per hour.
       Let $t =$ travel time (in hours). Then $t = x/55$.
    c. Find the number of cents in Q quarters, D dimes, and N nickels.
       Let $C =$ number of cents. Then $C = 25Q + 10D + 5N$. ▪

Example 2
A Revenue Model   A merchant purchased a group of designer jeans for \$1800 and sold all but 5 pairs at a profit of \$6 each. Write an algebraic equation for the revenue in terms of the variables: cost per pair of jeans and the number purchased. What is the relationship between these variables?

Solution   To begin, let us define x to be the cost of each pair and y the number purchased. Since revenue = selling price times the number sold, we have

Selling price $= 6 + x$
Number sold $= y - 5$
Revenue $= (6 + x)(y - 5)$

Since the total group cost is \$1800, we have $1800/x =$ total cost divided by the cost per pair = number purchased, y. Thus, the relationship between x and y is

$$\frac{1800}{x} = y$$

ALGEBRA REVIEW 2: ALGEBRAIC EXPRESSIONS

An **algebraic expression** is an expression consisting of sums, differences, products, quotients, powers, and/or roots of variables and constants. When values are assigned to the variables, an algebraic expression becomes a number. Some examples of algebraic expressions are the following:

$$x + 2y \qquad 3x^2 - 5xy$$
$$\sqrt{3x} - \frac{4}{\sqrt[3]{y}} \qquad 3x^2 + 2xy + \frac{6z - 4xy^3}{2a^2 + bz}$$

Expressions consisting simply of products, quotients, powers, or roots of variables and constants are called **terms**. For example, the expression $3x^2 - 5xy$ has two terms, namely $3x^2$ and $-5xy$. Similarly, the terms of $\sqrt{3x} - 4/\sqrt[3]{y}$ are $\sqrt{3x}$ and $-4/\sqrt[3]{y}$.

Algebraic expressions satisfy certain properties that allow you to manipulate or simplify them just as you can simplify similar numerical expressions. These properties are (where A, B, and C are any terms):

Substituting this expression for y, we can also write the revenue equation as

$$\text{Revenue} = (6 + x)\left(\frac{1800}{x} - 5\right)$$

Developing appropriate algebraic expressions is often easier if you break the problem down into smaller pieces and concentrate on each individual piece. In the next section we present several examples of mathematical models. You should study these carefully in order to understand how the statement of the problem is translated into a model. The key ideas to remember are the following:

Guidelines for Mathematical Problem Solving

1. Read the problem carefully.
2. When appropriate, draw a picture to help organize your thoughts.
3. Identify the constants and variables in the problem.
4. List any relationships that are explicitly stated or implied.
5. Translate the terms of each relationship into mathematical expressions.
6. Combine the terms into the final model.

Mathematical modeling is very much an art. The more you practice, the better you will become at it. Throughout this book we will see many more examples of mathematical modeling in practice, particularly in the opening section of each chapter.

Commutative properties: $A + B = B + A$

$\quad\quad AB = BA$

Associative properties: $A + (B + C)$
$\quad\quad = (A + B) + C$

$\quad\quad A(BC) = (AB)C$

Distributive property: $A(B + C) = AB + AC$

Thus, for example, we have

$$14x(x + y) + 2x^2 = (14x^2 + 14xy) + 2x^2$$
$$= (14x^2 + 2x^2) + 14xy$$
$$= 16x^2 + 14xy$$

An equation involving algebraic expressions is called an **algebraic equation.**

EXERCISES

Simplify the following algebraic expressions:

a. $(2x)(5x + 2)$
b. $6\sqrt{x}(x + 4\sqrt{y})$
c. $(2x - 3)(4x) + 6x^2$
d. $(6 + 2x)x + 5x^2 - 3x$
e. $4x(x + y) + 2y(x - y)$

PROBLEMS (Section 1-1)

In Problems 1–10, define the unknowns and express the relation as an algebraic equation.

1. Mike is 11 years older than Tom.

2. Sue is 5 years younger than Kim.

3. The time (in hours) to travel x miles at a speed of 40 miles per hour

4. The speed per hour to travel y miles in 6 hours

5. The number of cents in q quarters, n nickels, and p pennies

6. The number of dollars in t tens, f fives, and d one-dollar bills

7. Bob is 3 years older than twice Mary's age.

8. Tom is 2 years older than Bill. The sum of their ages is 38.

9. The total time (in hours) to take a trip if someone travels the first x miles at a speed of 55 miles per hour and the last y miles at a speed of only 45 miles per hour

10. The number of cents if someone has 2 more dimes than nickels and 4 more quarters than nickels

11. In Example 2, suppose the selling price of a pair of jeans is $42. How many jeans are sold?

12. A store purchased a group of record albums and sold all but 10 at a profit of $3 each. The total revenue received was $1350.
    a. Write an equation relating the total revenue to the number of albums purchased and the cost of each.
    b. If the total cost was $900, write an equation for total revenue in terms of the number purchased only.

13. A store purchased a group of radios. They sold 2 of them to employees at cost. All but 4 of the remaining radios were sold at a profit of $10 each. The total revenue received was $592.
    a. Write an equation relating total revenue to the number of radios purchased and the cost of each.
    b. If the total cost was $480, write an equation for total revenue in terms of the number of radios purchased only.

1-2 FURTHER EXAMPLES OF MATHEMATICAL MODELING

In this section several examples of modeling will illustrate the general process and, at the same time, present certain practical guidelines that can help to make modeling an easier task.

Example 3
A Profit Model

A large consumer goods manufacturer is interested in introducing a new product. The price that will be charged to retail stores (the wholesale price) is $2.50 per unit. Annual costs for cleaning and maintenance of the machines that produce the product are $8000. The cost for raw materials, labor, and other expenses is determined to be $1.55 for each unit produced. The company would like to compute the profit for any quantity that may be manufactured and sold.

Solution   In order to develop an appropriate model, we first identify the constants and variables in this problem statement. The constants in this problem are the wholesale price, $2.50; annual costs of $8000; and the unit production costs of $1.55. The only variable is the quantity that is manufactured and sold; let this be denoted by x. The fundamental economic relationship between cost, revenue, and profit is

$$\text{Profit} = \text{Revenue} - \text{Cost}$$

Revenue is found by multiplying the price per unit by the number of units sold. Thus, we have

$$\text{Revenue} = (2.50 \text{ dollars/unit}) (x \text{ units})$$
$$= 2.50x \text{ dollars}$$

It is a good habit to include the **dimensions,** or units of measurement, of all quantities in the model. This helps us to determine whether we are multiplying or dividing the correct quantities. If the units of measurement for the revenue equation we set up turned out to be something other than dollars, then we have made a mistake.

The total cost is the fixed cost of $8000 plus the cost of the units produced, or

$$\text{Cost} = \$8000 + (1.55 \text{ dollars/unit}) (x \text{ units})$$
$$= 8000 + 1.55x \text{ dollars}$$

The profit P is therefore

$$\text{Profit} = \text{Revenue} - \text{Cost}$$
$$P = 2.50x - (8000 + 1.55x)$$
$$= 0.95x - 8000 \text{ dollars}$$

■

Example 4
Fuel Consumption   Commuters who drive to work have automobiles with different fuel economies. Also, they drive different distances to work, and often drive a different number of days each month. Suppose that the Department of Transportation (DOT) is interested in measuring the average monthly fuel consumption of commuters in a certain city. The DOT might survey a sample of commuters and collect information about the number of miles per day driven to and from work, the number of days driven per month, and the fuel economy of their automobiles, measured in miles per gallon. From these data the DOT may wish to determine how many gallons of gasoline per month are consumed by a particular commuter.

Solution   We can construct a model for fuel consumption by carefully examining the dimensions of these quantities. Let us first define some symbols:

g = Gallons of fuel consumed per month
x = Miles per day driven to and from work
d = Number of days per month driving to work
y = Fuel economy, miles per gallon

We seek an expression for g in terms of the variables x, d, and y. By writing down the dimensions of each, we can help to verify the correct expression by observing how the dimensions "cancel out":

$$
\begin{aligned}
g &= \frac{(x \text{ miles/day}) (d \text{ days/month})}{(y \text{ miles/gallon})} \\
&= \frac{xd}{y} \frac{(\text{miles/day})(\text{days/month})}{\text{miles/gallon}} \\
&= \frac{xd}{y} \left(\frac{\cancel{\text{miles}}}{\cancel{\text{day}}}\right) \left(\frac{\cancel{\text{days}}}{\text{month}}\right) \left(\frac{\text{gallons}}{\cancel{\text{mile}}}\right) \\
&= \frac{xd}{y} \text{ gallons/month}
\end{aligned}
$$

Sometimes, this may take some trial and error, but attention to the dimensions of quantities usually makes the task of modeling easier and more reliable.   ■

Example 5
Monthly Driving Cost

Van pooling has become increasingly popular as a method for reducing transportation costs and saving energy. Many companies purchase 9- or 12-passenger vans for their employees to use in lieu of driving individually to and from work. In promoting van pooling, it is helpful to show the comparative costs of driving alone versus pooling. Let us develop a model to compute monthly costs of operating an automobile.

Solution

The variables we consider are the number of miles driven per day, denoted by x, and the miles per gallon obtained by the vehicle, denoted by y. Assume the following values for the constants in this problem:

a. There are 21 working days per month.
b. Fuel cost is $1.50 per gallon.
c. The cost of lubrication and oil change is $15, and this must be done every 8000 miles.
d. Miscellaneous maintenance amounts to $0.025 per mile.
e. The cost of parking is $2 per day.

To develop a model for total monthly costs MC, we must sum the average monthly costs of fuel C_f, lube and oil C_l, miscellaneous maintenance C_m, and parking C_p. That is,

$$MC = C_f + C_l + C_m + C_p$$

We shall express all terms in dollars per month; attention to dimensions is critical in this problem.

First, consider the monthly fuel cost C_f. This is the fuel cost per gallon times the number of gallons of fuel used, or

$$C_f = (1.50 \text{ dollars/gallon}) (g \text{ gallons/month}) = 1.50g \text{ dollars/month}$$

By using the results of Example 4, we see that g is given by

$$g = \frac{21x}{y} \text{ gallons/month}$$

We may now write the fuel cost as

$$C_f = (1.50)\left(\frac{21x}{y}\right) = \frac{31.5x}{y} \text{ dollars/month}$$

The average monthly cost of oil changes is also a little tricky. The cost is $15 per oil change, but this will not occur every month. We need to find the average number of oil changes made per month. Again, by paying close attention to the dimensions we can write the number of oil changes per month as

$$\left(\frac{1 \text{ oil change}}{8000 \text{ miles}}\right)\left(x \text{ } \frac{\text{miles}}{\text{day}}\right)\left(21 \text{ } \frac{\text{days}}{\text{month}}\right) = \frac{21x}{8000} \text{ oil changes/month}$$

Therefore,

$$C_l = (15)\left(\frac{21x}{8000}\right) = 0.039375x \text{ dollars/month}$$

In a similar manner we can compute the monthly miscellaneous cost by

$$C_m = (21x \text{ miles/month})(0.025 \text{ dollars/mile}) = 0.525x \text{ dollars/month}$$

The monthly parking cost is simply $2 per day times 21 days/month, or

$$C_p = 42 \text{ (dollars/month)}$$

The total average monthly cost is given by the sum of these terms and results in the model

$$MC = \frac{31.5x}{y} + 0.039375x + 0.525x + 42$$

$$= \frac{31.5x}{y} + 0.564375x + 42 \text{ dollars}$$

Thus, if a car gets 25 miles/gallon and travels 30 miles daily, the average monthly cost is

$$MC = \frac{31.5(30)}{25} + 0.564375(30) + 42 \approx \$96.73$$ ∎

This was a pretty complicated example, but it illustrated two important lessons. First, you must pay very careful attention to the dimensions in a problem; and second, it is much easier to develop a model if it is broken down into small pieces which are analyzed separately.

The previous model is an example of a model that contains more than one variable. The applications of calculus to such models will be studied in a later chapter. Until then, we will deal only with models that can be written (possibly after a little manipulation) in terms of a single variable.

Example 6
Container Design   Let us reconsider Problem C listed in the Introduction to this chapter. Recall that a manufacturer wishes to design an open cardboard box by cutting small square corners out of a 28 × 28-inch piece of cardboard and folding up the sides. Let us construct a mathematical model for the volume of the box.

Solution   The variables are the volume of the box and the size of the corners that are removed from the cardboard. The only constant we know is the size of the cardboard. A useful procedure for mathematical problem solving is, whenever possible, to draw a picture. This often helps in defining variables and establishing the relations among the variables and constants. An illustration of this problem is given in Figure 1-2. From the figure, we see that the length and width of the box will be $28 - 2x$ inches and the height will be x inches. Therefore, the volume V will be

$$V = (\text{Length})(\text{Width})(\text{Height})$$
$$= (28 - 2x)(28 - 2x)(x) \text{ cubic inches}$$

The problem is to find the value of x that *maximizes* this quantity. This is an example of an *optimization problem*. One of the most important uses of calculus is for solving such problems. (We shall see this in Chapter 3.) We may simplify this expression by multiplying the factors (see Algebra Review 3):

$$V = (28 - 2x)(28 - 2x)(x)$$
$$= (784 - 56x - 56x + 4x^2)(x)$$
$$= 784x - 56x^2 - 56x^2 + 4x^3$$
$$= 784x - 112x^2 + 4x^3$$

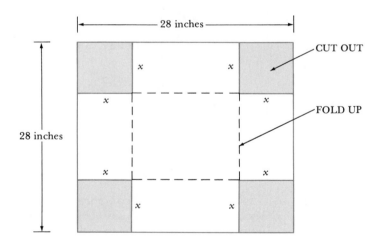

Figure 1-2
Container design

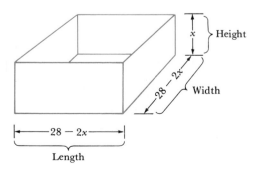

ALGEBRA REVIEW 3: MULTIPLYING ALGEBRAIC EXPRESSIONS

A very common algebraic operation is the multiplication of two or more expressions involving several terms. In the simplest case we wish to multiply

$$(A + B)(C + D)$$

where A, B, C, and D are algebraic terms. To do this systematically, we multiply each term in the first pair of parentheses by each term in the second as follows:

$$(A + B)(C + D) = AC + AD + BC + BD$$

For example, consider the expression

$$\underbrace{(3x}_{A} + \underbrace{2x^2)}_{B}\underbrace{(4}_{C} + \underbrace{x)}_{D}$$

This is simplified as

$$(3x)(4) + (3x)(x) + (2x^2)(4) + (2x^2)(x)$$
$$= 12x + 3x^2 + 8x^2 + 2x^3$$
$$= 12x + 11x^2 + 2x^3$$

EXERCISES

Multiply the following algebraic expressions:

a. $(6x + y)(y + 4x)$ b. $(3y^3 + \sqrt{x})\left(\dfrac{y}{x} + 2\right)$

c. $(xy - x)(1 - y)$ d. $(x - y)(x + y)$

e. $(\sqrt{x} + 1)(\sqrt{x} - 3)$

PROBLEMS (Section 1-2)

1. A store bought 10 television sets for x dollars each. Each one was sold for $100 above cost. Write an equation expressing the revenue received in terms of the cost x.

2. It costs a company $5000 per year to run a certain machine plus $0.90 for each manufactured unit the machine produces. Write an equation for the total yearly cost in terms of the number of units produced.

3. A food company is producing a new product. They will charge retail stores $1.25 per unit. The annual cost for the equipment needed is $12,000 plus $0.75 for each unit produced. Write an equation for profit in terms of the number of units produced.

4. Steve drives x miles per day to work. He pays $1.30 per gallon of gas. If he works 22 days per month and averages y miles per gallon, what is his monthly cost for fuel in terms of x and y?

5. Steve, in Problem 4, pays $20 for an oil change every 5000 miles and encounters other monthly expenses totaling $35. What is his total average monthly cost for driving to work?

6. A backyard patio has a rectangular shape. If the length is 3 feet less than twice the width, write an equation for the area of the patio:
   a. In terms of the width
   b. In terms of the length

7. A park district has a wading pool and a regular pool at their swimming complex. The wading pool is square-shaped. The regular pool is a rectangle whose width is the same as that of the wading pool and is 30 feet less than its length. Write an equation for the total area of both pools in terms of the side of the wading pool.

8. A container corporation wants to design an open cardboard box by cutting small square corners out of a 30 × 24-inch piece of rectangular cardboard and folding up the sides. Write an equation for the volume of the container in terms of the length of the side of the cut-out portion.

9. In Problem 8, write an equation for the surface area of the container.

10. A homeowner wishes to add a rectangular addition of 120 square feet to one end of her house.

The cost of building the exterior wall is $80 per linear foot, and the cost of removal of the interior wall is $50 per linear foot. Express the area of the addition in terms of the length and width of the addition. Express the total cost in terms of the length of the addition.

11. A closed rectangular box with a square bottom and top is to contain 8 cubic feet. Express the surface area of the box in terms of the side of the square bottom.

1-3 FUNCTIONS

In the models presented so far, we have expressed a mathematical relationship between two or more variables. For instance, the fire department cost model was

$$TC = 24,000 + 0.60x$$

and the driving cost model was

$$MC = \frac{31.5x}{y} + 0.564375x + 42$$

In the fire department model, TC is written in terms of x and we say that TC is a *function* of x. Given a value for the variable x, a unique value of TC is determined. Similarly, in the driving cost model, MC is a *function* of x and y. When values are given to x and y, a unique value of MC is determined. The variables TC and MC are called *dependent variables* while x (in the fire department model) and x and y (in the driving model) are called *independent variables*.

We will study functions of several variables in detail in a later chapter. For now we are content to consider only functions of a single variable.

> A **function** is a rule that assigns to each value of an independent variable exactly one value of a dependent variable.

Functions are important in mathematical applications in all disciplines. In physics, the motion of an object that is thrown into the air can be described as a function of time. In psychology, the analysis of data can lead psychologists to determine learning rates as a function of time or of the IQ of an individual. In business, many functional relationships exist between costs, profits, and design variables.

Functions, or rules associating dependent variables with independent variables, are usually symbolized by lowercase letters such as *f, g, h,* and so on. The independent variable is traditionally called x and the dependent variable is usually called y, although other choices are common in specific application areas. We can therefore visualize the association between the dependent and independent variables which a function establishes in the following way:

$$x \qquad\qquad \xrightarrow{\quad f \quad} \qquad\qquad y$$

Independent variable   *Function*   *Dependent variable*

If y is a function of x, then we usually write "$y = f(x)$" (pronounced "y equals f of x"). The notation $f(x)$, which incorporates the symbol for the function as well as the independent variable, will be used to designate a general function. The value of the function $f(x)$ when the independent variable x takes on a specific value a will be denoted "$f(a)$." This is found by *substituting a* everywhere x appears in the formula for $f(x)$. Thus, if

$$f(x) = 4x^2 - 2x$$

we have

$$f(-2) = 4(-2)^2 - 2(-2) = 20$$

$$f(a) = 4a^2 - 2a$$

$$f(a + b) = 4(a + b)^2 - 2(a + b)$$

$$f(z^3) = 4(z^3)^2 - 2(z^3)$$

Not all relationships between two variables are functions. For instance, consider the equation $y^2 = x^2$. If x is chosen as 2, then $y^2 = 4$ and hence, $y = 2$ *or* $y = -2$ will satisfy this equation. That is, to the value $x = 2$ this equation allows the assignment of *two* values of y. We therefore cannot say that this equation defines y as a function of x, since the definition of function requires that to each value of the independent variable x only one value of the dependent variable y is assigned.

The rule that assigns the relationship between the independent and dependent variables can be expressed in many ways, such as a single formula, or by groups of formulas, charts, tables, or graphs. The following examples illustrate some different ways of expressing functions that are useful in many practical situations.

Example 7
Empirical Data — Meteors

An amateur astronomy group observed meteors during the hour from 11:00 P.M. to midnight each evening from January 7 through January 12, 1986. Their observations are given in the table at the top of the next page. If we let x stand for the day of the month and $f(x)$ stand for the number of meteors sighted on date x, then we see that the table defines a function. Note that each value of x is associated by the table with a unique value of $f(x)$. However, it is not the case in this example that each function value $f(x)$ is associated with only a single value of

Date, x	Sightings, $f(x)$
7	14
8	19
9	10
10	14
11	22
12	26

the independent variable x. For example, $f(7) = 14$ and $f(10) = 14$. The definition of function requires only that each value of x be assigned to a unique value of $y = f(x)$, not that each value of $y = f(x)$ be associated with only a single value of x. ∎

Example 8
Salary Structure with Commissions

A salesperson's income is based upon a fixed salary, $15,000, plus commission. The commission schedule is set up so that she receives 7.5% for the first $100,000 in sales, and 9% for any amount above this. Express her total yearly income as a function of sales.

Solution

Let x = yearly sales. We see that if $0 \le x \le 100,000$, then her income will be $15,000 + 0.075x$. If $x > 100,000$, then she will make 7.5% commission on the first $100,000 plus 9% on the amount above $100,000, or

$$15,000 + 0.075(100,000) + 0.09(x - 100,000)$$

We may express the salesperson's income as a function of sales in the following manner:

$$f(x) = \begin{cases} 15,000 + 0.075x & \text{if} \quad 0 \le x \le 100,000 \\ 22,500 + 0.09(x - 100,000) & \text{if} \quad x > 100,000 \end{cases}$$

We have defined this function by two different formulas, depending on the value of the independent variable x. This is perfectly acceptable mathematically and is often necessary. In order to evaluate $f(50,000)$ we use the first formula, since $0 \le 50,000 \le 100,000$:

$$f(50,000) = 15,000 + 0.075(50,000)$$
$$= 18,750$$

To find $f(125,000)$, we must use the second formula, since $125,000 > 100,000$:

$$f(125,000) = 22,500 + 0.09(125,000 - 100,000)$$
$$= 24,750$$ ∎

Example 9
Sales of a New Product

Many companies carefully collect and analyze sales data for new products over time. The sales of a new product often follow a curve similar to the one shown in Figure 1-3. This is called a *life cycle curve*, and represents sales as a function of time. Thus, time is the independent variable and sales is the dependent variable. Even if the formula for the function is unknown, the curve (which can be

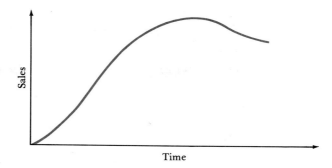

Figure 1-3
Graph of new product sales as a
function of time

estimated from historical data) provides some useful information. For example, we see that sales grow rapidly at the beginning of the life cycle, then begin to level off, and eventually fall as competitive products are introduced into the marketplace. We might be interested in the rate at which sales increase or when the maximum value is attained. Techniques of calculus are necessary to study these issues and will be addressed in Chapters 2 and 3. ■

Example 10
The Vending Machine Function

A common type of vending machine has various items hanging from numbered hooks behind a glass panel. To buy an item, a coin is inserted and a button corresponding to the number of the desired item is pushed. The item then drops from the hook into a bin for retrieval. This machine is an example of a function, since there is a correspondence between two sets (buttons and items) such that for each button there is a unique item associated with it. In this case, we cannot write the function as a formula. A convenient way of specifying such a function is to list the independent variables along with their corresponding dependent variables as ordered pairs of the form $(x, f(x))$, where $f(x)$, the value of the function at x, is the product received when button x is pushed. For instance, we might have

(1, Potato chips)
(2, Potato chips)
(3, Hershey bar)
(4, Three Musketeers)
(5, Corn chips)

It is important to note that although "potato chips" appears twice, the definition of a function is still satisfied. What is *not* permissible for a function would be the following:

(1, Potato chips)
(2, Potato chips)
(2, Corn chips)
(3, Hershey bar)
(4, Three Musketeers)

since when the independent variable equals 2, *two* distinct values are assigned. No vending machine gives two for the price of one! The same is true for functions. ■

Expressing a function in terms of ordered pairs is useful if the independent variable takes on only a few values. In most cases, formulas and/or graphs are used to express functions.

In dealing with a function $y = f(x)$, we are often interested in the values that x and $f(x)$ may assume.

The **domain of a function** is the set of permissible values that the independent variable may take on.

The domain may be explicitly defined, implicitly defined by the context of the problem, or defined by what makes sense mathematically. For example, the domain of the function in Example 7 is the set of values $\{7, 8, 9, 10, 11, 12\}$. In the salesperson's salary model, x is restricted to values greater than or equal to zero in the definition of $f(x)$. In the profit model, permissible values of x are not explicitly defined; however, since the number of items produced and sold cannot be negative, it is implied that $x \geq 0$. (In real life, of course, the number of units must take on integer values, but we relax this assumption in many mathematical models.) To illustrate a mathematically defined domain, consider the function $f(x) = 1/(x - 2)$. Since division by 0 is not permitted in mathematics, the domain of $f(x)$ consists of all values of x not equal to 2. Thus, any values that lead to illegal mathematical operations in the real number system, such as division by 0 or taking square roots of negative numbers, cannot be in the domain of a function. In cases where the domain is not explicitly given, we will understand it to be the largest set of values of the independent variable for which the function makes sense.

We will also be interested in the set of values that a function $f(x)$ assumes.

The **range of a function** is the set of all possible values that the dependent variable may assume.

For example, the range of the function in Example 7 is the set of values $\{10, 14, 19, 22, 26\}$. (Check this!) For the function $f(x) = 1/(x - 2)$, as x varies over all values not equal to 2, $f(x)$ will take on all real numbers except 0. It is often easier to determine the range of a function by analyzing the graph of the function. This topic will be considered in the next section.

Functions are often expressed as sums, $f(x) + g(x)$; differences, $f(x) - g(x)$; products, $f(x)g(x)$; or quotients, $f(x)/g(x)$, of other functions. This is sometimes done for convenience when it is difficult to simplify an expression algebraically. For example, if $f(x) = (x + 2)^5$ and $g(x) = (2x + 3)^4$, then the product $f(x)g(x) = (x + 2)^5(2x + 3)^4$ would be very time-consuming to expand, so we leave it in the form $f(x)g(x)$. In other cases, it simply makes the problem easier to interpret. For instance, in Example 3 we could have defined the revenue function as $R(x) = 2.50x$, the cost function as $C(x) = 8000 + 1.55x$, and the profit function as $P(x) = R(x) - C(x)$. Likewise in Example 2, we could have defined functions for the selling price $p(x) = 6 + x$ and the number of pairs of jeans sold $n(x) = (1800/x) - 5$. The revenue function would then be the product $p(x)n(x)$. Thus, we see that functions are useful in building more complex mathematical models.

PROBLEMS (Section 1-3)

1. Let $f(x) = 2x - 5$. Compute:

   a. $f(0)$
   b. $f(6)$
   c. $f(a + b)$
   d. $f(z^3)$

2. Suppose that $g(x) = 6x^2 - 2x + 1$. Find:

   a. $g(-1)$
   b. $g(4)$
   c. $g(y + z)$
   d. $g(a^2)$

3. Let $f(x)$ be defined as follows:

$$f(x) = \begin{cases} 3x + 2 & \text{if } x \le 4 \\ 4 - x^2 & \text{if } x > 4 \end{cases}$$

   Compute $f(0)$, $f(4)$, and $f(10)$.

4. A function is defined as

$$g(x) = \begin{cases} 1 & \text{if } x < 0 \\ 0 & \text{if } x \ge 0 \end{cases}$$

   Find $g(-1)$, $g(0)$, and $g(1)$.

In Problems 5–12, find the domain and range for each function.

5. $f(x) = x^2$

6. $g(x) = 3x^3$

7. $f(x) = \dfrac{1}{x^3}$

8. $g(x) = \dfrac{1}{x^2}$

9. $f(x) = \sqrt{x + 4}$

10. $g(x) = \sqrt{3x - 12}$

11. $f(x) = \dfrac{1}{x(x - 2)}$

12. $g(x) = \dfrac{1}{(x - 1)(x + 3)}$

13. Each of the following pairs of objects bears some relationship to one another. Which are functions and which are not? If not, tell why, and if so, define the independent and dependent variables.

   a. Grocery items and prices
   b. Names and telephone numbers
   c. Students and grade-point averages
   d. Fingerprints and people
   e. Husbands and wives

14. An advertising company needs workers to help with a big mass mailing project. They will pay a worker $600 plus a nickel for each pamphlet he or she mails. Express the worker's salary as a function of the number of pamphlets mailed.

15. The value F of P dollars held in a savings account for t years and earning simple annual interest at a rate of r percent is given by

$$F = P\left(1 + \frac{rt}{100}\right)$$

   If the interest rate is 9%, that is, $r = 9$, what will be the value of $5000 after 3 years? Why is F a function of t?

In Problems 16–18, express the desired function mathematically.

16. In electrical engineering, power equals resistance times the square of current.

17. IQ is equal to mental age times 100 divided by 12.

18. Degrees Celsius is equal to $\frac{5}{9}$ times degrees Fahrenheit minus 32.

19. The cost of labor and materials (variable cost) for manufacturing a color monitor for a personal computer is $250. Research and development costs (fixed costs) amount to $90,000.

a. If x monitors are produced, find the function that gives the total cost.
b. What is the total cost for production of 100, 1000, and 5000 monitors?
c. Find the function that gives the cost per monitor as a function of x.
d. What is the cost per monitor for production volumes of 100, 1000, and 5000?

20. A physiology experiment on the exertion expended by rats in seeking food when held back by a harness attached to a spring yielded the following function:

$$f(x) = -0.2x + 70 \qquad 30 \le x \le 170$$

where x is the distance (in centimeters) from the food and $f(x)$ is the pull (in dynes) of the rat on the spring.
a. What is the domain and range of the function?
b. Find $f(30)$, $f(100)$, and $f(170)$.
c. From your answers to part b, what general conclusions might be reached?

21. The reaction times of humans for selecting among several responses when given a certain

Number of Possible Responses	Reaction Time (Seconds)
2	0.35
3	0.40
4	0.45
5	0.50
6	0.60

stimulus are given in the table. What are the independent variable, dependent variable, domain, and range for this function?

22. The WATS (Wide Area Transmission Service) monthly telephone rates from Boston to Denver depend on the total duration of calls. The rates are given in the table.

Monthly Call Duration	Cost per Hour (Dollars)
First 15 hours	27.90
Next 25 hours	20.37
Next 40 hours	17.86
Additional hours	15.10

a. Write a mathematical function representing cost in terms of the number of hours of calls.
b. If there is an average of 4.3 weeks per month, what would be the monthly WATS cost for a weekly call volume of 5 hours? 15 hours? 25 hours?

23. For cities a distance of 1000 miles or more apart, the cost of private line telephone service is $1098.92 per month plus $0.58 for each additional mile over 1000.
a. Write the monthly cost as a function of the distance.
b. If Boston and Denver are 1757 miles apart, what would the monthly cost be? Compare this with the monthly cost of WATS in the previous problem. Which service is better?

1-4 GRAPHS OF EQUATIONS AND FUNCTIONS

One of the most useful ways to represent an equation in two variables is by means of a graph on a Cartesian coordinate system (see Algebra Review 4). This gives a picture of the equation and allows us to learn a great deal about it.

GRAPHS OF EQUATIONS

The graph of an equation is the set of points (x, y) that satisfy the equation. The usual method for constructing a graph of an equation is to pick values for one variable, usually x, and use the equation to determine values for the other variable. We then plot these points on the coordinate system and try to draw a smooth curve through them. Generally, the more x values taken, and the closer their spacing, the more accurate will be the sketch of the graph.

Example 11   Let us graph the equation $y - 2x + 6 = 0$. Choosing several values for x, we find the corresponding values for y that satisfy this equation. For example, if $x = 0$, then we have

$$y - 2(0) + 6 = 0$$
$$y + 6 = 0$$
$$y = -6$$

If $x = 1$, then we find that $y = -4$. A table of values of x and y is given in the margin. The graph of this equation is shown in Figure 1-4 on page 22.

x	y
0	-6
1	-4
2	-2
3	0
4	2
5	4

ALGEBRA REVIEW 4: RECTANGULAR COORDINATES

The *rectangular*, or *Cartesian, coordinate system* consists of two perpendicular lines, called the *x-axis* (horizontal) and *y-axis* (vertical). Their intersection is called the *origin*. Any point on the coordinate system is represented by a pair of numbers (a, b) where a is the x value and b is the y value, as shown in Figure a. The plane is divided into four parts, called *quadrants*.

Figure a
Cartesian coordinate system

Values of x to the right of the origin and

values of y above the origin are positive; values of x to the left of the origin and values of y below the origin are negative, as illustrated in Figure b.

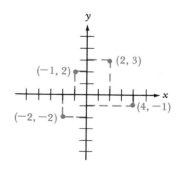

Figure b
Four points in the xy-plane

EXERCISES

Plot the following pairs of points on the Cartesian coordinate system:

a. $(6, 3)$ and $(2, -4)$ b. $(-4, 1)$ and $(0, -6)$
c. $(2, 0)$ and $(-3, -1)$

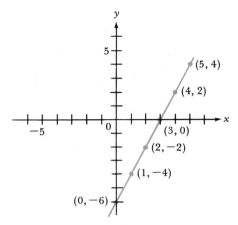

Figure 1-4
Graph of the equation
y − 2x + 6 = 0

In graphing an equation, it is often useful to find the points (if any) where the graph crosses the coordinate axes. These points may be described by numbers called the *intercepts* of the graph.

A **y-intercept** of a graph is the *y*-coordinate of any point on the graph whose *x*-coordinate is 0.

An **x-intercept** of a graph is the *x*-coordinate of any point on the graph whose *y*-coordinate is 0.

Thus, in Example 11, $y = -6$ is the only *y*-intercept since this is the only *y* value corresponding to $x = 0$. To find *x*-intercepts, we would set $y = 0$ and solve for *x*. What value(s) do you obtain? Identify the intercepts of the graph in Figure 1-4.

Example 12   Graph the equation $2x^2 + x - y = 4$.

Solution   We choose several values for *x* and find the values of *y* that satisfy the equation. For example, if $x = 0$, then we have $2(0)^2 + 0 - y = 4$. Therefore, $y = -4$. If $x = 1$, then $2(1)^2 + 1 - y = 4$, and $y = -1$. A table of values of *x* and *y* is given in the margin. These points are plotted in Figure 1-5.

x	y
0	−4
1	−1
−1	−3
2	6
−2	2

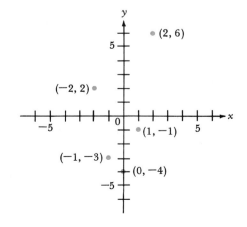

Figure 1-5
Points satisfying the equation
2x² + x − y = 4

Most curves are smooth, and it is important to determine carefully where changes in direction occur. From these points, it appears that the curve reaches its lowest value somewhere between $x = -1$ and $x = 1$. If we try additional points in this range (and are a little lucky), we will find that the curve "bottoms out" near the point $x = -\frac{1}{4}, y = -4\frac{1}{8}$. Later, we will develop calculus techniques for determining exactly such minimum points of curves. The graph of $2x^2 + x - y = 4$ is shown in Figure 1-6.

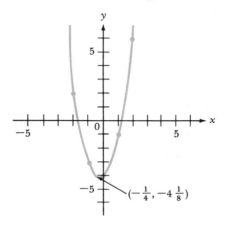

Figure 1-6
Graph of $2x^2 + x - y = 4$

From this example we see that graphing equations is not always easy, especially if we wish to be exact. Fortunately, techniques of calculus will help us to do this more easily. This will be discussed in Chapters 3 and 4.

Example 13

Consider the equation $y - 3 = 1/(x - 2)$. When $x = 2$, we see that y is not defined since we would be dividing by 0. Choosing values for x other than 2, we obtain the table shown in the margin. Notice that as x gets close to 2, values of y get large in magnitude, and as x gets large in magnitude, y gets close to 3. The graph of this equation is shown in Figure 1-7. In this figure, we see that the curve approaches the horizontal line $y = 3$ as x gets very large in either the positive or negative direction. Also, the curve approaches the vertical line $x = 2$ as x gets close to 2 from either direction. Such lines are called *asymptotes*.

x	y
-998	2.999
-48	2.98
-2	2.75
0	2.5
1	2
1.999	-997
2.001	1003
52	3.02
1002	3.001

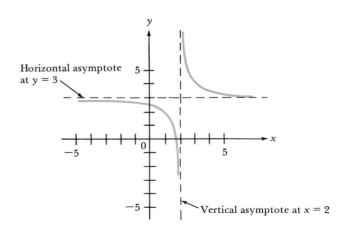

Figure 1-7
Graph of $y - 3 = \dfrac{1}{x - 2}$

> If the y-coordinates of points on a graph become arbitrarily large in magnitude as the x-coordinates approach a fixed value a, then the line $x = a$ is called a **vertical asymptote** of the graph.
>
> If the y-coordinates of points on a graph approach a fixed value b as the x-coordinates become arbitrarily large in magnitude, then the line $y = b$ is called a **horizontal asymptote** of the graph.

When graphing an equation, care must be taken to include enough points to describe the curve accurately. For instance, consider the equation $y = x^3$. If we plot (x, y) for $x = -1, 0$, and 1, we obtain Figure 1-8. This suggests that the graph is a straight line, but if we plot more points, we find that the curve appears as shown in Figure 1-9.

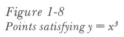

Figure 1-8
Points satisfying $y = x^3$

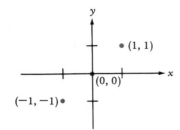

Figure 1-9
Graph of $y = x^3$

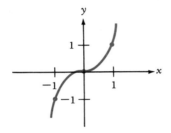

Example 14   Consider the equation $x^2 + y^2 = 9$. Again, we choose values for x and compute values for y that satisfy the equation. However, we must be careful, because for

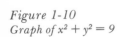

Figure 1-10
Graph of $x^2 + y^2 = 9$

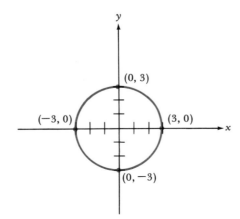

any value for y that satisfies the equation, the negative value will also satisfy it. If we try $x = 0$, we find that $y = 3$ and $y = -3$ will satisfy the equation. If $x = 2$, then $y = \sqrt{5}$ and $y = -\sqrt{5}$. For $x = 3$, $y = 0$. These points are plotted in Figure 1-10, and we see that the equation represents a circle. Note that this equation does not define y as a function of x, since, for example, *two* values of y are assigned to $x = 0$. ∎

GRAPHS OF FUNCTIONS

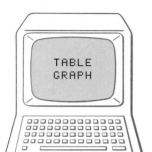

For any function $f(x)$, if we graph the equation $y = f(x)$, then we obtain the *graph of the function* $f(x)$.

> The **graph of the function $f(x)$** is the set of points $(x, f(x))$ where x is any value in the domain of the function.

One important property that distinguishes graphs of functions from more general graphs of equations is that *any vertical line drawn through the graph of a function can intersect the graph in at most one point.* This follows from the definition of a function. Figure 1-11 illustrates this property of functions. If a vertical line $x = a$ intersects a graph at more than one point, say at the points (a, y_1) and (a, y_2), where $y_1 \neq y_2$, then the value $x = a$ would be assigned to two distinct values y_1 and y_2 by the rule that defines the graph. This contradicts the definition of a function and is illustrated in Figure 1-12. This *vertical line property* holds because of the fact that any function can have only one value of the dependent variable y for any given value of the independent variable x. This also

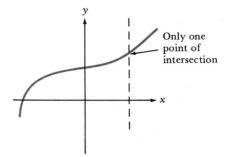

Figure 1-11
A graph of a function

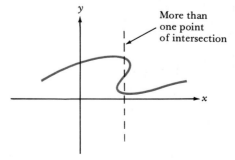

Figure 1-12
A graph that is not a function

makes it easier to graph functions than general equations, since any value of x we choose will give only one corresponding value for y.

Graphs of functions are very useful in problem solving since they give a problem solver a lot of information about the behavior of the function. To see this, consider the following example.

Example 15
Spread of Disease
Suppose that the percentage of students in a college dormitory that are infected by a certain stomach disorder t days after the first case was discovered is given by the function

$$f(t) = \frac{50t}{t^2 + 25}$$

Figure 1-13 illustrates the graph of $f(t)$. From this graph, we see that the percentage of students infected increases rapidly over the first 5 days, reaches a peak of 5% on the fifth day, and then starts to decline at a slower rate. In fact, when $t = 20$, $f(t)$ is still approximately 2.35%. The student health center might be interested in how the rate of infection is changing, or when it reaches its peak. This information can easily be obtained from the graph of the function, and we shall see how calculus helps us find this information.

Figure 1-13
Graph of $f(t) = \dfrac{50t}{t^2 + 25}$

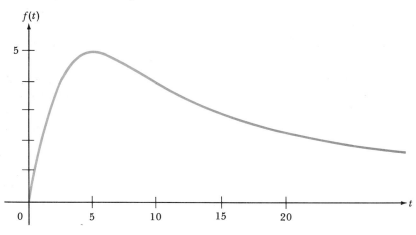

An important property of many functions is that they are *continuous*. You probably have an intuitive notion of continuity in terms of a graph. If you can draw the graph of a function without lifting your pencil from the paper, then the function is continuous. Of course, this is not a formal mathematical definition, and we cannot rigorously define continuity until we introduce the material in the next chapter. However, it is important at this point to at least be able to determine whether a function is continuous from its graph. For instance, the functions in Figures 1-11 and 1-13 are continuous while the function in Figure 1-7 is not. A function that is not continuous is called *discontinuous*.

PROBLEMS (Section 1-4)

In Problems 1–8, graph each equation and determine whether the graph represents a function.

1. $y + 3x - 5 = 0$

2. $y - x + 4 = 0$

3. $x^2 - 4x - y = -4$

4. $y^2 = 3x + 4$

5. $x = \dfrac{1}{y - 3}$

6. $y = \dfrac{2}{x + 1}$

7. $x^2 + y^2 = 4$

8. $4x^2 + y^2 = 36$

9. The cost of delivering a package from Indianapolis to another city varies with the distance according to the graph shown here.

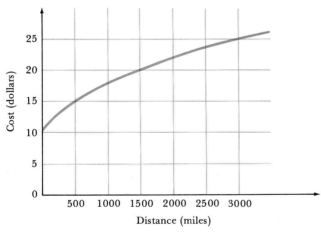

a. Why does this represent a function? Identify the independent and dependent variables.
b. Construct a table of costs for distances of 500, 1000, 1500, and 3000 miles.

10. The number of tons of pollutants emitted from an industrial complex depends on the amount spent on control as represented by the graph shown here.

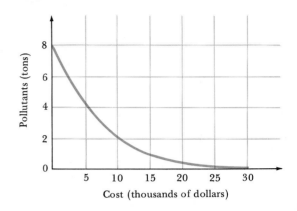

a. Identify the independent and dependent variables for this function.
b. Construct a table of the number of tons of pollutants emitted for control costs of $5000, $10,000, and $30,000.

11. One gram of potatoes contains 0.19 gram of carbohydrates while 1 gram of soybeans contains 0.35 gram of carbohydrates. If a person requires 300 grams of carbohydrates each day, determine all combinations of potatoes and soybeans that will meet this requirement by constructing an appropriate relation and drawing its graph. State three different combinations of potatoes and soybeans that equal 300 grams.

12. A company produces leather belts and leather gloves. One belt uses 40 square inches of leather, and one pair of gloves requires 60 square inches. If 12,000 square inches of leather are available, determine all combinations of belts and gloves that can be produced by constructing an appropriate relation and drawing its graph. Find three different combinations of belts and gloves that use all the leather available.

13. In photochemistry, the Bunsen–Roscoe law regarding the amount of light required to produce a sensory effect is:

Intensity $\times$ Time of action = A constant

or $It = c$. For example, in a bright light, a camera

requires a quicker shutter speed than in a sub-dued light to have the same exposure.
a. Express time of action t as a function of intensity I.
b. If $c = 4$, graph t as a function of I. How can you interpret this graph?

14. In physics, the formula for kinetic energy is $K = \frac{1}{2}mv^2$. If mass is $m = 20$, graph the kinetic energy as a function of the velocity v.

15. When driving, the required visibility for safe stopping depends on the speed of the vehicle. The distance is estimated to be

$$D = 1.23v + 0.0173v^2$$

where D is in meters and v is the velocity of the vehicle expressed in kilometers per hour. Graph the function for speeds from 0 to 80 kilometers per hour.

16. The number of reported incidents of sickness in a chemical plant is a function of the concentration of sulfur dioxide in the air. A study estimated that this function is

$$N = 100 + 0.03C$$

where C is the concentration of sulfur dioxide. Draw a graph of this function. What does it tell you?

17. In 1984, the consumption of coal in an industrial region was 3.5 million tons. Assuming a 5% increase per year, determine the consumption for 1994, 2000, 2010, and plot these points in a coordinate system. Draw a smooth curve through the points. What does this curve indicate about the usage of coal?

18. A government-supported study estimates that the percentage savings in gasoline consumption that would result from a 30¢ per gallon tax will be 34% over the next 5 years, 25% over the following 5 years, and 20% over the next following 5 years. Plot these points in a coordinate system and draw a smooth curve through them. Interpret the meaning of this graph.

19. Traffic engineers have found that the number of seconds, T, that a traffic light at a certain intersection should remain yellow is

$$T = 1 + \frac{v}{2.4} + \frac{7.0}{v}$$

where $v =$ the average speed (in miles per hour) of the vehicles approaching the intersection.
a. Graph the term $1 + (v/2.4)$.
b. Graph the term $7.0/v$.
c. Graph the function T.
d. Using inspection and trial-and-error calculations, estimate the minimum time that the light should be yellow.

20. Draw the graph of the WATS telephone cost function in Problem 22 of Section 1-3.

21. For each of the figures below, determine whether the graph of the function is continuous.
a.

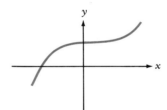

b.

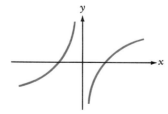

c.

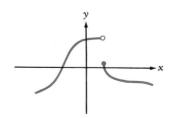

1-5 SOME SPECIAL FUNCTIONS

In this section we introduce several common types of functions that are useful in modeling and solving problems in a variety of applied areas.

LINEAR FUNCTIONS

One of the simplest but most useful functions we shall encounter is the *linear function.*

A **linear function** is a function that can be written in the form

$$f(x) = mx + b$$

where m and b are constants called the *slope* and *y-intercept,* respectively.

The graph of a linear function is defined by the equation $y = mx + b$ and is a nonvertical straight line. The constant b agrees with the definition of y-intercept given earlier, since when $x = 0$, $f(0) = m(0) + b = b$. That is, b indicates the point where the graph crosses the y-axis. The slope gives a measure of the angle of the line relative to the positive x-coordinate axis. The slope is the distance along the y-axis from one point to another on the same line, divided by the distance along the x-axis from the first point to the second. Thus, if (x_1, y_1) and (x_2, y_2) are two distinct points on the line, the slope is given by

$$m = \frac{y_2 - y_1}{x_2 - x_1}$$

and is independent of the two points chosen. This is so because if

$$y_1 = f(x_1) = mx_1 + b \qquad \text{and} \qquad y_2 = f(x_2) = mx_2 + b$$

then

$$\frac{y_2 - y_1}{x_2 - x_1} = \frac{(mx_2 + b) - (mx_1 + b)}{x_2 - x_1} = \frac{m(x_2 - x_1)}{x_2 - x_1} = m$$

irrespective of the distinct points (x_1, y_1) and (x_2, y_2) chosen on the graph. This is illustrated in Figure 1-14.

Figure 1-14
Illustration of the slope of a line

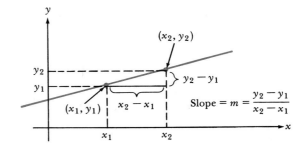

Example 16   The slope of the line joining $(x_1, y_1) = (-2, 6)$ and $(x_2, y_2) = (4, 1)$ is

$$m = \frac{1-6}{4-(-2)} = \frac{-5}{6}$$

This line is shown in Figure 1-15.

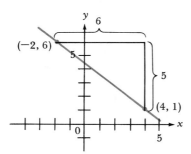

Figure 1-15
Line joining the points $(-2, 6)$
and $(4, 1)$ with slope of $-\frac{5}{6}$

The equation of a straight line can be represented in several ways. The form $y = mx + b$ is known as the **slope–intercept** form of the straight line. If the equation of a straight line is written in this form, then we immediately know its slope is the coefficient of the x term and its y-intercept is the value of the constant term.

Another way of writing the equation of a line is to consider one point (x_1, y_1) on the line to be fixed and let (x, y) represent any other point on the line. Then the slope is given by

$$m = \frac{y - y_1}{x - x_1}$$

which can be written as

$$y - y_1 = m(x - x_1)$$

This is known as the **point–slope** form of the equation of a straight line. If we write this in slope–intercept form, we get

$$y = mx + (y_1 - mx_1)$$

Therefore, the slope is m and the y-intercept is $y_1 - mx_1$. The point–slope form is useful when we know the slope of a line and some point through which it passes.

Example 17   Find the equation of the line passing through $(-3, 4)$ and having a slope of 2. Letting $(x_1, y_1) = (-3, 4)$, we have

$$y - y_1 = m(x - x_1)$$
$$y - 4 = 2[x - (-3)]$$
$$y - 4 = 2(x + 3)$$
$$y - 4 = 2x + 6$$

and finally,

$$y = 2x + 10$$

The graph of this equation is shown in Figure 1-16.

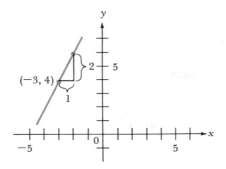

Figure 1-16
Line passing through (−3, 4) with
slope of 2

In general, the equation of any straight line can be written in the form

$$Ax + By + C = 0$$

where A and B are not both 0. This is called the **general form** of the equation of a straight line. If $B = 0$, the line is the vertical line

$$x = -\frac{C}{A}$$

If $B \neq 0$, the line is nonvertical and the equation can be written

$$y = \frac{-A}{B} x - \frac{C}{B}$$

Therefore, the slope is $-A/B$ and the y-intercept is $-C/B$.

A summary of these different characterizations of straight lines is given in the box.

Forms of the Equation of a Straight Line

> *Slope−intercept form:* $y = mx + b$
>   where slope $= m$, y-intercept $= b$
>
> *Point−slope form:* $y - y_1 = m(x - x_1)$
>   where (x_1, y_1) is any point on the line, slope $= m$, y-intercept $= y_1 - mx_1$
>
> *General form:* $Ax + By + C = 0$
>   where slope $= -A/B$, y-intercept $= -C/B$ (for $B \neq 0$)

APPLICATIONS OF LINEAR FUNCTIONS

Linear functions play an important role in modeling because in many models the change in the dependent variable is directly proportional to the change in the independent variable. That is, if $f(x)$ is a linear function, then as x increases by one unit, $f(x)$ changes by a constant amount, namely m. This fact can be demonstrated mathematically by examining the difference

$$f(x + 1) - f(x) = [m(x + 1) + b] - [mx + b]$$
$$= mx + m + b - mx - b$$
$$= m$$

The slope of a linear function represents the change in $f(x)$ when x is increased by one unit.

Many practical situations can be modeled by linear functions. For example, in business, costs and revenues are often linear functions of the number of units produced and sold; in physics, the pressure at a point in a liquid is a linear function of the depth in the liquid; in psychology, learning rates are often linear over certain domains. The following examples illustrate some applications of linear functions.

Example 18
Cost of Utilities

In a certain suburb, water is purchased from the city and is metered. The residential cost is $7.50 plus $1.36 for every metered unit of water used. Thus, if x units of water are used, the total cost will be the linear function

$$f(x) = 7.50 + 1.36x$$

Example 19
Automobile Rebates

To promote the sales of a new model, Anderson Motors proposes a rebate of at least $300. The Consumer Research Department has determined that they will sell 400 more cars for every $100 rebate over $200. The company is currently selling 5000 cars per month at a base price of $8000. Let x denote the size of the rebate and N be the number of cars sold. We wish to express N as a function of x. Since the number of cars sold increases by a constant amount for every $100 rebate over $200, we have a linear relationship between N and x. When $x = 300$, then $N = 5400$; when $x = 400$, then $N = 5800$. This gives us two points to establish a linear function: (300, 5400) and (400, 5800). The slope is

$$m = \frac{5800 - 5400}{400 - 300} = \frac{400}{100} = 4$$

Alternatively, we could have argued that the slope is the additional number of cars per $1 rebate, or $400/100 = 4$ cars per dollar.

Using the point (300, 5400), we can now obtain the intercept:

$$5400 = 4(300) + b$$
$$4200 = b$$

Thus, the number of cars sold as a function of the rebate is

$$N = f(x) = 4x + 4200 \qquad x \geq 200$$

Example 20
Quantity Discounts

It is not unusual for companies to offer discounts when products are purchased in large quantities. This type of pricing structure is called a *quantity discount*. Suppose that a company offers a corporate discount on a popular microcomputer software program. The first 10 copies are sold for $500 each, the next 10 copies are sold for $400 each, and any additional copies cost $300 each. Therefore, if 25 copies are purchased, the total cost will be $(10)(500) + (10)(400) + (5)(300) = \$10,500$. We can express the cost of any quantity x by the following function:

$$f(x) = \begin{cases} 500x & \text{if} \quad 0 < x \leq 10 \\ 5000 + 400(x - 10) & \text{if} \quad 10 < x \leq 20 \\ 9000 + 300(x - 20) & \text{if} \quad 20 < x \end{cases}$$

The graph of $f(x)$ is shown in Figure 1-17. Notice that the graph is composed of pieces of straight lines with different slopes. This is an example of a *piecewise linear* function.

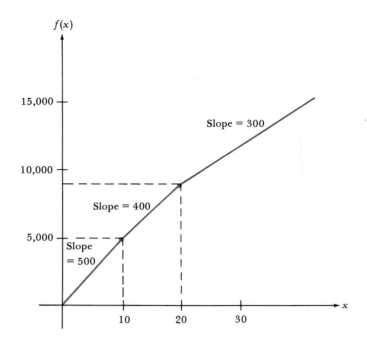

Figure 1-17
Graph of a piecewise linear function

NONLINEAR FUNCTIONS

Many mathematical models cannot be adequately described by linear functions. A **nonlinear function** is one which cannot be written as $f(x) = mx + b$. The quantity discount model presented in Example 20 is composed of pieces of linear functions, but as a whole, it is nonlinear. To see this more clearly, consider a related situation in which the unit price is reduced by 1¢ from a base price of $10 for every additional unit purchased up to 500 units. The total price as a function of the number of units purchased is given by the function

$$f(x) = \begin{cases} 10 & \text{if} \quad x = 1 \\ 10 + 9.99 = 19.99 & \text{if} \quad x = 2 \\ 19.99 + 9.98 = 29.97 & \text{if} \quad x = 3 \\ \quad \cdot & \quad \cdot \\ \quad \cdot & \quad \cdot \\ \quad \cdot & \quad \cdot \end{cases}$$

Such functions are difficult to work with mathematically. It would be more convenient to express this by a single expression in x — in other words, to draw a smooth curve that passes through these points. Such a function is given by (you need not worry about how we obtained this function)

$$f(x) = 10.005x - 0.005x^2$$

As a quick check, let us substitute $x = 2$ into this function:

$$f(2) = 10.005(2) - 0.005(2)^2$$
$$= 20.01 - 0.02 = 19.99$$

which agrees with the function definition above.

One of the simplest types of nonlinear functions is the *quadratic function*.

A **quadratic function** has the form

$$f(x) = ax^2 + bx + c$$

where a, b, and c are constants and a is not equal to 0.

The graph of a quadratic function is a *parabola*. Examples of parabolas are given in Figures 1-18 and 1-19.

Figure 1-18

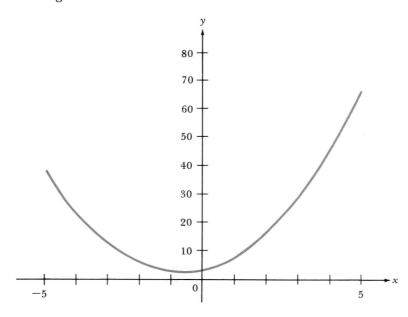

Example 21
Motion of Projectiles

A missile is shot into the air. Its coordinates (in kilometers) t seconds after it is launched are given by

$$x = 30t \qquad \text{Distance from launch site}$$
$$y = 240t - 16t^2 \qquad \text{Height above the ground}$$

We will show that the path of the missile is a parabola. We need to express y as a function of x. This can be done by solving for t in terms of x and substituting the result into the equation for y. Doing this, we have

$$t = \frac{x}{30}$$

and

$$y = 240\left(\frac{x}{30}\right) - 16\left(\frac{x}{30}\right)^2$$
$$= 8x - \tfrac{16}{900}x^2$$
$$= 8x - \tfrac{4}{225}x^2$$

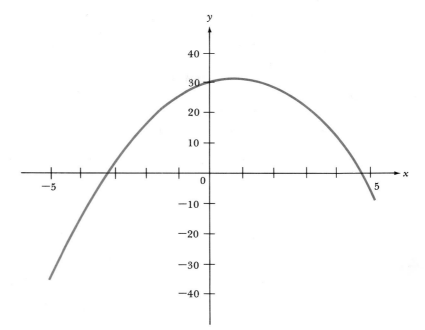

Figure 1-19

This is a quadratic function with $a = -\frac{4}{225}$, $b = 8$, and $c = 0$. Thus, the path followed by the missile is a parabola, pictured in Figure 1-20.

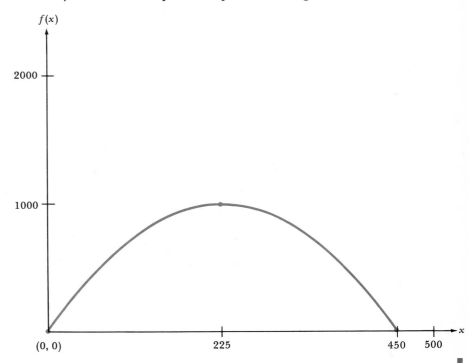

Figure 1-20
Graph of $f(x) = 8x - \frac{4}{225}x^2$

Example 22
Concert Ticket Pricing   A rock concert promoter knows that if ticket prices are \$10, a capacity crowd of 12,000 can be expected for a certain concert. However, at higher prices, fewer

people will be expected to attend. The promoter feels that for each $1 increase in price, 500 fewer tickets will be sold. The promoter would like to determine the revenue that will be received for various values of ticket prices.

Solution   To model this, define y to be the additional price per ticket over $10. The number of tickets sold will then be $12,000 - 500y$. Since the ticket price is $10 + y$, the total revenue TR will be the number of tickets sold multiplied by the price, or

$$TR = (12,000 - 500y)(10 + y)$$
$$= 120,000 - 5000y + 12,000y - 500y^2$$
$$= -500y^2 + 7000y + 120,000$$

Therefore, revenue is a quadratic function of ticket price.   ■

Both linear and quadratic functions are special cases of a general class of functions called *polynomials*.

A **polynomial** is a function of the form

$$P(x) = a_n x^n + a_{n-1}x^{n-1} + \cdots + a_1 x + a_0$$

where the a's are constants and n is a nonnegative integer.

For linear functions, $n = 1$; for quadratics, $n = 2$. Polynomials are useful in many applications but are rather difficult to graph if n is large.

There are many other types of nonlinear functions such as those involving exponents and logarithms. These will be studied in later chapters.

PROBLEMS (Section 1-5)

In Problems 1 – 4, find the slope of the line containing the two points given.

1. (4, 2) and (6, 8)

2. (−8, 12) and (0, −12)

3. (8, 3) and (14, 5)

4. (−7, −5) and (5, 4)

In Problems 5 – 14, find an equation for each line.

5. Slope = 2 and passing through the point (−1, 3)

6. Slope = −4 and passing through the point (−2, 5)

7. Slope = $\frac{1}{2}$ and passing through the point (6, 6)

8. Slope = $-\frac{3}{4}$ and passing through the point (4, −6)

9. Slope = $-\frac{2}{3}$ and y-intercept = 7

10. Slope = $\frac{1}{6}$ and y-intercept = −4

11. Passing through the points (−4, 2) and (6, 1)

12. Passing through the points (1, 3) and (−1, 2)

13. Passing through the points (−1, 4) and (2, 4)

14. Passing through the points (2, 3) and (2, 6)

15. In a certain town, electricity costs $6.75 per month plus $0.09 for every kilowatt-hour used. Write a linear function for the total monthly cost.

16. To promote more air travel, United States Airlines proposes a discount of at least $50. The airline has determined that they will sell 100 more tickets for every $20 discount over $50. The airline is currently selling 6000 tickets per month at a standard price of $250. Express the number of tickets sold as a function of the amount of the discount.

17. Students are planning a short trip that will include bus transportation. The bus company requires at least 50 students and charges $4 per student for the first 50 students. For each additional student, the price per student is reduced by $0.01 for all students. Express the total cost as a function of the number of students going on the trip. Assume a linear relationship between the cost per student and the number of students.

18. The pressure at any point under the ocean surface is a function of the depth below the surface. Pressure is 15 pounds per square inch at the surface and 45 pounds per square inch 60 feet below. Develop a function for pressure at any ocean depth assuming a linear increase. What is the pressure at 1000 feet? 10,000 feet?

19. A warehouse worker's energy expenditure depends on the number of cartons handled per minute. The energy expended in calories per minute has been found to be 100 if 6 cartons per minute are handled and 132 if 12 cartons per minute are handled. Assuming a linear relationship, determine a function relating energy expended to the number of cartons handled per minute.

20. The number of new mortgages at a bank is found to be a function of the interest rate. At 14%, 28 loans per month were processed. For each percentage decrease in the rate, 3.8 additional loans

per month were made. Determine the appropriate function. How many loans are expected to be processed if the rate is 11%?

21. The accident rate per million vehicle miles over a rural road is a function of the average speed over the road. If the average speed is 40 miles per hour, the accident rate is 1.08. For every additional unit increase in speed, the rate rises by 0.03. Develop the appropriate function for the accident rate.

22. A metal rod changes in length as a linear function of the temperature (in degrees Celsius). When the temperature is 15°C, the length is 65.45 centimeters; when the temperature is 85°C, the length is 65.52 centimeters. Construct a function relating length to temperature.

23. A firefighters' union contract calls for fixed salary increases of $1200 for each of the next 3 years. If the current salary (at $t = 0$) is $21,600, develop a function expressing the salary over the next 3 years.

24. In a given month, the consumer price index is determined to be 160. Economists estimate that over the next 6 months it will increase by $1\frac{1}{2}\%$ of this value each month. Write a function expressing the value of the consumer price index over the next 6 months, using $t = 0$ to represent the current month.

25. A machine was purchased for $20,000 in 1980. Its value in 1990 is estimated to be $5000. Assuming a linear decrease in value over time, write a function expressing the machine's value as a function of time (let $t = 0$ represent 1980).

26. The cost per unit of a military helicopter is estimated to be $f(x) = 1.2/\sqrt{x}$, where x is the number of helicopters produced. Is $f(x)$ a polynomial? Why or why not? Is $f(x)$ linear or nonlinear? Sketch the graph of $f(x)$. How would you interpret the meaning of this function?

1-6 CONCLUSION: THE NEED FOR CALCULUS

In this chapter we have discussed the importance of mathematics in solving problems, and we have presented a variety of situations for which mathematical models can be developed. Let us return to the four problems posed at the beginning of this chapter. We saw in Section 1-1 how to set up and solve the Forest Hills Fire Department problem. The biologist's problem can also be answered using straightforward algebraic methods. Can you do it? We'll give you some hints. Compute the number of bacteria in the population after 2 hours and the number present after 3 hours. The rate of population growth will be expressed in dimensions of bacteria per hour.

The container design problem was set up in Example 6 (Section 1-2). However, unless you graph the function carefully, there does not seem to be a way of finding the dimensions that give the largest capacity. In the last problem relating to the spread of a flu virus, you probably cannot figure out how to "sum up" the total number of people affected. For instance, if $t = 1$, we can calculate y to be 784.6 people per hour; when $t = 5$, $y = 975.6$ people per hour. The rate at which the virus spreads changes over time, and this makes the problem difficult. These problems cannot be solved with the tools of algebra. Nevertheless, such problems arise in many important application areas. This is why calculus is necessary. In later chapters we will learn how to solve these types of problems.

1-7 CHAPTER REVIEW

1. Mathematical modeling is the process of translating a problem statement into a mathematical form. A model is a representation of a problem using symbols and mathematical relationships.

2. The components of a mathematical model are variables, constants, and relationships among them.

3. The key ideas in developing models are *(1)* reading the problem carefully, *(2)* drawing pictures whenever possible, *(3)* breaking down the problem into small manageable pieces, and *(4)* paying close attention to dimensions of all quantities.

4. A function is a rule associating a dependent variable with an independent variable such that for every value of the independent variable, there is assigned one and only one value for the dependent variable.

5. Functions can be expressed by formulas, tables of ordered pairs, or graphs, depending on the particular application.

6. The domain of a function is the set of possible values that the independent variable can assume;

the range is the set of values that the dependent variable can assume.

7. A graph of a function has the property that any vertical line that crosses the graph can do so at only one point.

8. Linear functions can be expressed as

$$y = mx + b$$
$$y - y_1 = m(x - x_1)$$
$$Ax + By + C = 0$$

and are characterized by the slope and intercepts. Linear functions are used when the change in the dependent variable varies directly with (or is proportional to) the change in the independent variable.

9. Nonlinear functions that arise in many applications include quadratic functions, which are special cases of the general class of functions known as polynomials.

10. Techniques of algebra cannot solve all mathematical problems that commonly occur; calculus is a useful and necessary branch of mathematics for solving some of these problems.

REVIEW PROBLEMS (Section 1-7)

In Problems 1–4, define the unknowns and express the relations as algebraic equations.

1. Jane is 3 years younger than Bill.

2. The number of cents in d dimes and twice as many quarters as dimes

3. A store purchased a group of bicycles and sold all but 2 at a profit of $35 each. Write an algebraic equation for the revenue in terms of the cost per bicycle and the number purchased.

4. It costs a dairy $3500 per year to run a skim milk machine plus $0.85 for each half gallon produced. What is the total cost?

5. A farmer has a rectangular plot of land. If the length is 50 feet more than three times the width, write an expression for the area of the land in terms of the width.

6. Given the function $f(x) = -3x + 5$, find $f(0)$, $f(-4)$, and $f(10)$.

7. The number of mutations in a certain organism depends on the dosage of x-rays. A function for the percent mutation is

   $$p(x) = 3x \qquad 0 \le x \le 10$$

   where x is the dosage of x-rays.
   a. What is the domain and range of this function?
   b. Find $p(0)$, $p(5)$, and $p(10)$.
   c. What conclusions can be drawn about the behavior of the mutations?

8. The 1984 Ohio income tax for taxpayers with incomes over $40,000 was computed with the following table:

Taxable Income	Tax
$40,000–80,000	$1710.00 + 6.65% of the excess over $40,000
$80,000–100,000	4370.00 + 7.6% of the excess over $80,000
Over $100,000	5890.00 + 9.5% of the excess over $100,000

a. Construct a function for tax in terms of income.
b. Graph the function.
c. If a corporate executive made $85,000 in 1983, what was his tax?

In Problems 9–12, graph each equation. Determine whether the graph represents a function.

9. $y + 2x + 3 = 0$

10. $y^2 = 2x + 1$

11. $x^2 = y^2 - 1$

12. $y - x^2 = 0$

13. Natural gas consumption (in millions of cubic feet) in North America was 672.8 in 1970, 638.7 in 1975, 619.4 in 1980, and 611.3 in 1985. Plot these points on a graph and estimate the consumption in 1990 by drawing a smooth curve through the points.

Find the slope of the line containing the given points.

14. (2, 1) and (5, 4)

15. (3, 4) and (6, −5)

Find the equation for each line in slope–intercept form.

16. Slope $= 1$ and passing through the point (3, 2)

17. Slope $= \frac{1}{6}$ and passing through the point (−1, 5)

18. What are the values of A, B, and C in the general form of the equation of a straight line for:
a. The slope–intercept form?
b. The point–slope form?

19. To mail a certain package, the cost is $1.80 plus $0.20 per ounce. Write a linear function for the total cost.

20. The change in the length of a spring is proportional to the weight attached to it. If there is no weight, the length is 5 inches. For every 5 pounds of weight, the spring gets $\frac{1}{2}$ inch longer. Construct a function relating the number of pounds of weight on the spring to the length. Graph the function you develop.

21. There is never just one way to model a given problem mathematically. Variables can often be defined in different ways. In Example 22, suppose we define x to be the price to charge per ticket. Develop a model for total revenue in terms of x. What is the relationship between x and y? Can you show that your model and the one given in Example 22 are equivalent?

22. A personal computer was purchased in January 1986 for $4000. If the accounting value of the computer (for tax purposes) is 0 in January 1989, write a linear function expressing its value over time (in years).

23. A special university loan fund allows faculty to borrow $5000 at 5% simple interest. If the total interest of $(0.05)(5000) = \$250$ is paid in equal monthly installments along with equal portions of the amount borrowed, write a function expressing the cumulative amount paid back over a 25 month period.

THE NATURE AND LANGUAGE OF MATHEMATICS

If you are a student of business, economics, the social sciences, or the life sciences, you will probably not use calculus as part of your daily routine. However, this does not mean that the study of calculus and other mathematical subjects will not be useful in your future profession. Throughout this book we will see numerous examples of realistic situations where concepts of calculus are necessary for problem solving. On a more general level, everyone uses mathematics to some extent for balancing a checkbook, presenting a graph of sales estimates, analyzing psychological or biological data, and so on. Mathematics is the understanding of structure and relationships, the develop-

ment of theories and their application to real problems. Every discipline — business, economics, the sciences — has its own structure, relationships, theories, and applications. To be successful in any profession, you have to be able to understand these structures and relationships, perhaps develop your own theories, and then be able to apply them to situations that arise. Mathematics provides a framework for developing these thought processes that are so crucial to career success. In this supplement, we briefly discuss some aspects of the nature of mathematics and its language.

1S-1 PURE AND APPLIED MATHEMATICS

Mathematicians often have heated debates on the difference between pure mathematics and applied mathematics. The distinction is not always clear, and we certainly do not wish to stir up any controversy. However, this text is primarily devoted to applied mathematics, and it is important to understand the role that pure mathematics plays.

Pure mathematics is "mathematics for mathematics' sake." The focus is on the understanding of logical and geometrical structures and relationships, the creation of new concepts and ideas with no immediate application in mind. *Applied mathematics,* on the other hand, is "mathematics for something else." It is focused on problem solving, on better understanding our world and environment so that it may be improved or so that new vistas may be explored.

In pure mathematics, one begins with a system of **axioms,** or assumptions that are accepted to be true, and proceeds to develop a collection of **theorems,** or mathematical truths, that logically follow from the axioms. No real-world applications are intended. The pure mathematician's satisfaction is derived from the beauty of the theory and the elegance of the reasoning behind it. In applied mathematics, one usually begins with a problem to be solved. A model is created that represents the problem, and mathematical tools, often taken directly from pure mathematics, are used to solve the problem.

For example, we are used to counting in a *base 10* number system using the ten digits $0-9$. In this system, $1 + 1 = 2$, $9 + 1 = 10$, and so on. Suppose, however, that we begin with a different assumption and define a number system to have only two elements, 0 and 1. We may begin with the assumptions that $0 + 0 = 0$, $0 + 1 = 1$, $1 + 0 = 1$, and $1 + 1 = 0$. What properties does such a number system have? What truths logically follow from these assumptions? These are some questions that a pure mathematician might ask.

An applied mathematician might be concerned with developing mathematical models of electrical current flow in a complex system of switches. How might this be done? One way might be to associate the numbers 0 and 1 with the off and on status of the switches. Such models might require a theory that deals only with a number system composed of 0's and 1's. The applied mathematician might attempt to create such a theory, but the ultimate goal is clearly to solve some well-defined problem. Or, if such a theory already exists, the applied mathematician might recognize the analogy and be able to apply it immediately. Thus, we see that practice often breeds theory and that theory can also breed practice.

Theory is important in understanding applications. Tools of mathematics, or any discipline for that matter, are ineffective unless properly applied, and to properly apply any tool, the underlying assumptions and theory must be understood. Hence, a certain level of rigor is necessary when dealing with a mathematical subject. The language of mathematics is very precise, and it will be important for you to distinguish between general English usage and the mathematical language that is used in this book.

1S-2   THE LANGUAGE OF MATHEMATICS

To understand mathematics, you have to be able to read mathematical statements carefully. Consider the following:

"Let f be a function and let u and v be in the domain of f. If $f(u) \neq f(v)$, then $u \neq v$."

This begins with a statement of *assumptions* that govern everything that follows; that is, f is a function, and u and v are in the domain of f. Next, we state a *premise:* (If) $f(u) \neq f(v)$. We are saying, suppose that u and v are points in the domain of f such that $f(u) \neq f(v)$. Finally, we state a *conclusion:* (Then) $u \neq v$. We are making the proposition that the conclusion will always be true whenever the assumptions and premise are true.

Can we prove that this proposition is true? Recall the definition of a function as a rule that assigns a unique value $f(x)$ to every value of x in its domain. If $u = v$ and $f(u) \neq f(v)$, we are saying that there must be assigned two different values of $f(x)$ to the same value of x. This, however, contradicts the assumption that $f(x)$ is a function. Therefore, the proposition must be true.

Any proposition that can be proved from accepted assumptions and premises is called a **theorem.** Thus, we might say that the proposition given above is a theorem about functions. Throughout this book we state a number of theorems. Although we prove only a few of them in the supplements, you should keep in mind that a theorem is not a definition, but a result that can be shown to always be true. Many theorems are expressed as follows:

If "statement A," then "statement B."

This means that whenever statement A is true, then statement B logically follows. It does *not* mean that if statement B is true, then statement A must also be true. In many cases, a *counterexample* can be constructed that demonstrates that even if statement B is true, then statement A can be false. For example, in the proposition stated above, statement A is "$f(u) \neq f(v)$" and statement B is "$u \neq v$." Can you construct a counterexample for the proposition: "Let f be a function and let u and v be in the domain of f. If $u \neq v$, then $f(u) \neq f(v)$"? When reading a theorem, try to understand what assumptions are made and why, what the premise is, and what conclusion follows.

While this is an applied text, it is impossible to learn mathematics properly without an adequate knowledge of its nature and language. Thus, you should always keep these ideas in mind in your study of mathematics.

INTRODUCTION TO THE DERIVATIVE

INTRODUCTION

In the previous chapter we showed how mathematical models are developed by translating verbal statements of problems into mathematical expressions. In doing so, we used logic and experience to propose the correct relations. For instance, we know that profit = revenue − cost; therefore, we only needed to develop mathematical expressions for revenue and cost in order to find profit. However, not all mathematical models are this easily constructed. For example, suppose that a high school typing teacher is interested in predicting how fast an average student will be able to type after spending a certain amount of time in class. There does not seem to be any "common sense" relationship between the amount of time spent learning to type and the speed or performance achieved.

One way to obtain an idea of how such learning takes place is to collect some data. Figure 2-1 is an actual plot of typing speed measured in words per

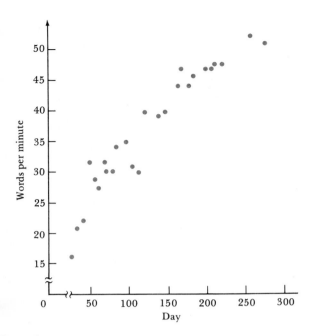

Figure 2-1
Plot of typing speed over time

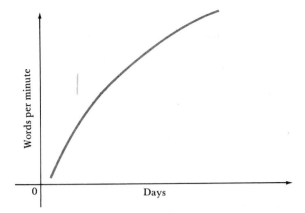

Figure 2-2
A learning curve for typing speed

minute versus the number of days into the school year for a high school student.* Data in this form are difficult to work with mathematically. However, you can see that there is a definite relationship between typing speed and the time spent learning.

It would be useful to be able to construct a function that expresses this relationship and is easy to manipulate mathematically. Such a function might be similar to the one shown in Figure 2-2. This is an example of a *learning curve*. Learning curves are often used to explain and predict progress when learning a new skill. The form of the function seems to make sense, because when a new skill is first tried, learning occurs rapidly. However, everyone has a maximum level of proficiency, and learning becomes progressively more difficult over time. Therefore, the curve begins to level off.

How to find an appropriate function that "fits" a set of data is a subject that is beyond the scope of this

book. Let us suppose that this has been done and that the following function has been proposed:

$$f(t) = 8t^{0.3} + 0.04t$$

where t is the number of days into the school year and $f(t)$ is the number of words per minute, or typing speed, achieved. To determine whether this is an appropriate function, we might try some values for t and see if the model predicts the typing speed closely. (You will need a calculator with the capability of computing exponents to do this.) For instance, if $t = 32$, then $f(32) = 8(32)^{0.3} + 0.04(32) \approx 24$. Other values are given in the table. By visually checking these points, you can see that this function appears to be an appropriate model for this situation.

t	$f(t)$
32	24
160	43
256	52

Suppose that the typing teacher is interested in *how fast* learning is taking place; that is, *at what rate* is a

* Adapted from J. R. Evans, "A Case Study in Mathematical Modeling," *International Journal of Mathematical Education in Science and Technology*, vol. 12, no. 4, pp. 393–398 (1981).

student increasing his or her typing speed? This rate is usually expressed in words per minute per day. The change in typing speed per unit of time is called the *rate of change.*

The **rate of change of a function** describes how fast the dependent variable changes as the value of the independent variable changes.

Using the data from the table, we see that between day 32 and day 160, the speed increased from 24 to 43 words per minute. Therefore, the *average change* in speed over this time period was

$$\frac{(43-24) \text{ words per minute}}{(160-32) \text{ days}} = \frac{19}{128}$$

$$\approx 0.15 \text{ word per minute per day}$$

Note that the dimensions of this rate of change are the dimensions of the dependent variable (words per minute) divided by the dimensions of the independent variable (days). We interpret this by saying that, *on the average,* typing speed increased at a rate of about 0.15 word per minute per day for each day of the 128 days

considered. During the period of time from day 160 to day 256, the *average rate of change* was

$$\frac{52-43}{256-160} = \frac{9}{96}$$

$$\approx 0.09 \text{ word per minute per day}$$

The average rate of change is different for the two different time periods and seems to depend on the time period chosen. Suppose we looked only at the rates of change from day 160 to 165 and from day 251 to 256. Would we expect these rates to be the same? Probably not, based on our previous observations.

What if we tried to answer the question "On the 100th day of the school year, what is the rate of change of typing speed?" We run into a major problem. In the previous examples, we were interested in the rate of change between *two distinct points in time.* We were able to calculate the rate of change by dividing the actual change in speed by the number of days between the two points in time. If we are interested in the "instantaneous" rate of change on a single day, this cannot be done. How, then, can we answer this question? New methods beyond algebra are needed, and this is where we actually begin to study calculus. The basic concepts of calculus depend strongly on a new idea called a *limit.*

2-1 LIMITS

The concept of a limit is fundamental to the understanding of calculus and its applications. In this section we introduce the notion of a limit of a function and present some computational methods for evaluating limits. Further discussions about limits and their applications to ideas in this chapter are presented in the supplement to this chapter.

Suppose a company manufactures metal shafts that are subjected to heavy stress, such as in automobiles or airplanes. The manufacturer would like to determine the maximum safe load on the shaft before it reaches its breaking point. To determine this maximum, shafts are subjected to increasingly higher loads as shown in Figure 2-3 (page 48) until the breaking point is reached.

Suppose that the data in the table on page 48 have been collected. It appears that the load that the shaft can handle safely before it breaks is any value less than 318 pounds. Observe that the load can never *equal* 318 pounds, but can get very close to 318 without breaking.

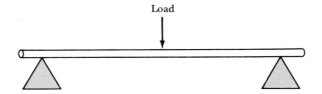

Load

Figure 2-3
Subjecting a metal shaft to load
stress

Load (Pounds)	Break?
300	No
310	No
315	No
317	No
317.5	No
317.9	No
317.95	No
318	Yes

We may represent the data as a function by letting x denote the load in pounds and defining

$f(x) = 0$ if the part does not break with a load of x

$f(x) = 1$ if the part does break with a load of x

Mathematically, we have

$$f(x) = \begin{cases} 0 & \text{if } x < 318 \\ 1 & \text{if } x \geq 318 \end{cases}$$

The graph of this function is shown in Figure 2-4. Imagine yourself as a bug walking along this graph toward $x = 318$ from the *right*. What value of the function $f(x)$ do you get close to? Since the function is always equal to 1 when $x > 318$, your answer should be 1. Eventually you will reach $x = 318$ and stop (or else take a tumble!). We would say that the limit of $f(x)$ as x approaches 318 from the right is 1. Now suppose you walk along the graph toward $x = 318$ from the *left*. The value of the function remains at 0 as long as you do not reach $x = 318$. We would say that the limit of $f(x)$ as x gets close to 318 from the left is 0. You can get as close as you wish to $x = 318$ and still $f(x) = 0$. Thus, the limit of a function is the value that the function approaches (but does not necessarily have to equal) as x gets close to a particular value (but does not equal that value). In this example, even though $f(318) = 1$, the limit as x approaches 318 from the left is still 0.

Figure 2-4
Graph of the breaking function for
the metal shaft load stress example

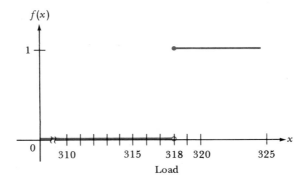

We may speak of limits from the left or limits from the right of a particular value of x, say $x = a$. *If the limits from both sides are the same, then we say that the limit exists as x approaches a; otherwise, we say that the limit does not exist.* In the example we have just considered, the limit does not exist as x approaches 318.

> If the limit of a function as x approaches a exists and is equal to L, we write
>
> $$\lim_{x \to a} f(x) = L$$

The symbol "lim" stands for "limit" and "$x \to a$" represents "x approaches a." We read $\lim_{x \to a} f(x) = L$ as "the limit of the function $f(x)$ as x approaches a is L."

Example 1   Determine whether the limit of $f(x) = 3x - 5$ exists as x approaches 1, and if so, determine its value.

Solution   We write this limit as

$$\lim_{x \to 1} (3x - 5)$$

Let us select values of x greater than 1 and getting close to 1:

x	$f(x)$
2	1
1.5	-0.5
1.1	-1.7
1.01	-1.97
1.001	-1.997

For values of x less than 1 and getting close to 1, we have:

x	$f(x)$
0	-5
0.5	-3.5
0.9	-2.7
0.99	-2.03
0.999	-2.003

In either case, the value of $f(x) = 3x - 5$ appears to be getting closer to -2, so we write

$$\lim_{x \to 1}(3x - 5) = -2 \qquad \blacksquare$$

A more rigorous way of defining a limit is to say that we can make the difference between $f(x)$ and L as small as we wish by taking x close enough to a.

That is, *the limit of $f(x)$ as x approaches a is L, if for all values of x close enough to a but distinct from a, the magnitude of the difference between $f(x)$ and L is no greater than a prescribed value.*

Example 2   For the function $f(x) = 3x - 5$ given in Example 1, how close must x be to $a = 1$ in order for the difference between $f(x)$ and $L = \lim_{x \to 1} f(x)$ to be less than 0.006?

Solution   Since $f(x) = 3x - 5$ and $L = -2$, we want

$$-0.006 < f(x) - L < 0.006$$

or

$$-0.006 < (3x - 5) - (-2) < 0.006$$

This is equivalent to saying that

$$(3x - 5) - (-2) < 0.006 \quad \text{and} \quad (-2) - (3x - 5) < 0.006$$

Solving for x in the first inequality, we obtain

$$3x - 3 < 0.006$$
$$3x < 3.006$$
$$x < 1.002$$

For the second inequality, we have

$$-3x + 3 < 0.006$$
$$-3x < -2.994$$
$$x > 0.998$$

(When multiplying or dividing an inequality by a negative number, remember that you must reverse the inequality.) Therefore, if $0.998 < x < 1.002$, then the difference between $f(x) = 3x - 5$ and the limit $L = -2$ will be less than 0.006. Stated another way, whenever x is within 0.002 of $a = 1$, $f(x)$ is within 0.006 of the limit $L = -2$.   ∎

In the above example, the function $f(x)$ is defined at $x = a$. However, in order for $\lim_{x \to a} f(x)$ to exist, it is not necessary that $f(x)$ be defined at $x = a$. This is illustrated in the next example.

Example 3   Consider the function

$$f(x) = \begin{cases} x^2 & \text{if } x < 5 \\ 3x + 10 & \text{if } x > 5 \end{cases}$$

Does the limit of $f(x)$ as $x \to 5$ exist, and if so, what is its value?

Solution   When $x < 5$, $f(x) = x^2$. Taking some values less than 5, but close to 5, we get:

x	$f(x) = x^2$
4.9	24.01
4.99	24.9001
4.999	24.990001

Approaching $x = 5$ from the right, we have:

x	$f(x) = 3x + 10$
5.1	25.3
5.01	25.03
5.001	25.003

In both cases, it appears that the function approaches the value of 25. Thus, we conclude that the limit of $f(x)$ is 25 as x approaches 5. The graph of this function is given in Figure 2-5.

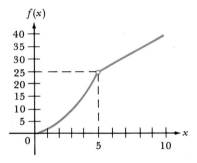

Figure 2-5
Graph of f(x) for Example 3

In Example 3, $f(x)$ was not defined at $x = 5$. We therefore see that $\lim_{x \to a} f(x)$ may exist even if $f(a)$ is not defined. As a general guideline, in computing $\lim_{x \to a} f(x)$, we do not care what the value of $f(a)$ is or whether it even exists. All we care about are values of $f(x)$ as x gets *close to a,* but is distinct from a.

Example 4   Let us modify the previous example slightly. Let

$$f(x) = \begin{cases} x^2 & \text{if } x < 5 \\ 30 & \text{if } x = 5 \\ 3x + 10 & \text{if } x > 5 \end{cases}$$

Even though $f(5) = 30$, we still see that $f(x)$ gets close to 25 as x approaches 5 from either the left or the right. Thus,

$$\lim_{x \to 5} f(x) = 25$$

The graph of this function and the limit are shown in Figure 2-6 on page 52.

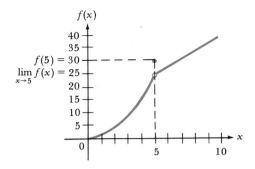

Figure 2-6
Graph of f(x) for Example 4

Example 5   Consider the function

$$f(x) = \frac{1}{x}$$

Does the limit as x approaches 0 exist?

Solution   As $x \to 0$ from the left, the value of the function will always be negative. Some values close to 0 are given in the table:

x	$f(x)$
-0.1	-10
-0.01	-100
-0.001	$-1,000$

As $x \to 0$ from the left, the values of the function become unbounded in the negative direction. In this case, we say that $f(x)$ approaches negative infinity, and we write this as $f(x) \to -\infty$.

When $x \to 0$ from the right, the value of the function will always be positive. For example, we have:

x	$f(x)$
0.1	10
0.01	100
0.001	1,000

In this case, the function values become unbounded in the positive direction, and we say that $f(x)$ approaches positive infinity, written $f(x) \to \infty$. Since the limits from each side of 0 do not equal a common finite value L, the limit of $f(x)$ does not exist as $x \to 0$. This is shown in Figure 2-7. The concept of infinite limits is treated more extensively in the supplement.

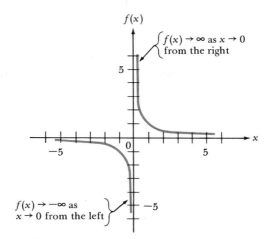

$f(x)$

$\begin{cases} f(x) \to \infty \text{ as } x \to 0 \\ \text{from the right} \end{cases}$

$f(x) \to -\infty$ as
$x \to 0$ from the left

Figure 2-7

Graph of $f(x) = \dfrac{1}{x}$

COMPUTATIONAL METHODS FOR LIMITS

Later in this chapter we will need to compute limits in order to find rates of change of functions. We now present some useful rules and techniques that enable us to compute limits of most functions rather easily. There are seven important rules that can be used to simplify the computations necessary for computing limits. These are given in the box. In all of these, we assume that $\lim_{x \to a} f(x)$ and $\lim_{x \to a} g(x)$ exist.

Basic Limit Theorems

1. If c is a constant, then $\lim_{x \to a} c = c$.

2. $\lim_{x \to a} x = a$

3. If c is a constant, then $\lim_{x \to a} c\, f(x) = c \lim_{x \to a} f(x)$.

4. $\lim_{x \to a}[f(x) + g(x)] = \lim_{x \to a} f(x) + \lim_{x \to a} g(x)$

   $\lim_{x \to a}[f(x) - g(x)] = \lim_{x \to a} f(x) - \lim_{x \to a} g(x)$

5. $\lim_{x \to a} f(x)g(x) = \left[\lim_{x \to a} f(x)\right]\left[\lim_{x \to a} g(x)\right]$

6. If $\lim_{x \to a} g(x) \neq 0$, then $\lim_{x \to a} \dfrac{f(x)}{g(x)} = \dfrac{\lim_{x \to a} f(x)}{\lim_{x \to a} g(x)}$.

7. $\lim_{x \to a}[f(x)]^n = \left[\lim_{x \to a} f(x)\right]^n$

How can these rules be useful? Suppose $f(x) = 5x + 3$ and we wish to find the limit as $x \to 2$. We may compute $\lim_{x \to 2}(5x + 3)$ as follows:

$$\lim_{x \to 2}(5x + 3) = \lim_{x \to 2} 5x + \lim_{x \to 2} 3 \qquad \text{Rule 4}$$

$$= 5 \lim_{x \to 2} x + \lim_{x \to 2} 3 \qquad \text{Rule 3}$$

$$= 5 \lim_{x \to 2} x + 3 \qquad \text{Rule 1}$$

$$= 5(2) + 3 \qquad \text{Rule 2}$$

$$= 13$$

Example 6   Find

$$\lim_{x \to 1} \frac{(3x + 1)^2}{2x^2 + 3}$$

Solution   Since $\lim_{x \to 1} 2x^2 + 3 \neq 0$, by rule 6 we can write

$$\lim_{x \to 1} \frac{(3x + 1)^2}{2x^2 + 3} = \frac{\lim_{x \to 1}(3x + 1)^2}{\lim_{x \to 1}(2x^2 + 3)}$$

$$= \frac{\left[\lim_{x \to 1}(3x + 1)\right]^2}{\lim_{x \to 1}(2x^2 + 3)} \qquad \text{Rule 7}$$

$$= \frac{\left[\lim_{x \to 1} 3x + \lim_{x \to 1} 1\right]^2}{\lim_{x \to 1} 2x^2 + \lim_{x \to 1} 3} \qquad \text{Rule 4}$$

$$= \frac{\left[3 \lim_{x \to 1} x + \lim_{x \to 1} 1\right]^2}{2\left(\lim_{x \to 1} x\right)^2 + \lim_{x \to 1} 3} \qquad \text{Rules 3 and 7}$$

$$= \frac{[(3)(1) + 1]^2}{(2)(1) + 3} \qquad \text{Rules 1 and 2}$$

$$= \frac{16}{5}$$

It would indeed be very tedious to have to employ these limit rules in such detail. Fortunately, they can be applied in general to show the following results:

THEOREM

If $f(x)$ is a polynomial, then

$$\lim_{x \to a} f(x) = f(a)$$

THEOREM

If $f(x) = g(x)/h(x)$ where both $g(x)$ and $h(x)$ are polynomials and $h(a) \neq 0$, then

$$\lim_{x \to a} f(x) = \frac{g(a)}{h(a)} = f(a)$$

Thus, for polynomials and functions that are ratios of polynomials (as long as the denominator is not zero), we can compute limits by simple substitution. In many other cases which we will not encounter in the remainder of this chapter, evaluating limits is not this easy, and certain "tricks of the trade" must be used. These are discussed in the chapter supplement.

In Section 1-4 we gave an intuitive notion of a *continuous* function in terms of its graph. Any function with the property that $\lim_{x \to a} f(x) = f(a)$ for each a in the domain of $f(x)$ is a continuous function. The theorems stated above show that polynomials and ratios of polynomials are continuous functions. This concept is addressed further in the supplement.

PROBLEMS (Section 2-1)

In Problems 1–10, construct a table of values for x and f(x) by letting x approach a from both the left and right side. Then determine whether the limit exists, and if so, determine its value.

1. $f(x) = 6 + \dfrac{4}{x}, \quad a = 4$

2. $f(x) = 3x^3 + 4, \quad a = -2$

3. $f(x) = \dfrac{2x^2 - 2x}{x - 1}, \quad a = 1$

4. $f(x) = \dfrac{x^2 + x}{x}, \quad a = 0$

5. $f(x) = \dfrac{1}{x + 3}, \quad a = -3$

6. $f(x) = \dfrac{3}{(2x - 1)^2}, \quad a = \dfrac{1}{2}$

7. $f(x) = \sqrt{5x - 5}, \quad a = 6$

8. $f(x) = \begin{cases} 3x - 1 & \text{if } x \neq 2 \\ 2 & \text{if } x = 2 \end{cases} \quad a = 2$

9. $f(x) = \begin{cases} x^2 & \text{if } x \leq 4 \\ 3x + 4 & \text{if } x > 4 \end{cases} \quad a = 4$

10. $f(x) = \begin{cases} 2x & \text{if } x < 0 \\ 1 - 2x & \text{if } x \geq 0 \end{cases} \quad a = 0$

For each of the following graphs, tell whether the limit exists as x approaches 2:

11.

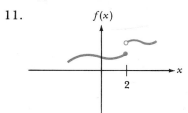

12.

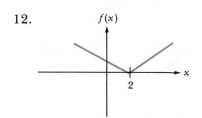

13.

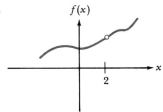

14.

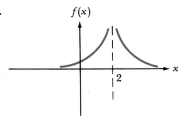

15.

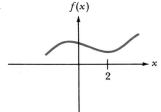

16.

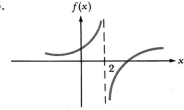

17. If

$$f(x) = \begin{cases} 2x + 3 & \text{if } x < 3 \\ x^2 - 1 & \text{if } x \geq 3 \end{cases}$$

show that $\lim_{x \to 3} f(x)$ does not exist.

Find the limits of the following functions or show that the limit does not exist:

18. $g(t) = \begin{cases} 3t - 4 & \text{if } t \leq 6 \\ 14 & \text{if } t > 6 \end{cases}$　as $t \to 6$

19. $g(y) = \begin{cases} 6y^2 - 4y & \text{if } y < 1 \\ 3 & \text{if } y = 1 \\ 3 - y & \text{if } y > 1 \end{cases}$　as $y \to 1$

20. $f(x) = \dfrac{1}{|x|}$　as $x \to 0$

21. $f(x) = \dfrac{x}{|x|}$　as $x \to 0$

In Problems 22–33, evaluate the limits using the limit rules and state reasons for each step.

22. $\lim_{x \to 2}(x^2 + 2x - 5)$

23. $\lim_{t \to 4}(t^2 - 16)$

24. $\lim_{x \to 2}(3x - 10)$

25. $\lim_{x \to -7}(3 - 7x)$

26. $\lim_{x \to 1} \dfrac{x^3 + 1}{x + 1}$

27. $\lim_{t \to -2} \dfrac{t^2 + 4}{t^3 + 7}$

28. $\lim_{h \to 1} \sqrt{\dfrac{2h^2 + 4h - 1}{h^2 + 3}}$

29. $\lim_{x \to -1} 7$

30. $\lim_{x \to 10} 52$

31. $\lim_{x \to 0} \dfrac{\sqrt{1 + x^2}}{x + 1}$

32. $\lim_{h \to 0}(x^2 + 4xh + 3h^2)$

33. $\lim_{h \to 0} \dfrac{\sqrt{x^2 + h}}{5}$

2-2 PROPERTIES OF SLOPES

In Chapter 1 we saw that linear functions are useful in developing mathematical models, and we reviewed the concept of a straight line. The slope of a linear function, to recall, is the ratio of the change in the y-coordinates to the change in the x-coordinates for two points (x_1, y_1) and (x_2, y_2) on the line; that is,

$$m = \frac{y_2 - y_1}{x_2 - x_1} \qquad x_1 \neq x_2$$

as illustrated in Figure 2-8. For a straight line, the slope is always the same regardless of the two points chosen to compute it.

Figure 2-8
The slope of a line

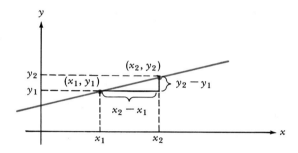

Example 7
Consumer Demand

A new product is being introduced in selected cities by a large consumer products company. By putting the product on display in supermarkets and drugstores, sales of 6000 units are expected during the first 3 months. However, by sending free samples to groups of residents, sales are expected to be increased. Research indicates that for every 10 samples distributed, 3 additional regular items will be purchased. We may use this information to develop a model of sales as a function of the number of samples distributed. Graphically, we shall represent sales (the dependent variable) on the y-axis, and the number of samples distributed (the independent variable) on the x-axis. If no samples are distributed, that is, $x = 0$, then sales will equal 6000. This is the y-intercept. Also, if x increases by 10, then y will increase by 3; that is, when $x = 10$, then $y = 6003$. Using these two points $(0, 6000)$ and $(10, 6003)$, we see that the slope is equal to

$$m = \frac{6003 - 6000}{10 - 0} = \frac{3}{10} = 0.3$$

Notice that it does not matter what value x is; a 1 unit increase in x will give a 0.3 unit increase in y. The model for sales as a function of the number of samples is

$$y = f(x) = mx + b$$
$$= 0.3x + 6000$$

This is illustrated in Figure 2-9 on page 58.

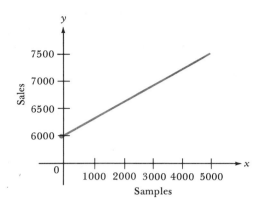

Figure 2-9
Graph of y = 0.3x + 6000

The value of the slope of a line has several important graphical interpretations. If the slope is positive, the line rises as x increases; if the slope is negative, the line falls as x increases (Figure 2-10). Also, the larger the magnitude of the slope, the steeper is the line (Figure 2-11). Some additional properties of slopes are given in the box.

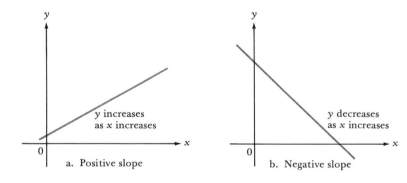

Figure 2-10
Illustration of positive and negative slopes

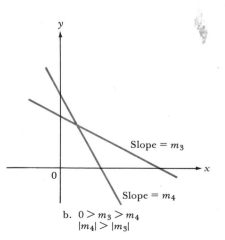

Figure 2-11
The magnitude of the slope affects the steepness of the line

Properties of Slopes

> 1. The slope of a vertical line (parallel to the y-axis) is undefined.
> 2. The slope of a horizontal line (parallel to the x-axis) is 0.
> 3. The slopes of any two parallel lines are equal.
> 4. Perpendicular lines have negative reciprocal slopes. That is, if one line has a slope of $m_1 \neq 0$ and the other has a slope of m_2, then $m_2 = -1/m_1$.

Note that parallel and perpendicular properties of lines depend only on the slope, and are independent of the value of the intercept.

Example 8   Consider the line $y = \frac{3}{2}x + 2$. The slope is $\frac{3}{2}$ and its negative reciprocal is $-\frac{2}{3}$. Therefore, the line $y = -\frac{2}{3}x + 8$ must be perpendicular to $y = \frac{3}{2}x + 2$. Also, the line $y = \frac{3}{2}x + 5$ is parallel to $y = \frac{3}{2}x + 2$, since both have the same slope. This is illustrated in Figure 2-12.

Figure 2-12
Perpendicular lines have negative reciprocal slopes. Parallel lines have the same slope.

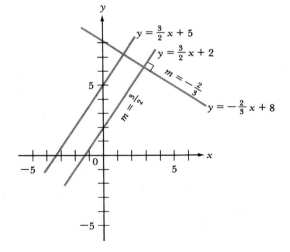

Since the slope of a straight line specifies how much the function changes as x is increased by 1 unit, the slope can be interpreted as the *rate of change* of the function with respect to x. Thus, in the consumer demand example, the rate of change of sales with respect to the number of samples distributed is 0.3. This means that if 1 additional sample is distributed, then sales will change by 0.3. Since this value is positive, it indicates an increase. A negative value would have indicated a decrease in sales.

PROBLEMS (Section 2-2)

In Problems 1–6, find the slope of each line. Write an equation of a different line that is parallel to the given line.

1. $y = 4x + 3$

2. $y = -\frac{2}{3}x - 5$

3. $2x + y = 5$

4. $y = -x$

5. $3y - 2x = 7$

6. $6x - 4y = 5$

In Problems 7–12, find the slope of each line. Write an equation of a line that is perpendicular to the given line.

7. $y = \frac{3}{2}x + 1$

8. $y = -3x + 2$

9. $y - 2x = -4$

10. $2x - 4y + 8 = 0$

11. $3x + 2y = -2$

12. $7y - 5x - 3 = 0$

13. How does the graph of a line with positive slope differ in appearance from the graph of a line with negative slope?

14. Find an equation of the line parallel to the y-axis and passing through the point $(4, -2)$.

15. Find an equation of the line parallel to the x-axis and passing through the point $(5, -3)$.

16. Find an equation of the line that passes through the point $(1, -4)$ and is parallel to the line $x + 5y - 3 = 0$.

17. Find an equation of the line that passes through the point $(-2, -3)$ and is parallel to the line $3x - 7y + 4 = 0$.

18. Find an equation of the line that passes through the point $(3, -2)$ and is perpendicular to the line $2x + 3y + 4 = 0$.

19. Find an equation of the line that passes through the point $(1, 5)$ and is perpendicular to the line $5x - 4y + 1 = 0$.

20. Use the definition of slope to explain why the slope of a vertical line is undefined and also why the slope of a horizontal line is 0.

21. Find an equation of the line that passes through the origin and is perpendicular to the line whose x-intercept is 6 and y-intercept is -2.

22. Find an equation of the line that passes through the point $(-1, -3)$ and is parallel to the line whose x-intercept is -4 and y-intercept is 8.

23. Suppose a person has a total of $100 to spend on two items. The first item X costs $2 per unit and the second item Y costs $3 per unit. Write an equation relating the number of units x of item X to the number of units y of item Y purchased, assuming that the entire $100 is spent. What happens to y if the number of units x is increased by 1 unit?

24. In accounting, the *book value* of an item purchased for D dollars having an expected life of N years and a salvage value of S dollars is often expressed as a linear function of the number of years in use, t. This is called *linear depreciation* and is important in tax computations. The book value is given by

$$V(t) = D - \left(\frac{D - S}{N}\right)t$$

If $D = \$8400$, $N = 5$, and $S = \$1400$, evaluate $V(t)$. What is the meaning of the slope of this equation? How does the slope change as D is increased? As S is increased? As N is increased?

2-3 THE RATE OF CHANGE OF A FUNCTION

Time Interval	Number of Cars
6:30–6:31	50
6:35–6:36	60
6:40–6:41	92
6:45–6:46	130
6:50–6:51	200
6:55–6:56	271

In the previous section we found that the rate of change of a linear function is given by its slope. However, what if the function that interests us is not linear? How can we determine the rate of change? In this section we address the problem of finding the rate of change of a nonlinear function.

To illustrate the rate of change of a nonlinear function, we consider the following situation. A traffic survey conducted on a particular segment of interstate highway during the morning rush hour resulted in the collection of the data in the table. Analysts found that the data can be rather accurately represented by the function

$$y = f(t) = \tfrac{1}{3}t^2 + 50$$

where t = the number of minutes past 6:30 and y = the number of cars on the segment of highway. Note that the domain of the function is $0 \le t \le 30$. Figure 2-13 illustrates the graph of this function.

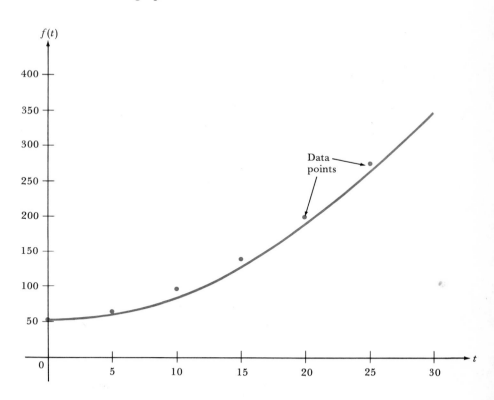

Figure 2-13
Graph of $y = f(t) = \tfrac{1}{3}t^2 + 50$

Since this function is not linear, we cannot properly define its slope. How, then, can we express the rate of change of traffic volume over time? One thing we can do is to compute the *average* rate of change over a specific time interval.

> The **average rate of change of a function** $f(x)$ **between** $x = a$ **and** $x = b$ **is given by**
>
> $$\frac{f(b) - f(a)}{b - a} = \frac{\text{Change in } f(x)}{\text{Change in } x}$$

Clearly, the points $(a, f(a))$ and $(b, f(b))$ lie on the graph of the function $f(x)$. The line joining two points on the graph of a function $f(x)$ is called a **secant line.** The slope of the secant line between $(a, f(a))$ and $(b, f(b))$ is computed in the usual fashion; that is, the change in the y-coordinates divided by the change in the x-coordinates gives the slope of the secant line:

$$\frac{f(b) - f(a)}{b - a}$$

Therefore, the average rate of change of a function between $x = a$ and $x = b$ is equal to the slope of the secant line between the points $(a, f(a))$ and $(b, f(b))$.

For the function $y = f(t)$ in the traffic survey example, the average rate of change between $t = 0$ and $t = 30$ is given by

$$\frac{f(30) - f(0)}{30 - 0} = \frac{[\frac{1}{3}(30)^2 + 50] - 50}{30}$$

$$= \frac{350 - 50}{30} = 10 \text{ vehicles per minute}$$

Notice that since the numerator is expressed in "vehicles" and the denominator in "minutes," the dimensions of the average rate of change for this example are "vehicles per minute." This can be interpreted as saying that, on the average, the number of vehicles on the highway increased by 10 each minute. But, as Figure 2-13 shows, this is not what really happens. The increase is much slower at the beginning of the time period than at the end. For instance, over the time interval $0 \le t \le 10$, the average rate of change is

$$\frac{f(10) - f(0)}{10 - 0} = \frac{[\frac{1}{3}(10)^2 + 50] - 50}{10}$$

$$= \frac{83\frac{1}{3} - 50}{10} = 3\frac{1}{3} \text{ vehicles per minute}$$

On the other hand, over the interval $20 \le t \le 30$, the average rate of change is

$$\frac{f(30) - f(20)}{30 - 20} = \frac{[\frac{1}{3}(30)^2 + 50] - [\frac{1}{3}(20)^2 + 50]}{10}$$

$$= \frac{350 - 183\frac{1}{3}}{10} = 16\frac{2}{3} \text{ vehicles per minute}$$

In Figure 2-14 we can clearly see that the slope of the secant line from $t = 0$ to $t = 10$ is less than the slope of the secant line between $t = 20$ and $t = 30$. Thus, we see that *the average rate of change depends on the time interval chosen.*

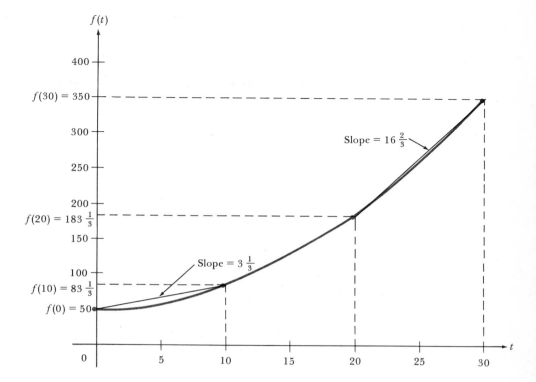

Figure 2-14
Average rates of change for
$y = f(t) = \frac{1}{3}t^2 + 50$

Notice that the line segments in Figure 2-14 are approximations to the actual curve. The smaller the time interval chosen, the better the approximation. We could use the slopes of these line segments as a measure of the "steepness" of the curve. In other words, the curve is steeper between $t = 20$ and $t = 30$ than it is between $t = 0$ and $t = 10$ since the slope is larger.

For a function $f(x)$, we often denote the average rate of change by the symbol $\Delta f/\Delta x$, that is,

$$\frac{\Delta f}{\Delta x} = \frac{f(b) - f(a)}{b - a}$$

The symbol Δ is the Greek letter delta and is usually used to denote the change in a particular quantity. Thus, Δf is read as "the change in (the function) f" and Δx as "the change in the variable x."

Example 9   Compute the average rates of change for the function $f(x) = -6x^2 + 4x + 2$ between $x = -2$ and $x = 0$, and between $x = 0$ and $x = 2$.

Solution   We first compute $f(-2) = -30, f(0) = 2$, and $f(2) = -14$. Between $x = -2$ and $x = 0$ the average rate of change is

$$\frac{\Delta f}{\Delta x} = \frac{f(0) - f(-2)}{0 - (-2)} = \frac{2 - (-30)}{2} = 16$$

This means that the value of the function has increased, on the average, 16 units for every unit increase in x between $x = -2$ and $x = 0$. Between $x = 0$ and $x = 2$ we compute the average rate of change as

$$\frac{\Delta f}{\Delta x} = \frac{f(2) - f(0)}{2 - 0} = \frac{-14 - 2}{2} = -8$$

In this case, the function has decreased (as indicated by the negative sign) an average of 8 units for every unit increase in x between $x = 0$ and $x = 2$. This is illustrated in Figure 2-15.

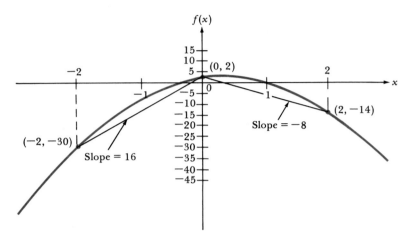

Figure 2-15
Average rates of change for
$f(x) = -6x^2 + 4x + 2$

PROBLEMS (Section 2-3)

1. Consider the empirical meteor data discussed in Example 7 in Chapter 1:

Date, x	Sightings, $f(x)$
7	14
8	19
9	10
10	14
11	22
12	26

What is the average rate of change between $x = 7$ and $x = 8$, between $x = 8$ and $x = 12$, and between $x = 7$ and $x = 12$?

2. The price of a share of common stock of a well-known computer firm over a 6 week period is given in the following table:

Week	Stock Price
1	50
2	49
3	$47\frac{1}{2}$
4	48
5	52
6	50

What is the average rate of change between week 1 and each successive week?

3. Consider the function $f(x) = 5x - 3$. What is the average rate of change between $x = -3$ and $x = 0$, and between $x = 0$ and $x = 4$?

4. Consider the function $f(x) = -x^2 + 2x - 1$. What is the average rate of change between $x = -2$ and $x = -1$, and between $x = -1$ and $x = 0$?

5. A recovery curve following treatment for a patient with acquired dyslexia is shown here. Using the graph, find the average rate of change between the 50th and 100th weeks, the 50th and 80th weeks, and the 50th and 60th weeks. Can you estimate the rate of change of the error score "at" the 50th week?

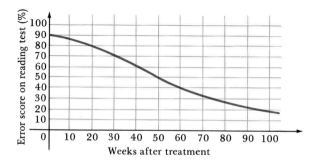

6. The temperature of a person during an illness follows the curve shown here. Find the average rate of change of temperature between the 10th and 20th hours, the 10th and 16th hours, and the 10th and 12th hours. Can you estimate the rate of change "at" the 10th hour?

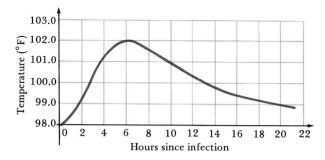

7. The number of employees in several industries is given in the table for the years 1947, 1968, and 1980. Find the average rate of change in each industry during the time periods 1947–1968 and 1968–1980.

	Employment (in Thousands)		
	1947	*1968*	*1980*
Agriculture, forestry, and fisheries	7,890	4,150	3,180
Mining	955	640	590
Transportation and utilities	4,160	4,500	5,000
Finance, insurance, and real estate	1,750	3,275	4,640
Services	5,050	15,000	21,000

8. Using a 1967 base of 100%, the costs of goods and services until 1980 are given in the table. Plot these points and find the average rate of change from 1967 to 1980 and between each time period (1967–1970, 1970–1972, etc.).

Year	*All Services*	*All Consumer Items*
1970	120	115
1972	134	126
1974	153	146
1976	180	170
1978	210	196
1980	270	247

9. In 1967, the average salary of industrial workers in the United States had increased to 143% of the 1956 value. Assuming a linear change, find the function expressing the percentage change over time. What is the rate of change during this time period? Assuming a continued linear increase, what would the estimate for 1985 be (as a percentage of the 1956 value)?

10. The book value of an item decreased by 15% of its original value of $5000 in 2 years. Assuming linear depreciation, find the function expressing the percentage change over time. How long will it be until the book value is zero, assuming a zero salvage value? (See Problem 24 in Section 2-2.)

11. In 5 years after 1980, the average starting salaries of marketing graduates in a business school

increased by 30%. Find the function expressing the percentage change over time from 1980, assuming a linear change. What is the average rate of change during these 5 years? What would the estimated percentage increase for 1987 be?

12. Let $f(x) = x^2$. Find the rate of change of $f(x)$ over the intervals from $x = 0$ to $x = 1$; $x = 0$ to $x = \frac{1}{10}$; $x = 0$ to $x = \frac{1}{100}$; and $x = 0$ to $x = \frac{1}{1000}$. Do you notice a trend? How does this relate to the concept of a limit?

13. Let $f(x) = x^2$. Find the rate of change of $f(x)$ over the intervals from $x = 1$ to $x = 1.1$; $x = 1$ to $x = 1.01$; and $x = 1$ to $x = 1.001$. Do you notice a trend? How does this relate to the concept of a limit?

2-4 THE SLOPE OF A TANGENT LINE

Suppose we were to observe traffic volume $f(t)$ on a highway segment at an instant t in time, and we ask "What is the rate of change in traffic volume at this instant?" Finding $\Delta f/\Delta t$ will not answer this question since we need two distinct points to compute this quantity. We would like to determine the rate of change at precisely one point $(t, f(t))$ on the curve. In order to find this, we need to introduce the notion of a tangent line.

You may have an intuitive notion of what we mean by a tangent line to a curve. However, the idea of a tangent line is rather subtle and most people, if pressed, would have difficulty defining the concept accurately. We leave a precise definition until later. Roughly speaking, the **tangent line to a curve at a given point** is a line that touches the curve at that point and approximates the curve well in a small region surrounding the point. Figure 2-16 illustrates some examples of lines tangent to curves at a point as well as examples of lines that are not tangent at that point.

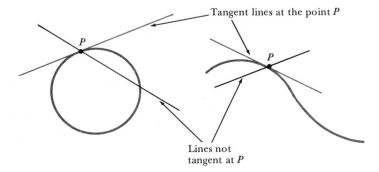

Figure 2-16
Illustration of tangent lines

The tangent line to a curve at a point is unique. It would seem reasonable to use the slope of the tangent line to a curve at a point as a measure of the "slope" or steepness of the curve itself at that point, even though the curve may not be linear. This is not difficult to imagine if one magnifies a small segment of a curve. Figure 2-17 shows a sequence of enlargements of a circle, and you can see that in a very small region around a point on the circle, the curve is approximately a

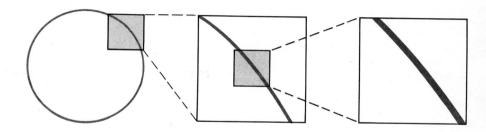

Figure 2-17
A portion of a circle magnified twice

straight line. The slope of the tangent line at a point, then, measures the rate of change of the function at that point. In Figure 2-18, several tangent lines are shown for the traffic volume function $f(t) = \frac{1}{3}t^2 + 50$. As t gets larger, the slopes of these lines increase and the curve becomes steeper. Thus, in view of our preceding discussion, if we can compute the slope of the tangent line to a curve at a point, we can use it as a measure of the rate of change of the function at that point.

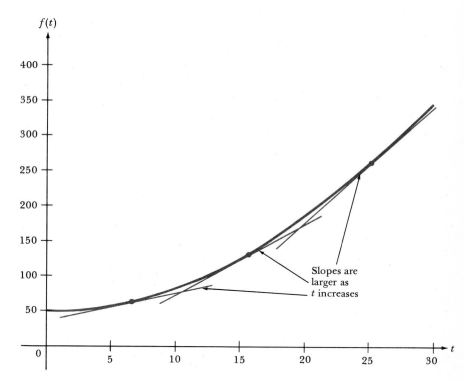

Figure 2-18

Tangent lines for the traffic volume function

Let $(a, f(a))$ be a point on the graph of a function $f(x)$. How can we determine the slope of the tangent line to the curve at this point? The difficulty we face is that we know only *one point* through which the line passes. The slope of a line is uniquely determined by *two points*. However, if we do know the rule defining the function $f(x)$, we can choose some other value for x, not equal to a, say $x = b$, and compute the slope of the secant line between the points $(a, f(a))$ and $(b, f(b))$. This secant line will be an *approximation* to the tangent line, as illustrated in

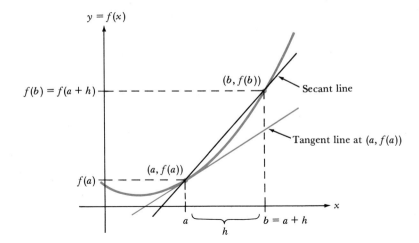

Figure 2-19
Approximating the tangent line at
(a, f(a)) by a secant line

Figure 2-19. Notice that we have defined h to be the distance along the x-axis between a and b. Therefore, $b = a + h$ so that $f(b) = f(a + h)$. The slope of this secant line is

$$\frac{f(b) - f(a)}{b - a} = \frac{f(a + h) - f(a)}{(a + h) - a} = \frac{f(a + h) - f(a)}{h}$$

We can approximate the slope of the tangent line at $(a, f(a))$ even better by choosing b closer to a, that is, letting the distance h get smaller. In Figure 2-20 we selected a point b_1 closer to a. Notice that the distance between a and b_1, h_1, is smaller than h. The slope of the new secant line between $(a, f(a))$ and $(b_1, f(b_1))$ is

$$\frac{f(b_1) - f(a)}{b_1 - a} = \frac{f(a + h_1) - f(a)}{h_1}$$

If we compare this to the formula for the slope of the first secant line, we see that

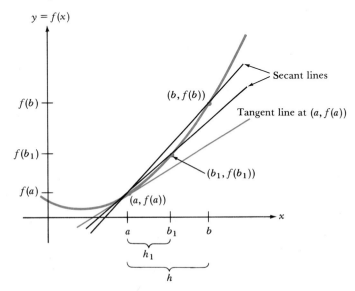

Figure 2-20
Finding a better approximation to
the tangent line

all that has changed is h. Intuitively, it would seem reasonable that if we select yet a smaller value for h, say h_2, then the value of

$$\frac{f(a + h_2) - f(a)}{h_2}$$

will be even closer to the true slope of the tangent line. We can choose h as small as we wish; however, we cannot let h be 0, since we cannot divide by 0.

We will illustrate this method of computing the slope of the tangent line at a point $(a, f(a))$ for the traffic volume example, $f(t) = \frac{1}{3}t^2 + 50$. Let $(a, f(a))$ be the point $(20, 183\frac{1}{3})$. Suppose we choose $b = 30$ so that $f(30) = 350$. Then h (the distance between a and b) is 10. The slope of the secant line between $(20, 183\frac{1}{3})$ and $(30, 350)$ is

$$\frac{350 - 183\frac{1}{3}}{10} = \frac{166\frac{2}{3}}{10} = 16\frac{2}{3}$$

Table 2-1 shows the values of the slopes of secant lines for various values of b, using decimal approximations where convenient.

Table 2-1

b	$f(b)$	$h = b - 20$	$f(b) - f(20)$	$[f(b) - f(20)]/h$
30	350	10	166.67	16.67
25	258.33	5	75	15
21	197	1	13.67	13.67
20.5	190.08	0.5	6.75	13.5
20.1	184.67	0.1	1.3367	13.367
20.01	183.4667	0.01	0.1334	13.34

Notice that as b gets closer to 20, h gets closer to 0. It appears that the slopes form a sequence that approaches $13\frac{1}{3}$. Choosing trial values in this way is quite tedious. Let us see if we can reach this same result mathematically. (You may wish to look over Algebra Review 5, page 70, at this point.)

We saw that the general expression for the slope of the secant line between $(a, f(a))$ and $(b, f(b))$, where $b = a + h$, is given by

$$\text{Slope} = \frac{f(a + h) - f(a)}{h}$$

If we let $a = 20$, we have

$$\text{Slope} = \frac{f(20 + h) - f(20)}{h}$$

Evaluating $f(20 + h)$ and $f(20)$, we obtain

$$\frac{f(20 + h) - f(20)}{h} = \frac{[\frac{1}{3}(20 + h)^2 + 50] - [\frac{1}{3}(20)^2 + 50]}{h}$$

We may simplify this as follows:

$$\frac{f(20+h)-f(20)}{h} = \frac{[\frac{1}{3}(400+40h+h^2)+50]-[183\frac{1}{3}]}{h}$$

$$= \frac{[183\frac{1}{3}+40h/3+h^2/3]-[183\frac{1}{3}]}{h}$$

$$= \frac{40h/3+h^2/3}{h}$$

$$= \frac{40h+h^2}{3h}$$

$$= \frac{h(40+h)}{3h}$$

$$= \frac{40+h}{3}$$

Now, we find the limit of this expression by using the sixth basic limit theorem stated in Section 2-1:

$$\lim_{h\to 0}\frac{f(20+h)-f(20)}{h} = \frac{\lim\limits_{h\to 0}(40+h)}{3} = \frac{40}{3} = 13\frac{1}{3}$$

We take this as the slope of the tangent line to $f(t)$ at the point $(20, f(20))$.

Since h can never equal 0, $[f(20+h)-f(20)]/h$ will never *equal* $13\frac{1}{3}$. However, for all h sufficiently small we can get as close to $13\frac{1}{3}$ as we like. For instance,

ALGEBRA REVIEW 5: SIMPLIFYING ALGEBRAIC FRACTIONS

We reduce algebraic fractions to lowest terms by identifying all common factors and dividing the numerator and denominator by these common factors. For example, consider the expression $12x^2/16x^3$. The common factor is $4x^2$. Therefore, we divide the coefficients 12 and 16 by 4 and we divide the variables x^2 and x^3 by x^2. This results in the following:

$$\frac{12x^2}{16x^3} = \frac{\overset{3}{\cancel{12}}x^{\cancel{2}}}{\underset{4}{\cancel{16}}x^{\cancel{3}}x} = \frac{3}{4x}$$

As another example, consider the expression

$$\frac{5x^2+2x}{3x}$$

Noting that x is a common factor of the numerator and denominator, we can factor and divide as follows:

$$\frac{5x^2+2x}{3x} = \frac{\cancel{x}(5x+2)}{\cancel{x}\cdot 3} = \frac{5x+2}{3}$$

Alternatively, we can separate the expression into

suppose we wish $[f(20 + h) - f(20)]/h$ to be within $\frac{1}{6}$ of $13\frac{1}{3}$, or no greater than 13.5. From Table 2-1 we see that whenever h is less than 0.5, the value of $[f(20 + h) - f(20)]/h$ will be less than 13.5. In a similar way, we can show that this quotient is greater than $13\frac{1}{3} - \frac{1}{6}$ for sufficiently small negative values of h.

Let us define a special symbol for limits of the type computed above. We will take such a limit as the *definition* of the slope of the tangent line at $(a, f(a))$, and hence, we are finally able to define unambiguously the notion of a tangent line.

Let $f(x)$ be a function and let a be in the domain of f. Define the symbol $f'(a)$ by

$$f'(a) = \lim_{h \to 0} \frac{f(a + h) - f(a)}{h}$$

assuming this limit exists. We then define the tangent line to the graph of $f(x)$ at the point $(a, f(a))$ to be the line through $(a, f(a))$ with slope $f'(a)$.

If the function $f(x)$ is itself linear, then its graph and its tangent line coincide, as you might expect.

individual fractions, and then reduce each fraction separately:

$$\frac{5x^2 + 2x}{3x} = \frac{5x^2}{3x} + \frac{2x}{3x}$$

$$= \frac{5x}{3} + \frac{2}{3}$$

$$= \frac{5x + 2}{3}$$

EXERCISES*

Simplify the following:

a. $\dfrac{7x^4}{49x}$ b. $\dfrac{64y^3}{4y^2 + 32y}$

c. $\dfrac{6xy^4}{18xy^2}$ d. $\dfrac{xy + y^2}{y(y^3 - 4y)}$

e. $\dfrac{2xh + h^2}{h(h + 2)}$

---

* Answers to algebra review exercises can be found at the end of the answer section in the back of the book.

Suppose, for example, that $f(x) = mx + b$. Then for any number a,

$$f'(a) = \lim_{h \to 0} \frac{f(a + h) - f(a)}{h}$$

$$= \lim_{h \to 0} \frac{[m(a + h) + b] - [ma + b]}{h}$$

$$= \lim_{h \to 0} \frac{\cancel{ma} + mh + \cancel{b} - \cancel{ma} - \cancel{b}}{h}$$

$$= \lim_{h \to 0} \frac{m\cancel{h}}{\cancel{h}}$$

$$= m$$

That is, the graph of $f(x)$ and its tangent line at $x = a$ are both lines with slope m. Therefore, these lines are parallel. However, since they both pass through the point $(a, f(a))$, the two lines in fact coincide.

The importance of the definition above is that it gives us a method for computing the slope of the tangent line to a curve at a specific point $(a, f(a))$.

Computing the Slope of the Tangent Line to $y = f(x)$ at $x = a$

Step 1.   Compute the expression

$$\frac{f(a + h) - f(a)}{h}$$

and simplify as much as possible.

Step 2.   Determine the limit, if it exists, of this expression as h approaches 0. The result is $f'(a)$, the slope of the tangent line at the point $(a, f(a))$.

Example 10   Determine the slope of the line tangent to the function $f(x) = 2x - 5x^2$ at $x = 1$.

Solution   When $x = 1$, $f(1) = 2(1) - 5(1)^2 = -3$. We compute the expression

$$\frac{f(a + h) - f(a)}{h} = \frac{f(1 + h) - f(1)}{h}$$

$$= \frac{[2(1 + h) - 5(1 + h)^2] - [-3]}{h}$$

$$= \frac{2 + 2h - 5(1 + 2h + h^2) + 3}{h}$$

$$= \frac{2 + 2h - 5 - 10h - 5h^2 + 3}{h}$$

$$= \frac{-8h - 5h^2}{h}$$

$$= -8 - 5h$$

Now,

$$\lim_{h \to 0} \frac{f(a+h) - f(a)}{h} = \lim_{h \to 0} (-8 - 5h)$$

$$= \lim_{h \to 0} (-8) - 5 \lim_{h \to 0} h = -8$$

Therefore, the slope of the tangent line at $x = 1$ is -8. ∎

Example 11 (You may wish to read Algebra Review 6, page 74, before studying this example.) Find an equation of the tangent line to the curve $g(t) = 3/t$ at $t = 5$.

Solution First, we find $g(5 + h) = 3/(5 + h)$ and $g(5) = \frac{3}{5}$. Next, compute

$$\lim_{h \to 0} \frac{g(a+h) - g(a)}{h} = \lim_{h \to 0} \frac{g(5+h) - g(5)}{h}$$

$$= \lim_{h \to 0} \frac{\dfrac{3}{5+h} - \dfrac{3}{5}}{h}$$

$$= \lim_{h \to 0} \frac{\dfrac{15 - 3(5+h)}{5(5+h)}}{h}$$

$$= \lim_{h \to 0} \frac{15 - 15 - 3h}{5h(5+h)}$$

$$= \lim_{h \to 0} \frac{-3h}{5h(5+h)}$$

$$= \lim_{h \to 0} \frac{-3}{5(5+h)}$$

$$= \lim_{h \to 0} \frac{-3}{25 + 5h} = \frac{-3}{25 + 0} = -\frac{3}{25}$$

This is the slope of the tangent line at $t = 5$. The equation of the tangent line in point–slope form is then

$$y - \tfrac{3}{5} = -\tfrac{3}{25}(t - 5)$$ ∎

It is important to realize that h can approach 0 from either the right or the left; that is, through either positive or negative values. If the limiting values are not the same as h approaches 0 from both sides, then the limit does not exist, and the function does not have a unique tangent line at that point.

Example 12 Consider the absolute value function

$$f(x) = |x|$$

which is defined by

$$|x| = \begin{cases} x & \text{if } x \geq 0 \\ -x & \text{if } x < 0 \end{cases}$$

(Note that $|x|$ is never negative; the $-x$ in the formula above is actually *positive*, since $x < 0$.) The graph of this function is shown in Figure 2-21.

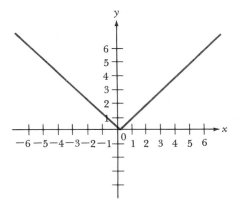

Figure 2-21
Graph of $y = |x|$

ALGEBRA REVIEW 6: ADDING AND DIVIDING FRACTIONAL TERMS

To add algebraic expressions involving fractional terms, it is necessary to obtain a common denominator similar to the manner in which numerical fractions are added in arithmetic. Thus, if a, b, c, and d represent algebraic expressions, then

$$\frac{a}{b} + \frac{c}{d} = \frac{a}{b} \cdot \frac{d}{d} + \frac{c}{d} \cdot \frac{b}{b}$$

$$= \frac{ad}{bd} + \frac{cb}{bd}$$

$$= \frac{ad + cb}{bd}$$

In this example, the common denominator is bd; we write each fraction using this denominator and then add the numerators together. Notice that when we multiply a/b by d/d and c/d by b/b we are simply multiplying each term by 1. For example, consider the expression

$$\frac{3x^2 + 2}{x + 1} + \frac{2x}{x - 1}$$

The common denominator is $(x + 1)(x - 1)$. So, we multiply the first term by $(x - 1)/(x - 1)$ and the second term by $(x + 1)/(x + 1)$. We obtain

$$\frac{3x^2 + 2}{x + 1} + \frac{2x}{x - 1}$$

$$= \frac{3x^2 + 2}{x + 1} \cdot \frac{x - 1}{x - 1} + \frac{2x}{x - 1} \cdot \frac{x + 1}{x + 1}$$

$$= \frac{(3x^2 + 2)(x - 1) + (2x)(x + 1)}{(x + 1)(x - 1)}$$

$$= \frac{(3x^3 - 3x^2 + 2x - 2) + (2x^2 + 2x)}{x^2 - 1}$$

$$= \frac{3x^3 - x^2 + 4x - 2}{x^2 - 1}$$

Suppose that $a = 0$. Now let us examine the limit

$$\lim_{h \to 0} \frac{f(0 + h) - f(0)}{h} = \lim_{h \to 0} \frac{|0 + h| - 0}{h}$$

$$= \lim_{h \to 0} \frac{|h|}{h}$$

If $h > 0$, then $|h|/h = h/h = 1$. If, however, $h < 0$, then $|h|/h = -h/h = -1$. As h approaches 0 from the positive side, the limiting value is $+1$, but as h approaches 0 from the negative side, the limiting value is -1. These "one-sided" limits exist but are not equal to each other. We therefore say that

$$\lim_{h \to 0} \frac{f(0 + h) - f(0)}{h}$$

does not exist, and hence, we cannot assign a slope to a tangent line to $f(x) = |x|$ at $x = 0$. From the graph we also see that there is no *unique* tangent line to the graph of the function at $x = 0$ that satisfies the definition presented on page 71. In general, tangent lines do not exist at places where the graph of the function is broken or has sharp points. ∎

To divide algebraic expressions involving fractional terms, we have

$$\frac{\dfrac{a}{b}}{\dfrac{c}{d}} = \frac{a}{b} \cdot \frac{d}{c} = \frac{ad}{bc}$$

For example,

$$\frac{\dfrac{2x + 4}{5x}}{\dfrac{3x}{4}} = \frac{2x + 4}{5x} \cdot \frac{4}{3x}$$

$$= \frac{(2x + 4)(4)}{(5x)(3x)} = \frac{8x + 16}{15x^2}$$

A special case we often see is when $d = 1$:

$$\frac{\dfrac{a}{b}}{c} = \frac{\dfrac{a}{b}}{\dfrac{c}{1}} = \frac{a}{b} \cdot \frac{1}{c} = \frac{a}{bc}$$

To illustrate this,

$$\frac{\dfrac{6x + 2h}{6 + h}}{h} = \frac{6x + 2h}{6 + h} \cdot \frac{1}{h}$$

$$= \frac{6x + 2h}{(6 + h)h} = \frac{6x + 2h}{6h + h^2}$$

EXERCISES

Simplify the following algebraic fractions:

a. $\dfrac{x}{y} + \dfrac{2x + 1}{y^2}$ b. $\dfrac{x + 2}{3x} + \dfrac{6x}{1 - x}$

c. $\dfrac{2}{x} + \dfrac{3}{y} + \dfrac{4}{x + y}$ d. $\dfrac{\dfrac{3x^2}{4x + 1}}{\dfrac{5 + x}{2}}$

e. $\dfrac{\dfrac{3}{5 + x} - \dfrac{3}{5}}{x}$

Rather than find the slope of the tangent line at a specific point, we can apply this procedure for any arbitrary point $(x, f(x))$. This leads to a new function which we call the *derivative* and define in the next section.

PROBLEMS (Section 2-4)

In Problems 1–12, find the slope of the tangent line to each function using the formula

$$f'(a) = \lim_{h \to 0} \frac{f(a + h) - f(a)}{h}$$

1. $f(x) = 7x - 2$ at $a = -1$

2. $f(x) = -3x + 4$ at $a = 2$

3. $f(x) = \frac{1}{2}x + 2$ at $a = 0$

4. $f(x) = -\frac{2}{3}x + \frac{4}{3}$ at $a = 1$

5. $f(x) = 3x^2 - 1$ at $a = 1$

6. $f(x) = 2x^2 + 3$ at $a = 3$

7. $f(x) = x^2 + 2x - 1$ at $a = 1$

8. $f(x) = 1 - x - x^2$ at $a = -2$

9. $f(x) = \dfrac{2}{x}$ at $a = 3$

10. $f(x) = \dfrac{4}{3x}$ at $a = 5$

11. $f(x) = \dfrac{3}{x^2}$ at $a = 2$

12. $f(x) = \dfrac{x + 2}{x - 1}$ at $a = 0$

In Problems 13–21, find an equation of the tangent line to the graph of each function at the given point $(a, f(a))$.

13. $f(x) = 6$ at $(-1, 6)$

14. $f(x) = 7x + 4$ at $(2, 18)$

15. $f(x) = x^2 + 1$ at $(1, 2)$

16. $f(x) = 3x^2 + 4$ at $(0, 4)$

17. $f(x) = -2x^2 + 4$ at $(1, 2)$

18. $f(x) = \dfrac{-2}{3x}$ at $(-2, \frac{1}{3})$

19. $f(x) = \dfrac{3}{x^2}$ at $(1, 3)$

20. $f(x) = x^2 + 3x$ at $(1, 4)$

21. $f(x) = 2x^2 - x$ at $(0, 0)$

The **normal line** to a curve at a point (a, b) on the curve is defined to be the line that passes through (a, b) and is perpendicular to the tangent line to the curve at (a, b). In Problems 22–27, find an equation of the normal line to the graph of $f(x)$ at the point $(a, f(a))$.

22. $f(x) = -3x + 7$ at $(1, 4)$

23. $f(x) = 8x - 14$ at $(2, 2)$

24. $f(x) = 7x^2 - 2$ at $(-1, 5)$

25. $f(x) = x^2 + x$ at $(1, 2)$

26. $f(x) = \dfrac{1}{x}$ at $(1, 1)$

27. $f(x) = \dfrac{2}{x^2}$ at $(2, \frac{1}{2})$

2-5 THE DERIVATIVE

We have seen that for some values a in the domain of a function $f(x)$, a unique number $f'(a)$ exists that represents the slope of the tangent line to the graph of $f(x)$ at the point $(a, f(a))$. We will formally define a new function, the *derivative* of

f(x), which associates with a the number $f'(a)$. In defining this function we will follow the custom of denoting the value of the independent variable by x.

The derivative of a function $y = f(x)$ is a function denoted by $y' = f'(x)$, which gives the slope of the tangent line at any point $(x, f(x))$ on the graph of $f(x)$. By examining the slope of the graph of a function at several points, we can make a rough sketch of the graph of the derivative function, as shown in Example 13.

Example 13 The graph of the function $y = f(x) = x^3 - 12x^2 + 45x - 50$ is shown in Figure 2-22. Using only this graph as a guide, sketch the graph of the derivative function $y' = f'(x)$.

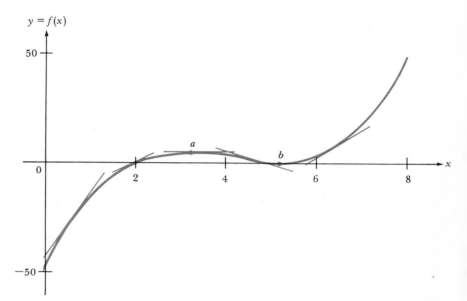

Figure 2-22
Graph of
$y = f(x) = x^3 - 12x^2 + 45x - 50$

Solution The derivative function gives the slope of the tangent line to the graph of $y = f(x)$ for any value of x. Two key points in Figure 2-22 are a and b. At these points, the tangent line is horizontal; therefore, the value of the derivative must be zero. Hence, the derivative function must cross the x-axis at these points. For values of x to the left of a, we see that the slopes of the tangent lines are positive but decrease as x gets closer to a. Between a and b, the slopes of the tangent lines are slightly negative, and to the right of b, the slopes are positive and increasing. Putting these facts together, we see that the derivative function must look something like that shown in Figure 2-23 (page 78). ∎

We now present a formal definition of the derivative function.

The **derivative of the function $f(x)$** is the function $f'(x)$ defined by

$$f'(x) = \lim_{h \to 0} \frac{f(x + h) - f(x)}{h}$$

provided this limit exists.

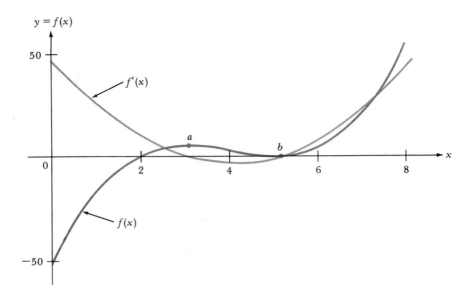

Figure 2-23
Graphs of $f(x)$ and $f'(x)$

The domain of the derivative function $f'(x)$ consists of those values of x for which the above limit exists. Since $f(x)$ appears in the definition of $f'(x)$, every value in the domain of $f'(x)$ must of necessity be in the domain of $f(x)$. However, as we have seen in Example 12, the domain of $f'(x)$ may be strictly smaller than the domain of $f(x)$. In that example, 0 is in the domain of $f(x) = |x|$, but not in the domain of $f'(x)$.

Notice that the definition of the derivative function is very similar to the definition of the slope of a tangent line at a specific point $(a, f(a))$. For the derivative, we let a be variable, thus defining a new function. Let us use this definition to find the derivatives of several functions.

Example 14   Find the derivative of the function $y = f(x) = \frac{1}{3}x^2 + 50$. (This is just the traffic flow function of Section 2-3 with the independent variable renamed.)

Solution   We compute

$$f'(x) = \lim_{h \to 0} \frac{f(x + h) - f(x)}{h}$$

First we evaluate the expression

$$\frac{f(x + h) - f(x)}{h} = \frac{[\frac{1}{3}(x + h)^2 + 50] - [\frac{1}{3}x^2 + 50]}{h}$$

$$= \frac{[\frac{1}{3}(x^2 + 2xh + h^2) + 50] - [\frac{1}{3}x^2 + 50]}{h}$$

$$= \frac{\frac{2xh}{3} + \frac{h^2}{3}}{h}$$

$$= \frac{2x}{3} + \frac{h}{3}$$

Taking the limit as h approaches 0, we have

$$f'(x) = \lim_{h \to 0} \left(\frac{2x}{3} + \frac{h}{3} \right)$$

$$= \frac{2x}{3}$$

Thus, the derivative is $f'(x) = 2x/3$. We can use this to say that when $x = 10$, the slope of the tangent line at the point $(10, f(10))$ is $f'(10) = 2(10)/3 = 6\frac{2}{3}$; when $x = 20$, the slope is $f'(20) = 2(20)/3 = 13\frac{1}{3}$. (Note that this is the same result found earlier in Section 2-4.) ∎

In Figure 2-24 we show graphs of both $f(x)$ and $f'(x)$. The scale used for $f(x)$ is given on the left vertical axis, while the scale used for $f'(x)$ is on the right vertical axis. At $x = 0$, we see that the tangent line to $y = f(x)$ is horizontal so that its slope, and the value of the derivative, is 0. When $x = 15$, the tangent line has a slope of 10; thus, the value of $f'(15) = 10$.

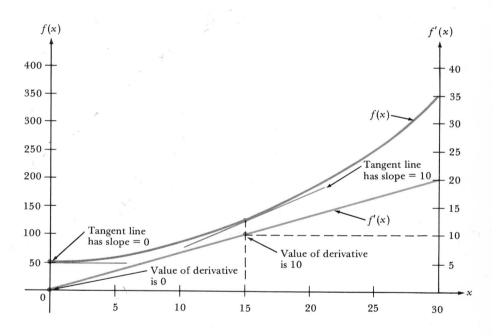

Figure 2-24
Graphs of $f(x) = \frac{1}{3}x^2 + 50$ and
$f'(x) = \dfrac{2x}{3}$

Example 15   Find the derivative of $f(x) = 2x - 5x^2$.

Solution   Proceeding according to the definition of $f'(x)$, we have

$$f'(x) = \lim_{h \to 0} \frac{f(x+h) - f(x)}{h}$$

$$= \lim_{h \to 0} \frac{[2(x+h) - 5(x+h)^2] - [2x - 5x^2]}{h}$$

$$= \lim_{h \to 0} \frac{[2x + 2h - 5(x^2 + 2xh + h^2)] - 2x + 5x^2}{h}$$

$$f'(x) = \lim_{h \to 0} \frac{2x + 2h - 5x^2 - 10xh - 5h^2 - 2x + 5x^2}{h}$$

$$= \lim_{h \to 0} \frac{2h - 10xh - 5h^2}{h}$$

$$= \lim_{h \to 0} \frac{h(2 - 10x - 5h)}{h}$$

$$= \lim_{h \to 0} (2 - 10x - 5h) = 2 - 10x$$

Thus, the derivative of $f(x) = 2x - 5x^2$ is $f'(x) = 2 - 10x$. When $x = 1$, the derivative is $f'(1) = 2 - 10(1) = -8$, which agrees with the result obtained in Example 10 (Section 2-4). ■

Several different types of notation are used to denote the derivative of a function $y = f(x)$:

Symbolic Notation for the Derivative

$$f'(x) \qquad y' \qquad \frac{dy}{dx} \qquad \frac{df}{dx}$$

Thus, the derivative in Example 15 could have been written in any of the following forms:

$$f'(x) = 2 - 10x$$

$$y' = 2 - 10x$$

$$\frac{dy}{dx} = 2 - 10x$$

$$\frac{df}{dx} = 2 - 10x$$

The notation dy/dx is somewhat appealing since we see that the derivative represents the slope of the tangent line, which in turn represents the change in the dependent variable per unit change in the independent variable. Recall our use of the $\Delta f/\Delta x$ notation. The symbol df/dx is an outgrowth of this notation. When we write a function using symbols other than $y = f(x)$, we modify the notation appropriately. For instance, the derivative of a function $s(t)$ can be written as $s'(t)$ or ds/dt.

The process of finding a derivative is called *differentiation*.

Example 16   **Differentiate the function**

$$y = f(x) = \frac{4}{x - 4}$$

Solution $$y' = \lim_{h \to 0} \frac{f(x + h) - f(x)}{h}$$

$$y' = \lim_{h \to 0} \frac{\dfrac{4}{(x+h)-4} - \dfrac{4}{x-4}}{h}$$

$$= \lim_{h \to 0} \frac{\dfrac{4(x-4) - 4(x+h-4)}{(x+h-4)(x-4)}}{h}$$

$$= \lim_{h \to 0} \frac{4x - 16 - 4x - 4h + 16}{h(x+h-4)(x-4)}$$

$$= \lim_{h \to 0} \frac{-4h}{h(x+h-4)(x-4)}$$

$$= \lim_{h \to 0} \frac{-4}{(x+h-4)(x-4)} = \frac{-4}{(x-4)(x-4)}$$

Therefore,

$$y' = \frac{-4}{(x-4)(x-4)} = \frac{-4}{(x-4)^2}$$

Note that y is not defined at $x = 4$ and neither is the derivative. Clearly, if the function is not defined at a point, then neither can the tangent line exist at that point. ∎

If the derivative of a function exists at a point a, then we say that the function is *differentiable* at a.

> A function $f(x)$ is **differentiable at a point $x = a$** if
>
> $$\lim_{h \to 0} \frac{f(a+h) - f(a)}{h}$$
>
> exists.

Generally, a function is not differentiable at places where the graph of the function is broken or has sharp points (at such points a tangent line cannot be unambiguously defined), or where the tangent line is vertical (since the slope is undefined).

Example 17   Which of the following functions are differentiable at $x = a$?

a.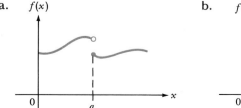

b.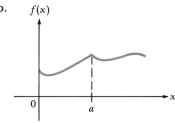

c. $f(x)$

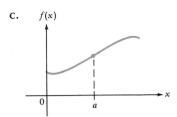

d. $f(x)$

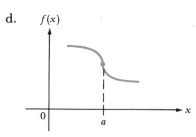

Solution   Only the function in part c is differentiable at $x = a$. The functions represented by the other three graphs are not differentiable, since in part a the graph is broken; the graph in part b has a sharp point at $x = a$; and in part d, the tangent line is vertical at $x = a$.

> A function $f(x)$ is said to be **differentiable** if it is differentiable at every point in its domain, that is, if $f'(x)$ exists for each x in the domain of $f(x)$.

ALGEBRA REVIEW 7: RATIONALIZING EXPRESSIONS INVOLVING RADICALS

If a quotient of two expressions involves square roots, its numerator can often be simplified by a process called *rationalizing the numerator*. The idea behind this is to eliminate the roots from the numerator by multiplying both the numerator and the denominator by the same expression. We know that $(a + b)(a - b) = a^2 - ab + ba - b^2 = a^2 - b^2$, and if either a or b is a square root, then the terms a^2 and b^2 will not contain any square roots. Thus, if the numerator is a *sum* of terms, we multiply both the numerator and the denominator by the *difference* of these terms. If the numerator is a *difference* of terms, we multiply by the *sum* of these terms. This technique is especially useful in many limit problems such as finding derivatives and others discussed in the supplement to this chapter.

For instance, consider the expression

$$\frac{\sqrt{x} - 2}{x}$$

Since the numerator is the difference of terms involving a radical, we multiply the numerator and denominator by the sum of these terms:

$$\frac{\sqrt{x} - 2}{x} \cdot \frac{\sqrt{x} + 2}{\sqrt{x} + 2} = \frac{x - 4}{x(\sqrt{x} + 2)}$$

EXERCISES

Rationalize the numerator for the following expressions:

a. $\dfrac{\sqrt{2x} + x}{x}$

b. $\dfrac{\sqrt{x^2 + 3} - 2}{x^2 - 1}$

c. $\dfrac{1 - \sqrt{x + 2x^2}}{x + 1}$

d. $\dfrac{3 + \sqrt{x}}{x - 9}$

e. $\dfrac{x - \sqrt{x}}{x - 1}$

We shall see later that every polynomial is a differentiable function, and we will develop a simple rule for differentiating any polynomial. In fact, we will find that many classes of functions (quotients of polynomials, radicals, exponentials, logarithms, and so on) are differentiable, and we will construct some general rules for differentiating such functions.

Example 18   Find the derivative of $f(x) = \sqrt{x}$. $x^{1/2}$

Solution   Using the limit definition, we have

$$f'(x) = \lim_{h \to 0} \frac{f(x+h) - f(x)}{h}$$

$$= \lim_{h \to 0} \frac{\sqrt{x+h} - \sqrt{x}}{h}$$

We may simplify this expression by rationalizing the numerator. (See Algebra Review 7.) This results in the following:

$$\frac{\sqrt{x+h} - \sqrt{x}}{h} \cdot \frac{\sqrt{x+h} + \sqrt{x}}{\sqrt{x+h} + \sqrt{x}}$$

$$= \frac{(\sqrt{x+h}\,\sqrt{x+h}) + (\sqrt{x+h}\,\sqrt{x}) - (\sqrt{x+h}\,\sqrt{x}) - (\sqrt{x}\,\sqrt{x})}{h(\sqrt{x+h} + \sqrt{x})}$$

$$= \frac{(\sqrt{x+h}\,\sqrt{x+h}) - (\sqrt{x}\,\sqrt{x})}{h(\sqrt{x+h} + \sqrt{x})}$$

$$= \frac{x+h-x}{h(\sqrt{x+h} + \sqrt{x})}$$

$$= \frac{h}{h(\sqrt{x+h} + \sqrt{x})}$$

$$= \frac{1}{\sqrt{x+h} + \sqrt{x}}$$

The derivative is therefore

$$f'(x) = \lim_{h \to 0} \frac{1}{\sqrt{x+h} + \sqrt{x}}$$

As h gets close to 0, $\sqrt{x+h}$ approaches $\sqrt{x}$. The derivative is then

$$f'(x) = \frac{1}{\sqrt{x} + \sqrt{x}}$$

$$= \frac{1}{2\sqrt{x}}$$

Figure 2-25 shows the graph of $f(x) = \sqrt{x}$. For any positive value of x, the derivative gives the slope of the tangent line at the point $(x, \sqrt{x})$ on the graph.

Figure 2-25
Graph of $f(x) = \sqrt{x}$

PROBLEMS (Section 2-5)

For Problems 1–16, find the derivative of each function using the limit definition of the derivative.

1. $g(x) = -10$

2. $f(x) = 36$

3. $g(x) = -2x + 4$

4. $f(x) = 5x - 1$

5. $f(x) = x^2$

6. $f(x) = 5x^2$

7. $g(x) = 3x^2 + 1$

8. $g(x) = \dfrac{2}{x}$

9. $f(x) = \dfrac{1}{x} + 3$

10. $h(x) = \dfrac{1}{x + 2}$

11. $f(x) = x^2 + 3x + 5$

12. $f(x) = 2x^2 + 5x + 4$

13. $g(x) = \sqrt{x} + 2$

14. $h(x) = \sqrt{x + 3}$

15. $f(x) = x^3$

16. $g(x) = 2x^3 - 1$

17. At what points do the graphs below fail to have tangent lines?

a.

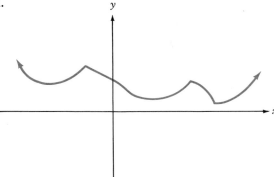

b.

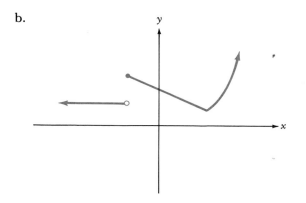

18. Where do the graphs in Problem 17 have tangent lines with positive slope? Negative slope?

19. The surface area of a sphere is $S(r) = 4\pi r^2$, where r is the radius. Using the limit definition of the derivative, find $S'(r)$. What does this derivative mean?

20. Many living cells are spherical in nature. The volume of a sphere is $V(r) = \frac{4}{3}\pi r^3$, where r is the

radius. Using the limit definition of the derivative, find $V'(r)$. How can this be interpreted? How might this result relate to the answer for Problem 19?

21. The velocity (in feet per second) of an object dropped from a building is given by the function $v(t) = 32t$, where t is time measured in seconds. Find dv/dt using the limit definition of the derivative.

22. The force (in pounds) exerted by a spring when it is stretched x feet past a certain position is given by $f(x) = 0.2x + 7$. Use the definition of the derivative to compute $f'(x)$.

23. The distance a solute travels through a biological fluid in time t is given by $f(t) = 2.5\sqrt{4t}$. Find $f'(t)$ using the limit definition of the derivative. How would $f'(t)$ be interpreted?

24. The total revenue for the rock concert promoter in Example 22 of Chapter 1 was $TR = g(y) = -500y^2 + 7000y + 120,000$. Find dg/dy using the limit definition of the derivative. (Note that y is the independent variable, hence the notation dg/dy.)

25. The cost function for producing x units of a product is $C(x) = 9000 + 60x$. Using the limit definition of the derivative, compute $C'(x)$, the marginal cost function.

26. The revenue function from the sale of x units of a product is determined to be $R(x) = 4000x - x^2$. Use the limit definition of the derivative to find $R'(x)$. What does $R'(x)$ mean?

27. The resistance of a certain electrical resistor is given by $R = 20(3 + 0.3t + 0.01t^2)$ ohms, where t is temperature (in degrees Celsius). Use the definition of the derivative to compute dR/dt.

2-6 THE DERIVATIVE OF A POWER FUNCTION

It would be very tedious to have to use the limit approach to evaluate a derivative every time one is needed. Fortunately, basic formulas have been developed for computing derivatives of many different functions. We shall first consider the simple case of a power function, $f(x) = x^p$, where p is any real number. (You may wish to consult Algebra Review 8 at this time.)

Let us start with $p = 0$, that is, a constant function $f(x) = x^0 = 1$. Applying the limit definition, we have

$$f'(x) = \lim_{h \to 0} \frac{f(x + h) - f(x)}{h}$$

$$= \lim_{h \to 0} \frac{1 - 1}{h}$$

$$= \lim_{h \to 0} \frac{0}{h} = 0$$

This confirms the fact that the horizontal line $y = 1$ has zero slope.

That was pretty easy, so let us try $p = 1$ so that $f(x) = x^1 = x$. In this case, we have

$$f'(x) = \lim_{h \to 0} \frac{f(x + h) - f(x)}{h}$$

$$= \lim_{h \to 0} \frac{(x + h) - x}{h}$$

$$= \lim_{h \to 0} \frac{h}{h}$$

$$= \lim_{h \to 0} 1 = 1$$

ALGEBRA REVIEW 8: EXPONENTS AND ROOTS

An *exponent* is a real number that is used to indicate the power of a number or variable. Exponents can be zero, positive, or negative. A positive integer exponent tells how many times to use the base as a factor:

$$x^n = \underbrace{x \cdot x \cdot x \cdot \cdots \cdot x}_{n \text{ factors}}$$

Thus, $x^3 = x \cdot x \cdot x$. By convention, $x^0 = 1$.

When an exponent is a positive fraction (which we assume to be in lowest terms), the denominator is the root of a radical and the numerator is the power of the base:

$$x^{m/n} = \sqrt[n]{x^m} = (\sqrt[n]{x})^m$$

Note that an even root of a negative number is impossible in the real number system. Thus, if n is even, the domain of $f(x) = x^{m/n}$ is the set of all nonnegative numbers.

A second rule of exponents that gives meaning to negative exponents is the following:

$$x^{-r} = \frac{1}{x^r}$$

We can use these rules to convert terms involving roots to powers or to convert terms involving the reciprocal of a power function to a power function. For example,

$$\frac{1}{x} = x^{-1}$$

This confirms the fact that the line $y = x$ has slope 1.
Let us try this again, this time for $f(x) = x^2$:

$$\lim_{h \to 0} \frac{f(x+h) - f(x)}{h} = \lim_{h \to 0} \frac{(x+h)^2 - x^2}{h}$$

$$= \lim_{h \to 0} \frac{x^2 + 2xh + h^2 - x^2}{h}$$

$$= \lim_{h \to 0} \frac{2xh + h^2}{h}$$

$$= \lim_{h \to 0} (2x + h) = 2x$$

One last time, let us find the derivative of $f(x) = x^3$:

$$\lim_{h \to 0} \frac{f(x+h) - f(x)}{h} = \lim_{h \to 0} \frac{(x+h)^3 - x^3}{h}$$

$$= \lim_{h \to 0} \frac{(x+h)(x+h)^2 - x^3}{h}$$

$$= \lim_{h \to 0} \frac{(x^3 + 3x^2h + 3xh^2 + h^3) - x^3}{h}$$

$$= \lim_{h \to 0} (3x^2 + 3xh + h^2) = 3x^2$$

$$\frac{5}{x^{3/2}} = 5x^{-3/2}$$

$$\sqrt{x} = x^{1/2}$$

$$\frac{1}{\sqrt[5]{x^3}} = \frac{1}{x^{3/5}} = x^{-3/5}$$

There are certain *laws of exponents* that allow the simplification of expressions involving exponents. These laws are:

$$x^r x^s = x^{r+s} \qquad \frac{x^r}{x^s} = x^{r-s}$$

$$(x^r)^s = x^{rs} \qquad (xz)^r = x^r z^r$$

$$\frac{1}{x^r} = x^{-r} \qquad \frac{x^r}{z^r} = \left(\frac{x}{z}\right)^r$$

EXERCISES

Convert the following terms to power functions:

a. $\dfrac{3}{2\sqrt{x}}$ b. $\dfrac{8}{\sqrt[3]{x^2}}$

c. $\sqrt[3]{x^7}\left(\dfrac{1}{x^{1/3}}\right)$

Convert the following to expressions involving radicals:

d. $x^{2/3}$ e. $(x+2)^{-3/2}$

A pattern seems to be emerging:

Function	Derivative
x^0	0
x^1	$1 = 1x^0$
x^2	$2x = 2x^1$
x^3	$3x^2$
x^4	?

Before reading further, try applying the limit formula yourself to find the derivative of $f(x) = x^4$. Can you see the pattern?

If we examine the exponent of x in each of these functions, we see that the derivative is found by multiplying the function by the exponent and then reducing the exponent by 1. For example, the derivative of $1 = x^0$ is $0x^{0-1} = 0$; the derivative of x is $1x^{1-1} = 1x^0 = 1$; the derivative of x^2 is $2x^{2-1} = 2x$; and the derivative of x^3 is $3x^{3-1} = 3x^2$. You should have discovered that the derivative of x^4 is $4x^3$. Although we have considered only cases in which the exponent is one of the first few nonnegative integers, the pattern turns out to be quite general and holds for *any* power of x. We may state this as a general rule:

THEOREM
Power Rule for Derivatives

If $f(x) = x^p$, then $f'(x) = px^{p-1}$ for any real number p.

Example 19   Find the derivatives of the following functions using the power rule:

a. $y = x^{14}$ b. $f(x) = x^{2/3}$ c. $g(t) = t^{-3/4}$ d. $y = \sqrt{x}$

Solution   Using the power rule, we obtain:

a. $y' = 14x^{14-1} = 14x^{13}$

b. $\dfrac{df}{dx} = \tfrac{2}{3}x^{(2/3)-1} = \tfrac{2}{3}x^{-1/3}$

c. $\dfrac{dg}{dt} = -\tfrac{3}{4}t^{(-3/4)-1} = -\tfrac{3}{4}t^{-7/4}$

d. Since $y = x^{1/2}$, then

$$y' = \tfrac{1}{2}x^{-1/2} = \frac{1}{2\sqrt{x}}$$

which confirms the result of Example 18 (Section 2-5).   ■

TWO IMPORTANT
THEOREMS

Recall that a polynomial function has the form

$$f(x) = a_n x^n + \cdots + a_1 x + a_0$$

where n is a nonnegative integer. In Chapter 1 we saw that polynomials are among the most useful functions in a variety of applications. Thus, knowing how to find the derivatives of such functions will be quite useful. We present two theorems which, along with the power rule, are useful in finding derivatives of polynomials. The first theorem is concerned with the derivative of a constant times a function.

THEOREM
Derivative of a Constant Times a Function

If $f(x) = cg(x)$, where $g(x)$ is any differentiable function and c is any constant, then $f'(x) = cg'(x)$.

This theorem states that the derivative of a constant times any function is the constant times the derivative of the function. We can use this result to show that the derivative of any constant function is 0. Suppose that $f(x) = c$. We may write $f(x)$ as cx^0 since $x^0 = 1$. We have already shown that the derivative of x^0 is 0. Therefore, by the above theorem, the derivative of $f(x) = c$ is $f'(x) = c \cdot 0 = 0$.

Example 20　Find the derivative of:

　　a. $f(z) = 5z^2$　　b. $y = 13t^{-7}$ ＝ $-91t^{-8}$

　　　　　　 $10z$

Solution　　a. We have $f'(z) = 5$ times the derivative of z^2. Since the derivative of z^2 is $2z$, the derivative of $f(z)$ must be $5(2z) = 10z$.
　　　b. $y' = 13(-7t^{-7-1}) = 13(-7t^{-8}) = -91t^{-8}$ ■

The second result that will be useful to us is the following:

THEOREM
Derivative of a Sum or Difference of Functions

If $f(x)$ and $g(x)$ have derivatives $f'(x)$ and $g'(x)$, respectively, then the derivative of $f(x) + g(x)$ is $f'(x) + g'(x)$ and the derivative of $f(x) - g(x)$ is $f'(x) - g'(x)$.

In other words, the derivative of a sum or difference of two (or more) functions is the sum or difference of the derivatives of the functions. (Proofs of both of the above theorems can be found in the supplement to this chapter.)

Example 21　Find the derivative of:

　　a. $y = x^5 + 4$　　b. $f(t) = 2.5t^{-3} - 4t$

　　　　 $5x^4$　　　　　　 $-7.5t^{-4} - 4$

Solution　The derivatives are:

　　a. $\dfrac{dy}{dx} = 5x^4 + 0 = 5x^4$　　b. $f'(t) = 2.5(-3t^{-4}) - 4 = -7.5t^{-4} - 4$ ■

We may use these theorems to find the derivatives of any polynomial; they are also useful in many other applications.

Example 22   Find the slope of the line tangent to the function $f(x) = 3x^3 + 4x - 3$ when $x = 2$.

Solution   According to the above theorem, the derivative of $f(x)$ is the sum or difference of the derivatives of the three terms that make up $f(x)$. Therefore,

$$f'(x) = 3(3x^{3-1}) + 4(1) - 0 = 9x^2 + 4$$

This gives the value of the slope of the tangent line to the graph of the function for any x in the domain of f. Specifically, if $x = 2$, then $f'(2) = 9(2^2) + 4 = 40$. Therefore, the slope of the tangent line to the function at $x = 2$ is 40. ∎

Example 23   Find the equation of any tangent line to the curve $y = -4x^2 + 3x + 9$ that also passes through the point $(2, 0)$. (Note that this point is *not* on the curve.)

Solution   The slope of the tangent line is given by the derivative $y' = -8x + 3$. But we do not know the value of x at the point of tangency. We do know, however, that the slope of the tangent line is given by

$$\frac{y - 0}{x - 2}$$

since (x, y) is the point of tangency and $(2, 0)$ is the point on the tangent line given in the problem statement. We set this equal to y' since a line has a unique slope, no matter how it is expressed:

$$\frac{y - 0}{x - 2} = -8x + 3$$

Since $y = -4x^2 + 3x + 9$, we then have

$$\frac{-4x^2 + 3x + 9}{x - 2} = -8x + 3$$

Rearranging this equation, we have

$$-4x^2 + 3x + 9 = (-8x + 3)(x - 2)$$
$$-4x^2 + 3x + 9 = -8x^2 + 19x - 6$$
$$4x^2 - 16x + 15 = 0$$

Using the quadratic formula (see Algebra Review 9), we find

$$x = \frac{-(-16) \pm \sqrt{(-16)^2 - 4 \cdot 4 \cdot 15}}{2 \cdot 4} = \frac{16 \pm 4}{8} = 2 \pm \tfrac{1}{2}$$

It appears that there are two solutions and indeed there are, namely

$$x = \tfrac{3}{2} \quad \text{and} \quad x = \tfrac{5}{2}$$

These are shown in Figure 2-26.
    To find the points of tangency, we substitute these values for x into the equation of the curve and obtain

$$x = \tfrac{3}{2} \quad \text{and} \quad y = \tfrac{9}{2}$$

and also,

$$x = \tfrac{5}{2} \quad \text{and} \quad y = -\tfrac{17}{2}$$

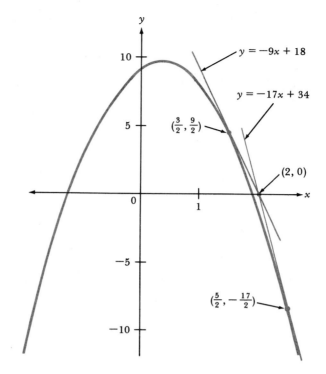

Figure 2-26
Graph of $y = -4x^2 + 3x + 9$

ALGEBRA REVIEW 9: THE QUADRATIC FORMULA

To solve a quadratic equation $ax^2 + bx + c = 0$, we may apply the *quadratic formula*. The quadratic formula is

$$x = \frac{-b \pm \sqrt{b^2 - 4ac}}{2a}$$

For example, if we have the equation $x^2 + 3x + 1 = 0$, then $a = 1$, $b = 3$, and $c = 1$. The solutions to this equation are

$$x = \frac{-3 \pm \sqrt{3^2 - 4(1)(1)}}{2(1)} = \frac{-3 \pm \sqrt{5}}{2}$$

Since $\sqrt{5} \approx 2.236$, we have

$$x = \frac{-3 + \sqrt{5}}{2} \approx -0.382$$

or

$$x = \frac{-3 - \sqrt{5}}{2} \approx -2.618$$

Many quadratic equations can be more easily solved by factoring. However, the quadratic formula can be used to solve *any* quadratic equation.

EXERCISES

Find the roots of the following quadratic equations using the quadratic formula:
a. $-2x^2 + 3x + 2 = 0$
b. $6x^2 + x + 1 = 0$
c. $3x^2 + 2x - 3 = 0$
d. $x^2 - 8x + 16 = 0$
e. $x^2 + x = 0$

Since $y' = -8x + 3$, the slope of the first line is $-8(\frac{3}{2}) + 3 = -9$ and the slope of the second line is $-8(\frac{5}{2}) + 3 = -17$.

The point–slope equation of the first line is therefore

$$\frac{y - \frac{9}{2}}{x - \frac{3}{2}} = -9 \qquad \text{or} \qquad y = -9x + 18$$

Similarly, the point–slope equation of the second line is

$$\frac{y - (-\frac{17}{2})}{x - \frac{5}{2}} = -17 \qquad \text{or} \qquad y = -17x + 34$$

Therefore, the equations of the two lines that pass through $(2, 0)$ and are tangent to the curve $y = -4x^2 + 3x + 9$ are

$$y = -9x + 18 \qquad \text{and} \qquad y = -17x + 34 \qquad \blacksquare$$

Examples 22 and 23 have illustrated how we can use derivatives to solve problems involving tangent lines to curves. There are more important uses of the derivative, however. As we indicated earlier in this chapter, the slope of a straight line can be interpreted as a rate of change. In the next section we examine applications of the derivative involving rates of change.

PROBLEMS (Section 2-6)

In Problems 1–26, find the derivative of each function using the rules given in this section.

1. $y = 8$

2. $f(x) = 15$

3. $f(x) = 5x$

4. $f(x) = -11x$

5. $y = 3x^4$

6. $y = -5x^3$

7. $f(x) = \frac{1}{2}x^6$

8. $f(x) = -\frac{3}{4}x^8$

9. $f(x) = 3x^3 + 4x$

10. $y = -x^2 - 5x$

11. $y = x^3 + 6x^2 - 9x$

12. $f(x) = -4x^4 + 3x^3 - 5$

13. $f(x) = x^5 + 3x^4 - 2x^2 + 1$

14. $y = x^6 - 2x^4 + x^3 - 5x^2 + 3x$

15. $f(x) = -4\sqrt{x}$

16. $y = 2\sqrt[3]{x}$

17. $f(x) = 7x^2 - \dfrac{1}{x}$

18. $f(x) = 2 + \dfrac{1}{x^2} - \dfrac{1}{x^3}$

19. $f(x) = 8x^2 + \dfrac{1}{x^2} - \dfrac{3}{2}$

20. $g(x) = \frac{1}{5}x^5 + 2x^3 - \dfrac{3}{x}$

21. $f(x) = 7\sqrt{x} - 2x$

22. $y = \dfrac{2}{\sqrt{x}} + x^2$

23. $g(x) = \dfrac{4}{\sqrt[3]{x^2}}$

24. $f(x) = -6x^{1/3} + 2x^2$

25. $f(x) = 3x^{2/3} - x^{1/5}$

26. $g(x) = -2x^{3/2} + \dfrac{4}{\sqrt{x}}$

In Problems 27–31, find the slope of the line tangent to the function at the given value of the independent variable.

27. $f(x) = x^2 + 3x$ at $x = 1$

28. $g(x) = 2x^4 + 4x^3 + x^2 - 2x + 3$ at $x = 0$

29. $y = x^3 + 3x - 8$ at $x = 2$

30. $f(x) = \sqrt{x} + \frac{5}{2}x^2$ at $x = 4$

31. $f(x) = \sqrt[3]{x} - \dfrac{1}{x}$ at $x = 8$

In Problems 32–37, find an equation of the tangent line to each function at the indicated point.

32. $f(x) = x^2 - 6x$ at $(-1, 7)$

33. $f(x) = 3x^2 - x$ at $(1, 2)$

34. $h(x) = 3x - \sqrt{x}$ at $(9, 24)$

35. $g(x) = x^3 - 2x^2 + 3x + 1$ at $(2, 7)$

36. $f(x) = 2x^{3/2}$ at $(4, 16)$

37. $g(x) = -2\sqrt[3]{x} + x^2$ at $(1, -1)$

38. Find $f'(x)$ if
$$f(x) = \begin{cases} 2x & \text{if } x < 0 \\ 4x + 1 & \text{if } x \ge 0 \end{cases}$$

Is the domain of $f'(x)$ the same as the domain of $f(x)$? [*Hint:* Sketch the graphs.]

39. Find $g'(x)$ if
$$g(x) = \begin{cases} x + 3 & \text{if } x < 1 \\ x^2 - 2x + 4 & \text{if } x \ge 1 \end{cases}$$

Is the domain of $g'(x)$ the same as the domain of $g(x)$? [*Hint:* Sketch the graphs.]

40. Find the equation of the two tangent lines to the curve $y = x^2 + 3$ that pass through the point $(1, 0)$.

41. The intensity of radiation received at a distance r from the source is
$$I = \frac{E}{4\pi} r^{-2}$$

where E is a constant. Find dI/dr using the constant times a function rule and the power rule.

42. The growth of an organism depends on the transfer of ingested food that diffuses through its surface. It can be shown that the rate of growth can be expressed as the function
$$r = am^{2/3} - bm$$

where a and b are constants and m represents the mass of the organism. Find dr/dm using the rules of this section.

43. An ecological formula relating the per capita effectiveness of predators A to the number of predators P is
$$A = 4P^{-0.25}$$

Compute the derivative dA/dP.

44. The revenue obtained from renting x apartments in a certain section of a city was found to be
$$R = 16{,}000 + 500x - 20x^2$$

Compute the derivative dR/dx.

45. A small business owner calculates that the total cost of manufacturing x units is
$$C(x) = 5x^2 + \frac{200}{x} + 350$$

What is the derivative of $C(x)$? Compute $C'(10)$.

46. In a biology experiment on the chemical breakdown of sugars, the mass M of glucose remaining t minutes after treatment was found to be given by the function
$$M(t) = 3.4 - 0.05t^2$$

Compute $M'(t)$. What is $M'(5)$?

2-7 INTERPRETING THE DERIVATIVE AS A RATE OF CHANGE

Let us return to the traffic flow example that we introduced in Section 2-3. Recall that the function is
$$f(t) = \tfrac{1}{3}t^2 + 50$$

and represents the volume of traffic over a road segment during a certain time interval. We saw that for any time interval $[a, b]$, the slope of the line between the points $(a, f(a))$ and $(b, f(b))$ represents the average rate of change of traffic volume over that interval. This is illustrated in Figure 2-27. As b approaches a, this line approaches the tangent line to the curve at the point $(a, f(a))$.

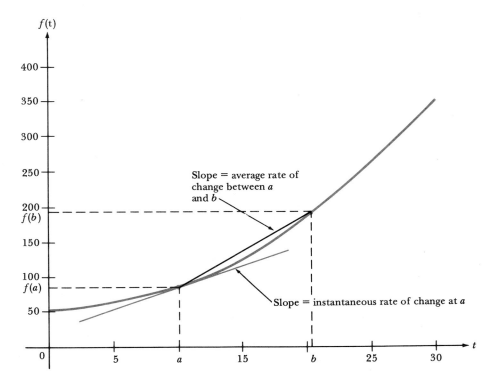

Figure 2-27
Rates of change for the traffic
volume example

What does this mean? The slope of the tangent line represents the *instantaneous rate of change* of the function at a. By "instantaneous" we mean that the rate of change is defined at a single point, not averaged over some interval as we previously computed. This corresponds with the notion of a limit — "instantaneous" implies a rate of change over a time interval that approaches zero, namely, a single point in time. The derivative of $f(t) = \frac{1}{3}t^2 + 50$ is $f'(t) = \frac{2}{3}t$. So, for example, at $t = 10$, $f'(10) = \frac{2}{3}(10) = 6\frac{2}{3}$. This means that at *precisely* 6:40 A.M., the model states that traffic volume is increasing at a rate of $6\frac{2}{3}$ vehicles per minute. At $t = 11$, $f'(11) = 7\frac{1}{3}$. Therefore, 1 minute later, the rate of change has increased to $7\frac{1}{3}$ vehicles per minute.

The **instantaneous rate of change of a function $f(x)$ at a point $x = a$** is given by $f'(a)$, the value of the derivative at $x = a$.

In order to use and interpret the derivative correctly, it is important to be able to associate the correct dimensions (units of measurement) with it. Since the definition of the derivative is a ratio of a change in the function values $f(x)$ (the dependent variable) to a change in the independent variable x, we can state the following rule:

> The dimensions of the derivative of $f(x)$ are always the dimensions of $f(x)$ divided by the dimensions of x.

Example 24   Interpret the derivative for the following functional relationships:

    a. Profit $P(x)$ is a function of x, the number of units produced and sold.
    b. The number of bacteria $N(t)$ in a culture is a function of t, the number of hours the culture has grown.

Solution   a. The derivative $P'(x)$ represents the instantaneous rate of change of profit for a specified level of production and sales, x. The dimensions are dollars per unit.
    b. The derivative $N'(t)$ represents the instantaneous growth rate of the population at the tth hour in dimensions of number of bacteria per hour. ∎

We conclude this section with some additional examples of applying the derivative to practical problems.

Example 25
Quantity Discounts   In Chapter 1 (Section 1-5), when we introduced nonlinear functions, we presented the following quantity discount model:

$$f(x) = 10.005x - 0.005x^2$$

This model gives the total cost of purchasing x units of some product. The derivative is

$$\frac{df}{dx} = 10.005 - 2(0.005)x = 10.005 - 0.01x$$

since

$$\frac{df}{dx} = f'(x) \approx \frac{f(x+1) - f(x)}{1}$$

This represents the approximate cost per unit of the $(x + 1)$st unit when x units of product are purchased. Equivalently, it tells us the approximate rate of change of the total cost if an additional unit is purchased. In economics, this is referred to as *marginal cost*. That is, the marginal cost when $x = a$, $f'(a)$, is the approximate cost of the $(a + 1)$st unit. For instance, if we purchase 100 units, the marginal cost is $f'(100) = 10.005 - 0.01(100) = 9.005$. Thus, the approximate cost of the 101st unit is \$9.005. ∎

In Example 25, let us compute the *actual* cost of purchasing the 101st unit. This is given by

$$f(101) - f(100)$$
$$= [10.005(101) - 0.005(101)^2] - [10.005(100) - 0.005(100)^2]$$
$$= 959.5 - 950.5$$
$$= 9.000$$

The marginal cost was found to be 9.005. Why the difference? This can be explained by examining Figure 2-28. Since the derivative is the slope of the tangent line, the y-coordinate of the tangent line will change by the value of the derivative for a 1 unit increase in x. However, $f(x)$ is nonlinear, so the tangent line is only an approximation at the point $(100, f(100))$. The y-coordinate on the tangent line at $x = 101$ approximates $f(101)$. It overestimates the value of the function by 0.005. We may therefore state the following:

The value of the derivative of a function $f(x)$ at $x = a$, $f'(a)$, represents the approximate change in $f(x)$ for a 1 unit increase in the value of x. That is,

$$f'(a) \approx \frac{f(a + 1) - f(a)}{1}$$

or

$$f(a + 1) \approx f(a) + f'(a)$$

Thus, the derivative $f'(x)$ at $x = a$, $f'(a)$, represents the **marginal value of the function $f(x)$ at $x = a$.** This is a fundamental concept in economics.

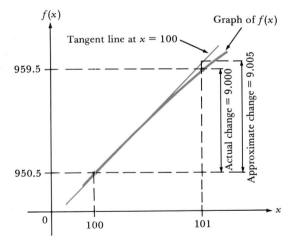

Figure 2-28
Approximation of actual change in
$f(x)$ by the tangent line

Example 26   Suppose that the demand x for a certain product is related to its price p by

$$x = 4000 - p$$

That is, as price increases, demand decreases. Since revenue is equal to price per unit times quantity sold, we can determine the revenue from the sale of x units as

$$R(x) = px = (4000 - x)x = 4000x - x^2$$

Further assume that the cost of producing x units is given by

$$C(x) = 9000 + 60x$$

Develop expressions for the marginal revenue, marginal cost, and marginal profit. Interpret these values when $x = 1970$.

Solution   The marginal revenue is the derivative

$$R'(x) = 4000 - 2x$$

Marginal cost is

$$C'(x) = 60$$

Since profit equals revenue minus cost, we have

$$P(x) = R(x) - C(x) = (4000x - x^2) - (9000 + 60x)$$
$$= 3940x - x^2 - 9000$$

Therefore, the marginal profit is

$$P'(x) = 3940 - 2x$$

Note that this is simply equal to $R'(x) - C'(x)$. (Can you state why this is true in general?)

When $x = 1970$, $R'(1970) = C'(1970) = 60$, and thus $P'(1970) = 0$. This means that the revenue and cost realized from the production and sale of the 1971st unit are equal and approximately \$60. Since the marginal profit is 0, the firm cannot increase profits by producing more than 1970 units. ∎

Example 27
A Population Model

The population of a small Kentucky town has increased rapidly since several major corporations have located their headquarters in a nearby city. The total population in thousands can be represented by the equation $P(t) = t^2 + 20$, where t is the number of years since 1980. How is the population changing in 1985?

Solution   First, we see that $t = 1985 - 1980 = 5$ years. The derivative is

$$P'(t) = 2t$$

Thus, when $t = 5$, $P'(5) = 2(5) = 10$ thousand people per year. This is the rate of population growth in 1985. ∎

Example 28
Spread of Disease

A new flu strain is spreading throughout the northeast. Doctors estimate that the number of people who will contract the flu over the next 30 days is given by

$$n(t) = 450t^3 - 40t^2 \qquad 0 \le t \le 30$$

How many people will have caught the flu by the 15th day? How fast is the flu spreading 20 days from now?

Solution   To answer these questions, we must first think about what we are looking for. The first question relates to the total number of people who have caught the flu. This is given by the function $n(t)$. Therefore, we only have to evaluate $n(15)$ to answer this question:

$$n(15) = 450(15)^3 - 40(15)^2 = 1,509,750 \text{ people}$$

To determine how fast the flu is spreading, we seek the number of people per day who catch the flu. Notice that the dimensions of what we are looking for give us a clue that we need to find a derivative. We have

$$\frac{dn}{dt} = 1350t^2 - 80t$$

At $t = 20$, $dn/dt = 538,400$. We interpret this to mean that, 20 days from now, 538,400 people per day are catching the flu.   ■

Example 29
Velocity

In many applications, particularly in science and engineering, a relation is developed that represents distance as a function of time. The derivative represents the rate of change of distance per unit of time, which we commonly call *velocity*. For example, suppose the distance of a Toronto subway train from the station is given by

$$s(t) = t^3 - 8t^2 + 40t + 4$$

where t is measured in hours. At $t = \frac{1}{2}$ hour, the train is

$$s(\tfrac{1}{2}) = (\tfrac{1}{2})^3 - 8(\tfrac{1}{2})^2 + 40(\tfrac{1}{2}) + 4 = 22\tfrac{1}{8}$$

miles from the station. The velocity of the train (rate of change of distance per unit time) is given by

$$v(t) = s'(t) = 3t^2 - 16t + 40$$

Therefore, at $t = \frac{1}{2}$ hour, the velocity is

$$v(\tfrac{1}{2}) = s'(\tfrac{1}{2}) = 3(\tfrac{1}{2})^2 - 16(\tfrac{1}{2}) + 40 = 32.75 \text{ miles per hour}$$ ■

PROBLEMS (Section 2-7)

In Problems 1–4, interpret the derivative for the given functional relationship.

1. Revenue is a function of the number of units produced and sold.

2. The population of a city is a function of the number of years it has existed.

3. The demand for an item is a function of the quantity available.

4. The area of a circle is a function of its radius.

5. The Dunn Co. finds that the total cost (in dollars)

to manufacture x items is represented by the function $C(x) = 0.02x^2 + 16x + 800$.
a. What is the total cost of manufacturing 50 items?
b. What is the marginal cost of the 50th item?
c. What is the actual cost of the 51st unit? Why is there a difference?
d. What is the marginal cost when 60 items are being manufactured?

6. The enrollment at a small college is approximated by the formula

$$e(t) = 7000 - 10(t^2 - 25t + 280)$$

where t represents the number of years since 1980.
  a. What is the rate of growth of the enrollment?
  b. What will be the rate of growth in 1990? In 1995?

7. A colony of bacteria is sprayed with a chemical. Let N represent the number of bacteria per milliliter remaining after t minutes, and suppose that $N(t) = 10,000(1600 - 80t + t^2)$.
  a. What is the average rate of change in the number of bacteria during the first 10 minutes? 20 minutes?
  b. What is the instantaneous rate of change in N at the end of 10 minutes? 20 minutes?
  c. Explain the differences in your answers to parts a and b.
  d. At what time will all the bacteria be dead? What is the derivative at this time?

8. A truck starts at town A and travels along a straight highway. The distance from A is given by $s(t) = 32t$, where $s(t)$ is measured in miles and t in hours.
  a. What distance did the truck travel after 5 hours? 10 hours?
  b. What is the velocity of the truck at $t = 5$? $t = 10$?

9. The cost (in dollars) to an appliance firm of producing x food processors is given by $C(x) = 23 + 10x + 15\sqrt{x}$.
  a. What is the total cost of making 25 food processors?
  b. What is the marginal cost of the 25th food processor?
  c. What is the marginal cost of the 100th food processor?
  d. How does the marginal cost seem to be changing as the number produced increases?

10. An analysis of the daily production of a toy factory assembly line shows that about $80t + t^2 - \frac{1}{12}t^3$ units are produced after t hours of work when $0 \le t \le 8$.
  a. How many units are produced after 2 hours? After 4 hours?
  b. What is the rate of production at $t = 2$ hours? At $t = 4$ hours?

11. Suppose that the revenue (in dollars) from producing x units of a product is given by $R(x) = 0.02x^2 - 5x$.
  a. What is the revenue from producing 10,000 units? 20,000 units?
  b. What is the marginal revenue at a production level of 10,000 units? 20,000 units?

12. Suppose that N is the number of gallons of water in a pool being drained for the winter and t is the number of minutes after draining starts. $N(t)$ is given by the function
$$N(t) = 40(50 - t)^2 \qquad 0 \le t \le 50$$
  a. Find the instantaneous rate at which the water is running out at 5 minutes, at 10 minutes, and at 20 minutes after the pool has started to drain. How do these rates compare to the *average* rates of change over the intervals [0, 5], [0, 10], and [0, 20]?
  b. How many minutes after the pool has started to drain will the pool be empty?

13. A company knows that its sales $S(x)$ are related to the advertising budget x by the function
$$S(x) = 100,000 + 800x - x^2$$
where x is in hundreds of thousands of dollars.
  a. What is the company's sales when the advertising budget is $300,000? $500,000?
  b. What is the rate of change of sales when the advertising budget is $300,000? $500,000? What do these figures mean?
  c. Should the company continue to increase the budget when it is $300,000? $500,000? Why or why not?

14. A toy rocket fired straight up into the air reaches a height $h(t) = 160t - 16t^2$ feet after t seconds.
  a. What is the rocket's initial velocity (when $t = 0$)?
  b. What is the velocity after 2 seconds?
  c. At what time will the rocket hit the ground?
  d. At what velocity will the rocket be traveling just as it smashes into the ground?

15. The volume V of a spherical raindrop is given by $V(r) = \frac{4}{3}\pi r^3$, where r is its radius. Suppose the raindrop is evaporating. At what rate does the volume change with respect to the radius?

16. The surface area S of a spherical raindrop is given by $S(r) = 4\pi r^2$, where r is its radius. Suppose the raindrop is evaporating. At what rate does the surface area change with respect to the radius?

17. The blood pressure P within a cylindrical vein or artery is given by $P(r) = Tr^{-1}$, where T is a constant and r is the radius of the vein or artery. What is the rate of change of the blood pressure with respect to the radius? Interpret your answer.

18. The area of a rectangular field of width w and perimeter 1000 feet is given by $A(w) = w(500 - w)$. At what rate does the area change with respect to the width?

19. Suppose the length $L(t)$ (in inches) of a metal bar is related to its temperature t (in degrees Fahrenheit) by the formula $L(t) = 100 + 0.001\sqrt{t}$. What is the rate of change of the length with respect to the temperature when the temperature is $400°$F?

20. The number of bacteria in a colony at time t days is found to be $N(t) = 1000 + 17t^3$. At what rate is the colony growing when $t = 3$?

21. The volume of a certain cylindrical grain silo is given by $V(r) = 7\pi r^2 + \frac{2}{3}\pi r^3$, where r is the radius of its base. At what rate does the volume change with respect to r when $r = 2$?

22. The total revenue function (in dollars) when Q units of a certain commodity are sold is given by

$$TR(Q) = 56 + 0.1Q + 2Q^{1/2}$$

Find the marginal revenue when $Q = 9$.

23. In Problem 22, suppose that the total cost function is given by

$$TC(Q) = 500 + 0.05Q + 0.1Q^2$$

Find the marginal cost and the marginal profit when $Q = 9$.

2-8 CONCLUSION: FINDING THE RATE OF CHANGE OF TYPING SPEED

We are now in a position to answer the questions posed by the typing teacher in the introduction to this chapter. How fast is the student learning to type on the 100th day? The typing speed function was given as

$$f(t) = 8t^{0.3} + 0.04t$$

The rate of change of typing speed with respect to time is given by the derivative,

$$f'(t) = 0.3(8)t^{0.3-1} + 0.04 = 2.4t^{-0.7} + 0.04$$

Therefore, on the 100th day, the student's typing speed is changing at a rate of $f'(100)$ words per minute per day (dimensions of the dependent variable divided by the dimensions of the independent variable), or

$$f'(100) = 2.4(100)^{-0.7} + 0.04$$
$$\approx 0.1355 \text{ word per minute per day}$$

2-9 CHAPTER REVIEW

1. If the limit of a function $f(x)$ as x approaches a exists and is equal to L, we write

$$\lim_{x \to a} f(x) = L$$

This means that the function $f(x)$ can be made arbitrarily close to L by taking x close enough to a (but distinct from a).

2. The basic limit theorems can be used to compute limits of functions that are algebraic combinations of simpler functions.

3. If $f(x)$ is a polynomial or ratio of polynomials with nonzero denominator, then

$$\lim_{x \to a} f(x) = f(a)$$

4. The rate of change of a function describes how fast the dependent variable changes as the value of the independent variable changes.

5. The slope of a straight line can be interpreted as the rate of change of y with respect to a change in x.

6. The average rate of change of $f(x)$ between $x = a$ and $x = b$ is given by

$$\frac{f(b) - f(a)}{b - a}$$

7. The slope of the tangent line to a curve at a point is a measure of the rate of change of the function at that point.

8. The slope of the tangent line to the graph of $y = f(x)$ at $(a, f(a))$ is

$$f'(a) = \lim_{h \to 0} \frac{f(a + h) - f(a)}{h}$$

9. The derivative function $f'(x)$ of the function $f(x)$ is defined by

$$f'(x) = \lim_{h \to 0} \frac{f(x + h) - f(x)}{h}$$

10. A function is differentiable if the derivative exists at all points in the domain of the function.

11. For simple power functions $y = x^p$, the derivative is $y' = px^{p-1}$ for any real number p.

12. The derivative of a constant times a function is equal to the constant times the derivative of the function.

13. The derivative of the sum or difference of two functions is equal to the sum or difference of the derivatives.

14. In applications, the derivative is interpreted as the instantaneous rate of change and always has the dimensions of the dependent variable divided by the dimensions of the independent variable.

15. The marginal value of a function $f(x)$ at $x = a$ is given by the derivative $f'(a)$, and represents the approximate change in $f(x)$ for a 1 unit increase in the value of x when $x = a$.

REVIEW PROBLEMS (Section 2-9)

In Problems 1 – 4, construct a table of values for x and f(x) by letting x approach a from both the left and the right sides. Then determine whether the limit exists, and if so, determine its value.

1. $f(x) = \dfrac{1}{x^2} + 3, \quad a = -1$

2. $f(x) = \dfrac{x^2 - 1}{x - 1}, \quad a = 1$

3. $f(x) = \begin{cases} 3x & \text{if } x < 1 \\ 2.9 & \text{if } x \geq 1 \end{cases} \quad a = 1$

4. $f(x) = \begin{cases} x^3 & \text{if } x \leq -1 \\ 3x + 2 & \text{if } x > -1 \end{cases} \quad a = -1$

In Problems 5 – 10, evaluate the limits using the limit rules. Explain each step.

5. $\lim\limits_{x \to 5} 3x^2 + x - 1$

6. $\lim\limits_{x \to 0} \dfrac{2x - 1}{x + 4}$

7. $\lim\limits_{x \to -1} \dfrac{1}{x}$

8. $\lim\limits_{t \to 16} (-16)$

9. $\lim\limits_{h \to 0} \dfrac{\sqrt{h + 5}}{x}$

10. $\lim\limits_{h \to 0} (3xh + 2x^2 + h^3)$

In Problems 11–14, find the slope of each line. Then write the equation of a line that is parallel to the given line and one that is perpendicular to the given line.

11. $y = 2x - 7$

12. $y = -4x + 1$

13. $y = -\frac{3}{5}x + 2$

14. $y + 3x + 2 = 0$

15. Find an equation of the line through the point $(3, 3)$ and parallel to the line $y - x + 1 = 0$.

16. Find an equation of the line through the point $(6, 4)$ and perpendicular to the line $y + 3x - 3 = 0$.

In Problems 17–20, find the slope of the tangent line to each function using the limit definition of the derivative.

17. $f(x) = \dfrac{4}{x}$ at $x = 5$

18. $g(x) = 2x^2 + 1$ at $x = 2$

19. $f(x) = 3x^2 + 2x$ at $x = -2$

20. $s(t) = t^2 + t - 2$ at $t = 3$

For Problems 21–24, find the derivative of each function using the limit definition of the derivative.

21. $f(x) = 3x^2$

22. $g(x) = 2x^2 - 5$

23. $h(x) = \dfrac{5}{x}$

24. $f(x) = \sqrt{3x}$

25. The velocity (in feet per second) of an object thrown from a tall tower is given by the function $v = 32t + 4$, where t is measured in seconds. Find dv/dt using the limit definition of the derivative.

In Problems 26–37, find the derivative of each function using the rules given in Section 2-6.

26. $f(x) = 3x^2$

27. $f(x) = -2x^2 + 5$

28. $y = x^3 - 3x$

29. $y = -\frac{2}{3}x^6$

30. $g(x) = 2x^2 - 4x + 1$

31. $y = 2x^4 - 5x^3 + 6x^2 - 3x + 7$

32. $f(x) = 3\sqrt{x}$

33. $y = 2\sqrt[4]{x}$

34. $y = 8x^2 + \dfrac{1}{x}$

35. $g(x) = 5x^2 + \dfrac{2}{x} - \dfrac{1}{x^2}$

36. $f(x) = \dfrac{8}{\sqrt[3]{x^2}}$

37. $g(x) = 2x^{3/5} - 3x^2$

For the functions given in Problems 38–41, find an equation of the tangent line to each function at the indicated point.

38. $f(x) = x^2 + 3x$ at $(-1, -2)$

39. $g(x) = x^3 - 2x^2 + 3$ at $(2, 3)$

40. $f(x) = 2\sqrt{x}$ at $(4, 4)$

41. $g(x) = 3x^{2/3} + x^2$ at $(8, 76)$

42. Interpret the derivative for the following functional relationships:
    a. Commission is a function of the number of items sold.
    b. The cost of a plane ticket in a group plan is a function of the number of people going on the trip.

43. The mass of protein in a cell increases over time t according to the function $M(t) = 32 + 17t + 11t^2$. Find $M'(t)$ and $M'(1)$. What does $M'(1)$ mean?

44. The cost of a computer system is proportional to the square root of its speed of access. A formula for estimating the cost is given by $C(s) = 1400\sqrt{s}$. What is dC/ds and how is it interpreted?

45. A candy company has determined that the total daily cost for producing x boxes of chocolate candy is $C(x) = \frac{1}{2}x^2 + x + 3$ dollars. Find the average rate of change of cost from 0 to x and also the instantaneous rate of change at x. Explain the difference.

46. Suppose that t hours after being placed in a freezer, the temperature of a piece of meat is given by

$$T(t) = 70 - 12t + \frac{4}{t} \text{ degrees} \qquad 0 < t \leq 5$$

a. What is the temperature after 1 hour?
b. How fast is the temperature falling after 2 hours?

47. The total cost (in dollars) to a firm for manufacturing x air filters is given by

$$C(x) = \frac{x^2}{2} + 20x + 60$$

a. What is the total cost of producing 50 filters?
b. Determine the marginal cost function.
c. Compute the marginal cost when $x = 50$.
d. What is the actual cost of manufacturing the 51st part? Why does this differ from your answer in part c?

MORE ON LIMITS

In Chapter 2 we introduced the concept of a limit of a function and its important application to differentiation in calculus. In this supplement we present addi- tional computational techniques for evaluating limits and discuss further applications of limits in mathematics and calculus.

2S-1 INDETERMINATE FORMS AND LIMITS INVOLVING INFINITY

In Chapter 2 we stated two important theorems regarding limits, namely:

If $f(x)$ is a polynomial, then $\lim_{x \to a} f(x) = f(a)$.

And:

If $f(x) = g(x)/h(x)$, where both $g(x)$ and $h(x)$ are polynomials with $h(a) \neq 0$, then $\lim_{x \to a} f(x) = g(a)/h(a)$.

For many functions, direct substitution of the value that x approaches yields a meaningless mathematical quantity such as $0/0$, ∞/∞, $\infty - \infty$, or $(\infty)(0)$. Such meaningless quantities are called **indeterminate forms.** When these forms arise in trying to compute a limit, it is usually necessary to perform some type of algebraic manipulation (factoring, for example) on the function. Such simplifi- cation will usually enable us to apply the limit rules from Chapter 2 and find the limit, if it exists.

Example 1   Find the limit of the function

$$f(x) = \frac{x^2 - 9}{x - 3}$$

as $x \to 3$.

Solution   If we substitute $x = 3$, we obtain $0/0$, which is meaningless, and although both the numerator and the denominator are polynomials, the rules given above cannot be applied. However, a limit does exist. You should verify that when $x = 3.01$, then $f(x) = 6.01$; when $x = 3.001$, then $f(x) = 6.001$. Thus, it appears that the limit is 6. To show this more rigorously, we can simplify the function by observing that the numerator is the difference of two squares (see Algebra Review 10, page 106):

$$f(x) = \frac{x^2 - 9}{x - 3} = \frac{(x + 3)(x - 3)}{x - 3}$$

$$= x + 3$$

We are allowed to divide the numerator and denominator by $x - 3$ as long as it is not zero. Note that this will be true as x approaches 3 *but does not equal 3.* It is now easy to see that the limit is 6.   ∎

Example 2   Find

$$\lim_{x \to 0} \frac{2x^2 + 3x}{4x}$$

Solution   As $x \to 0$, we obtain the indeterminate form $0/0$. In this case, we wish to eliminate any powers of x in the denominator so that it will not be 0 as $x \to 0$. We can accomplish this by multiplying both the numerator and the denominator by the reciprocal of the *lowest* power of x. This gives

$$\lim_{x \to 0} \frac{2x^2 + 3x}{4x} \cdot \frac{1/x}{1/x} = \lim_{x \to 0} \frac{2x + 3}{4} = \frac{3}{4}$$

since $2x \to 0$ as $x \to 0$.   ∎

Example 3   Find

$$\lim_{x \to 1} [(x - 1)\sqrt{x - 2}]$$

ALGEBRA REVIEW 10: DIFFERENCE OF TWO SQUARES

In Algebra Review 3 we reviewed how to multiply two algebraic expressions. Suppose that A and B are terms. Then

$$(A + B)(A - B) = A^2 + BA - AB - B^2$$
$$= A^2 - B^2$$

In general, the product of $A + B$ and $A - B$ is a difference of two squares, $A^2 - B^2$.

Viewing this fact in reverse, whenever we have an expression $C - D$ where both C and D are perfect squares, we are able to write it as a product of two algebraic expressions, called *binomials*. For example,

$$x^2 - 4 = (x + 2)(x - 2)$$
$$16x^2 - y^2 = (4x + y)(4x - y)$$

This result is often useful in simplifying algebraic fractions (see Algebra Review 5).

EXERCISES*

a. Write $(6x + 5)(6x - 5)$ as the difference of two squares without formal multiplication.
b. Write $(9 - 49x^2)$ as the product of two binomials.
c. Write $(4x^4 - 9y^2)$ as the product of two binomials.

---

* Answers to algebra review exercises can be found at the end of the answer section in the back of the book.

Solution   We might think that if we substitute $x = 1$, we will obtain a value of 0. However, if x is very close to 1, but not equal to it, then $\sqrt{x - 2}$ is not a real number. That is, the expression $(x - 1)\sqrt{x - 2}$ does not make sense for x near 1. Hence, the limit does not exist.   ∎

Example 4   Find

$$\lim_{x \to -1} \frac{1}{(x + 1)^2}$$

Solution   When $x = -1$, we have the meaningless expression $1/0$. Consider what happens to the function when x is close to -1. If x is greater than -1 but very close to it, then the term $(x + 1)^2$ will be positive and very small. Hence, $1/(x + 1)^2$ will be very large. For instance, if $x = -0.999$, then $x + 1 = 0.001$, $(x + 1)^2 = 0.000001$, and $1/(x + 1)^2 = 1,000,000$. If x is less than -1 and very close to it, then $x + 1$ will be a very small negative number. However, $(x + 1)^2$ will be positive. Thus, $1/(x + 1)^2$ will be a large positive number. In either case, we see that the value of the function grows infinitely large. Thus, we have

$$\lim_{x \to -1} \frac{1}{(x + 1)^2} = \infty$$

This is called an *infinite limit*.   ∎

We wish to stress that the symbol ∞ used above does not stand for a real number. In discussing infinite limits we simply find it convenient to use the expression

$$\lim_{x \to a} f(x) = \infty$$

to convey the meaning that as x approaches a, the function values $f(x)$ become unbounded in the positive direction. In a similar way, if the function values are negative and become unbounded as x approaches a, we write

$$\lim_{x \to a} f(x) = -\infty$$

In either of these cases, note that the line $x = a$ is a vertical asymptote of the graph of $f(x)$. (See Figure 2S-1, for example.)

In all examples we have studied thus far, we have considered a to be a finite number. Let us now examine some problems in which x becomes unbounded or "tends toward infinity," that is,

$$\lim_{x \to \infty} f(x)$$

Such limits are called *limits at infinity*. We write $\lim_{x \to \infty} f(x) = L$ if the function values $f(x)$ can be made arbitrarily close to some finite number L by taking x sufficiently large. A similar meaning is attached to the symbol $\lim_{x \to -\infty} f(x) = L$. If either of these conditions holds, we say that $y = L$ is a horizontal asymptote of the graph of $f(x)$. (See, for example, Figure 2-7 in Chapter 2, where $L = 0$.)

Example 5   Find

$$\lim_{x \to \infty} \frac{3x^2 - 1}{4x^2 + 2}$$

Solution   We see that as x becomes very large, both the numerator and the denominator approach infinity. This would result in the indeterminate form ∞/∞. We can simplify this function by multiplying both the numerator and the denominator by $1/x^2$ (equivalently multiplying the whole function by 1). This results in

$$\frac{3x^2 - 1}{4x^2 + 2} \cdot \frac{1/x^2}{1/x^2} = \frac{3 - (1/x^2)}{4 + (2/x^2)}$$

Now as $x \to \infty$, the terms $1/x^2$ and $2/x^2$ approach 0. (Notice that they never actually reach 0 but get as close to 0 as we wish.) Therefore, the function approaches the limit $\frac{3}{4}$.   ∎

Why did we choose to multiply the numerator and denominator by $1/x^2$? We wanted to eliminate any powers of x in the numerator and to get terms of the following form: constant divided by a power of x. Terms in this form approach 0 as x approaches ∞. We might state this as a rule of thumb: If $x \to \infty$ and the function assumes the indeterminate form ∞/∞, then multiply the numerator and denominator by the reciprocal of the *highest* power of x in the function. This is only a rule of thumb and caution must be taken in some cases, as the following example illustrates.

Example 6   Find

$$\lim_{x \to \infty} \frac{3x^2 + 4x + 5}{2x + 1}$$

Solution   If we multiply the numerator and denominator by the reciprocal of the highest power of x, namely $1/x^2$, we obtain

$$\lim_{x \to \infty} \frac{3 + (4/x) + (5/x^2)}{(2/x) + (1/x^2)}$$

Now as $x \to \infty$, the terms in the denominator, $2/x$ and $1/x^2$, both approach 0, as do the terms $4/x$ and $5/x^2$ in the numerator. Thus, the fraction approaches $3/0$. A nonzero constant divided by 0 is not an indeterminate form; rather, we can view this in the following fashion. As $x \to \infty$, the numerator approaches 3 and the denominator approaches 0 from the right. For very small values close to 0 in the denominator, the value of the whole fraction will become very large. For example, $3/0.001 = 3000$; $3/0.0001 = 30,000$; and $3/0.000001 = 3,000,000$; and so on. Therefore, we may argue that the limit tends toward infinity.

We may show this more easily by multiplying the numerator and denominator by the reciprocal of the *lowest* power of x, namely $1/x$. This yields

$$\lim_{x \to \infty} \frac{3x + 4 + (5/x)}{2 + (1/x)}$$

Now as $x \to \infty$, the terms $5/x$ and $1/x$ approach 0, but the denominator approaches 2. Since the numerator contains the term $3x$, this will approach infinity as $x \to \infty$. Thus, the limit of the function is positive infinity. ∎

Although both approaches in Example 6 resulted in the same answer, using the lowest power of x was simpler. The choice is not always easy to determine beforehand, however.

2S-2 AN APPLICATION OF LIMITS TO FINANCE

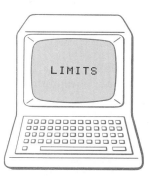

When money is placed in a bank or other savings institution, it earns *interest*, that is, an amount paid for the use of the money. There are several ways of computing interest. *Simple interest* is a fixed percentage of a sum of money. If the interest rate is 12% per year, for example, then $100 will earn $0.12(100) = \$12.00$ in interest after 1 year. More commonly, interest is *compounded*. That is, savings earn a portion of the annual interest over a shorter period of time; this allows the interest earnings to be added to the account and also earn interest. To illustrate this, suppose that the 12% interest rate is compounded quarterly. This means that every 3 months the amount in the savings account earns $\frac{1}{4}$ of the annual rate, or 3%. We compute the total amount of money in the account over 1 year as indicated in Table 2S-1. As you can see in the table, compounding effectively produces a higher rate than simple interest.

Table 2S-1

Month	Interest Earned	Total Savings
0	—	$100.00
3	$(100.00)(0.03) = 3.00$	103.00
6	$(103.00)(0.03) = 3.09$	106.09
9	$(106.09)(0.03) = 3.18$	109.27
12	$(109.27)(0.03) = 3.28$	112.55

Let us develop a formula for the amount of savings when an arbitrary number of compounding periods is considered. Let $r =$ annual interest rate expressed as a decimal, $P =$ initial amount in savings, and $n =$ number of compounding periods per year (for example, $n = 4$ denotes quarterly compounding, $n = 365$ denotes daily compounding). The amount of interest earned for one compounding period is the amount in savings times r/n. Therefore, after the first period, the initial sum of P dollars earns $P(r/n)$ in interest, and the total amount in the account will be

$$P + P\left(\frac{r}{n}\right) = P\left(1 + \frac{r}{n}\right)$$

This amount will earn a rate of r/n in the second period. Thus, we will have

$$P\left(1+\frac{r}{n}\right)+P\left(1+\frac{r}{n}\right)\left(\frac{r}{n}\right)=P\left(1+\frac{r}{n}\right)\left(1+\frac{r}{n}\right)$$
$$=P\left(1+\frac{r}{n}\right)^2$$

Continuing in this fashion, we see that after n periods, the total savings will be

$$P\left(1+\frac{r}{n}\right)^n$$

When $P = 100$, $r = 0.12$, and $n = 4$, we have

$$(100)\left(1+\frac{0.12}{4}\right)^4 = 112.55$$

If $n = 12$, we get

$$(100)\left(1+\frac{0.12}{12}\right)^{12} = 112.68$$

The amount in the savings account becomes larger as more compounding periods are used.

It is currently common practice to have interest compounded *continuously*. This means that the number of compounding periods in 1 year grows infinitely large. Mathematically, this is equivalent to saying that n approaches infinity. We would like to be able to compute the value of $P(1 + r/n)^n$ as n approaches infinity, that is,

$$\lim_{n\to\infty} P\left(1+\frac{r}{n}\right)^n$$

Since P is a constant, from one of the basic rules on limits, we know that this is equal to

$$P\lim_{n\to\infty}\left(1+\frac{r}{n}\right)^n$$

Now suppose that $r = 1$. In Table 2S-2, we evaluate some sample values of $(1 + 1/n)^n$ as n gets large. This sequence appears to be approaching a constant value. It turns out that this value is known as the number $e \approx 2.71828$. The number e is like π in that it does not terminate as a decimal fraction. The number e is the base of *natural logarithms*, which will be discussed more extensively in Chapter 6. For any value of r, it can be shown that

$$\lim_{n\to\infty}\left(1+\frac{r}{n}\right)^n = e^r$$

Thus, the value of P dollars compounded continuously at a rate r is Pe^r. For $P = 100$ and $r = 0.12$, we have

$$100e^{0.12} \doteq \$112.75$$

This is the formula used by financial institutions to compute continuous compound interest. Surprisingly, this everyday formula is based on limit concepts!

Table 2S-2

n	$\left(1 + \dfrac{1}{n}\right)^n$
1	2.00000000
10	2.593742469
100	2.704813829
1,000	2.716923932
10,000	2.718145926
100,000	2.718268237
1,000,000	2.718280469

2S-3 CONTINUITY

In Chapter 1 we introduced the notion of continuity strictly from an intuitive point of view. In this section we wish to define this concept from a formal mathematical viewpoint.

A function $f(x)$ is said to be *continuous at a point a* if it satisfies the following conditions:

1. $f(a)$ exists; that is, a is in the domain of $f(x)$

2. $\lim_{x \to a} f(x)$ exists

3. $\lim_{x \to a} f(x) = f(a)$

A function is *continuous over an interval* if it is continuous at every point in the interval. A function is **continuous** if it is continuous at every point in its domain.

To determine whether a function is continuous at a point $x = a$, three questions must be addressed: First, is $f(a)$ a real number? Second, does the limit of $f(x)$ as $x \to a$ exist? Third, if the limit exists, does it equal $f(a)$? Obviously, it is impossible to test every point in an interval for these three conditions, since there are an infinite number of them. It is usually sufficient to identify and examine potential "trouble points" where discontinuities might arise. Typically, these are points that involve division by 0 or where the definition of the function changes. The examples that follow will illustrate these cases.

Before presenting these examples, we should note that most functions encountered in practical applications are continuous. For example, all polynomial functions are continuous at every real number; $\sqrt{x}$ is continuous for nonnegative values of x; and the sums, differences, and products of continuous functions are also continuous (do you see how this follows from the limit rules?). Rational functions (quotients of two polynomials) are also continuous where the denominator is not equal to 0.

Example 7   Is the function

$$f(x) = \frac{1}{x - 2}$$

continuous for all real numbers x?

Solution   Since $f(x)$ is not defined at $x = 2$, we see that condition 1 is violated at this point. The function is not continuous at $x = 2$; therefore, it is not continuous for all the real numbers. This is easily seen from the graph in Figure 2S-1.

Figure 2S-1

Graph of $f(x) = \dfrac{1}{x-2}$

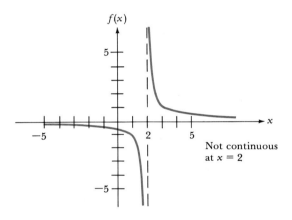

Example 8   Consider the function

$$h(x) = \begin{cases} \dfrac{x^2 - 1}{x - 1} & \text{if } x \neq 1 \\ 3 & \text{if } x = 1 \end{cases}$$

Solution   We have $h(x)$ defined for all real numbers x; therefore, condition 1 is satisfied. A potential point of discontinuity is $x = 1$, since the function is defined differently around this point. To verify conditions 2 and 3, we must show that the limit of the function as $x \rightarrow 1$ exists and is equal to $f(1) = 3$.

If $x \neq 1$, then the function is defined by $(x^2 - 1)/(x - 1)$. We may compute the limit as $x \rightarrow 1$ as follows:

$$\lim_{x \to 1} \frac{x^2 - 1}{x - 1} = \lim_{x \to 1} \frac{(x - 1)(x + 1)}{x - 1}$$

$$= \lim_{x \to 1} x + 1 = 2$$

Although the limit exists, it does not equal $f(1)$ and therefore the function is not continuous at $x = 1$. This is seen from the graph in Figure 2S-2.

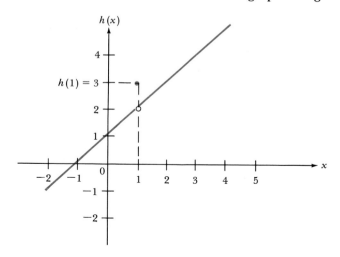

Figure 2S-2
Graph of h(x)

■

Example 9   Test the following function for continuity:

$$f(x) = \begin{cases} 2x - 1 & \text{if } x < 2 \\ x^2 - 1 & \text{if } x \geq 2 \end{cases}$$

Solution   A possible trouble point is where the function changes definition. This occurs at $x = 2$, since the function is defined by different formulas on either side of $x = 2$. We note that $f(2) = 2^2 - 1 = 3$. To evaluate the limit of $f(x)$ as $x \to 2$, we must check that both the left-hand and right-hand limits exist and are equal to each other. As $x \to 2$ from the left, $f(x) = 2x - 1$. This limit is clearly equal to 3. As x

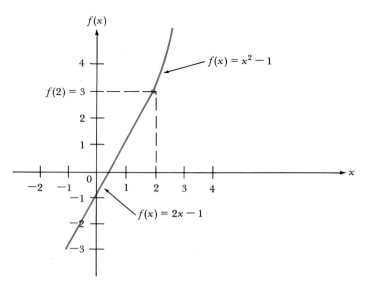

Figure 2S-3
Graph of f(x)

approaches 2 from the right, $f(x) = x^2 - 1$. Again the limit is 3. Therefore, we can say that

$$\lim_{x \to 2} f(x) = 3 = f(2)$$

and $f(x)$ is continuous at $x = 2$. This is shown in Figure 2S-3.   ■

The relationship between continuity and differentiability is summarized in the following theorem:

THEOREM

If a function is differentiable at a point, then it is continuous at that point.

This theorem can be stated in a different, but logically equivalent, way: If a function is not continuous at a point, then it cannot be differentiable at that point.

We hasten to point out that the converse of the theorem above is not valid. That is, if a function is continuous at a point, then it does not necessarily follow that it is differentiable at that point. For example, the function $f(x) = |x|$ is continuous at $x = 0$, but we showed in Chapter 2 that it is not differentiable at $x = 0$.

2S-4 SOME PROOFS OF DIFFERENTIATION RULES

We may also use the limit theorems presented in Chapter 2 to prove theorems about derivatives. In Chapter 2 we stated two important theorems regarding the derivatives of a constant times a function and the sum or difference of two functions. We will prove these now.

We stated that if $f(x) = cg(x)$, then $f'(x) = cg'(x)$. Using the limit definition of the derivative, we have

$$f'(x) = \lim_{h \to 0} \frac{f(x + h) - f(x)}{h}$$

$$= \lim_{h \to 0} \frac{cg(x + h) - cg(x)}{h}$$

$$= \lim_{h \to 0} c \left[\frac{g(x + h) - g(x)}{h} \right]$$

Now by rule 3, this last expression is equal to

$$c \lim_{h \to 0} \frac{g(x + h) - g(x)}{h} = cg'(x)$$

Next, consider the sum $h(x) = f(x) + g(x)$. In Chapter 2 we stated that the derivative of this sum is $h'(x) = f'(x) + g'(x)$. We prove this as follows:

$$h'(x) = \frac{d}{dx}[f(x) + g(x)] = \lim_{h \to 0} \frac{[f(x + h) + g(x + h)] - [f(x) + g(x)]}{h}$$

$$= \lim_{h \to 0} \frac{f(x + h) - f(x) + g(x + h) - g(x)}{h}$$

$$= \lim_{h \to 0} \frac{f(x + h) - f(x)}{h} + \lim_{h \to 0} \frac{g(x + h) - g(x)}{h} \qquad \textit{By rule 4}$$

$$= f'(x) + g'(x)$$

The proof for the difference of two functions is similar.

PROBLEMS (Supplement to Chapter 2)

In Problems 1 – 16, evaluate each limit.

1. $\lim\limits_{x \to 2} \dfrac{(2x - 1)^2}{3x^2 + 5}$

2. $\lim\limits_{x \to 1} \dfrac{4x + 3}{(2x + 1)^3}$

3. $\lim\limits_{x \to 2} \dfrac{x^2 - 4}{x - 2}$

4. $\lim\limits_{x \to -5} \dfrac{x^2 - 25}{x + 5}$

5. $\lim\limits_{x \to 1} (3x^3 - 2x^2 + 4x + 1)$

6. $\lim\limits_{x \to 1} [(x^2 + x + 1)(x - 2)]$

7. $\lim\limits_{t \to 0} \dfrac{4}{t^2}$

8. $\lim\limits_{y \to 0} \dfrac{3y + 2y^2}{4y - y^3}$

9. $\lim\limits_{x \to \infty} \dfrac{3x + 5}{x - 2}$

10. $\lim\limits_{x \to \infty} \dfrac{4 + 2x}{5 - 3x}$

11. $\lim\limits_{x \to \infty} \dfrac{x^2 + 2x + 3}{3x^2 - 5x - 4}$

12. $\lim\limits_{x \to \infty} \dfrac{2x^3 + 3x^2 - 5}{5x^3 + 2x + 3}$

13. $\lim\limits_{x \to -1} \dfrac{2x^2 - x - 3}{x + 1}$

14. $\lim\limits_{t \to 2^+} \sqrt{t - 2}$ (The notation 2^+ means to approach 2 from the right.)

15. $\lim\limits_{t \to 2^-} \sqrt{t - 2}$ (The notation 2^- means to approach 2 from the left.)

16. $\lim\limits_{t \to 2} \sqrt{t - 2}$

17. A manufacturer has developed a model for the amount of time necessary to produce t standard machine parts as

$$f(t) = \frac{\sqrt{t^3} - 9\sqrt{t}}{t - 9}$$

However, this is not defined at $t = 9$. What value should be assigned to $f(9)$ so that the function is continuous?

In Problems 18 – 29, find points (if any) where the given functions are not continuous and state why they are not continuous at these points.

18. $f(x) = 2x^3 - 3x - 1$

19. $f(x) = 3x^2 - 9x$

20. $f(x) = \dfrac{1}{x - 5}$

21. $f(x) = \dfrac{3}{x - 1}$

22. $f(x) = \dfrac{4x}{x + 4}$

23. $f(x) = \dfrac{3}{x^2 + 1}$

24. $f(x) = \sqrt{x + 2}$

25. $f(x) = \sqrt{x^2 + 2}$

26. $f(x) = \dfrac{x - 2}{x^2 - 9}$

27. $f(x) = \dfrac{x^2 - 4}{x - 2}$

28. $f(x) = \begin{cases} x - 3 & \text{if } x \le 1 \\ 2x & \text{if } x > 1 \end{cases}$

29. $f(x) = \begin{cases} \dfrac{x^2 - 4}{x - 2} & \text{if } x \ne 2 \\ 4 & \text{if } x = 2 \end{cases}$

30. Is it possible for a function to have a limit at a point when the function is discontinuous at that point? Why or why not?

31. A business firm estimates that the interest rate it has to pay is a function of the amount borrowed, x (in millions of dollars). The function is defined as

$$f(x) = \begin{cases} 0.07 & \text{if } 0 \le x \le 2 \\ 0.05 + 0.01x & \text{if } 2 < x \le 6 \\ 0.12 & \text{if } 6 < x \end{cases}$$

Graph this function. At what points is the function not continuous?

3

APPLICATIONS OF THE DERIVATIVE

OUTLINE

INTRODUCTION

A forest fire is burning down a narrow valley 3 miles wide at a speed of 40 feet per minute. The fire can be contained by cutting a firebreak through the forest across the valley. It takes 30 seconds for one person to clear 1 foot of the firebreak. The value of lost timber is $3000 per square mile. Each person hired is paid $10 per hour, and it costs $25 to transport and supply each person with the appropriate equipment. How many people should be sent to contain the fire, and where should the firebreak be located?*

To set up a mathematical model for this problem, it is first helpful to draw a picture. Figure 3-1 illustrates this situation. Next, we define the unknowns in the problem. From Figure 3-1 we denote the distance from the fire front to the location of the firebreak by the variable y. The other unknown in this problem is

---

* This problem is a simplified version of a real application that was investigated by G. M. Parks in "Development and Application of a Model for Suppression of Forest Fires," *Management Science*, vol. 10, pp. 760–766 (1964).

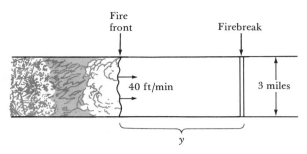

Figure 3-1
Forest fire problem

the number of people to hire, which we denote by the variable x.

The constants in this problem are listed below:

a. Width of the valley, 3 miles
b. Speed of the fire:

$$40 \frac{ft}{min}$$

$$= \left(40 \frac{ft}{min}\right) \times \left(\frac{1}{5280} \frac{mi}{ft}\right) \times \left(60 \frac{min}{hr}\right)$$

$$= \frac{2400}{5280} \left(\frac{ft}{min}\right)\left(\frac{mi}{ft}\right)\left(\frac{min}{hr}\right)$$

$$\approx 0.4545 \text{ mi/hr}$$

c. Rate (per person) at which the firebreak is cleared:

$$30 \frac{sec}{ft}$$

$$= \left(30 \frac{sec}{ft}\right) \times \left(\frac{1}{3600} \frac{hr}{sec}\right) \times \left(5280 \frac{ft}{mi}\right)$$

$$= \frac{(30)(5280)}{3600} \left(\frac{sec}{ft}\right)\left(\frac{hr}{sec}\right)\left(\frac{ft}{mi}\right)$$

$$= 44 \text{ hr/mi per person}$$

That is, one person can clear 1 mile in 44 hours.
d. Value of lost timber, $3000 per square mile
e. Wages paid, $10 per hour per person
f. Transportation and supply cost, $25 per person

Notice that we have changed all dimensions to units of miles and hours; in other words, we have *standardized*

the dimensions. The modeling task is usually made easier if this is done early in the process, that is, when variables and constants are first defined.

The next phase in constructing a model is to determine whether there are any appropriate relations among the variables and constants. Although it is not explicitly stated in the problem definition, you should realize that the firebreak must be completed before the fire reaches it. This can be stated in a different fashion: "The time it takes for the fire to reach the firebreak must not exceed the time to clear the firebreak." In the ideal case (remember that models are only approximations to reality), we could argue that these times must be equal in order not to lose any unnecessary timber. Obviously, these times depend on how many people are hired, where the firebreak is located, the speed of the fire, the width of the valley, and the rate at which people can cut the firebreak.

One of the keys to modeling discussed in Chapter 1 is to break down the problem into small pieces (recall the driving cost example). The time relation stated above can be broken down into three components:

a. Time for the fire to travel y miles
b. "Must equal"
c. Time to clear the firebreak

By considering each of these components separately, it is easier to develop the correct relation. The time for the fire to travel y miles at 0.4545 mile per hour is

$$\frac{y}{0.4545} \text{ hr}$$

Since one person can clear 1 mile of the firebreak in 44 hours, then x people can do it in

$$\frac{44}{x} \text{ hr}$$

The width of the valley is 3 miles; therefore, it will take 3 times this amount of time, or

$$(3)\left(\frac{44}{x}\right) = \frac{132}{x} \text{ hr}$$

We now have mathematical expressions for both components a and c. Since these must equal each other (component b), we have the relation

$$\frac{y}{0.4545} = \frac{132}{x}$$

What are we trying to accomplish in this problem? If no people are hired, then the fire will destroy a large amount of timber (at a cost of $3000 per square mile). On the other hand, if we hire many people, then little timber will be lost, but it may cost a lot to pay and supply these people. Thus, we seek a happy medium, a solution that *minimizes* the total cost incurred.

The cost of lost timber is

$$(3000 \text{ dollars/sq mi})(3y \text{ sq mi}) = 9000y \text{ dollars}$$

The labor cost to pay x people $10 per hour is

$$\left(\frac{10 \text{ dollars/hr}}{\text{person}}\right)(x \text{ people})\left(\frac{132}{x} \text{ hr}\right)$$

$$= 1320 \text{ dollars}$$

Finally, the transportation and supply costs are

$$(25 \text{ dollars/person})(x \text{ people}) = 25x \text{ dollars}$$

Summing these, we want to minimize the quantity

$$9000y + 1320 + 25x$$

and satisfy the restriction

$$\frac{y}{0.4545} = \frac{132}{x}$$

at the same time.

As in Chapter 1, we see that modeling with mathematics requires careful thought and analysis. This skill can only be developed through practice and more practice!

Let us try to minimize the total cost by selecting some values for x and y by trial and error. If $x = 100$ people, then using the restriction that $y/0.4545 = 132/x$, we have

$$y = \frac{(0.4545)(132)}{100} \approx 0.6$$

The total cost will be

$$9000(0.6) + 1320 + 25(100) = \$9220$$

If we choose $x = 200$, then you may verify that $y \approx 0.3$ and the total cost will be $9020; if $x = 150$, then $y \approx 0.4$ and the total cost is $8670. Clearly, such trial and error is time-consuming and makes it very difficult to find the best solution. In this chapter, we shall see how techniques of differentiation will assist us in solving problems like the forest fire problem which involve the minimization or maximization of a function. We will also see how calculus helps in drawing graphs of functions. First, however, we need to develop some additional mathematical concepts.

3-1 CONCEPTS OF OPTIMIZATION

The forest fire problem we have presented is an example of an optimization problem.

> An **optimization problem** is one in which we want to *minimize* or *maximize* some quantity. The expression that we wish to minimize or maximize is called the **objective function.**

In some applications, for instance, we may wish to minimize cost, time, or distance; in others, we might maximize profit, savings, or value. Optimization problems arise in many areas of business, engineering, and the social and natural sciences, and represent some of the most important applications of calculus.

The minimum and maximum of a function are best described in reference to the graph of the function. Consider, for example, the graph of the function $f(x) = -x^2 + 4x + 5$, as shown in Figure 3-2. From this graph we see that when $x = 2$, $f(x)$ assumes its largest value among all points in the domain of f (in this case, all real numbers). We say that $f(x)$ assumes its *absolute maximum* at $x = 2$, since the point $(2, f(2))$ is the highest point on the graph. This leads to the following definition:

A function $f(x)$ is said to reach its absolute maximum at $x = c$ if $f(c) \geq f(x)$ for all x in the domain of $f(x)$. We call the number $f(c)$ the **absolute maximum** of $f(x)$.

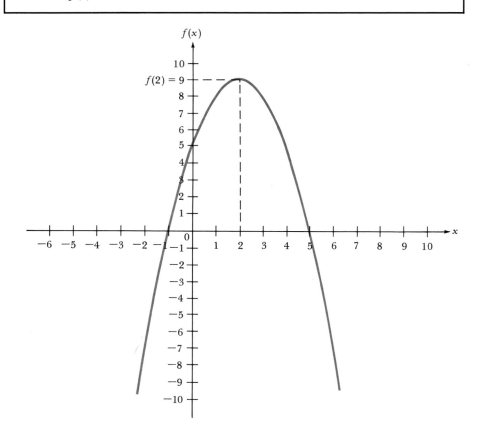

Figure 3-2
Graph of $f(x) = -x^2 + 4x + 5$

In a similar manner, we may define the *absolute minimum* of a function:

A function $f(x)$ is said to reach its absolute minimum at $x = c$ if $f(c) \leq f(x)$ for all x in the domain of $f(x)$. We call the number $f(c)$ the **absolute minimum** of $f(x)$.

An absolute maximum or minimum is sometimes called a *global* maximum or minimum. Observe that $f(x) = -x^2 + 4x + 5$ does not have an absolute minimum, since the function values can be made smaller than any prescribed number by taking x sufficiently large.

The absolute maximum value and absolute minimum value of a function are called the *extreme values* of the function. The following theorem guarantees that a continuous function on a closed interval actually has extreme values (see Algebra Review 11, page 122 for a discussion of intervals).

Extreme Value Theorem

> A continuous function defined on a closed interval [a, b] has an absolute maximum value and an absolute minimum value on [a, b].

This theorem is illustrated in Figure 3-3a. Figures 3-3b and 3-3c show that if

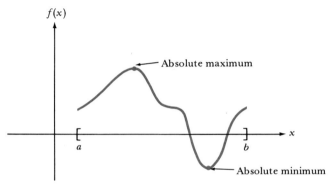

a. A continuous function on a closed interval has extreme values

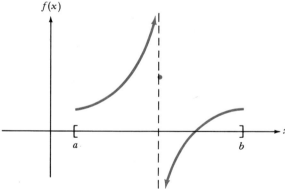

b. A discontinuous function on a closed interval with no extreme values

Figure 3-3
Extreme value theorem

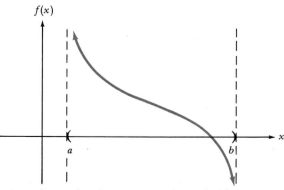

c. A continuous function on an open interval with no extreme values

either of the hypotheses that the function is continuous or the interval is closed is relaxed, then the conclusion of the theorem does not necessarily hold.

Next, consider the function $f(x) = x^3 - 6x^2$, as shown in Figure 3-4. The domain of $f(x)$ is the set of all real numbers. As x gets large, the function values get large without bound; therefore, no absolute maximum exists. Also, as x gets small (more negative), the function values become smaller (more negative) without bound. Consequently, there is no absolute minimum either. However, suppose that we look only at values of x close to $x = 0$. Although we can see that $f(0) = 0$ is not the largest value of the function over the entire domain, we do see that within a small open interval around $x = 0$, for example, $(-1, 1)$, the function assumes its largest value at $x = 0$. We then say that the function assumes a *relative maximum* value at $x = 0$.

A function $f(x)$ is said to have a **relative maximum** value $f(c)$ at $x = c$ if $f(c) \geq f(x)$ for all x in some open interval (a, b) in the domain of $f(x)$ containing c.

ALGEBRA REVIEW 11: INTERVALS

An **open interval (a, b)** is defined to be the set of real numbers $\{x | a < x < b\}$. (This notation is read "the set of all real numbers x such that x is greater than a and x is less than b.") A **closed interval $[a, b]$** is defined to be the set of all real numbers $\{x | a \leq x \leq b\}$. The points a and b are called **endpoints** of the intervals. The difference between an open interval and a closed interval is that while an open interval is defined with respect to the points a and b, it does not include these points. A closed interval, however, includes its endpoints.

For example, the interval $(0, 1)$ consists of all numbers *greater than* 0 and *less than* 1, that is, $\{x | 0 < x < 1\}$. The interval $[0, 1]$ consists of all real numbers *greater than or equal to* 0 and *less than or equal to* 1, that is, $\{x | 0 \leq x \leq 1\}$. We may also define *half-open* intervals such as $[3, 6)$. This is the set of all real numbers greater than or equal to 3 and less than 6, that is, $\{x | 3 \leq x < 6\}$. Similarly, $(3, 6] = \{x | 3 < x \leq 6\}$.

EXERCISES*

a. Describe the following intervals as sets of numbers:
   $(2, 7)$, $[3, 4]$, $(0, 1]$
b. Write the following sets as intervals:
   $\{x | -3 \leq x \leq 5\}$, $\{x | 2 < x\}$, $\{x | 0 < x \leq 1\}$
c. Write the following in both interval and set notation:
   The set of all real numbers less than 18 and greater than or equal to -1.

---

* Answers to algebra review exercises can be found at the end of the answer section in the back of the book.

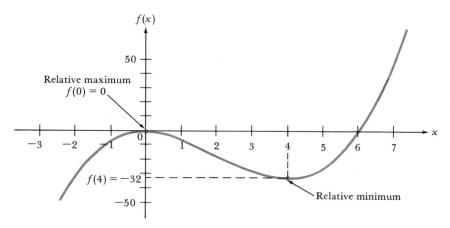

Figure 3-4
Graph of $f(x) = x^3 - 6x^2$

In such a case we also say that a relative maximum occurs at the point $(c, f(c))$ on the graph of the function. We will call $(c, f(c))$ a *relative maximum point* and the number $f(c)$ a *relative maximum.* In other words, a relative maximum occurs at a point $(c, f(c))$ if the value of $f(c)$ is no smaller than the value of $f(x)$ for any x close to, and on either side of, $x = c$. This is illustrated in Figure 3-5. Notice that $f(c) \geq f(x)$ need not hold for *every* interval around $x = c$; rather, we only need to be able to find *some* interval for which this holds. Also, observe that the interval (a, b) must contain c; that is, the function must be defined on both sides of $x = c$.

In Figure 3-4, we can also see that if we look only at values of x close to $x = 4$, the function assumes its smallest value at $x = 4$, $f(4) = -32$. We may therefore call $f(4)$ a *relative minimum.* We define a relative minimum as follows:

A function $f(x)$ is said to have a **relative minimum** value $f(c)$ at $x = c$ if $f(c) \leq f(x)$ for all x in some open interval (a, b) in the domain of $f(x)$ containing c.

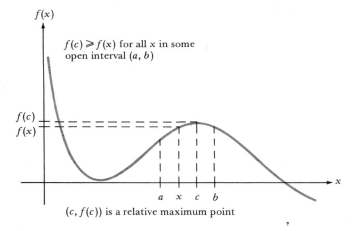

Figure 3-5
Illustration of relative maximum

A relative minimum is illustrated in Figure 3-6. We often call a relative minimum or relative maximum a *local minimum* or *local maximum,* respectively. The terms *optimum* and *extremum* are often used to denote either a minimum or a

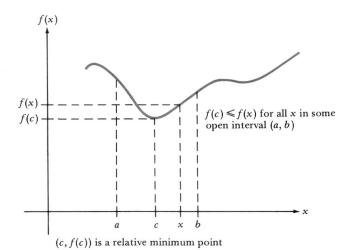

Figure 3-6
Illustration of relative minimum

maximum. (Keep in mind that the plurals of minimum, maximum, optimum, and extremum are *minima, maxima, optima,* and *extrema,* respectively.)

Example 1   For the graph in Figure 3-7, determine whether the points $(a, f(a))$, $(b, f(b))$, $(c, f(c))$, and $(d, f(d))$ are relative optimum points, absolute optimum points, or neither.

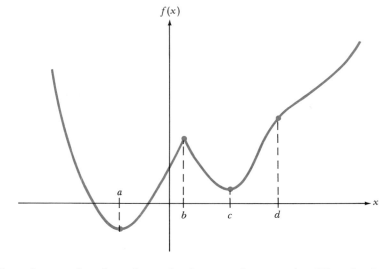

Figure 3-7
Graph for Example 1 (domain of f is all real numbers)

Solution   First observe that there is no absolute maximum point. The absolute minimum occurs at the point $(a, f(a))$; this point is also a relative minimum point. The point $(c, f(c))$ is another relative minimum point, and the function has a relative maximum at the point $(b, f(b))$. At the point $(d, f(d))$, there is neither a relative optimum nor an absolute optimum.   ∎

Suppose that we now modify the function in Figure 3-4 by restricting its domain as follows:

$$f(x) = x^3 - 6x^2 \qquad -2 \leq x \leq 7$$

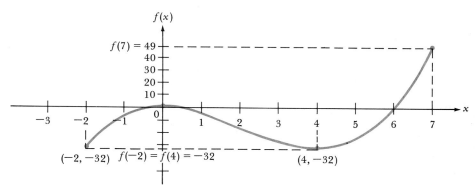

Figure 3-8
Graph of
$f(x) = x^3 - 6x^2, -2 \le x \le 7$

The domain of $f(x)$ is the closed interval $[-2, 7]$. From Figure 3-8, we see that this function assumes its largest value of 49 at $x = 7$ and its smallest value of -32 when $x = -2$ and when $x = 4$. The points $(0, f(0))$ and $(4, f(4))$ are also relative maximum and relative minimum points, respectively. We summarize these observations in Table 3-1.

Table 3-1

Type of Optimum	At	Value
Absolute maximum	$x = 7$	49
Absolute minimum	$x = -2$	-32
	$x = 4$	-32
Relative maximum	$x = 0$	0
Relative minimum	$x = 4$	-32

From this example we see that an absolute maximum or minimum can occur at an endpoint of a closed interval that defines the domain of the function. However, a relative optimum cannot occur at an endpoint by definition; since the function is not defined on both sides of the endpoint, there cannot be any open interval within the domain that contains an endpoint.

Example 2 For the function in Figure 3-9, suppose the domain is defined by the interval $a \le x \le b$. Identify all points at which relative or absolute optima occur.

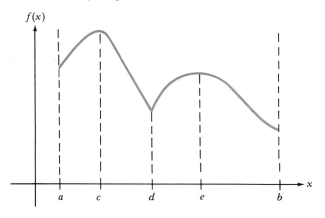

Figure 3-9
Graph for Example 2

Solution   The solution is given in Table 3-2.

Table 3-2

Type of Optimum	At
Absolute maximum	$x = c$
Absolute minimum	$x = b$
Relative maximum	$x = c$ and $x = e$
Relative minimum	$x = d$

In this section we have described the nature of relative and absolute optimum points for functions. It would certainly be convenient to be able to find such points mathematically. We shall do this in the following sections and then see how these procedures can be applied to practical problems.

PROBLEMS (Section 3-1)

For each graph, identify all points in the interval [a, b] at which relative or absolute optima occur.

1.

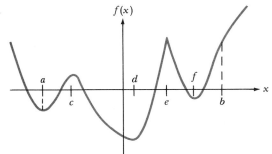

2.

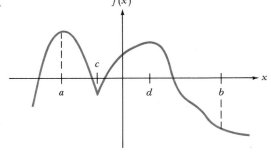

3.

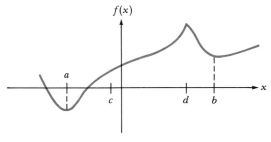

4.

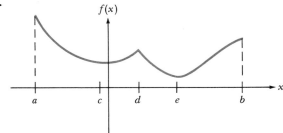

5.

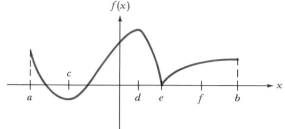

$f(x)$

a c d e f b x

6. Can a relative optimum or absolute optimum, or both, occur at an endpoint? Why?

7. Sketch a graph with a relative maximum point and absolute minimum point.

8. Sketch a graph with an absolute minimum point, absolute maximum point, relative minimum point, and relative maximum point.

9. Sketch a graph over an open interval whose absolute minimum point is also a relative minimum point and which does not have an absolute maximum point.

10. Sketch a graph with two relative maximum points and absolute optima occurring at the endpoints of a closed interval.

3-2 CRITICAL POINTS

NEWTON

Let us reconsider the function $f(x) = x^3 - 6x^2$ shown in Figure 3-4. We saw that at $x = 0$ we have a relative maximum, and at $x = 4$ we have a relative minimum. If you examine Figure 3-4 carefully, you can see that the tangent lines to the graph at these points are horizontal, and hence their slopes are both 0. Since the slope of the tangent line to a curve at any point is given by the value of the derivative at that point, we see that the derivative must be 0 at these points. This suggests a method for finding points that might be relative optima: Set the derivative of the function equal to 0 and solve for all values of x that satisfy this equation. If we apply this procedure to $f(x) = x^3 - 6x^2$, we have

$$f'(x) = 3x^2 - 12x = 0$$

Factoring out x in this equation (see Algebra Review 12, page 128) yields

$$x(3x - 12) = 0$$

Therefore, there are two solutions:

$$x = 0$$

and

$$3x - 12 = 0 \quad \text{or} \quad x = 4$$

At these points the function *potentially* has relative optima. We stress "potentially" since it is not necessarily true that every value x that satisfies $f'(x) = 0$ yields a relative maximum or minimum. This is easily shown by considering the function $f(x) = x^3$. Setting the derivative to 0 yields

$$f'(x) = 3x^2 = 0$$

The only solution is $x = 0$. However, from the graph of this function in Figure 3-10 we can see that at $x = 0$ there is neither a relative maximum nor a relative minimum. We shall soon discover how to recognize this particular case.

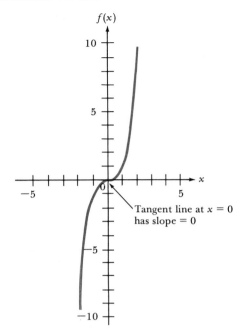

Figure 3-10
Graph of $f(x) = x^3$

Tangent line at $x = 0$ has slope = 0

ALGEBRA REVIEW 12: SOLVING ONE-VARIABLE EQUATIONS

One of the most fundamental algebraic techniques used in calculus problems is that of solving an equation in one variable, $f(x) = 0$. Any value of x such that $f(x) = 0$ is called a **root,** or **solution,** of the equation. Two equations are **equivalent** if one can be derived from the other by the following operations:

1. Addition or subtraction of the same quantity on both sides of the equation
2. Multiplication or division of both sides by the same nonzero quantity

Equivalent equations have the same roots.

Consider the equation

$$6x + 3 = 0$$

We subtract 3 from both sides, obtaining

$$6x = -3$$

Next, divide both sides by 6:

$$\frac{6x}{6} = -\frac{3}{6}$$

$$x = -\frac{1}{2}$$

A second complication arises in that a point can be a relative optimum where the derivative does not even exist. Such points cannot be identified by setting the derivative to 0. To illustrate this, consider the function

$$f(x) = \begin{cases} -x^2 + 6x & \text{if } x \le 6 \\ -x^2 + 16x - 60 & \text{if } x > 6 \end{cases}$$

The graph of this function is shown in Figure 3-11. Note that when $x = 6$, the

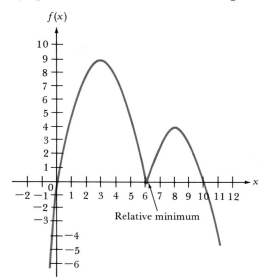

Figure 3-11
Illustration of relative minimum
where derivative does not exist

Many equations having more than one solution can be *factored*. For example, consider

$$8x + 4x^2 = 0$$

We may factor $4x$ from each term, giving

$$4x(2 + x) = 0$$

Setting each factor equal to 0 yields roots of the equation. Thus, we have

$$4x = 0 \qquad \text{or} \qquad x = 0$$

and

$$(2 + x) = 0 \qquad \text{or} \qquad x = -2$$

Therefore, the roots of the equation are $x = 0$ and $x = -2$. We shall discuss the process of factoring other types of equations in subsequent algebra reviews.

EXERCISES

Find the roots of each equation and explain all steps.
a. $3x + 4 = 8x - 1$
b. $5x^2 + 2x = 5x - x^2$
c. $x^2 - 18 = -9 - 2x^2$
d. $2(x - 3) = -3(4 - 2x)$
e. $\dfrac{x + 3}{6} - \dfrac{x + 4}{2} = 1$

function assumes a relative minimum. However, the derivative of $f(x)$ does not exist at $x = 6$, as the sharp point on the graph at $x = 6$ indicates. To see this, we note that when $h < 0$, then $6 + h < 6$, and so

$$\frac{f(6 + h) - f(6)}{h} = \frac{-(6 + h)^2 + 6(6 + h) - 0}{h}$$

$$= \frac{-36 - 12h - h^2 + 36 + 6h}{h}$$

$$= -6 - h$$

However, when $h > 0$, then $6 + h > 6$, and we have

$$\frac{f(6 + h) - f(6)}{h} = \frac{-(6 + h)^2 + 16(6 + h) - 60}{h}$$

$$= \frac{-36 - 12h - h^2 + 96 + 16h - 60}{h}$$

$$= \frac{4h - h^2}{h}$$

$$= 4 - h$$

Therefore,

$$\lim_{h \to 0} \frac{f(6 + h) - f(6)}{h}$$

does not exist, since the limiting value is -6 when $h \to 0$ from the left, while the limiting value is 4 when $h \to 0$ from the right. Since the left-hand and right-hand limits are not the same, the derivative does not exist at $x = 6$.

We may summarize these arguments by defining a *critical point* of a function.

A **critical point** of a function $f(x)$ is a point $(c, f(c))$ where either:

1. The derivative has a value of 0, that is, $f'(c) = 0$.

Or:

2. The function is defined but the derivative does not exist at $x = c$.

Thus, critical points are points that may be relative maximum points or minimum points. However, a given critical point need not be a relative optimum point. To find relative optima, we must first identify all critical points of the function. The examples that follow illustrate these different cases.

Example 3   Identify all critical points of the function

$$f(x) = x^3 - 9x^2 + 24x$$

Solution   (You may wish to read Algebra Review 13 on page 132 at this time.) First, there are no points where the derivative does not exist since the function is a simple polynomial and hence may be differentiated by the rules developed in Chapter 2. The derivative is

$$f'(x) = 3x^2 - 18x + 24$$
$$= 3(x^2 - 6x + 8)$$
$$= 3(x - 4)(x - 2)$$

Setting this expression to 0, we see that $f'(x) = 0$ at $x = 4$ and $x = 2$. Therefore, $(4, f(4))$ and $(2, f(2))$ are both critical points. These critical points might be relative maximum or minimum points. ∎

Example 4   Find all critical points of

$$f(x) = \begin{cases} x^2 - 16x + 77 & \text{if } x < 7 \\ 2x & \text{if } x \geq 7 \end{cases}$$

Solution   A point where the derivative may not exist is at $x = 7$, since the function changes definition at $x = 7$. Note that

$$\frac{f(7 + h) - f(7)}{h} = \begin{cases} \dfrac{(7 + h)^2 - 16(7 + h) + 77 - 14}{h} & \text{if } h < 0 \\ \dfrac{2(7 + h) - 14}{h} & \text{if } h > 0 \end{cases}$$

$$= \begin{cases} \dfrac{14h + h^2 - 16h}{h} = -2 + h & \text{if } h < 0 \\ \dfrac{2h}{h} = 2 & \text{if } h > 0 \end{cases}$$

Therefore, the limiting value as $h \to 0$ from the left is -2, while the limiting value as $h \to 0$ from the right is 2. Since these are not equal, the derivative does not exist at $x = 7$. Thus, $(7, f(7))$ is a critical point.

Next, setting $f'(x) = 0$ for $x < 7$, we have

$$f'(x) = 2x - 16 = 0$$
$$x = 8$$

This is clearly incompatible with the condition $x < 7$, and hence there are no critical points for $x < 7$. Since $f'(x) = 2$ for $x > 7$, there are no critical points for $x > 7$. Therefore, the only critical point is $(7, f(7)) = (7, 14)$. ∎

Example 5   Let $f(x) = 3x^5 - 8$. Identify any critical points.

Solution   Since $f'(x) = 15x^4$, there are no points where the derivative does not exist. Setting $f'(x) = 0$, we have

$$f'(x) = 15x^4 = 0$$

The only solution to this equation is $x = 0$. Thus, $(0, f(0)) = (0, -8)$ is the only critical point. ∎

PROBLEMS (Section 3-2)

Identify all critical points for each function in Problems 1–18.

1. $f(x) = x^2 - 4x + 1$

2. $f(x) = 3x^2 + 2x + 7$

3. $f(x) = -7x^2 + 2x - 12$

4. $f(x) = 3x + 4$

5. $f(x) = x^4$

6. $f(x) = 4x^3$

7. $f(x) = x^2 - 10x$

8. $f(x) = x^3 + 3x^2$

9. $f(x) = x^3 - 3x^2 - 9x$

10. $f(x) = x^4 + 4x^3 + 4x^2$

11. $f(x) = \frac{1}{3}x^3 - \frac{5}{2}x^2 + 6x$

12. $f(x) = \frac{2}{3}x^3 - \frac{7}{2}x^2 + 3x$

13. $f(x) = \begin{cases} x^2 - 6x + 15 & \text{if } x < 5 \\ \frac{2}{3}x + \frac{11}{3} & \text{if } x \geq 5 \end{cases}$

ALGEBRA REVIEW 13: SOLVING QUADRATIC EQUATIONS BY FACTORING

A **quadratic equation** is one that has the form

$$ax^2 + bx + c = 0$$

If the quadratic expression can be factored, the equation can be written as

$$(Ax + B)(Cx + D) = 0$$

where $AC = a$, $BD = c$, and $BC + AD = b$. The solution to this equation can now be found by setting each factor equal to 0. That is, we set

$$Ax + B = 0$$

and

$$Cx + D = 0$$

and solve each equation for x.

Let us illustrate this for a simple case. Consider the equation

$$x^2 - 4x - 12 = 0$$

In this example, $AC = a = 1$, $BD = c = -12$, and $BC + AD = b = -4$. Setting $A = C = 1$, we have $BD = -12$ and $B + D = -4$. We now examine $BD = -12$ to find products of numbers that satisfy this relation. The possibilities are 12 and -1, -12 and 1, 6 and -2, -6 and 2, 4 and -3, and -4 and 3. Next, we check the relation $B + D = -4$ to see if this can be satisfied with any of these possibilities. We see that if $B = -6$ and $D = 2$, then both relations hold. Therefore, the equation can be factored as

$$(x - 6)(x + 2) = 0$$

Now setting each factor equal to 0, we have

$$x - 6 = 0$$

and

$$x + 2 = 0$$

14. $f(x) = \begin{cases} x^2 + 4x & \text{if } x < 0 \\ x^2 - 10x & \text{if } x \geq 0 \end{cases}$

15. $f(x) = \begin{cases} x^3 - 3x^2 & \text{if } x < 3 \\ x^2 - 8x + 15 & \text{if } x \geq 3 \end{cases}$

16. $f(x) = \begin{cases} x^3 - 3x^2 - 9x & \text{if } x < 6 \\ x^2 - 10x + 78 & \text{if } x \geq 6 \end{cases}$

17. $f(x) = x^{2/3}$

18. $f(x) = |x|$

19. Are all relative optimum points critical points? Why or why not?

20. How do we recognize a critical point?

21. Does a linear function have any critical points? If so, how many?

22. How many critical points can a quadratic function have?

23. Find all critical points of the function $f(x) = -|x - 1|$. Does a relative maximum or relative minimum occur at these critical points?

24. Show that the function $f(x) = x^7$ has only one critical point and that this critical point is neither a relative maximum point nor a relative minimum point.

25. Find all critical points of the function $f(x) = 3x^{2/3} - x$.

26. Find all critical points of the function $f(x) = x^{2/5} - \frac{2}{5}x$.

Thus, $x = 6$ and $x = -2$ are solutions to the equation $x^2 - 4x - 12 = 0$.

If $a \neq 1$, then the process is more complicated. Consider

$$6x^2 + x - 40 = 0$$

We have $AC = 6$, $BD = -40$, and $BC + AD = 1$. Potential values for A and C are 6 and 1, and 3 and 2, and the respective negative values. Potential values for B and D are 8 and -5, 10 and -4, 20 and -2, and 40 and -1, and their negatives. If we try $A = 6$ and $C = 1$, we cannot find values for B and D that satisfy $BC + AD = 1$. However, if $A = 3$ and $C = 2$, we can choose $B = -8$ and $D = 5$, obtaining

$$6x^2 + x - 40 = (3x + 8)(2x - 5)$$

Again, setting each factor equal to 0, we solve

for x:

$$3x + 8 = 0$$
$$x = -\tfrac{8}{3}$$

and

$$2x - 5 = 0$$
$$x = \tfrac{5}{2}$$

EXERCISES

Solve the following quadratic equations by factoring:

a. $x^2 - 10x + 21 = 0$
b. $12x^2 - 23x + 5 = 0$
c. $2x^2 + 9x + 4 = 0$
d. $x^2 + 4x + 4 = 0$
e. $6x^2 + x - 15 = 0$

3-3 IDENTIFYING RELATIVE OPTIMA

In the previous section we saw how to identify points that are potentially relative maximum or minimum points. However, we did not show how to determine which were optima and how to distinguish between them. In this section we address these issues.

Through intuition, you would probably state that a function is *increasing* if, as we move to the right, the graph of the function rises. Mathematically this can be stated as follows:

> A function is **increasing** on an open interval if whenever a and b are in the interval and $a < b$, then $f(a) < f(b)$.

In a similar manner, you would probably say that a function is *decreasing* if, as we move to the right, the graph of the function falls. In mathematical terms, this is stated as:

> A function is **decreasing** on an open interval if whenever a and b are in the interval and $a < b$, then $f(a) > f(b)$.

These definitions are illustrated in Figure 3-12.

Figure 3-12
Illustration of increasing and decreasing functions

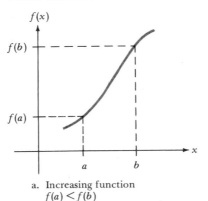

a. Increasing function
$f(a) < f(b)$

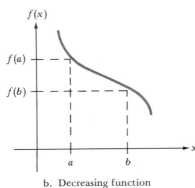

b. Decreasing function
$f(a) > f(b)$

Suppose that $(c, f(c))$ is the relative maximum point illustrated in Figure 3-13. Then in an interval to the left of $x = c$ the function is increasing, and in an interval to the right of $x = c$ the function is decreasing. For the relative minimum point $(d, f(d))$, we have just the opposite. In an interval to the left of $x = d$ the function is decreasing, and in an interval to the right of $x = d$ the function is increasing. From Figure 3-13 we also see that if a function is increasing, the slope of the tangent line is always positive, whereas if a function is decreasing, the

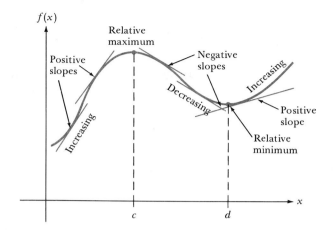

Figure 3-13
Relationships of increasing and
decreasing functions to relative
optima

slope of the tangent line is always negative. These observations provide a simple test for determining whether a critical point is a relative maximum, relative minimum, or neither:

First Derivative Test for
Relative Optima

> Suppose $(c, f(c))$ is a critical point of the continuous function $f(x)$. If the value of the derivative is positive in an interval (a, c) to the left of c and negative in an interval (c, b) to the right of c, then the critical point is a relative maximum point. If the value of the derivative is negative in an interval (a, c) to the left of c and positive in an interval (c, b) to the right of c, then the critical point is a relative minimum point. Finally, if the derivative has the same sign on both sides of c, then the critical point is neither a relative maximum point nor a relative minimum point.

Example 6

For the graph in Figure 3-14, identify all critical points and intervals where the function is increasing and decreasing. Show how this information is used to determine the type of critical point.

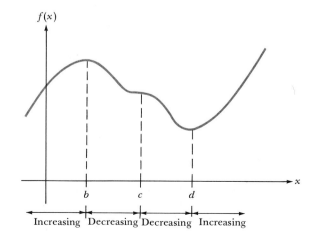

Figure 3-14
Graph for Example 6

Solution   The critical points occur when x equals b, c, and d, since the value of the derivative is 0 at these points. The function is increasing on the intervals $(-\infty, b)$ and (d, ∞), and is decreasing on the intervals (b, c) and (c, d). The point $(b, f(b))$ is a relative maximum point since the function is increasing to the left and decreasing to the right. The point $(d, f(d))$ is a relative minimum point since the function is decreasing to the left and increasing to the right. Finally, the point $(c, f(c))$ is neither since the function is decreasing on both sides.   ∎

Example 7   Determine whether the critical points found in Examples 3 – 5 (Section 3-2), are relative maxima, relative minima, or neither.

Solution   *Example 3:*

$$f(x) = x^3 - 9x^2 + 24x$$

$$f'(x) = 3x^2 - 18x + 24$$
$$= 3(x - 4)(x - 2)$$

The critical points were found to be at $x = 4$ and $x = 2$. Checking values of the derivative at nearby points on either side of both points, we have:

	x	$f'(x)$
Left of $x = 4$:	3.9	−0.57
Right of $x = 4$:	4.1	0.63
Left of $x = 2$:	1.9	0.63
Right of $x = 2$:	2.1	−0.57

Therefore, we conclude that there is a relative minimum at $x = 4$ and there is a relative maximum at $x = 2$.

Example 4:   The only critical point is at $x = 7$. The derivative does not exist at this point, but the first derivative test still applies. For $x < 7$, $f'(x) = 2x - 16$ and hence, for example, $f'(6.5) = -3$. On the other hand, $f'(x) = 2$ for $x > 7$. The derivative is negative just to the left of $x = 7$ and positive to the right of $x = 7$; thus, $(7, f(7))$ is a relative minimum point.

Example 5:   The only critical point is at $x = 0$. Since $f'(x) = 15x^4 > 0$ for $x \neq 0$, we see that $f'(x)$ is positive on both sides of $x = 0$. Therefore, $(0, f(0))$ is not a relative maximum point nor is it a relative minimum point.   ∎

Checking points in this fashion can be quite tedious. At critical points for which $f'(x) = 0$, there is usually a much easier way to determine whether the point is a relative minimum, maximum, or neither. This is based on the *second derivative*.

THE SECOND DERIVATIVE

The idea of the second derivative is easy: We simply take the derivative of the first derivative. For example, suppose that $f(x) = 4x^3 - 3x^2 + 2$. The first derivative is

$$f'(x) = 12x^2 - 6x$$

Taking the derivative of this function, we obtain $24x - 6$. This is another function which we call the *second derivative*.

Several different types of notation are used to denote the second derivative. The two most common are $f''(x)$ (read "f double prime of x"), which is analogous to the first derivative notation $f'(x)$, and the symbol d^2y/dx^2, which is an extension of the symbol dy/dx for the first derivative. The second notation is used when the function is written as $y = f(x)$. Notice that the exponent "2" comes after the d in the numerator and after the x in the denominator in the symbol d^2y/dx^2. However, it does not mean "squared."

We can continue to take derivatives of derivatives and obtain the third derivative, fourth derivative, and so on. We normally do not use the prime notation beyond the second derivative. Notations used for higher derivatives are: $f^{(3)}(x), f^{(4)}(x), \ldots$, or $d^3y/dx^3, d^4y/dx^4, \ldots$, and so on. For example, if $f(x) = x^5 + 3x^3 + x^2$, then

$$f'(x) = 5x^4 + 9x^2 + 2x$$
$$f''(x) = 20x^3 + 18x + 2$$
$$f^{(3)}(x) = 60x^2 + 18$$
$$f^{(4)}(x) = 120x$$

etc.

Applications seldom require us to take a derivative beyond the second.

Let us consider the meaning of the second derivative in order to see how it might be useful in determining whether a critical point is a relative maximum, minimum, or neither. Recall that the first derivative was defined as a rate of change of the function and represents the slope of the tangent line at a point on the graph of the function $f(x)$. Since the second derivative is the derivative of $f'(x)$, it measures the *rate of change of the slope of the tangent line to* $f(x)$. That is, $f''(x)$ represents how fast the slope is changing and also in which direction it is changing. Let us take a simple example. Suppose $f(x) = x^2$. Then $f'(x) = 2x$ and $f''(x) = 2$. Since the second derivative is *positive*, the slope of the tangent line must get larger as x increases. Moreover, the value of 2 tells us that for a unit change in x, the rate of change of the slope of the tangent line is 2, that is, the slope gets larger by 2 units. For example, when $x = -1$, then $f'(x) = -2$, and when $x = 0$, then $f'(x) = 0$. The slope increased by 2 units. If $x = 1$, then $f'(x) = 2$. Again we see a 2 unit increase, just as $f''(x) = 2$ predicted.

As another example, suppose that $f(x) = -x^2$. The first derivative is $f'(x) = -2x$ and the second derivative is $f''(x) = -2$. Since this is negative, we conclude that the slope must be decreasing; the value of -2 tells us that the rate of decrease is 2 units for every unit increase in x. For instance, if $x = -1$, then $f'(x) = 2$, and if $x = 0$, then $f'(x) = 0$. Finally, if $x = 1$, then $f'(x) = -2$, as expected.

Consider a critical point $(c, f(c))$ with $f'(c) = 0$. If the second derivative is positive, then as x moves from the left of c to the right of c, the slope must continually increase. Therefore, the slope must be less than 0 to the left of c and greater than 0 to the right of c. From the first derivative test, we must have a relative minimum at $x = c$. In a similar manner, if the second derivative is negative, then as x moves from the left of a critical point to the right of it, the slope must continually decrease. Therefore, the slope must be greater than 0 to the left of the critical point and less than 0 to the right. Again by the first derivative test, we must have a relative maximum at the critical point. This is precisely what happened for the functions $f(x) = x^2$ and $f(x) = -x^2$ discussed above.

If the second derivative equals 0, we cannot conclude that the point is *not* a relative maximum or minimum. For instance, let $f(x) = x^4$. The first derivative is $f'(x) = 4x^3$. The only solution to $f'(x) = 0$ is $x = 0$. The second derivative is $f''(x) = 12x^2$. Thus, $f''(0) = 0$. However, as shown in Figure 3-15, $(0, f(0))$ is a relative minimum point of the function $f(x) = x^4$. In general, the sign of the second derivative at the critical point indicates the nature of the critical point. We can summarize this as follows:

Second Derivative Test for Relative Optima

> Let $(c, f(c))$ be a critical point where $f'(c) = 0$. Then:
>
> 1. $(c, f(c))$ is a relative maximum point if $f''(c) < 0$
> 2. $(c, f(c))$ is a relative minimum point if $f''(c) > 0$
> 3. If $f''(c) = 0$, then the test is inconclusive.

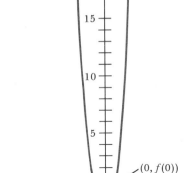

Figure 3-15
Graph of $f(x) = x^4$

When the second derivative test is inconclusive, we use the first derivative test to determine the nature of the critical point.

Example 8   Apply the second derivative test to Examples 3 – 5 (Section 3-2) to determine the nature of the critical points.

Solution   *Example 3:*   The first derivative is

$$f'(x) = 3x^2 - 18x + 24$$

and the critical points are at $x = 4$ and $x = 2$. The second derivative is

$$f''(x) = 6x - 18$$

At $x = 4$, $f''(4) = 6(4) - 18 = 6$. Since this is positive, we must have a relative minimum at the point $(4, f(4))$. When $x = 2$, $f''(2) = 6(2) - 18 = -6$. Therefore, the point $(2, f(2))$ is a relative maximum point since the second derivative is negative.

Example 4: We saw that there is a critical point at $x = 7$. However, the derivative does not exist at this point, and therefore the second derivative test cannot be used. (Nevertheless, we saw in Example 7 that the first derivative test shows that $(7, f(7))$ is a relative minimum point.)

Example 5: The only critical point occurs at $x = 0$. The second derivative is $f''(x) = 60x^3$; thus, $f''(0) = 0$. We cannot determine the nature of this critical point without examining the value of the first derivative on either side of $x = 0$ (as we did in Example 7). ∎

When either the first or second derivative tests cannot be used to determine the type of critical point, we can check values of $f(x)$ a small distance away from the point in order to determine its nature. For instance, in Example 4, $f(7) = 14$. If we compare this to $f(6.9)$ and $f(7.1)$, we get

$$f(6.9) = 14.21 \quad \text{and} \quad f(7.1) = 14.2$$

The value of the function is larger than $f(7)$ at a small distance away from both sides; therefore, we can conclude that $(7, 14)$ is a relative minimum point.

The procedure for identifying relative optima is summarized as follows:

Identifying Relative Optima

1. Compute the derivative $f'(x)$.
2. Find the critical points, that is, points $(c, f(c))$ such that $f'(c) = 0$ or $f'(c)$ does not exist.
3. Test each critical point using the first derivative test or the second derivative test.

PROBLEMS (Section 3-3)

For each graph in Problems 1 and 2, identify all critical points and intervals where the function is increasing and decreasing. Use this information to determine the type of critical point.

1.

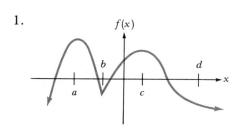

2.

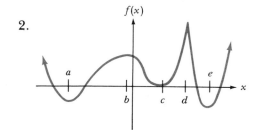

In Problems 3–10, find the relative optima of each function. Use the first derivative test to justify your answers.

3. $f(x) = x^2 - 4x + 1$

4. $f(x) = x^2 - 10x$

5. $f(x) = 4x^3$

6. $f(x) = x^3 + 3x^2$

7. $f(x) = x^3 - 3x^2 - 9x$

8. $f(x) = x^4 + 4x^3 + 4x^2$

9. $f(x) = \frac{1}{4}x^3 - \frac{5}{2}x^2 + 6x$

10. $f(x) = \frac{2}{3}x^3 - \frac{7}{2}x^2 + 3x$

In Problems 11–20, find the second derivative.

11. $f(x) = 2x^2 + x^5$

12. $f(x) = 5x - 3x^4$

13. $f(x) = x^3 + 4x^2 + 12x + 16$

14. $f(x) = x^3 - 5x^2 + 9x - 11$

15. $f(x) = (x + 2)^3 + 6$ [*Hint:* Expand the cube first.]

16. $f(x) = 3 - (x - 2)^2$

17. $f(x) = x + \dfrac{3}{x}$

18. $f(x) = \dfrac{5}{x} + 2x$

19. $f(x) = 3x^{2/3} - 5x$

20. $f(x) = 2\sqrt{x} + 3x$

21. Use the information given in the table below to replace the question marks in the last column with "Relative minimum," "Relative maximum," or "Neither."

$f'(c)$	$f'(x)$ over (a, c)	$f'(x)$ over (c, b)	$(c, f(c))$
a. 0	+	−	?
b. 0	+	+	?
c. 0	−	+	?
d. 0	−	−	?

22. For the odd-numbered problems 1–17 in Section 3-2, determine the nature of each critical point by using the second derivative test if applicable; otherwise, use the first derivative test if applicable.

In Problems 23–32, find all relative maxima or relative minima. Use the first or second derivative test to justify your answers.

23. $f(x) = 9x - x^2$

24. $f(x) = x^2 - 5x + 6$

25. $f(x) = (x - 3)^2$

26. $f(x) = (x + 2)^3$

27. $f(x) = x + \dfrac{4}{x}$

28. $f(x) = |x + 1|$

29. $f(x) = -|x - 2|$

30. $f(x) = \begin{cases} x^2 - x + 4 & \text{if } x < 2 \\ 3x & \text{if } x \geq 2 \end{cases}$

31. $f(x) = \begin{cases} -2x^2 + x & \text{if } x \leq 1 \\ 2x - 3 & \text{if } x > 1 \end{cases}$

32. $f(x) = x^{4/5}$

33. Let $f(x) = -x^4$. Show that the second derivative test is inconclusive, yet $f(x)$ has a relative maximum at $x = 0$.

34. Let $f(x) = x^7$. What is $f''(0)$? Does $f(x)$ have a relative maximum or relative minimum at $x = 0$?

35. Compute the fifth derivative, $f^{(5)}(x)$, of the polynomial

$$f(x) = -2x^4 - 3x^2 + 7$$

36. What is the $(n + 1)$st derivative of a polynomial of degree n?

3-4 IDENTIFYING ABSOLUTE OPTIMA

Many practical applications require us to find the absolute optimum of a function. In Section 3-1 we saw that when a continuous function is defined over a closed interval $[a, b]$, an absolute maximum or minimum point can occur at either:

1. A relative optimum

Or:

2. An endpoint of the interval

If a point is identified as being a relative maximum or relative minimum point by the first or second derivative test, we still have not determined whether it is an *absolute* optimum. However, we do know that the only points that can possibly be absolute optima are critical points and endpoints. Thus, we can restrict our search to this set of points. To locate the absolute maximum (or minimum) of the function, we need only evaluate the function at the critical points and endpoints and then choose the one for which the function value is largest (or smallest).

Finding Absolute Optima on a Closed Interval

> To find the absolute optima of a continuous function on a closed interval, first locate all critical points in the interval and compute the function values at these points. Next, compute the values of the function at the endpoints. The largest and smallest values computed are the absolute maximum and absolute minimum, respectively.

Example 9   Consider the function

$$f(x) = x^2 - 4x + 6 \qquad 0 \le x \le 3$$

To find the absolute maximum and minimum we first find all critical points:

$$f'(x) = 2x - 4 = 0$$
$$x = 2$$

The endpoints are $x = 0$ and $x = 3$. Therefore, the only points we need to consider are at $x = 0$, 2, and 3. Next, we evaluate $f(x)$ at each of these points:

x	$f(x)$
0	6
2	2
3	3

Since the largest value of $f(x)$ occurs at $x = 0$, then $(0, 6)$ is the absolute maximum point. The absolute minimum point is $(2, 2)$, corresponding to the smallest value of $f(x)$. Therefore, on the interval $[0, 3]$, the absolute maximum of the function is $f(0) = 6$ and the absolute minimum is $f(2) = 2$.   ∎

Example 10   Suppose that

$$f(x) = \frac{x^4}{4} - \frac{x^3}{3} - x^2 + 5 \qquad -1 \le x \le 1$$

Determine the absolute maximum and minimum.

Solution   The first derivative is $f'(x) = x^3 - x^2 - 2x$. Factoring and setting this to 0, we obtain

$$x^3 - x^2 - 2x = x(x^2 - x - 2)$$
$$= x(x - 2)(x + 1) = 0$$

The roots of this equation are $x = 0$, $x = 2$, and $x = -1$. However, $x = 2$ does not fall within the closed interval $[-1, 1]$ that defines the domain of the function. Consequently, we do not need to consider it. The only values at which absolute optima can occur are $x = 0$, $x = -1$ (which is also an endpoint), and $x = 1$. Checking the values of $f(x)$ at these points, we have

$$f(0) = 5$$
$$f(-1) = 4\tfrac{7}{12}$$
$$f(1) = 3\tfrac{11}{12}$$

Thus, the absolute maximum is $f(0) = 5$ and the absolute minimum is $f(1) = 3\tfrac{11}{12}$.   ∎

If the function is defined on an open interval, there may be no absolute optima (refer back to Figure 3-3c). However, if there are absolute optima, then any absolute optimum must also be a relative optimum. In such cases, we need only identify the critical points. Whether a point is a minimum or maximum can usually be determined by the second derivative test. However, if there are critical points where the derivative is undefined or if there is more than one relative minimum or maximum, then we must compare their function values in order to find the absolute optimum. When the function is defined on an infinite interval, we must also look at the behavior of the function as x approaches positive or negative infinity to determine whether absolute optima exist.

Example 11   Let $f(x) = 3x^{2/3}$. Determine the absolute optima.

Solution   The first derivative is

$$f'(x) = \tfrac{2}{3}(3)x^{-1/3}$$
$$= 2x^{-1/3} = \frac{2}{x^{1/3}}$$

If we set $f'(x) = 0$, we find that there is no solution. However, when $x = 0$, the function is defined but the first derivative is not. Thus, $(0, 0)$ is a critical point.

Since the derivative is not defined at $x = 0$, we cannot use the second derivative test. However, $f'(x) = 2/x^{1/3} > 0$ for $x > 0$ and $f'(x) = 2/x^{1/3} < 0$ for $x < 0$. Therefore, $f(x)$ has a relative minimum at $x = 0$ by the first derivative test. Since $f(0) = 0$, the point $(0, 0)$ is a relative minimum point. Notice that this is the only relative minimum point and the function is defined over all real numbers; hence, this point must also be the absolute minimum point. There is no absolute maximum since $f(x) = 3x^{2/3} = 3(x^{1/3})^2$ grows arbitrarily large as x approaches either positive or negative infinity. ∎

Example 12   Find the absolute optima for the function

$$y = x^3 + \tfrac{1}{2}x^2 - 2x$$

Solution   If we check the behavior of the function as x approaches positive infinity, the x^3 term dominates all others and so the value of y grows indefinitely. As x approaches negative infinity, the value of x^3 also approaches negative infinity and dominates all other terms. Therefore, there are no absolute optima. ∎

PROBLEMS (Section 3-4)

In Problems 1–12, determine the absolute maximum and minimum if they exist.

1. $f(x) = x^2 - 4x + 6$

2. $f(x) = -2x^2 + 8x - 7$

3. $f(x) = x^3 + x$

4. $f(x) = -3x^3 + 6x$

5. $f(x) = 4x^{1/2}$

6. $f(x) = -5x^{2/5}$

7. $f(x) = x^3 + 3x^2 - 9x + 5$

8. $f(x) = x^3 + 4x^2 - 3x + 7$

9. $f(x) = \dfrac{x^3}{3} - \dfrac{3x^2}{2} - 6x + 15$

10. $f(x) = \dfrac{2x^3}{3} - 3x^2 + 3x + 8$

11. $f(x) = \dfrac{4}{x^2} + 2$

12. $f(x) = \dfrac{2}{x} + 1$

In Problems 13–30, find the absolute maximum and minimum of the given function on the specified interval.

13. $f(x) = 6x + 2, \quad -10 \le x \le 1$

14. $f(x) = 2x^2 - 6x + 14, \quad -3 \le x \le 2$

15. $f(x) = 3x^2 + 2x + 1, \quad 0 \le x \le 2$

16. $f(x) = x^3 - 3x, \quad 0 \le x \le 3$

17. $f(x) = (x + 1)^2, \quad -2 \le x \le 2$

18. $f(x) = \sqrt{x} - x^2, \quad 0 \le x \le 10$

19. $f(x) = x^3 - 6x^2, \quad -1 \le x \le 3$

20. $f(x) = 2x^3 + 9x^2 - 24x + 3, \quad -2 \le x \le 2$

21. $f(x) = \dfrac{1}{x} + x^2, \quad -1 \le x \le 3, \quad x \ne 0$

22. $f(x) = \dfrac{1}{x^2} + 4, \quad -2 \le x \le -1$

23. $f(x) = x + \dfrac{1}{x}, \quad \tfrac{1}{2} \le x \le 4$

24. $f(x) = x^{1/3} + 1, \quad -27 \le x \le 64$

25. $f(x) = 8x + \frac{1}{2}x^{-2}, \quad \frac{1}{4} \le x \le 2$

26. $f(x) = 2x^5 - 5x^2, \quad -1 \le x \le 2$

27. $f(x) = x^5 - 10x^4 + 7, \quad 0 \le x \le 3$

28. $f(x) = 2x^8 - 6x^7, \quad -1 \le x \le 2$

29. $f(x) = -|x - 1| + 3, \quad 0 \le x \le 4$

30. $f(x) = 2|x + 2| + 4, \quad -3 \le x \le 0$

31. What kind of points should be tested when looking for absolute optima?

3-5 SOLVING OPTIMIZATION PROBLEMS

Optimization problems, like the forest fire problem developed in the Introduction to this chapter, represent some of the most important applications of calculus in business and the social and natural sciences. In this section we will solve a variety of problems using the methods we have studied in this chapter.

Example 13
Facility Sizing

A large retail drug chain has made an extensive study to determine the best size for a new store at a particular location. Based on population, per capita income, and competition, a model was developed which relates the average weekly dollar sales, S, to the size of the store, x (in 1000 square feet). It was found that

$$S(x) = -100x^2 + 1000x - 1500 \qquad 2 \le x \le 8$$

What store size will maximize sales?

Solution

Our objective is to find the absolute maximum of S(x) on the interval [2, 8]. We begin by finding all critical points of S(x). Taking the first derivative of S with respect to x and setting it equal to 0, we find

$$\frac{dS}{dx} = -200x + 1000 = 0$$

Therefore, the only critical point is at

$$x = \tfrac{1000}{200} = 5$$

and

$$\begin{aligned} S(5) &= -100(5^2) + 1000(5) - 1500 \\ &= -2500 + 5000 - 1500 \\ &= \$1000 \text{ per week} \end{aligned}$$

Checking the endpoints, we find that $S(2) = \$100$ and $S(8) = \$100$. Therefore, S(5) is the absolute maximum. Hence, maximum sales of $1000 per week is predicted by this model for a store size of 5000 square feet. ∎

Example 14
Production Cost

The total cost (in dollars) of producing x tons of a chemical used for water purification in swimming pools is determined to be

$$C(x) = 4x^4 - 9x^3 + 6x^2 + 15x \qquad 0.3 \le x < \infty$$

What amount should be produced in order to minimize the cost per ton?

Solution   In applied optimization problems, it is crucial to recognize the proper objective that is to be optimized. It is certainly tempting to take $C(x)$ and forge ahead and differentiate it without carefully reading the problem. The cost $C(x)$ represents the *total cost* for producing x tons, while the objective is to minimize the *cost per ton*. Hence, the cost per ton is $C(x)/x$, and we wish to minimize the function

$$f(x) = \frac{C(x)}{x} = 4x^3 - 9x^2 + 6x + 15 \qquad 0.3 \le x < \infty$$

Differentiating, we have

$$f'(x) = 12x^2 - 18x + 6$$
$$= 6(2x^2 - 3x + 1)$$
$$= 6(2x - 1)(x - 1)$$

The critical points of $f(x)$ therefore occur at $x = \frac{1}{2}$ and $x = 1$. To determine the character of the critical points, we check the second derivative:

$$f''(x) = 24x - 18$$

Since $f''(\frac{1}{2}) = -6$, the cost per ton function has a relative maximum of $f(\frac{1}{2}) = 16.25$. At $x = 1, f''(1) = 6$ and hence $f(1) = 16$ is a relative minimum. At the left endpoint, $f(0.3) = 16.098$, and as x approaches infinity, so does $f(x)$. Therefore, an absolute minimum cost of \$16 per ton occurs when 1 ton is produced.   ∎

Creating a mathematical model is probably the most difficult phase of solving optimization problems. You may wish to review Chapter 1 for some hints on setting up mathematical models. Some problems require relatively simple model building skills; others are more complicated. However, remember that the more examples you study and the more problems you attempt, the better you will become at developing mathematical models.

Example 15
Health Care Delivery   The number of patients treated per year at a psychiatric clinic is given by $N = P/T$, where T = average patient treatment time (in years) and P = number of in-resident patients. Based on patient records, the following relation was established:

$$T = \frac{50}{160 - P} \qquad 1 \le P \le 150$$

How many in-resident patients should the clinic serve in order to maximize the number of patients treated per year?

Solution   We seek a value of P that will maximize $N = P/T$ and satisfy the relation given above. First, we must express N as a function of one variable. Since we seek a value of P to maximize N, we express N in terms of P:

$$N = \frac{P}{T} = \frac{P}{50/(160 - P)}$$
$$= \frac{P(160 - P)}{50}$$

Hence,

$$N(P) = \frac{160P}{50} - \frac{P^2}{50}$$
$$= 3.2P - 0.02P^2$$

Differentiating N with respect to P yields

$$N'(P) = 3.2 - 0.04P$$

Setting this equal to 0 gives

$$3.2 - 0.04P = 0$$
$$P = \frac{3.2}{0.04} = 80$$

Thus, the only critical point occurs at $P = 80$ and $N(80) = 128$. Since $N(1) = 3.18$ and $N(150) = 30$, the endpoints yield smaller values of N. Thus, we find that if 80 in-patients are treated, then the absolute maximum of 128 patients can be treated per year. ■

Example 16
Recreational Mathematics

The product of two positive numbers is 81. Find the numbers so that their sum is a minimum.

Solution

Let x and y be the two unknown numbers. We know that $xy = 81$, and we wish to minimize the sum $S = x + y$. We may choose to express this sum in terms of either x or y alone. Solving the relation $xy = 81$ for x, we obtain $x = 81/y$. Substituting $81/y$ for x in the sum yields

$$S = \frac{81}{y} + y \qquad 0 < y < \infty$$

as the function we want to minimize. Setting the first derivative equal to 0, we have

$$\frac{dS}{dy} = \frac{-81}{y^2} + 1 = 0$$
$$y^2 = 81$$
$$y = 9$$

Since $d^2S/dy^2 = 162/y^3 > 0$, $y = 9$ gives a relative minimum for the sum S. Since S approaches infinity as y approaches 0 or ∞, it follows that $y = 9$ yields an absolute minimum. The corresponding value of x is $x = 81/y = 81/9 = 9$, and the minimum value of the sum is $S = x + y = 9 + 9 = 18$. Can you generalize this result to any arbitrary value of the product of two positive numbers? ■

Many optimization problems involve finding the best design for some physical object. Engineers solve optimization problems similar to the ones that follow in a variety of industrial and service applications. Table 3-3 gives a list of some of the simpler geometric formulas required in many of these types of problems.

Example 17
Container Design

A manufacturer of powdered lemonade is considering redesigning a new container in order to cut packaging costs. The container must be cylindrical and hold 201 ($\approx 64\pi$) cubic inches of lemonade by volume. The cost is proportional to the surface area of the container. What are the best dimensions?

Table 3-3
Some geometric formulas

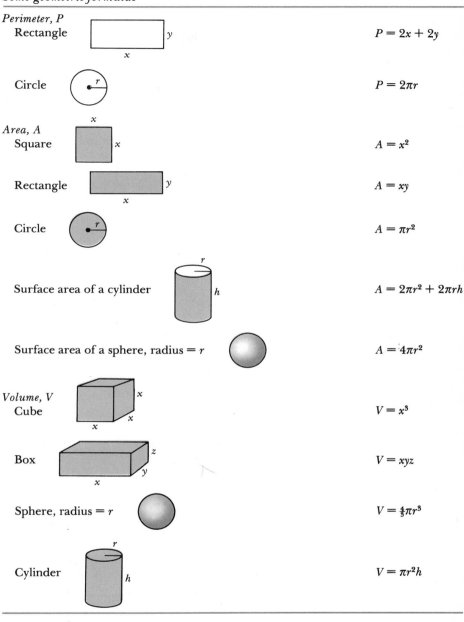

Perimeter, P		
Rectangle		$P = 2x + 2y$
Circle		$P = 2\pi r$
Area, A		
Square		$A = x^2$
Rectangle		$A = xy$
Circle		$A = \pi r^2$
Surface area of a cylinder		$A = 2\pi r^2 + 2\pi rh$
Surface area of a sphere, radius $= r$		$A = 4\pi r^2$
Volume, V		
Cube		$V = x^3$
Box		$V = xyz$
Sphere, radius $= r$		$V = \frac{4}{3}\pi r^3$
Cylinder		$V = \pi r^2 h$

Solution The goal is to minimize the surface area (and hence the cost) of the container, which is given by

$$A = 2\pi r^2 + 2\pi rh$$

Since the volume must be 64π cubic inches, we must also satisfy the constraint

$$V = \pi r^2 h = 64\pi$$

As in previous examples, we need to express the area function in terms of a single variable. We can use the relation for V to eliminate either r or h. Note, however, that if we eliminate r, we have

$$\frac{64\pi}{\pi h} = r^2 \quad \text{or} \quad r = \sqrt{\frac{64}{h}}$$

which involves a radical. In this case, it is easier to eliminate h:

$$h = \frac{64\pi}{\pi r^2} = \frac{64}{r^2}$$

Substituting this into the expression for the surface area, we have

$$A = 2\pi r^2 + 2\pi r \left(\frac{64}{r^2}\right)$$

$$= 2\pi r^2 + \frac{128\pi}{r}$$

Taking the derivative and setting it equal to 0, we get

$$\frac{dA}{dr} = 4\pi r - \frac{128\pi}{r^2} = 0$$

$$4\pi r^3 = 128\pi$$

$$r^3 = 32$$

$$r \approx 3.17 \text{ inches}$$

The second derivative is

$$\frac{d^2A}{dr^2} = 4\pi + \frac{256\pi}{r^3} > 0 \qquad \text{for } r > 0$$

and hence, the value of r that we have found yields a relative minimum of A. Since A grows without bound as r approaches 0 or infinity, we have actually found the value of r that yields an absolute minimum for A. Using the relation between r and h, we can find h as

$$h = \frac{64}{r^2} \approx \frac{64}{(3.17)^2} \approx 6.37 \text{ inches} \qquad \blacksquare$$

Example 18
Room Addition

A homeowner wishes to add a room of 120 square feet to his house. The cost of building the exterior wall is \$80 per linear foot, and the cost of removing the interior wall is \$50 per linear foot. What are the optimal dimensions?

Solution

Figure 3-16 shows the definitions of the unknown variables x and y. The goal is to minimize cost, which equals \$80 times the length of the exterior wall plus \$50 times the length of the interior wall to be removed. From Figure 3-16 we see that these lengths are $x + 2y$ and x, respectively. Therefore, the total cost function is

$$C = 80(x + 2y) + 50x = 130x + 160y$$

In order to have 120 square feet, we must satisfy the constraint

$$xy = 120$$

Solving for x and substituting into the cost function, we have

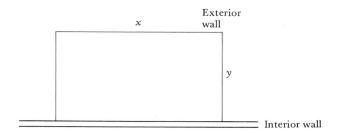

Figure 3-16
Room addition dimensions

$$C(y) = 130\left(\frac{120}{y}\right) + 160y$$

Then

$$C'(y) = \frac{-130(120)}{y^2} + 160$$

The critical points are then given by

$$160y^2 = 15,600$$
$$y^2 = 97.5$$
$$y \approx 9.87 \qquad \textit{Note that we need not consider the negative solution.}$$

The second derivative is

$$C''(y) = \frac{(130)(120)(2)}{y^3} = \frac{31,200}{y^3}$$

and hence, at the critical point the value of the second derivative is

$$C''(9.87) = \frac{31,200}{(9.87)^3} \approx 32.45 > 0$$

The second derivative test indicates that we have a relative minimum.

By considering the behavior of the function $C(y)$ as y approaches 0 or infinity, we find that there is an absolute minimum of the total cost function C at $y \approx 9.87$ feet. The other dimension of the optimal room is then

$$x = \frac{120}{y} \approx \frac{120}{9.87} \approx 12.15 \text{ feet}$$ ∎

The basic steps used in solving applied optimization problems are summarized as follows:

Solving Applied
Optimization Problems

1. Identify the constants and variables in the problem, and assign symbols to the variables.
2. Express the dependent variable to be optimized in terms of the other (independent) variables of the problem.
3. Use the constraint equations relating the independent variables to eliminate all but one independent variable.
4. Optimize the resulting function of a single variable.

PROBLEMS (Section 3-5)

1. The fungi concentration at a local swamp depends on the oxygen concentration in the water, x, according to the function

$$f(x) = x^3 - 7x^2 - 160x + 2100$$

   a. Find the oxygen concentration that will lead to the minimum fungi concentration.
   b. What is the minimum concentration?

2. The number of young that survive in a breeding colony of birds depends on the density of the nests. If the density of the nests is low, the young birds are easy prey for predators, and if it is high, there is a food shortage. The number surviving is given by the function $N(p) = 100p(4 - p)$, where p is the density of nests (nests per square foot), $0 \le p \le 4$. Find the optimum density of the nests.

3. In physiology, static strength is the maximum force that muscles can exert isometrically in a single voluntary effort. Let T be the maximum time (in seconds) that a subject can hold $100p\%$ of maximum force (endurance), where p is the decimal fraction of maximum force held (strength). Given the function

$$T = -90 + \frac{216}{p} - \frac{36}{p^2} - \frac{6}{p^3}$$

   find the maximum time that the force can be held.

4. The air flow through the respiratory system during coughing is given by the function

$$v = \frac{kr^2}{\pi}(r_0 - r)$$

   where k is a constant, r_0 is the radius of the windpipe under normal conditions, r is the radius when the windpipe is constricted by coughing, and v is the velocity of the air through the windpipe. At what radius is the air flow maximal?

5. The product of two positive numbers is 225. Find the numbers so that their sum is a minimum.

6. The difference between two numbers is 24. Find the numbers so that their product is a minimum.

7. The sum of one number and three times a second number is 30. What numbers should be selected so that their product is as large as possible?

8. Find numbers x and y such that $x + y = 60$ and $x^2 y$ is maximized.

9. Find the number that exceeds twice its fourth power by the largest amount.

10. Of all rectangles with fixed perimeter 400, which has the largest area?

11. Of all rectangles with area equal to 100, which has the smallest perimeter?

12. Find the dimensions of a rectangle with perimeter 120 inches that has maximum area. What is the maximum area?

13. A manufacturer of frozen orange juice is reconsidering the size of the container in order to reduce costs. The container must be cylindrical and hold 8π (≈ 25) cubic inches of orange juice by volume. The cost is proportional to the surface area of the container. What are the best dimensions?

14. A rectangular field is to be enclosed on all four sides with a fence. Fencing materials cost $4 per foot for two opposite sides and $8 per foot for the other two sides. Find the maximum area that can be enclosed for $3200.

15. The owner of a small beauty shop wishes to expand her shop by 200 square feet by building a rectangular addition. The cost of building the exterior wall is $70 per linear foot, and the cost of removing the current exterior wall is $40 per linear foot. What are the optimal dimensions of the addition?

16. What is the maximum area of a rectangle that has sides along the positive x-axis and y-axis and lies below the line $2y + 3x = 1$?

17. A cylindrical container is to be made with a capacity of 100 cubic inches. If the material used for the top and bottom of the container costs twice as much per square inch as the material for the sides, what is the most economical design?

18. A playing field in the shape of a rectangle with a semicircle at each end is surrounded by a track of length 880 yards. Find the dimensions of the field that maximize the rectangular area in the center.

19. A rectangular box with square bottom and top is to contain 8 cubic feet. What dimensions of the box will yield the least surface area?

20. A container company wants to manufacture a box with a volume of 288 cubic inches that is open on top and is twice as long as it is wide. Find the dimensions of the box produced from the minimum amount of material.

21. A company wishes to construct a rectangular box with square bottom and top that has a volume of 96 cubic feet. The material for the sides and bottom costs 40¢ per square foot, but that for the top costs only 20¢ per square foot. If the labor cost is

$2 per square foot, what are the dimensions of the most economical box?

22. A grocery chain plans to construct a new store. The site of the store requires the following: (1) the front of the store should have 100 feet for parking; (2) each of the two sides of the store should have 40 feet of clearance for driving through; (3) the back of the store should have 50 feet of clearance for deliveries; and (4) the store space should be at least 25,000 square feet (see the figure). What is the smallest amount of property that the grocery chain can buy?

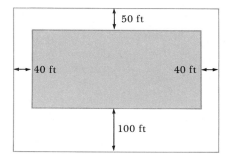

3-6 APPLICATIONS TO BUSINESS AND ECONOMICS

Many useful applications of optimization occur in business and economics. For instance, a firm seeks to maximize profit or minimize cost through manipulation of production levels, advertising budgets, and so on.

Example 19
Maximization of Profit

The production cost of a small leather key chain is a function of the number of key chains produced. The total cost for producing and selling x hundred units is estimated to be

$$C(x) = \frac{x^2}{2} - x + 15$$

One hundred key chains sell for $30. What level of production and sales will maximize profit?

Solution   The basic relation between profit, sales revenue, and cost is

Profit = Revenue − Cost

Since the selling price is constant, the revenue received from the sale of x hundred units will be $30x$. Therefore, profit is given by the function

$$P(x) = 30x - \left(\frac{x^2}{2} - x + 15\right)$$

$$= -\frac{x^2}{2} + 31x - 15$$

The derivative is $P'(x) = -x + 31$. Setting this equal to 0 yields $x = 31$. The second derivative is

$$P''(x) = -1$$

and hence, P has a relative maximum at $x = 31$. The behavior of P for small or large values of x shows that $(31, P(31))$ is actually an absolute maximum point. The optimal production is therefore 3100 units.   ∎

One of the fundamental relationships in economics is that as price increases, demand usually falls, and as price falls, demand rises. Many problems involve setting prices in order to maximize revenue given a certain demand equation.

In several of the examples below we find that a continuous function has only a single relative optimum on a semi-infinite interval, that is, an interval (a, ∞) or $(-\infty, a)$. When this happens, the single relative optimum is also the absolute optimum. (Sketch a couple of examples to see why this is true!)

Example 20
Maximization of Revenue   The annual demand for a personal computer disk drive is estimated to be $-0.4p^2 + 500p + 200,000$, where p is the price of the disk drive. What price should the company set if revenue maximization is their primary goal?

Solution   Revenue equals price times quantity sold. Thus, multiplying the demand function by p will give an expression for revenue:

$$R = (-0.4p^2 + 500p + 200,000)p$$
$$= -0.4p^3 + 500p^2 + 200,000p$$

The critical points are found from the equation

$$\frac{dR}{dp} = -1.2p^2 + 1000p + 200,000 = 0$$

Using the quadratic formula to solve this equation, we find

$$p = \frac{-1000 \pm \sqrt{(1000)^2 + (-4)(-1.2)(200,000)}}{-2.4}$$

$$= \frac{100(-10 \pm \sqrt{196})}{-2.4}$$

$$= 100\left(\frac{-10 \pm 14}{-2.4}\right)$$

Therefore,

$$p = -\tfrac{1000}{6} \qquad \text{or} \qquad p = +1000$$

Clearly, we may disregard the negative root. (The domain of R is implicitly $p \geq 0$; no manufacturer would *pay* a customer to take his product!) We leave it to the reader to compute $R''(p)$ and verify that $R''(1000) < 0$. Therefore, the optimal price (that is, the price giving the maximum revenue) is

$$p = \$1000$$ ∎

Example 21
Maximization of Revenue

During the off-season at a North Carolina resort hotel, on the average 50 rooms are booked when the price is $100 per night. If the price is decreased to $80, then an average of 80 rooms can be booked. Assuming that the demand function is linear, what price should be charged in order to maximize revenue?

Solution

We must first find the demand function that relates the number of rooms booked to the price charged. Once this is determined, revenue can be computed by multiplying the price by the number of rooms booked. Let x be the price charged and y be the number of rooms booked. The data provided in the problem statement give us two points on the demand curve $y = mx + b$: (100, 50) and (80, 80). We leave it as an exercise for you to verify that the correct linear equation passing through these points is $y = -1.5x + 200$. Now we can express the revenue as

$$\begin{aligned} R &= (\text{Price})(\text{Quantity sold}) \\ &= x(-1.5x + 200) \\ &= -1.5x^2 + 200x \end{aligned}$$

Setting the first derivative equal to 0 yields

$$\frac{dR}{dx} = -3x + 200 = 0$$
$$x = \frac{200}{3} \approx 66.67$$

Since $R''(x) = -3 < 0$, $x \approx \$66.67$ yields an absolute maximum for the revenue. ∎

Often, there is more than one way to set up a particular problem. A different choice of variable definition will lead to a different model, but, hopefully, the same result. We emphasize that there is no one "correct" way to do modeling problems. To illustrate this, we shall now solve Example 21 in a slightly different manner.

Example 21
Continued

Let us suppose that a price of $100 per night is currently charged. Let w represent the *change* in price from $100. Then, if $w = 0, y = 50$, and if $w = -20$, $y = 80$, where y is still the number of rooms booked. Using these two points and the point–slope formula, we find that the linear demand function relating y and w is

$$y = -1.5w + 50$$

The total revenue is

$$R = (100 + w)(-1.5w + 50) = -1.5w^2 - 100w + 5000$$

Differentiation yields

$$\frac{dR}{dw} = -3w - 100 = 0$$

$$w = -\tfrac{100}{3} \approx -33.33$$

We leave it to the reader to verify that this value of w actually gives a maximum for R. The optimal price is then $100 + w \approx \$66.67$, which agrees with the previous result.   ∎

One of the most common types of problems encountered in business organizations is that of finding a cost-minimizing tradeoff between conflicting objectives. For instance, consider the problem of determining the number of bank tellers to staff. If there are too many tellers, labor costs will be high, but customers will not have to wait very long for service. On the other hand, too few tellers will cause long lines (and probably lost business), but labor costs will be low. The bank manager's problem is to select a staff level that will balance the objectives of low labor cost and high customer service. This general relationship is shown in Figure 3-17.

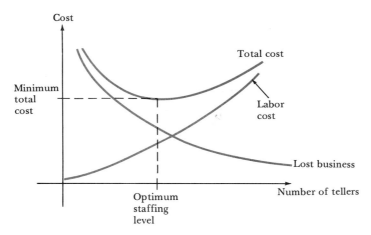

Figure 3-17
Cost as a function of number of bank tellers

Example 22
Labor Staffing

The manager of a downtown fast-food restaurant wishes to estimate the number of employees to have on the job during the busy lunch hour. Labor cost is $5 per hour per employee. The manager estimates that the lost profit per hour by having only x employees is $300/x$ dollars. What staffing level will minimize the total cost per hour?

Solution

The total cost per hour is given by

$$C = 5x + \frac{300}{x}$$

Taking the derivative and setting it equal to 0, we have

$$\frac{dC}{dx} = 5 - \frac{300}{x^2} = 0$$

Solving for x, we find that

$$5x^2 = 300$$
$$x^2 = 60$$
$$x = \sqrt{60} \approx 7.76$$

Since $d^2C/dx^2 = 600/x^3 > 0$ for $x > 0$, this yields a minimum for the total cost. Therefore, approximately 8 employees should be on the job. ∎

Many other situations are similar to this example. Perhaps the most famous is the problem of determining the optimum lot size for ordering inventory. This problem was modeled in the early 1900's and still is successfully in use today to control inventories in many companies. The following example illustrates this problem.

Example 23
Economic Order Quantity

The Holton Drug Company purchases approximately 24,000 cases per year of a brand of toothpaste from a supplier. Each time an order is placed, it costs Holton $38, regardless of the quantity requested. From accounting records, it is estimated that the cost of maintaining an inventory of the toothpaste is $2.16 per case per year. The purchasing manager would like to determine the appropriate order quantity that will minimize the total cost of maintaining inventory and ordering.

Solution

Since demand is constant at 24,000 cases per year, we see that if large orders are placed, they do not have to be placed very often. Thus, ordering costs will be small, but a large amount of inventory will be maintained at a high cost. Small orders will reduce the average amount of inventory that is held in stock, but orders will have to be placed more frequently, thus increasing the order cost. Let Q be the amount ordered each time an order is placed. The number of orders per year will be

$$\frac{24,000 \text{ cases per year}}{Q \text{ cases per order}} = \frac{24,000}{Q} \text{ orders per year}$$

The ordering cost will be

$$\left(\frac{24,000}{Q}\right)(38) = \frac{912,000}{Q} \text{ dollars}$$

The inventory cost will be $2.16 times the average amount of inventory that is maintained. If Q units are purchased at one time and are used at a constant rate, the average amount held in inventory will be $Q/2$, since half the time there will be $Q/2$ or more units in stock and half the time there will be less than $Q/2$. Therefore, the inventory cost will be

$$(2.16)\left(\frac{Q}{2}\right) = 1.08Q \text{ dollars}$$

The total cost to Holton Drugs is then

$$C = 1.08Q + \frac{912,000}{Q} \text{ dollars}$$

Setting the derivative equal to 0, we see that the optimal order quantity (called the *economic order quantity*) is found as follows:

$$\frac{dC}{dQ} = 1.08 - \frac{912,000}{Q^2} = 0$$

$$1.08Q^2 = 912,000$$

$$Q^2 = \frac{912,000}{1.08}$$

$$Q \approx 919 \text{ cases per order}$$

Since the second derivative,

$$\frac{d^2C}{dQ^2} = 1,824,000\, Q^{-3}$$

is positive for positive values of Q, this quantity yields a minimum for the total cost. The total cost to the company will be

$$1.08(919) + \frac{912,000}{919} \approx \$1985 \qquad \blacksquare$$

PROBLEMS (Section 3-6)

1. It was found that the average manufacturing cost (in dollars) for a leather key chain was given by the function

   $$C(x) = x^2 - 6x + 11$$

   where x is the number of key chains manufactured (in thousands).
   a. How many key chains should be manufactured in order to minimize the average cost per chain?
   b. What is the minimum average cost per key chain?

2. Market researchers have determined that the fraction of market share of a new product x months after introduction is estimated to be

   $$A(x) = \frac{2x - 4}{x^2} \qquad x \geq 2$$

   In how many months is there a maximum market share?

3. The total cost (in dollars) of producing x hundred pounds of a detergent is determined to be

   $$C(x) = 2x^4 - 27x^3 + 12x^2 + 18x$$

   What amount should be produced in order to minimize the *cost per pound?*

4. The number of people per year attending a health resort is given by $N = P/T$, where $T =$ average person's staying time (in weeks) and $P =$ total number of people over the year. Based on previous records, the following relation was established:

   $$T = \frac{50}{930 - 3P}$$

   How many people should the health resort serve in order to maximize the number of people per year attending?

5. The production cost of a small glass container is a nonlinear function of the number of contain-

ers produced. The total cost for producing x hundred units is estimated to be

$$C(x) = 0.03x^2 + 11.4x + 900$$

One hundred glass containers sell for $15. What level of production will maximize profit?

6. The annual demand for a microwave oven is estimated to be $-0.5p^2 + 200p + 100,000$, where p is the price of the unit. What price should the company set if revenue maximization is their primary goal?

7. The owner of two brand new movie theaters is trying to determine what price he should charge in order to maximize revenue. An average of 100 seats will be sold when the price is $4 per ticket. If the price is decreased to $3, then an average of 125 tickets will be sold. Assuming that the demand function is linear, what price should be charged in order to maximize revenue?

8. The manager of a large grocery store wishes to estimate the number of employees to have working at the checkout counters during the busy hours of the day. Labor cost is $4 per hour. The manager estimates that the lost profit per hour by having x employees is $160/x$ dollars. What is the total cost function? What staffing level will minimize the total cost?

9. A service station uses 300 cases of motor oil each year. The cost of storing 1 case for 1 year is $1.80, and the ordering fee is $20 per shipment. Assuming that the motor oil is used at a constant rate throughout the year, how many cases should be ordered each time to minimize the total cost?

10. A market has a steady annual demand for 15,000 cases of flour. It costs $3 to store 1 case for 1 year. The market pays $10 for each order that is placed. Find the number of cases per order that will minimize the total cost.

11. The total profit (in thousands of dollars) from the sale of x hundred thousand folding chairs is approximated by the function

$$P(x) = -x^3 + 6x^2 + 96x - 200$$

   a. Find the number of chairs that must be sold in order to maximize profit.
   b. Find the maximum profit.

12. The total profit (in dollars) from the sale of x thousand large pizzas is approximated by the function

$$f(x) = -x^3 + 6x^2 + 48x + 5000$$

   a. Find the number of pizzas that should be sold in order to maximize profit.
   b. Find the maximum profit.

13. A new car owner found that it costs $0.016x^2 - 1.344x + 35$ cents per mile to drive at an average speed of x miles per hour. Find the speed that will minimize the cost per mile.

14. The manager of a small electronics company has determined that it costs $0.003x^2 - 0.9x + 800$ dollars to produce x calculators. Find the production level that would minimize the cost per calculator.

15. A resort complex offers free gifts for potential customers to visit for a weekend and be subjected to a high-pressure sales pitch. The fraction of customers who will buy a membership after x hours of sales talk is given by the function

$$f(x) = -\tfrac{3}{500}x^2 + \tfrac{3}{50}x \qquad 0 \le x \le 10$$

After how many hours will the largest proportion of potential customers make a purchase?

16. The yield of McGloughin's Apple Orchard in Taylorsport, Kentucky, depends on the number of trees. If 25 trees are planted, the average yield per tree is 450 apples. For every additional tree, the average yield per tree is expected to decrease by 5. How many trees should be planted in order to maximize the total yield?

17. The director of a day-care center can take care of 20 children at $300 per child per month. For each $20 increase in the monthly rate, the center will lose 1 child. What fee will maximize revenue? If the center can accept no more than 25 children, what is the day-care center's optimal fee?

18. The value of a product is related to the percentage of defects observed in the product. However, the cost of the product is also related to the level of defects since it costs more to prevent them. Some minor defects are usually acceptable in a

manufactured good as long as the design performance is not affected. Suppose the cost is
given by the function

$$C = -\tfrac{9}{10}p + 90 \qquad 0 \le p \le 100$$

where p is the percentage of defects. The value of

the product is given by

$$V = -0.009p^2 + 90$$

At what level of defects is the difference between value and cost the maximum? (Graph these
functions first.)

3-7 CURVE SKETCHING

Sketching a graph simply by plotting points can be tedious, and as we pointed
out in Chapter 1, improper selection of points can easily lead to errors. Many of
the concepts that we introduced regarding optimization involved graphs of
functions. Specifically, we noted the following:

Some Properties of Graphs of
Functions and Derivatives

> 1. A function is increasing over an interval if $f'(x) > 0$ for all x in the
>   interval. It is decreasing on an interval if $f'(x) < 0$ for all x in the
>   interval.
> 2. Points $(c, f(c))$ on the graph are possible relative optima if $f'(c) = 0$ or if
>   the derivative does not exist at $x = c$.
> 3. If $f''(c) < 0$ at a critical point $(c, f(c))$, the function changes from in-
>   creasing to decreasing, indicating a relative maximum. If $f''(c) > 0$ at a
>   critical point $(c, f(c))$, the function changes from decreasing to increas-
>   ing, indicating a relative minimum.

These properties of the first and second derivatives can provide very useful
information for drawing graphs of functions. For example, consider the function

$$f(x) = x^3 - 12x^2 + 45x - 50$$

The first two derivatives are

$$\begin{aligned} f'(x) &= 3x^2 - 24x + 45 \\ &= 3(x^2 - 8x + 15) \\ &= 3(x - 3)(x - 5) \end{aligned}$$

and

$$f''(x) = 6x - 24$$

We can gain considerable insight about the shape of the graph by simply examining the critical points and the first and second derivatives. Setting the first
derivative equal to 0 gives critical points at $x = 3$ and $x = 5$. Since $f''(3) = -6$,
we know that a relative maximum occurs at the point $(3, f(3)) = (3, 4)$. Also,
$f''(5) = 6$, so a relative minimum occurs at the point $(5, f(5)) = (5, 0)$. Part of the
graph must appear as shown in Figure 3-18.

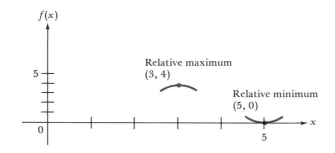

Figure 3-18
A portion of the graph of
$f(x) = x^3 - 12x^2 + 45x - 50$

What does the graph do to the left of $x = 3$? How do we fill in the gap between $x = 3$ and $x = 5$? What is the behavior to the right of $x = 5$? To answer these questions we consider the three intervals on the x-axis determined by the values $x = 3$ and $x = 5$.

Let us examine the intervals $(-\infty, 3)$, $(3, 5)$, and $(5, \infty)$. When $x < 3$, we also have $x < 5$, and hence, the derivative is positive:

$$f'(x) = 3\underbrace{(x - 3)}_{<0}\underbrace{(x - 5)}_{<0} > 0$$

Therefore, the function must be increasing over the interval $(-\infty, 3)$. Between $x = 3$ and $x = 5$, we have

$$f'(x) = 3\underbrace{(x - 3)}_{>0}\underbrace{(x - 5)}_{<0} < 0$$

Thus, $f(x)$ must be decreasing over this interval. Finally, if $x > 5$, we have

$$f'(x) = 3\underbrace{(x - 3)}_{>0}\underbrace{(x - 5)}_{>0} > 0$$

The function must be increasing over the interval $(5, \infty)$. The behavior of the function can be summarized in terms of the sign of the derivative with the following picture:

We conclude that $f(x)$ must look something like one of the graphs in Figure 3-19.

Figure 3-19
Some possible shapes for $f(x)$

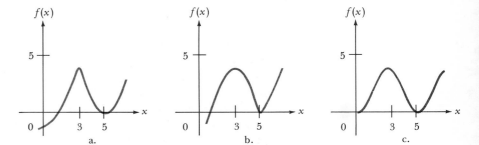

The process we have used to draw the graph of $f(x)$ is clearly superior to any trial-and-error, plug-and-chug method. There is, however, quite a bit more that we do not yet know about the function that will allow us to sketch a more accurate graph. For instance, does the function look more like Figure 3-19a, b, or c? Where does $f(x)$ cross the y-axis or x-axis? These are examples of questions that we address in this section.

CONCAVITY AND INFLECTION POINTS

In the example we have been discussing, we see that $f''(3) < 0$ and $f''(5) > 0$. Obviously, there should be some point in between $x = 3$ and $x = 5$ where $f''(x) = 0$. We can find this point by setting the second derivative equal to 0:

$$f''(x) = 6x - 24 = 0$$
$$x = 4$$

Recall that the second derivative measures the rate of change of the slope of the tangent line. If $x < 4$, then $f''(x) < 0$ and the slope (that is, the *first* derivative) is decreasing; when $x > 4$, then $f''(x) > 0$ and the slope is increasing. We use special terms to define these properties.

A graph is **concave up** over an interval if the first derivative is increasing over the interval.

A graph is **concave down** over an interval if $f'(x)$ is decreasing over the interval.

A point where the curve changes from concave up to concave down or vice versa is called a **point of inflection,** or an **inflection point.**

These concepts are illustrated in Figure 3-20.

Thus, for the function $f(x) = x^3 - 12x^2 + 45x - 50$, we may state that the graph is concave down for $x < 4$ and concave up for $x > 4$. Therefore, the point $(4, f(4))$ is a point of inflection.

If $f''(x)$ exists on an interval, then we can state a simple concavity test. Namely, if $f''(x) > 0$ on an interval, then the first derivative $f'(x)$ increases on that interval [its rate of change, $f''(x)$, is positive] and hence, the graph of the function $f(x)$ is concave up over the interval. Similarly, if $f''(x) < 0$ on an interval, then the graph of $f(x)$ is concave down on the interval.

Test for Concavity

If $f''(x) > 0$ on an interval, then the graph of $f(x)$ is concave up on the interval.

If $f''(x) < 0$ on an interval, then the graph of $f(x)$ is concave down on the interval.

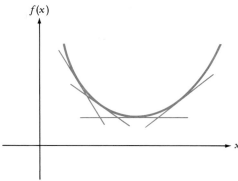

a. Graph is concave up; slope of tangent line increases from left to right

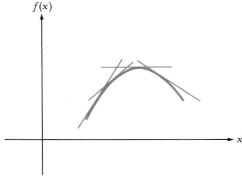

b. Graph is concave down; slope of tangent line decreases from left to right

Figure 3-20
Concavity and inflection points

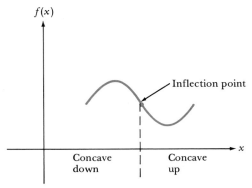

c. Concavity changes at an inflection point

If we examine tangent lines to the graph we see that at points where the graph is concave down, the curve lies below the tangent line; when the graph is concave up, the curve lies above the tangent line. Figure 3-21 (page 162) illustrates these properties for the function $f(x) = x^3 - 12x^2 + 45x - 50$, discussed earlier.

From this example, you might come to believe that a point of inflection will occur whenever $f''(x) = 0$. This, however, is not always the case. For example, consider the function $y = x^6$. The first derivative is $y' = 6x^5$, and the second

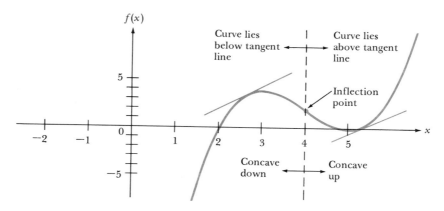

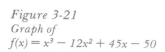

Figure 3-21
Graph of
$f(x) = x^3 - 12x^2 + 45x - 50$

derivative is $y'' = 30x^4$. The second derivative is 0 when $x = 0$, yet from Figure 3-22 we can see that $x = 0$ is clearly not an inflection point. How can we tell? One sure way is the following: If $f''(x) < 0$ for x in an interval (a, c) to the left of c and $f''(x) > 0$ for x in an interval (c, b) to the right of c, then the graph of $f(x)$ is concave down to the left of c and concave up to the right of c. Therefore, the concavity of the graph changes at $x = c$ and $(c, f(c))$ is an inflection point. A similar statement holds if $f''(x) > 0$ to the left of c and $f''(x) < 0$ to the right of c.

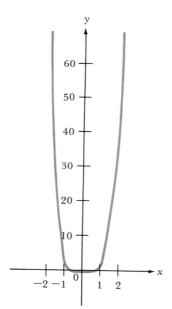

Figure 3-22
Graph of $y = x^6$

Test for Points of Inflection
(Second Derivative)

If the sign of $f''(x)$ changes at $x = c$, then $(c, f(c))$ is a point of inflection.

In the example above we found that $y'' = 30x^4$, and hence, $y''(x) > 0$ for $x < 0$ and $y''(x) > 0$ for $x > 0$. Therefore, the second derivative does not change sign at $x = 0$ and $(0, 0)$ is not an inflection point.

Alternatively, we can use first derivative information to test for a point of inflection.

Test for Points of Inflection
(First Derivative)

To determine whether a point $(c, f(c))$ where $f''(c) = 0$ is an inflection point, check the value of the first derivative on both sides. If the sign of the first derivative is the *same* on both sides, then the point is an inflection point.

In our example, since $y' = 6x^5$, the first derivative is negative for values of $x < 0$ and positive for values of $x > 0$. Therefore, the function is decreasing when $x < 0$ and increasing when $x > 0$. This indicates that a relative minimum occurs at $x = 0$ and that the function is concave up on both sides of $x = 0$. Notice also that we were unable to conclude that a relative minimum occurs at $x = 0$ from the second derivative test, since $y'' = 0$ at $x = 0$.

Inflection points also may occur when the second derivative does not exist, and it is important to check for this. When both the first and second derivatives exist, we can summarize their relationship to the shape of a graph as shown in Table 3-4.

Table 3-4
Graphical interpretation of first and second derivatives

		$f''(x)$		
		$+$	$-$	
	$+$	╱	╱	Increasing
$f'(x)$	0	╲╱	⋀	Relative optimum
	$-$	╲	╲	Decreasing
		Concave up	Concave down	

Example 24

(You should read Algebra Review 14 on page 164 before studying this example.) Determine the points of inflection and intervals where the following function is concave up and concave down:

$$y = \frac{x^4}{4} - 5x^3 + \frac{63x^2}{2} - 81x$$

Solution

The first derivative is

$$y' = x^3 - 15x^2 + 63x - 81$$
$$= (x^2 - 6x + 9)(x - 9)$$
$$= (x - 3)(x - 3)(x - 9)$$

Setting this equal to 0, we see that the critical points occur at $x = 3$ and $x = 9$. The second derivative is

$$y'' = 3x^2 - 30x + 63$$
$$= 3(x^2 - 10x + 21)$$
$$= 3(x - 3)(x - 7)$$

We have possible inflection points at $x = 3$ and $x = 7$, since $y'' = 0$ at these points.

The values $x = 3$, $x = 7$, and $x = 9$ therefore have special significance for the graph of this function. These values divide the x-axis into the four intervals $(-\infty, 3)$, $(3, 7)$, $(7, 9)$, and $(9, \infty)$. On the interval $(-\infty, 3)$ we find that $x - 3 < 0$ and $x - 9 < 0$. Therefore,

$$y' = \underbrace{(x - 3)}_{<0}\underbrace{(x - 3)}_{<0}\underbrace{(x - 9)}_{<0} < 0$$

while

$$y'' = 3\underbrace{(x - 3)}_{<0}\underbrace{(x - 7)}_{<0} > 0$$

on $(-\infty, 3)$. However, for x in the interval $(3, 7)$, we find that $x - 3 > 0$,

ALGEBRA REVIEW 14: FACTORING CUBIC EXPRESSIONS

There is a pattern for factoring the sum or difference of two cubes. Find the cube root of each term and then note the following:

$$a^3 + b^3 = (a + b)(a^2 - ab + b^2)$$
$$a^3 - b^3 = (a - b)(a^2 + ab + b^2)$$

For example, consider $x^3 + 64$. The cube root of x^3 is x; the cube root of 64 is 4. Thus,

$$x^3 + 64 = (x + 4)(x^2 - 4x + 16)$$

As a second example, consider $8x^3 - 27$. The cube roots of $8x^3$ and 27 are $2x$ and 3, respectively. Therefore,

$$8x^3 - 27 = (2x - 3)(4x^2 + 6x + 9)$$

In a general cubic polynomial, the process is more complicated since there is no simple formula to follow. When all the terms of the polynomial have integer coefficients and the coefficient of x^3 is 1, then $x - a$ is a factor if it exactly divides the polynomial. If $x - a$ is a factor of the polynomial, then a is a factor of the constant term in the polynomial. Consider the polynomial

$$x^3 + 7x^2 + 8x - 10$$

The constant term -10 has the factors $\pm 1, \pm 2, \pm 5$, and ± 10. If the polynomial is factorable, then $(x \pm 1)$, $(x \pm 2)$, $(x \pm 5)$, or $(x \pm 10)$ is a factor. Using long division with each of these expressions will determine whether any are factors.

$x - 7 < 0$, and $x - 9 < 0$. Therefore, on $(3, 7)$,

$$y' = \underbrace{(x - 3)}_{>0}\underbrace{(x - 3)}_{>0}\underbrace{(x - 9)}_{<0} < 0$$

and

$$y'' = 3\underbrace{(x - 3)}_{>0}\underbrace{(x - 7)}_{<0} < 0$$

Checking the signs of y' and y'' on each of the intervals in this way, we obtain the results in Table 3-5. We therefore see that at $x = 3$ and $x = 7$ there are inflection

Table 3-5

Interval	y'	y''	Concavity
$(-\infty, 3)$	<0	>0	Up
$(3, 7)$	<0	<0	Down
$(7, 9)$	<0	>0	Up
$(9, \infty)$	>0	>0	Up

We show below that $x + 5$ is a factor (it is the only one):

$$
\begin{array}{r}
x^2 + 2x - 2 \\
x + 5 \overline{)\, x^3 + 7x^2 + 8x - 10} \\
\underline{x^3 + 5x^2} \qquad\qquad\quad \\
2x^2 + 8x - 10 \\
\underline{2x^2 + 10x} \qquad\quad \\
-2x - 10 \\
\underline{-2x - 10} \\
0
\end{array}
$$

Therefore,

$$x^3 + 7x^2 + 8x - 10 = (x + 5)(x^2 + 2x - 2)$$

This process of factoring allows us to solve many cubic equations in a manner similar to that used for quadratic equations (see Algebra Review 13).

EXERCISES

Factor the following polynomials:
a. $125 - x^3$
b. $x^3 - 6x^2 + 11x - 6$
c. $x^3 - 2x^2 - x + 2$
d. $x^3 - 3x^2 + 3x - 1$
e. $x^3 + 3x^2 - x - 6$

points, since the concavity of the graph changes at these points. The graph of this function is given in Figure 3-23.

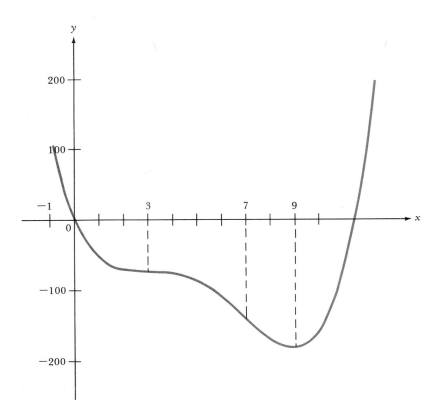

Figure 3-23
Graph of
$$y = \frac{x^4}{4} - 5x^3 + \frac{63x^2}{2} - 81x$$

Example 25   Determine inflection points and concavity properties for $f(x) = 6x^{2/3}$.

Solution   Taking derivatives, we have

$$f'(x) = 4x^{-1/3} = \frac{4}{x^{1/3}}$$

$$f''(x) = -\frac{4x^{-4/3}}{3} = -\frac{4}{3x^{4/3}}$$

Both the first and second derivatives do not exist at $x = 0$; therefore, $(0, f(0))$ may be a relative optimum point or an inflection point. Over the interval $(-\infty, 0)$, the first derivative is negative, and hence, the function is decreasing on this interval. The second derivative is also negative there (notice that the term $x^{4/3} = \sqrt[3]{x^4}$ can never be negative), indicating that $f(x)$ is concave down over this interval. Over the interval $(0, \infty)$, $f'(x) > 0$ and $f''(x) < 0$. Hence, the function is increasing and concave down over $(0, \infty)$. We conclude that $(0, 0)$ cannot be an inflection point and is a relative minimum point (in fact, an absolute minimum point). The graph of $f(x) = 6x^{2/3}$ is shown in Figure 3-24.

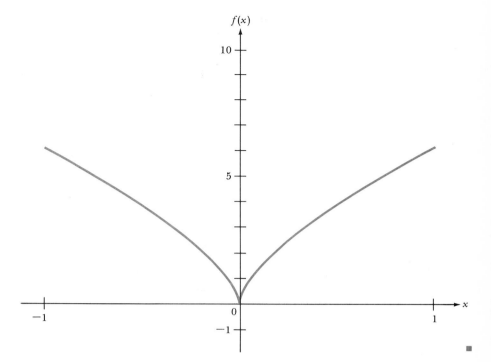

$f(x)$

Figure 3-24
Graph of $f(x) = 6x^{2/3}$

INTERCEPTS

Intercepts are the final aids to curve sketching that we discuss in this chapter. (Other concepts relating to new differentiation rules will be discussed in Chapter 4.) Intercepts were very briefly introduced in Chapter 1, but we will review them here. If we have the function $y = f(x)$, then an *x-intercept* is a number a such that $f(a) = 0$. That is, the graph crosses the x-axis at the point $(a, 0)$. The *y-intercept* of the graph of a function $y = f(x)$ is the number $f(0)$. That is, the graph of the function crosses the y-axis at the point $(0, f(0))$. The graph of a function $y = f(x)$ may have several x-intercepts, but can have only one y-intercept. Do you see why?

Example 26   Find the intercepts of the function $f(x) = x^3 - x^2 - 6x$.

Solution   Factoring this equation and setting $f(x) = 0$, we obtain

$$f(x) = x(x^2 - x - 6)$$
$$= x(x - 3)(x + 2) = 0$$

The roots are $x = 0$, $x = 3$, and $x = -2$. These are the x-intercepts. The y-intercept is found by computing $f(0) = 0$. ∎

**CURVE
SKETCHING
TECHNIQUES**

We are now in a position to put the facts we have discovered about graphs of functions together. Table 3-6 (page 168) gives a summary of the curve sketching concepts involving derivatives that we have discussed.

Table 3-6
Derivatives and curve sketching

Property of Graph	What to Look for in the Function	Test
Increasing/decreasing	Sign of $f'(x)$	Increasing if $f'(x) > 0$ Decreasing if $f'(x) < 0$
Relative optima	Points $(c, f(c))$ with $f'(c) = 0$ or $f'(c)$ nonexistent	First or second derivative test
Concavity	Sign of $f''(x)$	Concave up if $f''(x) > 0$ Concave down if $f''(x) < 0$
Inflection points	Points where $f''(c) = 0$ or $f''(c)$ nonexistent	Inflection point if $f''(x)$ changes sign at $x = c$
Intercepts	Points $(a, 0)$ where $f(a) = 0$; the point $(0, f(0))$	

When trying to graph a function, it is generally advisable to first identify the intercepts and any points of discontinuity. Next, locate and plot all relative optima and inflection points. Check for the concavity of the function over the intervals between these points and determine whether the function is increasing or decreasing. Finally (if necessary), plot some additional points in order to complete a smooth curve.

Example 27   Sketch the graph of the function $y = f(x) = x^2 - 4x - 12$.

Solution   The y-intercept is $f(0) = -12$. Factoring the equation and setting it equal to 0, we have $y = (x - 6)(x + 2) = 0$. Therefore, the x-intercepts are 6 and -2. The first derivative is $y' = 2x - 4 = 2(x - 2)$. Setting this equal to 0, we find that a

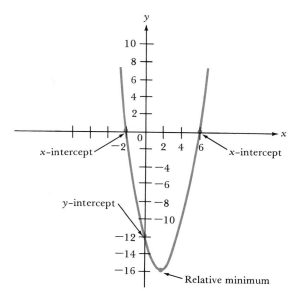

Figure 3-25
Graph of $y = x^2 - 4x - 12$

critical point occurs at $x = 2$. Since $y'' = 2 > 0$, the point $(2, f(2)) = (2, -16)$ is a relative minimum point. The first derivative is positive when $x > 2$ and negative when $x < 2$; thus, the function is increasing over the interval $(2, \infty)$ and decreasing over $(-\infty, 2)$. Since the second derivative is always positive, the function is always concave up. The graph, a parabola, is shown in Figure 3-25.   ∎

Example 28   Sketch the graph of the function

$$f(x) = -x^3 - 4x^2 + 35x$$

Solution   The y-intercept is $f(0) = 0$. Factoring and setting $f(x) = 0$, we obtain

$$f(x) = x(-x^2 - 4x + 35) = 0$$

One root is $x = 0$. Using the quadratic formula, we find that the other roots are $x \approx -8.25$ and $x \approx 4.25$. These three roots represent the x-intercepts. The first derivative of $f(x)$ is

$$f'(x) = -3x^2 - 8x + 35$$
$$= (-3x + 7)(x + 5)$$

Setting this equal to 0, we get

$$(-3x + 7)(x + 5) = 0$$

Therefore, the critical points occur at $x = \frac{7}{3}$, with $f(\frac{7}{3}) \approx 47.185$, and at $x = -5$, with $f(-5) = -150$.

The second derivative is $f''(x) = -6x - 8$. When $x = \frac{7}{3}$, the value of $f''(x)$ is -22. Thus, a relative maximum occurs at the point $(\frac{7}{3}, f(\frac{7}{3}))$. When $x = -5$, $f''(x)$ has a value of 22. Therefore, a relative minimum occurs at $(-5, -150)$. Checking the value of the first derivative over the intervals on the x-axis deter-

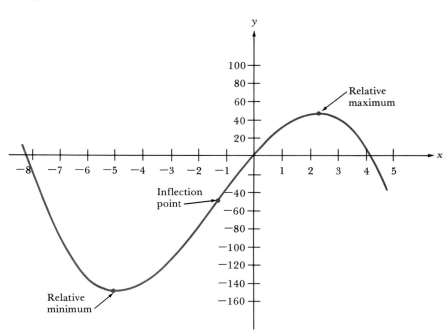

Figure 3-26
Graph of $f(x) = -x^3 - 4x^2 + 35x$

mined by the critical points, we see that $f'(x) < 0$ over the interval $(-\infty, -5)$, $f'(x) > 0$ over $(-5, \frac{7}{3})$, and $f'(x) < 0$ over $(\frac{7}{3}, \infty)$. Thus, $f(x)$ is decreasing, increasing, and decreasing, respectively, over these intervals.

Setting the second derivative equal to 0 yields the possible inflection points:

$$f''(x) = -6x - 8 = 0$$

$$x = -\tfrac{4}{3}$$

Since $f''(x) = 6(-x - \frac{4}{3})$, $f''(x) > 0$ for $x < -\frac{4}{3}$ and $f''(x) < 0$ for $x > -\frac{4}{3}$. Therefore, the function is concave up over $(-\infty, -\frac{4}{3})$ and concave down over $(-\frac{4}{3}, \infty)$, and hence, $(-\frac{4}{3}, f(-\frac{4}{3}))$ is an inflection point. The graph of $f(x)$ is given in Figure 3-26.   ∎

PROBLEMS (Section 3-7)

In Problems 1–8, determine intervals over which the functions are increasing or decreasing.

1. $f(x) = 3x + 7$

2. $f(x) = x^2 + 4x + 7$

3. $f(x) = 6x^2 - 3x + 4$

4. $f(x) = \frac{1}{3}x^3 + \frac{3}{2}x^2 + 2x - 4$

5. $f(x) = x^3 + 3x^2 + 3x - 2$

6. $f(x) = x + \dfrac{1}{x}$

7. $f(x) = x^{2/3}$

8. $f(x) = |x + 2|$

In Problems 9–16, determine the points of inflection and intervals where the functions are concave up and concave down.

9. $y = x^3 + 3x^2 + 3$

10. $y = \frac{2}{3}x^3 - 6x^2 + 3$

11. $y = x^3 - 3x^2 + 3x + 1$

12. $y = x^4 - 4x^3 + 12$

13. $f(x) = 9x^{2/3}$

14. $f(x) = -6x^{1/3}$

15. $f(x) = (x - 3)^3$

16. $f(x) = (x + 5)^3$

Find the intercepts of the following functions:

17. $f(x) = x^2 - 3x - 10$

18. $f(x) = x^2 - 12x + 32$

19. $y = x^3 + 9x^2 + 18x$

20. $g(x) = x^3 - 8x^2 - 33x$

21. $y = x^4 - 6x^3 + 8x^2$

22. $f(x) = x^4 - 13x^2 + 36$

23. $f(x) = x^3 + 4x^2 + x - 6$

24. $y = x^3 + x^2 - 16x - 16$

In Problems 25–37, sketch the graph of each function using the curve sketching techniques discussed in this section.

25. $y = x^2 + x - 20$

26. $y = x^2 - 9x + 18$

27. $y = -x^2 + 2x + 3$

28. $y = -x^2 + 5x$

29. $y = 2x^2 + 5x - 12$

30. $y = x^3 - 3x^2 + 3$

31. $y = x^3 - 3x + 1$

32. $y = x^3 - 9x^2 + 27x - 27$

33. $y = x^3 - 6x^2 + 3x + 10$

34. $y = 3x^4 + 4x^3$

35. $y = x + \dfrac{1}{x}$

36. $y = x^{1/3} + 2$

37. $y = x + x^{2/3}$

38. Sketch the graph of a function that has

the following properties: $f(-1) = 2$, $f(1) = -1$, $f'(x) > 0$ for $x < -1$, $f'(x) < 0$ for $-1 < x < 1$, $f'(x) > 0$ for $x > 1$.

39. Sketch the graph of a function that has the following properties: $f'(x) > 0$ for $x \neq 1$, $f'(1) = 0$, $f''(x) > 0$ for $x > 1$, $f''(x) < 0$ for $x < 1$.

3-8 CONCLUSION: SOLVING THE FOREST FIRE PROBLEM

The forest fire problem developed in the Introduction to this chapter can be stated as

Minimize $9000y + 1320 + 25x$

subject to the constraint

$$\frac{y}{0.4545} = \frac{132}{x}$$

To solve this problem, we first convert this function of two variables into a function of only a single variable. To do this, we solve the constraint for y:

$$y = \frac{(0.4545)(132)}{x} \approx \frac{60}{x}$$

Substituting this into the objective function, we obtain the following optimization problem:

Minimize $f(x) = 9000\left(\dfrac{60}{x}\right) + 1320 + 25x$

$$= \frac{540,000}{x} + 1320 + 25x$$

Now we may take the first derivative and set it equal to 0:

$$f'(x) = -\frac{540,000}{x^2} + 25 = 0$$

$$25 = \frac{540,000}{x^2}$$

$$25x^2 = 540,000$$

$$x^2 = \frac{540,000}{25} = 21,600$$

$$x \approx 147$$

The second derivative is $f''(x) = 1,080,000/x^3$. Since $f''(x) > 0$ for $x > 0$, we see that the value of the second derivative is positive at $x \approx 147$; hence, $(147, f(147))$ is a minimum point. We then find y using the expression $y \approx 60/x$ found above:

$$y \approx \frac{60}{x}$$

$$\approx \frac{60}{147} \approx 0.408$$

Therefore, about 147 people should be hired to construct the firebreak, and it should be located approximately 0.4 mile from the current fire front.

3-9 CHAPTER REVIEW

1. An optimization problem is one in which we seek to minimize or maximize some objective function.
2. A point $(c, f(c))$ is an absolute maximum (or mini-

mum) point of a function $f(x)$ if $f(c) \geq f(x)$ [or $f(c) \leq f(x)$] for all x in the domain of f.
3. A relative maximum (or minimum) point of a function $f(x)$ is a point $(c, f(c))$ on the graph of

$y = f(x)$ for which $f(c) \geq f(x)$ [or $f(c) \leq f(x)$] in some open interval (a, b) containing c.

4. Absolute optima can occur at points that are relative optima or at endpoints of a closed interval defining the domain of the function.

5. A critical point of a function $f(x)$ is a point $(c, f(c))$ where either $f'(c) = 0$ or $f'(c)$ does not exist.

6. A function is increasing (or decreasing) on an open interval if for every a, b in the interval, whenever $a < b$, then $f(a) < f(b)$ [or $f(a) > f(b)$].

7. If $(c, f(c))$ is a critical point, the first derivative test can be used to determine whether it is a relative minimum, relative maximum, or neither.

8. The second derivative measures the rate of change of the slope of the tangent line to the graph of a function.

9. The second derivative test can be used to determine whether a critical point $(c, f(c))$ with $f'(c) = 0$ is a relative maximum or minimum. However, if the second derivative is 0, the test is inconclusive and the first derivative test must be used.

10. Absolute optima can be determined by identifying all critical points and endpoints and comparing their function values.

11. A graph is concave up (or down) over an interval if $f'(x)$ is increasing (or decreasing) over the interval. This occurs if $f''(x) > 0$ [or $f''(x) < 0$] over the interval. A point where the concavity changes is an inflection point.

12. Calculus simplifies curve sketching by enabling one to identify critical points, inflection points, and other properties of functions.

REVIEW PROBLEMS (Section 3-9)

Identify all critical points for each function.

1. $f(x) = x^2 - 12x + 36$

2. $f(x) = 3x^3 - 4x + 5$

3. $f(x) = \frac{1}{3}x^3 - \frac{1}{2}x^2 - 30x$

4. $y = \begin{cases} x^3 - 5x^2 & \text{if } x < 5 \\ x^2 - 9x + 20 & \text{if } x \geq 5 \end{cases}$

5. $f(x) = x^4 - 18x^2 + 7$

Determine any relative maxima or minima for the following:

6. $f(x) = 2x^2 + 6x - 8$

7. $f(x) = x^3 - 6x^2 + 10$

8. $f(x) = x^4 - 4$

9. $g(x) = x + \dfrac{3}{x}$

10. $y = \dfrac{x + 1}{x}$

11. $f(x) = \dfrac{x^2 + 2x + 1}{x + 1}$

12. $f(x) = (x + 1)^2$

Find any absolute optima for each of the following functions:

13. $f(x) = x^2 + 2x - 4$

14. $y = -x^2 + 4x - 5$

15. $y = 3x^3 - 3x^2 + 5$

16. $f(x) = \frac{1}{5}x^5 - \frac{1}{3}x^3 + 4$

17. $f(x) = 2x^2 + 6x - 10$, $-3 \leq x \leq 6$

18. $y = -2x^3 + \frac{1}{2}x^2$, $-2 \leq x \leq 4$

19. $f(x) = x^3 - 3x^2$, $-1 \leq x \leq 3$

20. $g(x) = \dfrac{x - 2}{2x}$, $3 \leq x \leq 5$

21. $y = 3x^{2/3} - 2x$, $-1 \leq x \leq 1$

22. $f(x) = \dfrac{(x+1)^2}{x}, \quad 1 \le x \le 10$

23. The total profit (in dollars) from the sale of x units of a certain prescription drug is given by

$$P(x) = -x^3 + 3x^2 + 72x + 1280$$

   a. Find the number of units that should be sold in order to maximize total profit.
   b. What is the maximum profit?

24. A carpenter has been asked to build an open box with a square base. The sides of the box will cost $3 per square meter, and the base will cost $4 per square meter. What are the dimensions of the box having greatest volume that can be constructed for $48?

25. A rectangular play area of 600 square feet is to be fenced off. One side, which is along an existing wall, does not require fencing. The other three sides are to be fenced at a uniform cost of $4 per yard. What dimensions will minimize the cost of the fencing?

26. A travel agent offers a group tour to Europe only if the group has at least 60 people. Each member of a group of 60 pays $800. For each person in excess of 60 people, everyone's fare is reduced by $8. How many participants will maximize the agent's revenue, if at most 75 people can be accommodated?

27. If the price charged for a candy bar is $p(x)$ cents, then x thousand candy bars will be sold in a certain city, where

$$p(x) = 100 - \frac{x}{10}$$

   a. Find an expression for the total revenue from the sale of x thousand candy bars.
   b. Find the value of x that leads to maximum revenue.
   c. Find the maximum revenue.

28. The manager of an apartment complex with 80 units is trying to decide on the rent to charge. It is known from experience that if the rent is $200, all the units will be full. However, on the average, 1 additional unit will remain vacant for each $10 increase in the rent.

   a. Let x represent the number of $10 increases. Write an expression for the amount of rent per apartment.
   b. Write an expression for the number of apartments rented.
   c. Find the total revenue from all rented apartments.
   d. What value of x leads to the maximum revenue?
   e. What is the maximum revenue?

Determine the points of inflection and intervals where the following functions are concave up and concave down:

29. $f(x) = x^3 - 6x^2 + 12x - 8$

30. $f(x) = x^4 - 3x^3 + x^2$

31. $g(x) = (x+4)^3$

32. $y = \dfrac{x^4}{4} - 2x^2$

Find the intercepts of each of the following:

33. $y = x^2 + 4x - 77$

34. $y = x^3 + 5x^2 - 2x - 24$

35. $y = x^4 - 3x^3 - 10x^2 + 24x$

36. $f(x) = x^5 + 9x^4 + 27x^3 + 27x^2$

Sketch the graph of each function using curve sketching techniques presented in this chapter. Tell all you can about each function.

37. $y = x^2 - 6x - 27$

38. $y = 2x^2 - 10x + 12$

39. $y = 3x^3 - 4x + 5$

40. $y = 2x^3 + 3x^2 - 36x + 1$

41. $y = x^4 - x^3 + 1$

THE DIFFERENTIAL

The value of the derivative of a function $f(x)$ at $x = a$ represents the instantaneous rate of change of $f(x)$ at $x = a$. As we noted in Chapter 2, $f'(a)$ represents the *approximate* change in the value of the function when x is increased by one unit. In this supplement we show how to estimate the change in the value of $f(x)$ for any small change in x. This concept, called the *differential*, has several useful applications.

3S-1 DEFINITION OF THE DIFFERENTIAL

Recall that the derivative is defined as the limiting value of the slope of the secant line between $(x, f(x))$ and $(x + h, f(x + h))$:

$$f'(x) = \lim_{h \to 0} \frac{f(x + h) - f(x)}{h}$$

From Figure 3S-1, if we denote h by Δx and $f(x + h) - f(x)$ by Δy, we have

$$f'(x) = \lim_{\Delta x \to 0} \frac{\Delta y}{\Delta x}$$

Since $\Delta y / \Delta x$ equals $f'(x)$ only in the limit as Δx approaches 0, for any value $\Delta x \neq 0$, $\Delta y / \Delta x$ is an *approximation* to $f'(x)$, that is,

$$\frac{\Delta y}{\Delta x} \approx f'(x)$$

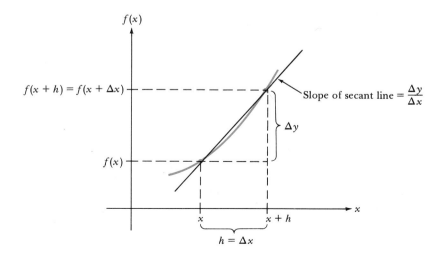

Figure 3S-1
The slope of the secant line between
$(x, f(x))$ and $(x + h, f(x + h))$

We may write this in a different fashion as

$$\Delta y \approx f'(x)\Delta x$$

This formula says that the change in the value of the function $f(x)$ at a point $(x, f(x))$, Δy, as x is changed by an amount Δx, is approximately equal to the value of the derivative at x times the change, Δx. We call $f'(x)\Delta x$ the *differential* and denote it as follows:

If $y = f(x)$ is a function with derivative $f'(x)$, the **differential** is defined as

$$dy = f'(x)\Delta x$$

Since $f'(x)$ is the slope of the tangent line at the point $(x, f(x))$, the differential provides an approximation to the change in the value of the function for a small change in x, Δx. This is illustrated in Figure 3S-2. We see that Δy is the actual change, and dy is only an approximation. The smaller Δx is, the better is the approximation. Thus, the differential is useful for investigating the effect on y of small changes in x.

Figure 3S-2
Differential approximation

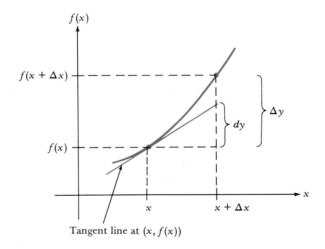

Tangent line at $(x, f(x))$

Example 1   Consider the function $y = f(x) = 2x^3 - x$. Approximate the change in y at $x = 1$ if x is increased by 0.1, 0.5, and 1.0. What percentage error is incurred for each of these values?

Solution   The derivative of $f(x)$ is

$$f'(x) = 6x^2 - 1$$

At $x = 1, f'(1) = 5$. Therefore, the differential of $f(x)$ is

$$dy = 5\Delta x$$

The approximate change in y is computed by

$$\Delta y = f(1 + \Delta x) - f(1) \approx dy = 5\Delta x$$

The *percentage error* is then computed by dividing the difference between the actual change and the estimated change by the actual change. The actual change is $\Delta y = f(1 + \Delta x) - f(1)$. Thus, we have

$$\text{Percentage error} = \left| \frac{\Delta y - dy}{\Delta y} \right| (100)$$

We may now construct the following table:

Table 3S-1

Δx	$dy = 5\Delta x$	$f(1 + \Delta x)$	Error $\Delta y = f(1 + \Delta x) - f(1)$	Percent Error $|(\Delta y - dy)/\Delta y|(100)$
0.1	0.5	1.562	0.562	11.03
0.5	2.5	5.25	4.25	41.18
1.0	5.0	14.0	13.0	61.54

We can see from Table 3S-1 that as Δx increases, the error also increases. Hence, the differential is only useful for relatively small values of Δx. ∎

3S-2 APPLICATIONS OF THE DIFFERENTIAL

Example 2
Economic Order
Quantity

In Example 23 of Chapter 3 we discussed the development and solution of the economic order quantity model for Holton Drug Company. In this problem, the derivative of the total cost function with respect to the purchase quantity was found to be

$$\frac{dC}{dQ} = 1.08 - \frac{912{,}000}{Q^2}$$

and the optimal purchase quantity was determined to be $Q \approx 919$. Suppose, however, that due to space limitations, the company cannot store more than 700 cases, but is considering expanding its capacity. How much savings will result if the capacity is expanded to accommodate an additional 10 cases?

Solution   At $Q = 700$, $dC/dQ = 1.08 - 912{,}000/(700)^2 \approx -0.7812$. Therefore, the differential of C is

$$dC \approx -0.7812\Delta Q$$

An increase in capacity of 10 cases would result in a change in total cost of approximately

$$dC \approx -0.7812(10) = -7.812$$

The negative sign indicates a decrease in total cost, or a savings of about $7.81

annually. The company might weigh these savings against the cost of increasing capacity.   ■

   We may use the differential to develop a formula for estimating the value of a function $f(x)$ for a small change in x. From Figure 3S-2, this is given by:

$$f(x + \Delta x) \approx f(x) + dy$$

or

$$f(x + \Delta x) \approx f(x) + f'(x)\Delta x$$

Example 3
Marginal Revenue

Economists have determined that the marginal revenue function for a product is estimated to be

$$R'(x) = 0.04x - 5 \text{ dollars per unit}$$

Currently, the company has an annual revenue of $1.950 million, and is producing at a level of $x = 10,000$ units per year in order to meet demand. How much revenue is expected if both demand and production are increased by 200 units? Decreased by 50 units?

Solution

At $x = 10,000$, the value of the derivative is

$$R'(10,000) = 0.04(10,000) - 5 = 395$$

Therefore, we may approximate $f(10,000 + \Delta x)$ by

$$f(10,000 + \Delta x) \approx 1,950,000 + 395\Delta x$$

If $\Delta x = 200$, we have

$$f(10,200) \approx 1,950,000 + 395(200)$$
$$= 2,029,000$$

This represents the expected revenue when the company is producing at a level of 10,200 units per year.
   If $\Delta x = -50$, we have

$$f(9950) \approx 1,950,000 + 395(-50)$$
$$= 1,930,250$$

Therefore, the expected revenue decreases by $19,750.   ■

   The differential has numerous applications in the physical sciences and engineering in the analysis of measurement errors, as the next example illustrates.

Example 4
Error Estimation

A thin gold plating in the shape of a square must be laid on the surface of an electronic component. Since this is a very expensive part of the component, the production engineer is interested in the magnitude of errors that may arise during production. If a 1% error is made on each side dimension of the square, by how much is the area affected?

Solution   Let x be the length of the side of the square. The area is then $A = x^2$. If x varies by Δx, then the change in the area is approximated by the differential

$$dA = \frac{dA}{dx}\,\Delta x$$

Since the derivative $dA/dx = 2x$, we have

$$dA = 2x\Delta x$$

Now, a 1% error in x means that $\Delta x = 0.01x$. Therefore, we have

$$dA = 2x(0.01x)$$
$$= 0.02x^2$$
$$= 0.02\,A$$

We can say that the area will be off by approximately 2% of its specifications. ∎

The differential can also be used to obtain quick approximations to roots of numbers.

Example 5
Numerical Approximation   Approximate $\sqrt{8}$ using differentials.

Solution   We note that the closest perfect square to 8 is 9. If we write $f(x) = \sqrt{x}$, then we can state the problem as one of estimating $f(x + \Delta x)$ by the differential approximation $f(x) + f'(x)\Delta x$, where $x = 9$ and $\Delta x = -1$. Thus, we have

$$f(8) \approx f(9) + f'(9)(-1)$$
$$= \sqrt{9} + \left(\frac{1}{2\sqrt{9}}\right)(-1)$$
$$= 3 - \tfrac{1}{6} \approx 2.833$$

The actual value of $\sqrt{8}$ to three decimal places is 2.828. Thus, the differential provides a fairly close approximation. ∎

PROBLEMS (Supplement to Chapter 3)

1. Consider the function $y = f(x) = x^2 - (x/2)$. Approximate the change in $f(x)$ at $x = 1$ if x is increased by 0.1, 0.5, and 1.0. What percentage error is incurred for each of these values?

2. Consider the function $y = f(x) = x^3 - 2x$. Approximate the change in $f(x)$ at $x = 2$ if x is increased by 0.1, 0.2, and 0.5. What percentage error is incurred for each of these values?

For each of the functions in Problems 3–7, first find $f(x + \Delta x)$ and $f'(x)$ for the given values of x and Δx. Then compute Δy by

$$\Delta y = f(x + \Delta x) - f(x)$$

and compute dy by

$$dy = f'(x)\Delta x$$

Finally, compare the results.

3. $f(x) = 5x - 4$; $x = 3$, $\Delta x = 0.1$

4. $f(x) = 3x^2$; $x = 2$, $\Delta x = 0.2$

5. $f(x) = x^2 + 1$; $x = 1$, $\Delta x = 0.1$

6. $f(x) = x^3 + x^2 - 2x$; $x = 2$, $\Delta x = 0.1$

7. $f(x) = \sqrt{2x}$; $x = 1$, $\Delta x = 0.2$

8. A factory finds that its total cost is determined to be

$$C(x) = \tfrac{1}{4}x^2 + 100 \text{ dollars}$$

What is the approximate change in cost as x, the number of units produced, increases from 50 to 55? As x decreases from 50 to 47?

9. A certain company's monthly profit, P, is estimated to be

$$P(x) = -\tfrac{3}{2}x^2 + 550x - 2000 \text{ dollars}$$

where x is the level of production and sales in units. What is the approximate change in profit as x increases from 100 to 105? As x decreases from 100 to 95?

10. Economists have determined that the marginal revenue function for a certain product is given by

$$R'(x) = 0.05x - 10$$

Currently, the company has an annual revenue of $1.5 million, and is producing at a level of $x = 20{,}000$ units per year. How much revenue is expected if both demand and production are increased by 500 units? Decreased by 1000 units?

11. If we are given the function $f(x) = x^{1/2}$, what is the rate of change of the function per unit change in x at $x = 0.04$? Estimate the change in the function if x increases from 0.040 to 0.042.

12. In Example 19 of Chapter 3 we discussed the optimal level of production of key chains in order to maximize profit. The profit function was given by

$$P(x) = -\frac{x^2}{2} + 31x - 15$$

and the optimum production level was 31 hundred units. Suppose, however, that the firm is currently producing at a level of $x = 27$. What would be the approximate increase in profit if the optimal number were produced?

13. In Example 21 of Chapter 3 we discussed the optimal price that a North Carolina resort hotel should charge for each room in order to maximize the revenue. The revenue function was given by

$$R(x) = -1.5x^2 + 200x$$

and the optimal price to charge was found to be $66.67. Suppose, however, that the owner is currently charging a round figure of $70.00. Approximately how much revenue could be gained if the optimal price is charged?

Find approximations for the numbers given in Problems 14–20 using differentials.

14. $\sqrt{15}$

15. $\sqrt{27}$

16. $\sqrt{62}$

17. $\sqrt{124}$

18. $\sqrt{1.23}$

19. $\sqrt[3]{26}$

20. $\sqrt[4]{9}$

21. The volume of a sphere is given by $V = \tfrac{4}{3}\pi r^3$. Find the approximate change in the volume of a snowball if the radius decreases from 6 centimeters to 5 centimeters.

22. An artificial lake is in the shape of a circle. Find the approximate increase in the area of the lake if its radius increases from 1.5 miles to 1.7 miles during the rainy season.

ADDITIONAL TECHNIQUES AND APPLICATIONS OF DIFFERENTIATION

INTRODUCTION

Many road tunnels in Great Britain suffer road congestion during busy work periods (normally 8–9 A.M. and 5–6 P.M.).* To maximize the flow of traffic during a busy period, many tunnel authorities prescribe recommended speeds and separation distances. We would like to develop a model that will provide such recommendations.

An important concept is the *flow rate,* which is defined as the number of cars passing a fixed point in a unit time interval. The flow rate depends on a number of factors such as the traffic speed, separation distance between cars, and the length of the cars. We will assume that traffic is uniform; that is, all cars travel at the same speed, v miles per hour, with an average separation distance of D feet and an average length of L feet. In order to standardize our dimensions, we shall convert the speed to feet per second as follows:

$$\left(\frac{v \text{ mi}}{\text{hr}}\right)\left(\frac{5280 \text{ ft/mi}}{3600 \text{ sec/hr}}\right) \approx 1.467v \text{ ft/sec}$$

Now suppose that a car has just passed a fixed

* Adapted and reproduced with permission from *Applying Mathematics: A Course in Mathematical Modelling,* by D. N. Burghes, I. Huntley, and J. McDonald, 1982, published by Ellis Horwood Limited, Chichester, England.

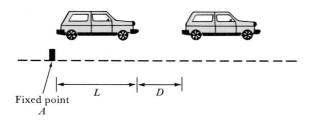

Figure 4-1
Uniform stream of traffic

point *A*, as shown in Figure 4-1. How long will it take for the next car to reach this point? It must travel a distance of $D + L$ feet at $1.467v$ feet per second. Therefore, the time to accomplish this will be

$$\frac{D + L}{1.467v} \text{ seconds per car}$$

The flow rate (in cars per second) is the reciprocal of this quantity, or

$$F = \frac{1.467v}{D + L} \text{ cars per second}$$

Therefore, we see that for cars of fixed length, the flow rate depends on the speed and the separation distance, both of which can be controlled by the tunnel authority traffic manager.

The separation distance *D* is composed of two parts: the thinking distance and the braking distance. Traffic studies have shown that both the thinking and braking distances are related to the speed as follows:

Thinking distance $= v$ feet

Braking distance $= \dfrac{v^2}{20}$ feet

Therefore, the total separation distance expressed as a function of speed is given by

$$D = v + \frac{v^2}{20}$$

We may now express the flow rate as a function of speed by

$$F(v) = \frac{1.467v}{v + v^2/20 + L}$$

For a value of $L = 12$ feet, the graph of $F(v)$ is shown in Figure 4-2.

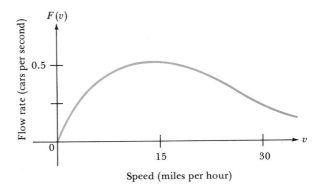

Figure 4-2
Graph of flow rate
$$F(v) = \frac{1.467v}{v + v^2/20 + 12}$$

This function clearly has a maximum value. In the previous chapter we showed that we can maximize a function by taking the derivative and setting it equal to 0. However, we have not yet encountered a function similar to $F(v)$. It is not a simple polynomial, and the power rule that we have learned for derivatives cannot be applied. In this chapter, we shall discuss several additional rules for differentiation that will enable us to tackle more difficult problems involving derivatives.

4-1 THE PRODUCT AND QUOTIENT RULES

In Example 6 of Chapter 1, we developed a model for the volume of a container that was made by cutting out squares with sides of length *x* from the corners of a 28 × 28-inch square sheet of cardboard. The volume of the container was shown to be $V = (28 - 2x)(28 - 2x)(x)$. Until now, in order to find the dimensions of the container that give the maximum volume, we would have to multiply the factors in this expression before differentiating. Thus, we can simplify this expression to $V = 4x^3 - 112x^2 + 784x$ so that we can apply the power rule for derivatives. The derivative is

$$\frac{dV}{dx} = 12x^2 - 224x + 784$$

For many problems, the algebra required to simplify a function in this fashion may be quite extensive. Can the derivative be computed from the original expression? Fortunately, the answer is yes, and the method to apply is given by the following rule:

THEOREM
Product Rule for Derivatives

Let $f(x)$ and $g(x)$ be two functions with derivatives $f'(x)$ and $g'(x)$, respectively. Then

$$\frac{d}{dx}[f(x)g(x)] = f(x)g'(x) + g(x)f'(x)$$

Stated simply, *the derivative of a product of two functions is equal to the first function times the derivative of the second function plus the second function times the derivative of the first function.* Note that the derivative of a product of two functions is *not* equal to the product of the derivatives! This rule is not too difficult to learn if you repeat it to yourself each time you use it. A proof of this theorem is given in the supplement to this chapter.

Example 1   Find the first derivative of $y = (x + 3)(2x - 4)$.

Solution   Using the product rule, we associate $x + 3$ with $f(x)$ and $2x - 4$ with $g(x)$. Then

$$\frac{dy}{dx} = \underbrace{(x + 3)}_{f(x)}\underbrace{(2)}_{g'(x)} + \underbrace{(2x - 4)}_{g(x)}\underbrace{(1)}_{f'(x)}$$
$$= (2x + 6) + (2x - 4)$$
$$= 4x + 2$$

We may check this by first multiplying out the factors as

$$y = 2x^2 + 2x - 12$$

Then, by the simple power rule, $dy/dx = 4x + 2$.

Example 2   Differentiate $s(t) = (3t^2 - 2t + 1)(t^3 + 4)$ using the product rule.

Solution

$$s'(t) = \overset{\substack{\textit{First} \\ \textit{function}}}{(3t^2 - 2t + 1)}\overset{\substack{\textit{Derivative} \\ \textit{of second} \\ \textit{function}}}{(3t^2)} + \overset{\substack{\textit{Second} \\ \textit{function}}}{(t^3 + 4)}\overset{\substack{\textit{Derivative} \\ \textit{of first} \\ \textit{function}}}{(6t - 2)}$$
$$= 15t^4 - 8t^3 + 3t^2 + 24t - 8$$

Let us see how we can apply the product rule to the container design problem. Suppose we consider the product of the first two factors, $(28 - 2x)(28 - 2x)$, to be the function $f(x)$ and the last term, x, to be $g(x)$; that is,

$$V = \underbrace{[(28 - 2x)(28 - 2x)]}_{f(x)}\underbrace{[x]}_{g(x)}$$

To find $f'(x)$, we can apply the product rule to $f(x)$ alone. This gives the following:

<pre>
First   Derivative   Second   Derivative
function   of second   function   of first
            function   function
    ↓   ↓   ↓   ↓
</pre>

$$f'(x) = (28 - 2x)(-2) + (28 - 2x)(-2)$$
$$= -56 + 4x - 56 + 4x$$
$$= -112 + 8x$$

Now we apply the product rule to V itself:

$$\frac{dV}{dx} = f(x)g'(x) + g(x)f'(x)$$

$$= [(28 - 2x)(28 - 2x)](1) + (x)(-112 + 8x)$$
$$= (28 - 2x)^2 - 112x + 8x^2$$
$$= 12x^2 - 224x + 784$$

This agrees with the result we obtained by first multiplying out the factors that comprise V. We have thus shown how the product rule can be extended to three or more functions.

Example 3 Find the derivative of the following using the product rule:

$$y = (x + 3)(x^2 - 2x + 1)(x^3 - 3x^2)$$

Solution Since this is the product of three factors, we use the product rule twice. In this case, we will associate the last two factors:

$$y = \underbrace{(x + 3)}_{\substack{First \\ function}}\underbrace{[(x^2 - 2x + 1)(x^3 - 3x^2)]}_{\substack{Second \\ function}}$$

$$y' = (x + 3)[(x^2 - 2x + 1)(3x^2 - 6x) + (x^3 - 3x^2)(2x - 2)]$$
$$+ [(x^2 - 2x + 1)(x^3 - 3x^2)](1)$$
$$= (x + 3)(5x^4 - 8x^3 - 3x^2 - 6x) + x^5 - 5x^4 + 7x^3 - 3x^2$$
$$= 6x^5 + 2x^4 - 20x^3 - 18x^2 - 18x \qquad ∎$$

Example 4 The change in body temperature (in degrees Fahrenheit) after x milligrams of a drug is injected into the body is given by

$$T = x^2\left(1 - \frac{x}{6}\right) \qquad x > 0$$

When is the change in temperature a maximum?

Solution   Using the product rule, the derivative of T with respect to x is

$$\frac{dT}{dx} = x^2 \left(-\frac{1}{6}\right) + \left(1 - \frac{x}{6}\right)(2x)$$

$$= \frac{-x^2}{6} + 2x - \frac{2x^2}{6}$$

$$= \frac{-3x^2}{6} + 2x$$

$$= \frac{-x^2}{2} + 2x$$

Setting this expression equal to 0 yields

$$\frac{-x^2}{2} + 2x = 0$$

$$x\left(\frac{-x}{2} + 2\right) = 0$$

The solutions to this equation are $x = 0$ and $x = 4$. Checking the second derivative,

$$\frac{d^2T}{dx^2} = -\frac{2x}{2} + 2$$

$$= -x + 2$$

we find that a relative maximum occurs when $x = 4$, since the second derivative has a value of $-4 + 2 = -2 < 0$ at this point. This relative maximum is an absolute maximum, since the function has only one critical point for positive x. ∎

Example 5   Use the product rule to differentiate

$$h(x) = \frac{x^2 - 3x + 2}{\sqrt[3]{x}}$$

Solution   First, we rewrite this as

$$h(x) = (x^2 - 3x + 2)(x^{-1/3})$$

Then,

$$\frac{dh}{dx} = (x^2 - 3x + 2)\left(-\frac{1}{3}x^{-4/3}\right) + x^{-1/3}(2x - 3)$$

$$= \frac{5}{3}x^{2/3} - 2x^{-1/3} - \frac{2}{3}x^{-4/3}$$

$$= \frac{5\sqrt[3]{x^2}}{3} - \frac{2}{\sqrt[3]{x}} - \frac{2}{3\sqrt[3]{x^4}}$$ ∎

In Example 5, we converted a ratio of functions to a product in order to apply the product rule. Sometimes it is not easy or desirable to do this. We next present a rule for differentiating a quotient directly. This is called the *quotient rule* for derivatives.

THEOREM
Quotient Rule for Derivatives

Let $f(x)$ and $g(x)$ be functions with derivatives $f'(x)$ and $g'(x)$, respectively. Then, for $g(x) \neq 0$,

$$\frac{d}{dx}\left[\frac{f(x)}{g(x)}\right] = \frac{g(x)f'(x) - f(x)g'(x)}{[g(x)]^2}$$

This is a pretty complicated expression. It is best remembered by saying that *the derivative of the quotient is equal to the denominator function times the derivative of the numerator function, minus the numerator function times the derivative of the denominator function, all divided by the square of the denominator function.* Like the product rule, if you repeat this to yourself each time you use it, you will quickly commit it to memory. A proof of this theorem is also given in the supplement to this chapter.

Let us illustrate this rule with the function given in Example 5:

$$h(x) = \frac{x^2 - 3x + 2}{x^{1/3}}$$

$$h'(x) = \frac{(x^{1/3})(2x - 3) - (x^2 - 3x + 2)(\frac{1}{3}x^{-2/3})}{(x^{1/3})^2}$$

Factoring out $x^{-2/3}$ in the numerator, we get

$$h'(x) = \frac{x^{-2/3}[x(2x - 3) - (x^2 - 3x + 2)(\frac{1}{3})]}{x^{2/3}}$$

$$= x^{-4/3}(2x^2 - 3x - \tfrac{1}{3}x^2 + x - \tfrac{2}{3})$$
$$= x^{-4/3}(\tfrac{5}{3}x^2 - 2x - \tfrac{2}{3})$$
$$= \tfrac{5}{3}x^{2/3} - 2x^{-1/3} - \tfrac{2}{3}x^{-4/3}$$
$$= \frac{5\sqrt[3]{x^2}}{3} - \frac{2}{\sqrt[3]{x}} - \frac{2}{3\sqrt[3]{x^4}}$$

which agrees with the result obtained in Example 5.

Example 6   Use the quotient rule to find the derivative of

$$y = \frac{x - 1}{x + 1}$$

Solution   Again, state the rule as it is being applied:

$$\frac{dy}{dx} = \frac{(x + 1)(1) - (x - 1)(1)}{(x + 1)^2}$$

$$= \frac{x + 1 - x + 1}{(x + 1)^2}$$

$$= \frac{2}{(x + 1)^2}$$

Example 7
Learning Curves
One of the earliest models for learning was proposed by L. L. Thurstone in 1919. This is given by the equation

$$y = \frac{ax}{x + b}$$

where y = number of successful acts accomplished when a task (for example, a multiple-choice test) has been performed x times, a is a constant relating to the limit of accomplishment, and b is a constant describing the rate of learning. The derivative dy/dx is the marginal increase in successful acts as an additional attempt is made. Applying the quotient rule, we obtain

$$y' = \frac{(x + b) \cdot a - ax \cdot 1}{(x + b)^2}$$

$$= \frac{ax + ab - ax}{(x + b)^2}$$

$$= \frac{ab}{(x + b)^2}$$

Example 8
Find the derivative of

$$y = \frac{1}{x^2 + 1}$$

Solution
$$\frac{dy}{dx} = \frac{(x^2 + 1)(0) - (1)(2x)}{(x^2 + 1)^2}$$

$$= \frac{-2x}{(x^2 + 1)^2}$$

Example 8 is an illustration of a particular case of the quotient rule, namely one in which the quotient is a *reciprocal*, that is, $f(x) = 1$. We may state this as a separate result:

THEOREM
Reciprocal Rule for
Derivatives

> Let $g(x)$ be a function with derivative $g'(x)$. Then, for $g(x) \neq 0$,
>
> $$\frac{d}{dx}\left[\frac{1}{g(x)}\right] = \frac{-g'(x)}{[g(x)]^2}$$

This theorem follows immediately from the quotient rule if we let $f(x) = 1$, noting that $f'(x) = 0$.

Example 9
Use the reciprocal rule to find the derivative of

$$y = \frac{4}{3x^4 + x}$$

Solution   Since the derivative of a constant times a function equals the constant times the derivative of the function, we have

$$y' = 4 \frac{d}{dx}\left(\frac{1}{3x^4 + x}\right)$$

$$= 4\left[\frac{-12x^3 - 1}{(3x^4 + x)^2}\right]$$

$$= \frac{-48x^3 - 4}{9x^8 + 6x^5 + x^2}$$ ∎

Example 10   The photosynthetic rate R of a leaf is related to light intensity x by the function
Photosynthesis

$$R = \frac{1}{4 + (0.5/x)}$$

How does the photosynthetic rate change as a function of light intensity?

Solution   The rate of change is given by the derivative dR/dx. We will find this by the reciprocal rule with $g(x) = 4 + (0.5/x) = 4 + 0.5x^{-1}$. Since $g'(x) = -0.5x^{-2} = -0.5/x^2$, the derivative of R with respect to x is given by

$$\frac{dR}{dx} = \frac{-(-0.5/x^2)}{[4 + (0.5/x)]^2}$$

$$= \frac{0.5/x^2}{[4 + (0.5/x)]^2}$$

$$= \frac{0.5}{x^2[4 + (0.5/x)]^2}$$

$$= \frac{0.5}{16x^2 + 4x + 0.25}$$

The function can also be written as

$$R = \frac{1}{4 + (0.5/x)} = \frac{x}{4x + 0.5}$$

We leave it to the reader to verify that the quotient rule applied to this expression yields the same result for dR/dx as that found above.   ∎

APPLICATIONS TO CURVE SKETCHING

In Chapter 3 we showed how useful calculus is in sketching graphs of functions. However, we have not yet considered graphs involving quotients of functions. We have just seen how to differentiate such functions using the quotient and reciprocal rules. Now we will use these rules to sketch graphs of functions involving quotients. However, before doing this, we introduce an important feature of such graphs, namely, asymptotes.

A **vertical asymptote** is a line $x = a$ such that $f(x)$ approaches positive or negative infinity as x approaches a from either the left or the right. (A formal treatment of infinite limits can be found in the supplement to Chapter 2.) For example, consider the function $y = 1/(x - 1)$. The function is not defined when $x = 1$. Let us choose values of x close to, but to the left of, 1:

x	y
0.900	−10
0.990	−100
0.999	−1000

We see that y becomes very large in the negative direction; that is, y approaches negative infinity. If we choose values of x close to, but to the right of 1, we obtain:

x	y
1.100	10
1.010	100
1.001	1000

In this case, y approaches positive infinity. Therefore, the line $x = 1$ is a vertical asymptote.

A vertical asymptote is *odd* if the function approaches positive infinity from one side of the asymptote and negative infinity from the other side. If the function approaches positive infinity from both sides or negative infinity from both sides of a vertical asymptote, we say that the asymptote is *even*.

For the function $y = 1/(x - 1)$, the line $x = 1$ is an odd asymptote. An example of an even asymptote is the line $x = 1$ for the function $y = 1/(x - 1)^2$. As x gets close to 1 from either side of 1, the value of the function approaches positive infinity. Figure 4-3 illustrates both of these cases.

Figure 4-3
Vertical asymptotes

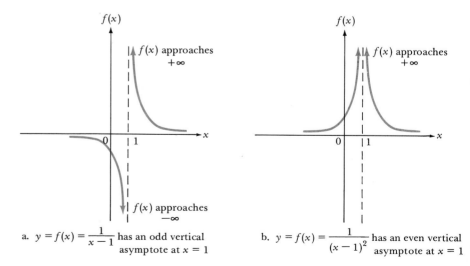

a. $y = f(x) = \dfrac{1}{x - 1}$ has an odd vertical asymptote at $x = 1$

b. $y = f(x) = \dfrac{1}{(x - 1)^2}$ has an even vertical asymptote at $x = 1$

Vertical asymptotes occur where a function would have a zero denominator and a nonzero numerator. To identify vertical asymptotes, first check for values of x that would cause the denominator to be zero, and then make sure that the numerator is not also zero for those values, assuming common factors have been canceled. To determine whether a vertical asymptote is even or odd, choose some points on either side of the asymptote and examine the behavior of the function.

Example 11   Determine whether the function

$$g(x) = \frac{x - 1}{(x + 2)^3}$$

has any vertical asymptotes.

Solution   The denominator is 0 when $x = -2$. At this value, the numerator is equal to -3; hence, the line $x = -2$ is a vertical asymptote. When x is slightly smaller than -2, the numerator is approximately -3 and the denominator is negative since the cube of a negative number is negative. Therefore, for $x < -2$, $g(x)$ is positive, and the function will approach positive infinity as x gets close to -2 from the left. If x is slightly larger than -2, the numerator is approximately -3 but the denominator is positive. Thus, as x approaches -2 from the right, the function will approach negative infinity. The asymptote $x = -2$ is therefore an odd asymptote.   ∎

A **horizontal asymptote** is a line $y = b$ such that $f(x)$ approaches b as x approaches either positive or negative infinity. For example, if $f(x) = 4/(x^2 + 1)$, we see that the denominator gets arbitrarily large as $|x|$ becomes very large. Since the numerator is constant, the value of $f(x)$ will get close to 0. Thus, the line $y = 0$ is a horizontal asymptote. The graph of $f(x) = 4/(x^2 + 1)$ is shown in Figure 4-4.

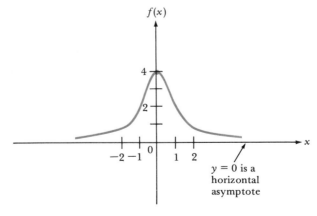

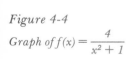

Figure 4-4

Graph of $f(x) = \dfrac{4}{x^2 + 1}$

Example 12
Average Cost   The cost of production often takes the form

$$C(x) = a + cx$$

where x is the number of units produced; a is the fixed cost of production (accounting for machine setup, for instance); and c is the unit cost to produce each item, or variable cost. The *average cost per unit* is $C(x)/x = (a/x) + c$. What happens as the number of units produced gets large?

Solution   The term a/x will approach 0 and c will remain constant. Therefore, the average cost per unit will approach the variable cost. The line $y = c$ is a horizontal asymptote for the average cost function. This is illustrated in Figure 4-5.

Figure 4-5

Average cost function,

$$f(x) = \frac{C(x)}{x} = \frac{a}{x} + c$$

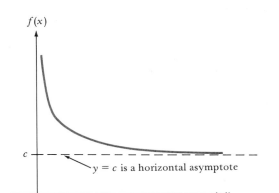

$f(x)$

c

$y = c$ is a horizontal asymptote

x

■

Example 13   Find all asymptotes for the function

$$y = 5 + \frac{3}{x}$$

Solution   First check for vertical asymptotes. When $x = 0$, the denominator of the second term is 0 and the numerator is 3. Therefore, the line $x = 0$ must be a vertical asymptote. As x gets close to 0 from the right, the term $3/x$ approaches positive infinity and, therefore, y approaches positive infinity. As x gets close to 0 from the left, $3/x$ approaches negative infinity and the function will also approach negative infinity. Thus, the line $x = 0$ is an odd asymptote.

Figure 4-6

Graph of $y = 5 + \dfrac{3}{x}$

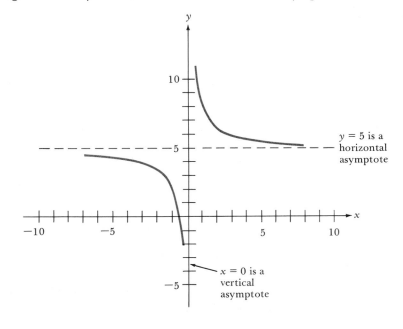

y

10

$y = 5$ is a horizontal asymptote

5

-10 -5 5   10 x

-5

$x = 0$ is a vertical asymptote

As x itself gets very large, the term $3/x$ tends toward 0 and y approaches the value 5. Also, if x approaches negative infinity, $3/x$ will again approach 0 (from the negative side, however) and y will approach 5. Thus, the line $y = 5$ is a horizontal asymptote. This is illustrated in Figure 4-6. ■

To summarize the procedure to identify asymptotes of graphs of functions: (1) Seek points at which the denominator is 0 and the numerator is nonzero; these yield vertical asymptotes. (2) To check for horizontal asymptotes, examine the behavior of the function as the independent variable approaches positive or negative infinity. If the value of the function approaches a constant, then we have identified a horizontal asymptote.

Example 14   Sketch the graph of

$$f(x) = \frac{x^2}{x + 5}$$

Solution   Using the curve sketching techniques discussed in Chapter 3, we note that the y-intercept is $f(0) = 0$ and the x-intercept is $x = 0$. The function has an odd vertical asymptote at $x = -5$. To find relative optima, we find $f'(x)$ using the quotient rule:

$$f'(x) = \frac{(x + 5)(2x) - x^2(1)}{(x + 5)^2}$$

$$= \frac{2x^2 + 10x - x^2}{x^2 + 10x + 25}$$

$$= \frac{x^2 + 10x}{x^2 + 10x + 25}$$

Setting $f'(x) = 0$ yields

$$x^2 + 10x = 0$$
$$x(x + 10) = 0$$

Therefore, critical points occur at the roots of this equation, namely at $x = 0$ and $x = -10$.

The second derivative is

$$f''(x) = \frac{(x^2 + 10x + 25)(2x + 10) - (x^2 + 10x)(2x + 10)}{(x^2 + 10x + 25)^2}$$

$$= \frac{50x + 250}{(x^2 + 10x + 25)^2}$$

$$= \frac{50(x + 5)}{[(x + 5)^2]^2}$$

$$= \frac{50}{(x + 5)^3}$$

When $x = 0$, $f''(0) = \frac{250}{625} > 0$. Thus, $(0, 0)$ is a relative minimum point. When $x = -10$, $f''(-10) = -\frac{250}{625} < 0$, so we have a relative maximum at the point $(-10, -20)$.

Since $f''(x) = 50/(x + 5)^3$ is never 0 the graph of $f(x)$ has no inflection points. We may check the signs of $f'(x)$ and $f''(x)$ over the intervals between critical points and the asymptotes, as shown in the table.

	Interval			
	$(-\infty, -10)$	$(-10, -5)$	$(-5, 0)$	$(0, \infty)$
$f'(x)$	+	−	−	+
$f''(x)$	−	−	+	+

This tells us that $f(x)$ is increasing over the intervals $(-\infty, -10)$ and $(0, \infty)$, and decreasing over $(-10, -5)$ and $(-5, 0)$. Also, the graph is concave down over $(-\infty, -5)$ and concave up over $(-5, \infty)$. The graph of this function is shown in Figure 4-7.

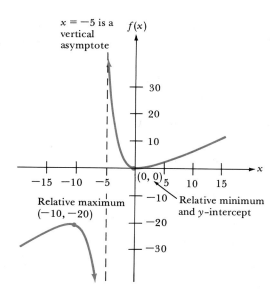

Figure 4-7

Graph of $f(x) = \dfrac{x^2}{x + 5}$

Example 15 Graph the function

$$y = \frac{2}{x^2 - 1}$$

Solution The y-intercept is $f(0) = -2$. Setting $y = 0$, we find that there are no x-intercepts; the curve does not cross the x-axis. Also, we see that the function is not defined when $x = 1$ or when $x = -1$. At these points we will have vertical asymptotes. The first derivative is

$$y' = \frac{-2(2x)}{(x^2 - 1)^2}$$

$$= \frac{-4x}{(x^2 - 1)^2}$$

Setting this equal to 0, we see that a critical point occurs at $x = 0$. Checking the second derivative, we find

$$y'' = \frac{(x^2 - 1)^2(-4) - (-4x)(2)(x^2 - 1)(2x)}{(x^2 - 1)^4}$$

$$= \frac{12x^2 + 4}{(x^2 - 1)^3}$$

At $x = 0$, $y'' = -4$; thus, the function has a relative maximum at $(0, -2)$. If we set $y'' = 0$, we see that there is no real solution. Hence, there are no points of inflection. Over the intervals $(-\infty, -1)$ and $(1, \infty)$, the second derivative is positive and the curve is concave up. Between $x = -1$ and $x = 1$, the second derivative is negative and the curve is concave down. Checking values of x on both sides of $x = -1$ and $x = 1$, we find that the function approaches positive infinity to the left of $x = -1$ and to the right of $x = 1$. As x gets close to -1 from the right or as x gets close to 1 from the left, the function approaches negative infinity. Both vertical asymptotes are therefore odd. Finally, as x approaches either positive or negative infinity, y approaches 0. Hence, $y = 0$ is a horizontal asymptote. The graph of this function is shown in Figure 4-8.

Figure 4-8

Graph of $y = \dfrac{2}{x^2 - 1}$

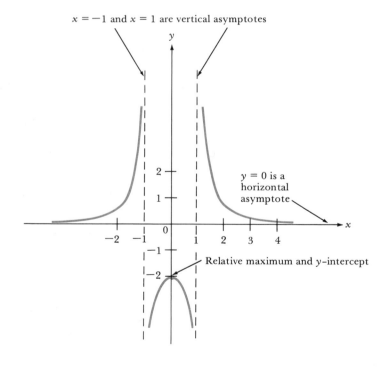

$x = -1$ and $x = 1$ are vertical asymptotes

$y = 0$ is a horizontal asymptote

Relative maximum and y-intercept

PROBLEMS (Section 4-1)

Find the derivative of each function using the product rule.

1. $f(x) = x^4(x + 3)$

2. $g(t) = 3t(t^2 + 5)$

3. $f(x) = (2x - 5)(4x + 1)$

4. $y = (x^2 - 2)(3x + 2)$

5. $f(p) = p^3(2p^2 + 3p - 1)$

6. $f(x) = (x + 6)(x^2 + 2x - 3)$

7. $y = (2x - 3)(2x^3 - x^2 + 1)$

8. $g(y) = (y^2 - 3y + 4)(y^3 + 2y^2 - 5)$

9. $f(z) = (z^2 - 3z + 2)(2z^3 - z^2 + 3)$

10. $y = \sqrt{x}(5x - 1)$

11. $f(x) = 3\sqrt{x}(x^3 + 7x^2)$

12. $f(x) = (x - 5)(x + 2)(x^2 - x + 6)$

Find the derivative of each function using the quotient rule.

13. $f(x) = \dfrac{1}{x^7 + 2}$

14. $y = \dfrac{x + 2}{x - 3}$

15. $g(t) = \dfrac{t^2}{t - 11}$

16. $f(x) = \dfrac{3x + 3}{1 - 4x}$

17. $f(x) = \dfrac{x^2 - 3x + 2}{x^7 - 2}$

18. $y = \dfrac{2x^2 + 3x}{x^2 + 9x - 2}$

19. $f(z) = \dfrac{\sqrt{z} + 5}{3z^2 + z + 6}$

20. $f(p) = \dfrac{3p^2 - 2p + 4}{p^2 + 3p - 1}$

21. $g(t) = \dfrac{t^2 + 5}{t(t + 6)}$

22. $y = \dfrac{3x^4 + 2x^2 + 7x - 5}{x^3 + 3x + 2}$

Find the derivative of each function using the reciprocal rule.

23. $f(x) = \dfrac{1}{x^6 - 3}$

24. $f(x) = \dfrac{5}{2x + 1}$

25. $f(x) = \dfrac{3}{x^2 + 2x}$

26. $f(x) = \dfrac{-1}{\sqrt{x} + 4x^3}$

27. The revenue R from the sales of x units of a product is given by the function

$$R(x) = \frac{3x}{x^2 + 1}$$

Compute the marginal revenue function.

28. If $p = (x + 13)/(x + 3)$ is a demand function, find the rate of change of the price p with respect to the quantity x.

29. The total receipts (in dollars) from the sale of x cameras is given by

$$T(x) = 150x(1 - 0.0001x)$$

Use the product rule to find the rate at which the total receipts are changing when $x = 100$.

30. The sales s of a product is related to the daily advertising expenditure x (in dollars) by the formula

$$s = \frac{5000x}{x + 100}$$

Determine the sensitivity of sales to advertising by finding ds/dx. Interpret the result.

Determine all asymptotes for each function in Problems 31–35.

31. $y = 5 + \dfrac{3}{x}$

32. $y = 4 + \dfrac{6}{x-3}$

33. $f(x) = \dfrac{7}{(x-2)^2} + 9$

34. $f(x) = \dfrac{5}{x^2} - 10$

35. $y = 8 - \dfrac{4}{x^3}$

In Problems 36–43, determine whether each function has any vertical asymptotes. If so, identify them as even or odd.

36. $f(x) = \dfrac{x-3}{(x+5)^3}$

37. $f(x) = \dfrac{x+1}{(x-3)^3}$

38. $g(x) = \dfrac{x+4}{(x-2)^2}$

39. $g(x) = \dfrac{(x-6)^2}{(x+7)^3}$

40. $f(x) = \dfrac{2x}{(x-3)^4}$

41. $y = \dfrac{(x+1)^2}{x^2 + 4x + 4}$

42. $f(x) = \dfrac{(x-5)^2}{15}$

43. $y = \dfrac{(x+3)^2}{x+3}$

Determine whether each function has any horizontal asymptotes.

44. $f(x) = \dfrac{5}{2x^2 + 1}$

45. $f(x) = -\dfrac{10}{x^3 + 9}$

46. $g(x) = \dfrac{3}{(1/x^2) + 1}$

47. $y = \dfrac{8}{-2 + (1/x^2)}$

48. $f(x) = \dfrac{2}{4x^4 + x^2}$

49. $g(x) = -\dfrac{15}{\sqrt{3x^2 + 6}}$

Use the techniques of Section 3-7 to sketch the graphs of the following functions:

50. $y = \dfrac{x+3}{x-2}$

51. $y = \dfrac{x-1}{x+2}$

52. $y = \dfrac{x}{x^2 + 9}$

53. $y = \dfrac{x^2}{(x+1)^2}$

54. $y = \dfrac{x}{\sqrt{x^2 - 4}}$

55. Suppose p is a positive number. Use the reciprocal rule to differentiate $f(x) = x^{-p} = 1/x^p$, and verify that the result is the same as that obtained by the power rule.

4-2 THE GENERAL POWER RULE AND CHAIN RULE

In all the examples and problems you have seen thus far, the basic rule used in finding derivatives has been the power rule, namely, that the derivative of x^p is px^{p-1}. When faced with the problem of differentiating, say $y = (3x + 2)^2$, you have had to expand the expression as $y = 9x^2 + 12x + 4$ in order to apply the power rule. It seems that it would be useful to know how to differentiate $(3x + 2)^2$ directly without expanding. In other problems, for example, those involving a fractional exponent, it might not even be possible to simplify such an expression. Such a situation is given in the next example.

Example 16
Design of a Gas Pipeline

A natural gas pipeline is to be run from a supply terminal across a river that is 1000 yards wide to a factory that is 2500 yards downstream. The construction cost is $12 per yard underground and $18 per yard under water. How much of the pipe should be constructed under water and how much underground?

Solution

A diagram of this situation is given in Figure 4-9. Obviously, the pipeline must cross the water, but there are several different ways the crossing can be done. Since the costs of placing the pipeline under water and underground differ, it is not clear how to best do this. In Figure 4-9, we let the variable x represent the distance from the point directly across the river from the supply terminal to the point at which the pipe reaches the opposite shore. (Note that $0 \leq x \leq 2500$.) The length of pipe that is underground must then be $2500 - x$. The cost for the underground portion of the pipeline will therefore be $12(2500 - x)$ dollars.

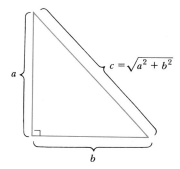

Figure 4-9
Pipeline problem

The length of the pipe under water can be found by geometry. Recall that the Pythagorean theorem states that for a right triangle, the length of the hypotenuse is equal to the square root of the sum of the squares of the other two sides. This is illustrated in Figure 4-10. Since the width of the river is 1000 yards, the length of the pipe under water will be $\sqrt{1000^2 + x^2}$. Hence, the cost of the

Figure 4-10
Illustration of the Pythagorean theorem

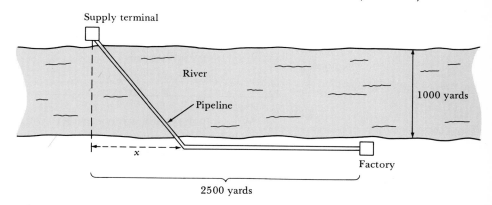

underwater portion of the pipeline will be $18\sqrt{1000^2 + x^2}$. The total cost is given by

$$C(x) = 12(2500 - x) + 18\sqrt{1000^2 + x^2}$$
$$= 30,000 - 12x + 18\sqrt{1,000,000 + x^2}$$

and we wish to minimize $C(x)$ on the interval $[0, 2500]$.

In order to minimize $C(x)$, we need to be able to differentiate this function. It is impossible to write $\sqrt{1,000,000 + x^2}$ as an ordinary polynomial. Thus, we cannot apply any of the differentiation rules we have previously learned. However, we can think of it as a generalization of the ordinary power function $f(x) = x^p$. We will complete this example after we develop a differentiation rule for general power functions.

We can express the cost of the underwater portion of the pipeline in Example 16 as a composition of two functions (see Algebra Review 15). Let $f(x) = 18\sqrt{x}$ and $g(x) = 1,000,000 + x^2$. Then $18\sqrt{1,000,000 + x^2} = f(g(x))$. Since $f(x)$ is essentially a power function with an exponent of $\frac{1}{2}$, we must differentiate a function of the form $[g(x)]^{1/2}$. The *general power rule for derivatives* applies to functions written in this form and can be stated as follows:

THEOREM
General Power Rule for Derivatives

> Suppose that $g(x)$ is a function with derivative $g'(x)$ and $f(x) = [g(x)]^p$. Then the derivative of $f(x)$ is
>
> $$f'(x) = p[g(x)]^{p-1}g'(x)$$

We apply the familiar power rule to the function $[g(x)]^p$ in the usual fashion: Multiply by the exponent and reduce the exponent by 1. However, the major difference is that we must also multiply by $g'(x)$. It is easy to see that this rule reduces to the ordinary power rule for simple functions like $y = x^p$. In such cases, $g(x) = x$, and the derivative of $g(x) = x$ is $g'(x) = 1$.

Example 17   Differentiate the function $y = (2 + 3x)^{100}$ using the general power rule.

Solution   Think of the nature of the composition of the functions. If $f(x) = x^{100}$ and $g(x) = 2 + 3x$, then $y = f(g(x)) = [g(x)]^{100}$. Using the general power rule, we have

$$y' = 100(2 + 3x)^{99}(3)$$
$$= 300(2 + 3x)^{99}$$

Example 18   Differentiate
$$y = \sqrt[5]{2x^2 + 3}$$

Solution   We rewrite this as $(2x^2 + 3)^{1/5}$. Letting $g(x) = 2x^2 + 3$, we have

$$\frac{dy}{dx} = \tfrac{1}{5}(2x^2 + 3)^{-4/5}(4x)$$

$$= \tfrac{4}{5}x(2x^2 + 3)^{-4/5}$$

Example 19   Find the derivative of

$$f(x) = (3x^2 - 2x)^4(x^3 + 3)^5$$

Solution   Using the product rule and the general power rule together,

$$f'(x) = (3x^2 - 2x)^4[5(x^3 + 3)^4(3x^2)] + (x^3 + 3)^5[4(3x^2 - 2x)^3(6x - 2)]$$
$$= (3x^2 - 2x)^4[15x^2(x^3 + 3)^4] + (x^3 + 3)^5[(24x - 8)(3x^2 - 2x)^3]$$ ■

Example 16
Continued   We can now use the general power rule to find the derivative of $C(x)$ in the pipeline problem of Example 16:

$$C(x) = 30,000 - 12x + 18(1,000,000 + x^2)^{1/2}$$

The derivative of the first two terms is -12. The derivative of the last term is

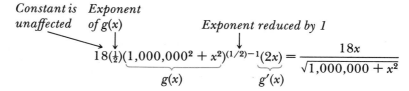

ALGEBRA REVIEW 15: COMPOSITION OF FUNCTIONS

If $f(x)$ is a function, we evaluate $f(a)$ by substituting a for x everywhere in the formula for $f(x)$. Often, it is useful to let a be another function of x. For example, to evaluate $f(g(x))$, we simply substitute $g(x)$ everywhere that x appears in the formula for $f(x)$. This is called a *composition of functions* $f(x)$ and $g(x)$ and can be thought of as a "function within a function." For example, if $f(x) = 3x^2 + 4x$ and $g(x) = 4x^3$, then $f(g(x))$ is found as follows:

$$f(g(x)) = f(4x^3)$$
$$= 3(4x^3)^2 + 4(4x^3)$$
$$= 3(16x^6) + 16x^3$$
$$= 48x^6 + 16x^3$$

We can also view this process in reverse. Given some function, can we write it as a composition of two functions? For instance, suppose $h(x) = (3x^2 + 4x + 1)^4$. If we let $f(x) = x^4$ and $g(x) = 3x^2 + 4x + 1$, then we see that $h(x) = f(g(x))$. Being able to view functions in this manner is a very useful technique in calculus.

EXERCISES*

a. Let $f(x) = 2x^2$ and $g(x) = 1/x$. Find $f(g(x))$ and $g(f(x))$.

b. Let $f(x) = x^4$ and $g(x) = \sqrt{x}$. Find $f(g(x))$ and $g(f(x))$.

c. Suppose $h(x) = 1/(x + 2)$. Write $h(x)$ as a composition of functions $f(g(x))$.

d. Consider the function $f(x) = (x^2 + 2x - 5)^3$. Write this as a composition of functions $h(g(x))$.

e. Let $h(x) = \sqrt[3]{2x + 1}$. Write this as a composition of functions $f(g(x))$.

---

*Answers to algebra review exercises can be found at the end of the answer section in the back of the book.

Therefore,

$$C'(x) = -12 + \frac{18x}{\sqrt{1,000,000 + x^2}}$$

We set this equal to 0 in order to find the optimum location for constructing the underwater portion of the pipeline:

$$-12 + \frac{18x}{\sqrt{1,000,000 + x^2}} = 0$$

$$-12\sqrt{1,000,000 + x^2} + 18x = 0$$

$$18x = 12\sqrt{1,000,000 + x^2}$$

$$324x^2 = 144(1,000,000 + x^2)$$

$$180x^2 = 144,000,000$$

$$x^2 = 800,000$$

$$x \approx 894.4 \text{ yards}$$

(Clearly, we do not consider the negative root.) We leave it to the reader to compare $C(894.4)$ with the values $C(0)$ and $C(2500)$ at the endpoints and thereby verify that $x \approx 894.4$ is indeed the value that yields the minimum cost. ∎

Example 20
Energy Expenditure   Scientists have estimated that the energy expenditure (in appropriate units) of a certain group of birds in flight is given by

$$E = \frac{0.35(v - 30)^2 + 100}{v}$$

where v is the speed (in miles per hour) of flight. What speed will minimize energy expenditure?

Solution   To take the derivative, both the quotient rule and the general power rule must be used:

$$\frac{dE}{dv} = \frac{v[2(0.35)(v - 30)] - [0.35(v - 30)^2 + 100](1)}{v^2}$$

$$= \frac{0.70v^2 - 21v - 0.35(v^2 - 60v + 900) + 100}{v^2}$$

$$= \frac{0.35v^2 - 215}{v^2}$$

Setting this expression equal to 0, we find

$$0.35v^2 - 215 = 0$$

$$v^2 = \frac{215}{0.35} \approx 614.29$$

$$v \approx 24.78 \text{ miles per hour}$$

Clearly, only the positive root is meaningful. When $v = 24.78$, the second derivative has a value of $430/(24.78)^3 > 0$. Therefore, a speed of 24.78 miles per hour will minimize the bird's energy expenditure. ∎

Example 21
Reaction Time

The reaction time to a stimulus in a psychology experiment was found to depend on the age x (in years) of the individual according to the formula

$$T = 0.05\sqrt{x^2 - 50x + 706} \qquad 16 \leq x \leq 65$$

At what age does the minimum reaction time occur?

Solution

Since the domain of T is defined by a closed interval, the minimum can occur at either a critical point or at an endpoint. Critical points are found by setting the first derivative equal to 0:

$$\frac{dT}{dx} = (\tfrac{1}{2})(0.05)(x^2 - 50x + 706)^{-1/2}(2x - 50)$$

$$= \frac{0.025(2x - 50)}{\sqrt{x^2 - 50x + 706}} = 0$$

Solving this equation, we see that the only critical point occurs when $2x - 50 = 0$, that is, at $x = 25$. Thus, we must evaluate the function at $x = 16$, 25, and 65. When $x = 16$, $T \approx 0.636$; for $x = 25$, $T \approx 0.450$; and when $x = 65$, $T \approx 2.05$. Therefore, the absolute minimum occurs at $x = 25$. ∎

THE CHAIN RULE

The general power rule provides a powerful method for computing derivatives of many common functions for which the ordinary power rule does not apply. The general power rule is a special case of a rule, called the *chain rule*, which applies to any composition of functions. Consider the composition of functions

$$y = f(g(x))$$

If we write $u = g(x)$, then by direct substitution, we have

$$y = f(u)$$

Since u is a function of x and y is a function of u, y is also a function of x. We would like to find a formula for dy/dx, the rate of change of y with respect to x.

Suppose x changes by an amount Δx. Then u will change by an amount Δu, where

$$\Delta u \approx \left(\frac{du}{dx}\right) \Delta x$$

This change Δu in u will give rise to a change Δy in y, where

$$\Delta y \approx \left(\frac{dy}{du}\right) \Delta u$$

Substituting from above, we find that a change of Δx in x results in a change Δy in y given by

$$\Delta y \approx \left(\frac{dy}{du}\right)\left(\frac{du}{dx}\right) \Delta x$$

This, of course, leads us to guess that the rate of change of y with respect to x, dy/dx, is given by

$$\frac{dy}{dx} = \left(\frac{dy}{du}\right)\left(\frac{du}{dx}\right)$$

This is, in fact, the case, as stated in the chain rule:

THEOREM
Chain Rule for Derivatives

Let $y = f(u)$ and $u = g(x)$. Then the derivative of y with respect to x is given by

$$\frac{dy}{dx} = \left(\frac{dy}{du}\right)\left(\frac{du}{dx}\right)$$
$$= f'(u)g'(x)$$
$$= f'(g(x))g'(x)$$

Let us see how this relates to the general power rule. The general power rule applies to functions of the form

$$y = f(g(x)) = [g(x)]^p$$

By defining $u = g(x)$, we have $y = f(u) = u^p$. Then

$$y' = f'(u)g'(x)$$
$$= p(u^{p-1})g'(x)$$

We must always replace u by $g(x)$ in order to get the answer solely in terms of x. This gives

$$y' = p[g(x)]^{p-1}g'(x)$$

Thus, when $f(u) = u^p$, the chain rule is the same as the general power rule. We illustrate the general power rule as an instance of the chain rule in the next example.

Example 22   Suppose $y = (4x^2 + 2x + 3)^4$. If we let $u = 4x^2 + 2x + 3$, then $y = u^4$. We take the derivatives of u with respect to x and y with respect to u, that is,

$$\frac{du}{dx} = 8x + 2$$

$$\frac{dy}{du} = 4u^3$$

Next, we substitute for u in the expression for dy/du:

$$\frac{dy}{du} = 4(4x^2 + 2x + 3)^3$$

Finally, we multiply du/dx by dy/du in order to obtain dy/dx:

$$\frac{dy}{dx} = \left(\frac{dy}{du}\right)\left(\frac{du}{dx}\right) = 4(4x^2 + 2x + 3)^3(8x + 2)$$

There are many other compositions of functions for which the general power rule does not apply. In such cases, the chain rule provides a simple method for differentiation. We shall see several examples where the chain rule is quite useful in Chapter 6. The chain rule is also crucial to the method of implicit differentiation and its application to related rate problems, which we study in the next section. In the following examples, we illustrate some applications of the chain rule.

Example 23   There exists a function $f(t)$ whose derivative is $f'(t) = 1/t$. What is dy/dt if $y = f(3t^2)$?

Solution   We do not even need to know what the actual function $f(t)$ is in order to determine dy/dt by the chain rule. Let $u = g(t) = 3t^2$. Then, clearly, $du/dt = 6t$. Also, for the function $y = f(u)$, we know that $dy/du = 1/u$. Therefore, by the chain rule,

$$\frac{dy}{dt} = \left(\frac{dy}{du}\right)\left(\frac{du}{dt}\right)$$

$$= \left(\frac{1}{u}\right)(6t)$$

$$= \left(\frac{1}{3t^2}\right)(6t)$$

$$= \frac{2}{t}$$ ∎

Example 24   There exists a function $f(x)$ such that $f'(x) = 1/(1 + x^2)$. Find dy/dx if $y = f(6x^3 + 2)$.

Solution   Let $u = g(x) = 6x^3 + 2$. Then $du/dx = 18x^2$. Since $dy/du = 1/(1 + u^2)$, we have

$$\frac{dy}{dx} = \left(\frac{dy}{du}\right)\left(\frac{du}{dx}\right)$$

$$= \left(\frac{1}{1 + u^2}\right)(18x^2)$$

$$= \frac{18x^2}{1 + (6x^3 + 2)^2}$$

$$= \frac{18x^2}{5 + 36x^6 + 24x^3}$$ ∎

PROBLEMS (Section 4-2)

Differentiate each function using the general power rule.

1. $y = (2x + 1)^{150}$

2. $f(x) = (5x^2 + 2x)^{500}$

3. $f(t) = (3t^2 - 5t + 6)^{10}$

4. $y = (5x^3 - 3x^2 + 9x - 5)^{25}$

5. $f(z) = \sqrt[3]{5z - 2}$

6. $y = \sqrt{(2x - 3)^3}$

7. $f(x) = (x^2 + 2)^{100}(x - 4)$

8. $f(p) = p^{2/3}(p^2 + 3)^8$

9. $y = (5x^2 - x)^3(x^4 + 2)^2$

10. $f(x) = \dfrac{x^2 - 1}{(x + 3)^4}$

Differentiate each function using the chain rule. Be explicit with all your derivatives and substitutions.

11. $y = (3x - 2)^{200}$

12. $f(t) = (4t^3 - 3t)^{75}$

13. $f(x) = (5x^2 + 2x - 7)^8$

14. $y = (x^4 - 2x^3 + 4x + 1)^{30}$

15. $f(p) = \sqrt[4]{6p - 5}$

16. $f(x) = \sqrt{3x^2 + x + 4}$

17. $f(z) = \sqrt[3]{(5z + 4)^2}$

18. $y = \dfrac{1}{(x - 15)^{25}}$

19. $f(x) = \dfrac{1}{(x^2 + 3x + 4)^4}$

20. $y = \left(\dfrac{2x}{x + 3}\right)^{10}$

21. Cost and activity are related by the function

$$C(a) = (30 - 2a + a^2)^{1/2}$$

Find the rate of change in cost, and then evaluate it at an activity level of $a = 10$.

22. A small manufacturing company estimates that the profit per unit is given by the function

$$P = \sqrt{0.05t^3 + 45.3t}$$

Find the marginal profit when $t = 2$ years.

23. A microwave company purchases microprocessors to install in their products. The cost per unit over the next few years is expected to be $C(t) = 5(1.6t + 10)^{3/2}$ dollars, where t is measured in years. Find the marginal cost when $t = 5$ years.

24. The reaction time in a simulated driving experiment was found to depend on the age x (in years) of the individual according to the formula

$$T = 0.02\sqrt{4000 + 40x - x^2} \qquad 16 \le x \le 60$$

At what age does the maximum reaction time occur?

25. There exists a function $f(x)$ whose derivative is $f'(x) = 3/(x + 1)$. Find dy/dx if $y = f(x^2 + x)$.

26. There is a function $f(t)$ such that $f'(t) = 3^t$. Find dy/dx if $y = f(8/x)$.

27. Suppose that $y = f(g(x))$, where $f(x)$ and $g(x)$ are differentiable functions. Find dy/dx at $x = 2$ if $g'(2) = -3$, $g(2) = 4$, and $f'(4) = 1.7$.

28. Let $y = f(g(x))$, where $f(x)$ and $g(x)$ are differentiable functions. Suppose that $dy/dx = 320$ at $x = 3$ and that $g'(3) = 4$ and $g(3) = 1$. What is $f'(1)$?

4-3 IMPLICIT DIFFERENTIATION

When we are given a function written as $y = f(x)$, we say that y is an *explicit* function of x. However, suppose that we are given the equation

$$x^2 - y^2 = 1$$

The graph of this equation is given in Figure 4-11. Note that x cannot assume any values in the interval $(-1, 1)$. (Why?) If we solve explicitly for y, we get two solutions:

$$y = +\sqrt{x^2 - 1} \qquad |x| \geq 1$$

and

$$y = -\sqrt{x^2 - 1} \qquad |x| \geq 1$$

We can think of the graph of the equation in Figure 4-11 as defining two functions, one represented by the portion of the graph above the *x*-axis, and the second corresponding to the portion of the graph below the *x*-axis. Thus, the equation $x^2 - y^2 = 1$ *implicitly* defines *y* as these two functions of *x*.

Let us take the derivative of each of these functions. For $y = +\sqrt{x^2 - 1} = (x^2 - 1)^{1/2}$, we have

$$\frac{dy}{dx} = \tfrac{1}{2}(x^2 - 1)^{-1/2}(2x)$$

$$= \frac{x}{\sqrt{x^2 - 1}}$$

Similarly, for $y = -\sqrt{x^2 - 1}$, we obtain

$$\frac{dy}{dx} = -\tfrac{1}{2}(x^2 - 1)^{-1/2}(2x)$$

$$= \frac{x}{-\sqrt{x^2 - 1}}$$

Notice that in both cases, the denominator is equal to *y*. Thus, we may write (in either case)

$$\frac{dy}{dx} = \frac{x}{y}$$

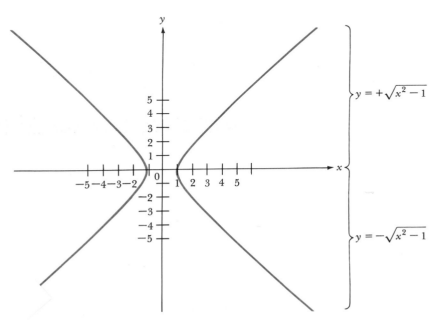

This is contrary to the usual way that we have been expressing derivatives. We defined the derivative of $y = f(x)$ as another function of x alone. Here, we are expressing the derivative in terms of both x and y. We must realize, however, that y is still a function of x, so that if we substitute an expression for y in terms of x, we would then have the derivative written only in terms of x.

Why did we do it this way? You might see one advantage to writing dy/dx as x/y. Even though the graph in Figure 4-11 represents two different functions, we do not need two different formulas for the derivatives. By knowing any point (x, y) on the curve, the slope of the tangent line will be given by x/y. For instance, the points $(2, \sqrt{3})$ and $(-3, -\sqrt{8})$ are on the curve. The slopes of the tangent lines to the curve at these points are given by $2/\sqrt{3}$ and $-3/(-\sqrt{8}) = 3/\sqrt{8}$, respectively. This is shown in Figure 4-12.

Figure 4-12

Slope of any point $= \dfrac{x}{y}$, $y \neq 0$

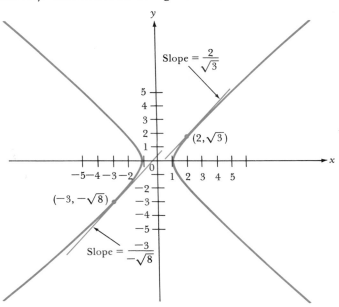

There is no obvious advantage to writing derivatives in this fashion if we first have to determine y as an explicit function of x every time. A technique called *implicit differentiation* allows us to find dy/dx directly from the original equation relating x and y.

Implicit differentiation works as follows: First, we differentiate each term on both sides of the equation individually with respect to x. Thus, for the example $x^2 - y^2 = 1$, we first consider the term x^2. The derivative, found by the power rule, is $2x$. Now consider $-y^2$. Remember that we are assuming that y is function of x, so we can think of this as $u = -y^2$ with $y = f(x)$. Therefore, u is composition of functions and the chain rule can be applied. We have

$$\frac{du}{dx} = \left(\frac{du}{dy}\right)\left(\frac{dy}{dx}\right)$$

Since $du/dy = -2y$, we have

$$\frac{du}{dx} = -2y\frac{dy}{dx}$$

Finally, differentiating the constant on the right side of the equation yields 0. The result of differentiating each term in the equation is

$$2x - 2y\frac{dy}{dx} = 0$$

We may now solve for dy/dx and obtain $dy/dx = x/y$. This agrees with our previous result.

Implicit differentiation is useful in many other situations where it is not even possible to solve an equation for y explicitly. Consider, for example, the equation

$$y^3x - x^2y = 2$$

We apply the same procedure, first differentiating each term on both sides of the equation individually with respect to x. The first term is y^3x. Since y is assumed to be a function of x, we have a product of two functions, y^3 and x, and the product rule must be used. This results in the following:

First function   *Derivative of second function*   *Second function*   *Derivative of first function*

$$(y^3)(1) + (x)\left(3y^2\frac{dy}{dx}\right) = y^3 + 3y^2x\frac{dy}{dx}$$

In a similar manner, for the term $-x^2y$, we obtain

$$(-x^2)\left(1\frac{dy}{dx}\right) + (y)(-2x) = -x^2\frac{dy}{dx} - 2xy$$

The derivative of the right-hand side of the equation is 0, and the result is

$$y^3 + 3y^2x\frac{dy}{dx} - x^2\frac{dy}{dx} - 2xy = 0$$

Solving for dy/dx yields

$$(3y^2x - x^2)\frac{dy}{dx} = 2xy - y^3$$

$$\frac{dy}{dx} = \frac{2xy - y^3}{3y^2x - x^2}$$

To summarize the method of implicit differentiation, we first differentiate each term on both sides of the equation *with respect to x*. For any terms involving y, we must use the chain rule. We collect all terms involving dy/dx, factor if necessary, and then solve for dy/dx.

Example 25   Find dy/dx for the equation $(3x + y)^2 = xy$ using implicit differentiation.

Solution   To differentiate the expression on the left, we must use the chain rule; for the term on the right, we use the product rule. Thus, differentiating both sides of the equation with respect to x, we obtain

$$2(3x + y)\left(3 + 1\frac{dy}{dx}\right) = x\frac{dy}{dx} + y(1)$$

$$18x + 6y + (6x + 2y)\frac{dy}{dx} = x\frac{dy}{dx} + y$$

This reduces to

$$(6x + 2y - x)\frac{dy}{dx} = -5y - 18x$$

$$\frac{dy}{dx} = \frac{-5y - 18x}{5x + 2y}$$

∎

Example 26   Use implicit differentiation to find the slope of the tangent line to the circle $x^2 + y^2 = 1$ at any point on the graph of the equation. What is the slope at the points $(1/2, -\sqrt{3}/2)$ and $(1/\sqrt{2}, 1/\sqrt{2})$? Show that your answer is correct using ordinary differentiation.

Solution   By implicit differentiation, we have

$$2x + 2y\frac{dy}{dx} = 0$$

$$2y\frac{dy}{dx} = -2x$$

$$\frac{dy}{dx} = \frac{-x}{y}$$

Therefore, the slope at $(1/2, -\sqrt{3}/2)$ is

$$-\frac{x}{y} = -\frac{1/2}{-\sqrt{3}/2} = \frac{1}{\sqrt{3}}$$

At $(1/\sqrt{2}, 1/\sqrt{2})$, the slope is

$$-\frac{1/\sqrt{2}}{1/\sqrt{2}} = -1$$

To see this using ordinary differentiation, we must first express y as a function of x:

$$y^2 = 1 - x^2$$
$$y = +\sqrt{1 - x^2} = (1 - x^2)^{1/2}$$

or

$$y = -\sqrt{1 - x^2} = -(1 - x^2)^{1/2}$$

The first function corresponds to the upper semicircle, and the second corresponds to the lower semicircle (see Figure 4-13). The derivatives of these functions are, respectively,

$$\frac{dy}{dx} = \frac{1}{2}(1 - x^2)^{-1/2}(-2x) = \frac{-x}{\sqrt{1 - x^2}}$$

and

$$\frac{dy}{dx} = -\frac{1}{2}(1-x^2)^{-1/2}(-2x) = \frac{x}{\sqrt{1-x^2}}$$

At the point $(1/2, -\sqrt{3}/2)$ on the lower semicircle we must use the second function. The slope is

$$\frac{dy}{dx} = \frac{1/2}{\sqrt{1-1/4}} = \frac{1}{\sqrt{3}}$$

At $(1/\sqrt{2}, 1/\sqrt{2})$, we use the first function since it lies on the upper semicircle:

$$\frac{dy}{dx} = \frac{-1/\sqrt{2}}{\sqrt{1-1/2}} = -1$$

These results are summarized in Figure 4-13.

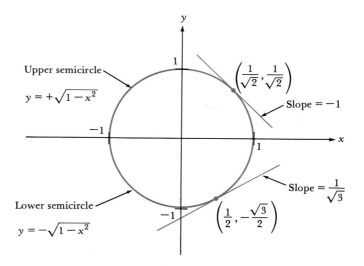

Figure 4-13
Graph of $x^2 + y^2 = 1$

Example 27
Conditioned Response in Psychology

For a group of mice, the average percentage p of conditioned responses resulting from x trials of a particular experiment is determined by the equation

$$p^2 = 64x - 64$$

What is the rate of change of p with respect to x, and how can this be interpreted?

Solution Since p is an implicit function of x, we have

$$2p \frac{dp}{dx} = 64$$

Therefore,

$$\frac{dp}{dx} = \frac{64}{2p} = \frac{32}{p}$$

The rate of change of the average percentage of conditioned responses is inversely proportional to the percentage p. That is, the higher the percentage response, the smaller the rate of change. ◼

RELATED RATES

In many situations, two variables are related to each other and also are functions of a third variable—for instance, time. As an example, suppose that profit is related to the rate of production x (in units per day) by

$$p = 25\sqrt{5x - x^2} - 20$$

Both x and p are implicit functions of time. Suppose the production rate is changing at a rate of 0.3 unit per day each day and we wish to determine the rate of change of profit with respect to time when $x = 4$ units per day.

The value of 0.3 unit per day each day represents the rate of change of the production rate with respect to time. Interpreted as a derivative, 0.3 unit per day each day is dx/dt. Since we seek the rate of change of profit with respect to time, we are trying to determine dp/dt. Using the chain rule, we have

$$\frac{dp}{dt} = \left(\frac{dp}{dx}\right)\left(\frac{dx}{dt}\right)$$

From the equation above, we can easily compute dp/dx. One way to do this is to differentiate the function p with respect to x directly. In order to emphasize that there is no one correct way of solving mathematical problems, we shall do it in a different fashion using implicit differentiation. Rewrite the equation as

$$p + 20 = 25\sqrt{5x - x^2}$$

Square both sides and obtain

$$(p + 20)^2 = 225(5x - x^2)$$

Now apply implicit differentiation:

$$2(p + 20)\frac{dp}{dx} = 225(5 - 2x)$$

$$\frac{dp}{dx} = \frac{225(5 - 2x)}{2(p + 20)}$$

When $x = 4$, $p = 25\sqrt{5(4) - 4^2} - 20 = 30$. Hence,

$$\frac{dp}{dx} = \frac{225[5 - 2(4)]}{2(30 + 20)} = \frac{-675}{100} = -6.75$$

We may now calculate dp/dt from the chain rule formula:

$$\frac{dp}{dt} = \left(\frac{dp}{dx}\right)\left(\frac{dx}{dt}\right)$$

$$\frac{dp}{dt} = (-6.75)(0.3) = -2.025$$

Therefore, the profit is decreasing at a rate of $2.025 per day.

Example 28
Radiation Therapy

The number of cancer cells N (in millions) destroyed by a radioactive treatment is given by

$$N = 2I^2$$

where I is the intensity of radioactivity. If the current intensity is 5 units and it is increasing at a rate of 0.4 unit per minute, what is the rate of cell destruction?

Solution

We seek the value of dN/dt and we know that $dI/dt = 0.4$. In other words, N and I are implicit functions of time. Using the chain rule, we have

$$\frac{dN}{dt} = \left(\frac{dN}{dI}\right)\left(\frac{dI}{dt}\right)$$

Since $dN/dI = 4I$, when $I = 5$, $dN/dI = 20$. Therefore,

$$\frac{dN}{dt} = 20(0.4) = 8 \text{ million cells per minute}$$

Example 29
Heat Loss

The amount of heat lost from the body depends on the external temperature. Physiologists have determined that the heat loss (in calories per minute) for a certain group of athletes is

$$L = 5 - 0.025T - \tfrac{1}{900}T^2 \qquad 30 \le T \le 80$$

where T is the external temperature (in degrees Fahrenheit). If the temperature is decreasing at a rate of 1°F per minute, how fast is the rate L changing when $T = 40°$F?

Solution

Since T is a function of time t, we have

$$\frac{dL}{dt} = -0.025\frac{dT}{dt} - \tfrac{1}{450}T\frac{dT}{dt}$$

Since dT/dt is given as -1, we have

$$\frac{dL}{dt} = -0.025(-1) - \tfrac{1}{450}(40)(-1)$$

$$\approx 0.1139$$

Thus, the heat loss is increasing by about 0.1139 calorie per minute.

Example 30
Relative Motion

A football is thrown at a speed of 6 yards per second toward a position 18 yards downfield from the quarterback to a receiver cutting across the field at a right angle to the throw. At the time of release, the receiver is 9 yards from the intended point of reception and running at a rate of 3 yards per second. How fast is the ball approaching the receiver after 2 seconds?

Solution

Figure 4-14 illustrates this situation. The distance between the receiver and the ball is given by the diagonal z. By the Pythagorean theorem, $z^2 = x^2 + y^2$. The speeds of the receiver and the ball are dx/dt and dy/dt. We seek dz/dt. Using

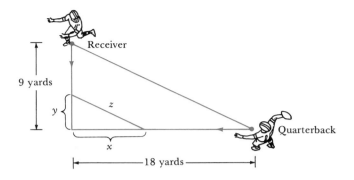

Figure 4-14

implicit differentiation,

$$2z \frac{dz}{dt} = 2x \frac{dx}{dt} + 2y \frac{dy}{dt}$$

$$\frac{dz}{dt} = \frac{x}{z} \frac{dx}{dt} + \frac{y}{z} \frac{dy}{dt}$$

After 2 seconds, the ball is 6 yards away from the intended point of reception, and the receiver is 3 yards away. Therefore, $x = 6$ and $y = 3$. This yields $z = \sqrt{45}$. Hence,

$$\frac{dz}{dt} = \left(\frac{6}{\sqrt{45}} \right) (6) + \left(\frac{3}{\sqrt{45}} \right) (3)$$
$$\approx 6.72 \text{ yards per second}$$ ∎

The procedure for solving related rate problems may be summarized as follows:

Solving Related Rate Problems

1. Draw a picture if possible, identifying and labeling all relevant variables and constants.
2. Find an equation relating the relevant dependent variables.
3. Differentiate implicitly with respect to the independent variable (usually time).
4. Substitute the appropriate values and solve for the required rate.

PROBLEMS (Section 4-3)

Find dy/dx by implicit differentiation in Problems 1–12.

1. $x^2 + y^2 = 4$

2. $x^2 = y^2 + 16$

3. $x^2 y = 20$

4. $2xy + y^2 = 9$

5. $5xy^2 + 4y + 1 = 0$

6. $3x^2 - 4xy = 10$

7. $3x^2 + 2xy + y^2 = 6$

8. $7x^2 = 3y^2 + 4xy$

9. $x^3 - 5y^2 = 10$

10. $xy^4 - x^2 - 3 = 0$

11. $x^2y + y^3 = 5$

12. $\sqrt{x} + \sqrt{y} = 9$

13. Find an equation of the tangent line to the curve $x^2 + 3y^2 = 12$ at the point $(1, 2)$. [*Hint:* Implicit differentiation will give you the slope, and substitution of the point will help you write the equation.]

14. Find an equation of the tangent line to the curve $x^2y^3 = 27$ at the point $(1, 3)$.

15. A demand function is given by the equation $p = 300 - \sqrt{q}$. Find the rate of change of q with respect to p when $q = 100$.

16. A window washer is on top of a ladder 25 feet long that is leaning against a vertical wall. If the bottom of the ladder slips away from the wall at the rate of 2 inches per minute, how fast is the ladder sliding down the wall when the bottom is 20 feet from the wall? [*Hint:* Let x be the height the ladder reaches and y be the distance from the foot of the ladder to the wall. Then $x^2 + y^2 = (25)^2$. What is dx/dt?]

17. A circular oil slick spreads out from a crippled supertanker so that its radius is increasing at a rate of 10 miles per hour. At what rate is the area of the slick changing when the radius is 2 miles?

18. A spherical balloon is being inflated at a rate of 3 cubic feet per minute. How fast is the radius increasing when the radius is 10 feet? [*Hint:* Recall that the volume V of a sphere of radius r is given by $V = \frac{4}{3}\pi r^3$.]

19. An ice cube is melting at a rate of 5 cubic centimeters per minute. How fast is the length of its side decreasing when the volume of the cube is 1000 cubic centimeters?

20. A man is standing $\frac{1}{2}$ mile from a railroad track as a train goes by at 60 miles per hour. When the front of the train is 1 mile away from the point on the track nearest to the man, at what speed is the front of the train moving away from him?

21. The annual consumption of poultry, p (in pounds), is related to annual income I by the equation

$$p = 85 \sqrt{\frac{20,000}{I}}$$

If a person's income is \$30,000 and is increasing at an annual rate of 5%, at what rate is p changing? That is, what is dp/dt?

22. A product currently sells for \$2.50 and has a demand function $D = 8000/P$. It is found that manufacturing costs are increasing at an annual rate of 12%. The company wishes to increase the selling price at the same rate. At what rate will the demand change?

23. Labor productivity is measured by the amount of output per labor hour. Suppose that output O is related to total labor hours L by the equation

$$O = 0.4L^2 + 2L - 4$$

If $L = 40,000$ and is increasing by 2% per year, at what rate is output changing?

4-4 CONCLUSION: SOLVING THE TRAFFIC FLOW PROBLEM

In the Introduction to this chapter, the flow of traffic during busy periods in Britain's road tunnels was considered. A model for the flow rate of cars was developed and given by

$$F(v) = \frac{1.467v}{v + v^2/20 + 12}$$

We seek to find the value of v (the speed of the vehicles) that will maximize the flow rate. We can now apply the quotient rule introduced in Section 4-1 to differentiate $F(v)$:

$$F'(v) = \frac{(v + v^2/20 + 12)(1.467) - (1.467v)(1 + 2v/20)}{(v + v^2/20 + 12)^2}$$

This expression is equal to 0 when the numerator is equal to 0. Therefore, we have

$$\left(v+\frac{v^2}{20}+12\right)(1.467)-(1.467v)\left(1+\frac{2v}{20}\right)=0$$

$$v+\frac{v^2}{20}+12-v-\frac{2v^2}{20}=0$$

$$-\frac{v^2}{20}+12=0$$

Solving for v^2 yields

$$v^2=240$$
$$v\approx15.49$$

We leave it to the reader to verify that $F'(15)>0$ and $F'(16)<0$, and hence, F has a relative maximum at $v\approx15.49$. The behavior of F for v near 0 or v very large shows that the flow rate is maximized when the speed through the tunnels is about $15\frac{1}{2}$ miles per hour.

4-5 CHAPTER REVIEW

1. The product rule is used to differentiate the product of two functions:

$$\frac{d}{dx}[f(x)g(x)]=f(x)g'(x)+g(x)f'(x)$$

2. The quotient rule is used to differentiate the quotient of two functions:

$$\frac{d}{dx}\left[\frac{f(x)}{g(x)}\right]=\frac{g(x)f'(x)-f(x)g'(x)}{[g(x)]^2}$$

3. The reciprocal rule is used to differentiate the reciprocal of a function:

$$\frac{d}{dx}\left[\frac{1}{g(x)}\right]=-\frac{g'(x)}{[g(x)]^2}$$

4. The general power rule applies to functions of the form $f(x)=[g(x)]^p$:

$$f'(x)=p[g(x)]^{p-1}g'(x)$$

5. The general power rule is a special case of the chain rule. The chain rule applies to any composition of functions $y=f(g(x))$. If $u=g(x)$, then $y=f(u)$ and we have

$$\frac{dy}{dx}=\left(\frac{df}{du}\right)\left(\frac{du}{dx}\right)$$

6. Implicit differentiation is a method of differentiating functions that are implicitly defined by an equation in x and y. We differentiate each term on both sides of the equation with respect to x, applying the chain rule to terms involving y, and then solve for dy/dx. This method is especially useful when it is not possible to solve an equation for y explicitly.

7. The chain rule is also useful in solving related rate problems, which occur when two variables are each related to one another and are also functions of a third variable (for example, time). If $y=f(x)$ and x is a function of t, then y is also a function of t. The chain rule implies that

$$\frac{dy}{dt}=\left(\frac{dy}{dx}\right)\left(\frac{dx}{dt}\right)$$

REVIEW PROBLEMS (Section 4-5)

Find the derivative of each function using the product rule.

1. $f(x)=4x(x^2+7)$

2. $f(x)=(5x+3)(3x^2-4)$

3. $y=(2\sqrt{x}+1)(4x^3+6x^2-x)$

4. $f(x)=\left(x^2-\frac{4}{x}+2\right)\left(2x^2+3x-\frac{8}{x^4}\right)$

Find the derivative of each function using the quotient rule.

5. $f(x) = \dfrac{x^2 - 3}{x^5}$

6. $y = \dfrac{5x^2 + 4}{4x^2 - 3x + 2}$

7. $f(p) = \dfrac{3p^2 - 5p + 3}{p^2 - 4p - 7}$

8. $y = \dfrac{20 + \sqrt{x}}{2x^2 + 3x + 1}$

Find the derivative of each function using both the product rule and the quotient rule. Show that the results are the same. Which method seems easier?

9. $f(x) = \dfrac{2x^3}{3x^2 + 1}$

10. $y = \dfrac{4x^2 - 2x + 5}{x^6}$

11. $g(t) = \dfrac{5t^2 - 4}{2t + 11}$

12. $y = \dfrac{2\sqrt{x} - 4}{x^2 + 6x + 2}$

13. The revenue R from the sales of a new product is given by the function

$$R(x) = \frac{5x}{2x^2 + 3}$$

Compute the marginal revenue.

14. The total cost for t tennis rackets is given by

$$C(t) = 125t(3 - 0.002t)$$

Find the marginal cost.

Differentiate each function using the general power rule.

15. $f(x) = (7x^2 + 4)^{60}$

16. $y = (2x^3 - 4x^2 - x)^{100}$

17. $f(x) = \left(3x^2 + 5x + \dfrac{1}{x}\right)^{205}$

18. $f(z) = \sqrt[3]{4z + 3}$

19. $y = \dfrac{7}{(x^5 - 6)^{10}}$

20. $f(p) = 3p^3(p^3 - 4p)^6$

Differentiate each function using the chain rule, and explain how the chain rule applies.

21. $f(x) = (2x^2 + 3)^{15}$

22. $y = (4x^3 - 5x)^{40}$

23. $f(t) = (2t^3 + 4t^2 - 11)^{13}$

24. $y = (6t^3 - 8t^2 + 15t)^{85}$

25. $f(x) = \sqrt[4]{5x^2 + x + 2}$

26. A record album company estimates that t years after an album is produced, the profit per album is given by the function $P = \sqrt{6t^3 - 2t}$, $0 \le t \le 5$. Find the marginal profit when $t = 3$ years.

27. The cost function $C(x)$ and revenue function $R(x)$ for a product are given by

$$C(x) = \sqrt{x + 2} \qquad R(x) = \frac{20x}{(x + 2)^2}$$

where x is the quantity produced and sold. Determine the marginal cost, marginal revenue, and the marginal profit.

Find dy/dx by implicit differentiation:

28. $y^2 + x^2 = 40$

29. $x^3y = 18$

30. $3xy + y^2 = 24$

31. $5x^2 - 2xy = 15$

32. $15x^2 = 5y^2 - 3xy$

33. $\sqrt[3]{x} + \sqrt{y} = 21$

34. Find the equation of the tangent line to the curve $x^2y^3 = 108$ at the point $(2, 3)$.

35. A man 6 feet tall is walking away from an illuminated flagpole 25 feet tall at a rate of 8 feet per second. How fast is the shadow of the man's head moving along the ground?

36. Differentiate the flow rate function in the Introduction to this chapter using the product rule and verify that it gives the same results as presented in Section 4-4.

Find all asymptotes for each function.

37. $y = 7 + \dfrac{2}{x}$

38. $y = \dfrac{3}{x} - 11$

39. $f(x) = 3 + \dfrac{4}{x - 2}$

40. $f(x) = \dfrac{6}{(x - 3)^2} + 8$

41. $y = \dfrac{4}{x^2} - 7$

42. $f(x) = 3 - \dfrac{2}{x^3}$

Sketch the graph of each function using techniques discussed in this chapter and in Chapter 3.

43. $y = \dfrac{4}{x^2 - 1}$

44. $y = \dfrac{8}{x^2 - 4}$

45. $y = \dfrac{x^2}{x + 4}$

46. $y = \dfrac{x^2}{x^2 + 3}$

47. $y = \dfrac{x - 1}{x + 3}$

48. The total cost C of producing a certain product is related to the quantity produced x by

$$C = 0.02x^3 - x^2 + 6x + 3$$

If $x = 10$ and if production is increasing at a rate of 20% per year, what is the rate of change in cost? That is, what is dC/dt?

49. Suppose that total output O is related to total labor hours L by the equation

$$O = 0.1L^2 + 4L - 6$$

If $L = 2500$ and is increasing by 3% per year, at what rate is output changing?

SOME PROOFS OF DIFFERENTIATION RULES

In Chapter 4 we stated three important theorems for differentiation: the product rule, quotient rule, and reciprocal rule. Using the limit theorems and the fact that differentiability implies continuity, as presented in the supplement to Chapter 2, we will prove the validity of these rules. All proofs are based on the fundamental limit definition of the derivative.

4S-1 PRODUCT RULE

We first prove the product rule: If $f(x)$ and $g(x)$ are functions with derivatives $f'(x)$ and $g'(x)$, respectively, then the derivative of $f(x)g(x)$ is $f(x)g'(x) + g(x)f'(x)$. Using the limit definition applied to the product $f(x)g(x)$, we have

$$\frac{d}{dx}[f(x)g(x)] = \lim_{h \to 0} \frac{f(x+h)g(x+h) - f(x)g(x)}{h}$$

If we subtract and add the term $f(x+h)g(x)$ to the numerator, we have

$$\frac{d}{dx}[f(x)g(x)] = \lim_{h \to 0} \frac{f(x+h)g(x+h) - f(x+h)g(x) + f(x+h)g(x) - f(x)g(x)}{h}$$

Now, using the limit theorems that state that the limit of a sum of terms is the sum of the limits and the limit of a product is the product of the limits, we may write the limit above as

$$\lim_{h \to 0} f(x+h) \frac{g(x+h) - g(x)}{h} + \lim_{h \to 0} g(x) \frac{f(x+h) - f(x)}{h}$$

$$= \lim_{h \to 0} f(x+h) \lim_{h \to 0} \frac{g(x+h) - g(x)}{h} + \lim_{h \to 0} g(x) \lim_{h \to 0} \frac{f(x+h) - f(x)}{h}$$

Since $f(x)$ is continuous, the limit of $f(x+h)$ as h approaches 0 is $f(x)$. Therefore, the last line is equal to

$$f(x)g'(x) + g(x)f'(x)$$

which proves the product rule given in Chapter 4.

4S-2 RECIPROCAL RULE

We prove the reciprocal rule next. Recall that this is stated as follows: If $g(x) \neq 0$, the derivative of $1/g(x)$ is equal to $-g'(x)/[g(x)]^2$. Using the limit definition of the derivative, we have

$$\frac{d}{dx}\left[\frac{1}{g(x)}\right] = \lim_{h \to 0} \frac{\dfrac{1}{g(x+h)} - \dfrac{1}{g(x)}}{h}$$

$$= \lim_{h \to 0} \frac{\dfrac{g(x) - g(x+h)}{g(x+h)g(x)}}{h}$$

$$= -\lim_{h \to 0}\left[\frac{g(x+h) - g(x)}{h} \cdot \frac{1}{g(x+h)g(x)}\right]$$

Using the fact that the limit of a product is equal to the product of the limits, this limit may be written as

$$\left[-\lim_{h \to 0}\frac{g(x+h) - g(x)}{h}\right]\left[\lim_{h \to 0}\frac{1}{g(x+h)g(x)}\right]$$

Since $g(x)$ is continuous, $g(x + h)$ approaches $g(x)$ as $h \to 0$, and we therefore have

$$\frac{d}{dx}\left[\frac{1}{g(x)}\right] = -g'(x)\frac{1}{[g(x)]^2}$$

$$= \frac{-g'(x)}{[g(x)]^2}$$

which completes the proof of the reciprocal rule.

4S-3 QUOTIENT RULE

Proof of the quotient rule follows simply from the product and reciprocal rules. We will first write the quotient $f(x)/g(x)$ as the product of $f(x)$ and $1/g(x)$ and apply the product rule:

$$\frac{d}{dx}\left[\frac{f(x)}{g(x)}\right] = \frac{d}{dx}\left[f(x)\frac{1}{g(x)}\right]$$

$$= f(x)\frac{-g'(x)}{[g(x)]^2} + \frac{1}{g(x)}f'(x)$$

$$= \frac{-f(x)g'(x)}{[g(x)]^2} + \frac{g(x)f'(x)}{[g(x)]^2}$$

$$= \frac{g(x)f'(x) - f(x)g'(x)}{[g(x)]^2}$$

This is the statement of the quotient rule. Note how we used the previous two theorems to prove this one.

5

INTEGRATION

INTRODUCTION

Sweigart and Associates, Architects, is in the process of designing a new office building. The design specifications must include the type and number of elevators in the building. Various types of elevators can be specified. In order to evaluate the performance of an elevator system in terms of its service to the building's occupants, the architects need to be able to compute the time it will take to travel between floors in the building. The building will be seven floors high with a distance of 14 feet between floors. Therefore, to travel from the first to the seventh floor, the elevator must go $6(14) = 84$ feet.

Suppose that the top speed of an elevator is 420 feet per minute. To travel 84 feet, it would take $\frac{84}{420} =$

0.2 minute, or 12 seconds. However, from your experience, you probably realize that an elevator does not start from a standstill and reach its maximum speed instantaneously; neither does it stop instantaneously. The speed gradually changes due to acceleration and deceleration. During acceleration, we will assume that the speed increases at a constant rate of 4 feet per second per second (ft/sec²). The relationship between speed v and time t (in seconds) is thus

$$v = 4t \text{ ft/sec}$$

until maximum speed is reached. Since the maximum speed is 420 feet per minute, or 7 feet per second, we may determine how long it takes to reach this speed by solving for t as follows:

$$7 = 4t$$

$$t = \tfrac{7}{4} = 1.75 \text{ seconds}$$

Similarly, if the deceleration rate is the same as the acceleration rate, the time to go from maximum speed to a stop will also be 1.75 seconds. Suppose that at $t = 8.25$, the speed begins to decrease at a constant rate of 4 feet per second until it reaches a value of 0 at time $t = 10$. This is described by a linear function passing through the points $(8.25, 7)$ and $(10, 0)$, which is given by $v = 40 - 4t$. Therefore, if the elevator starts at $t = 0$ and stops at $t = 10$, for instance, the speed will follow the function given below:

$$v(t) = \begin{cases} 4t & \text{if} \quad 0 \le t \le 1.75 \\ 7 & \text{if} \quad 1.75 < t \le 8.25 \\ 40 - 4t & \text{if} \quad 8.25 < t \le 10 \end{cases}$$

A graph of $v(t)$ is shown in Figure 5-1.

How far will the elevator actually travel during these 10 seconds? The basic relationship between speed, distance, and time is

$$\text{Distance} = \text{Speed} \times \text{Time}$$

When speed is constant, there is no problem. For

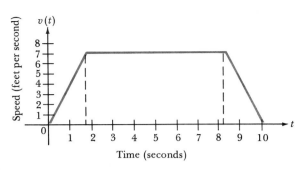

Figure 5-1
Graph of elevator speed

instance, in the interval $1.75 < t \le 8.25$, the distance traveled will be $7(8.25 - 1.75) = 45.5$ feet. However, during acceleration and deceleration, the speed is *constantly* changing. There does not seem to be any straightforward method of calculating the distance traveled.

Fortunately, calculus provides the answer. In this chapter, we introduce a new technique called *integration* which will allow us to answer questions such as this. We will also see that there is a very close relationship between integration and differentiation. Together, integration and differentiation are the essential elements of calculus.

5-1 ANTIDIFFERENTIATION

In the previous chapters you learned how to differentiate a variety of functions using several techniques. For example, using the simple power rule, you know that the derivative of the function $f(x) = x^2$ is $f'(x) = 2x$. Now consider this in reverse. Suppose we are given the function $g(x) = 2x$ and ask "What is a function whose *derivative* is $g(x)$?" From what we know, we would probably say that the answer must be $f(x) = x^2$. This is indeed correct, but it is not the only answer! Recall that the derivative of a constant is 0. Therefore, the functions $x^2 + 3$ and $x^2 - 5$ also have derivatives equal to $2x$. In general, if c is any constant, any function of the form $f(x) = x^2 + c$ would be a correct answer to the question posed above.

> The process of finding a function whose derivative is known is called **antidifferentiation**, and such a resulting function is called an **antiderivative.** That is, an antiderivative of a function $f(x)$ is a function $F(x)$ such that $F'(x) = f(x)$.

A common notation used to denote an antiderivative of a function $f(x)$ is $F(x)$.

Example 1   Find a function whose derivative is $f(x) = x^3$.

Solution   To solve this problem we must rely on our knowledge and experience in differentiation. When we differentiate a power function, we multiply the function by the exponent and reduce the exponent by 1. If this result is x^3, then the original function must have had an exponent of 4. Is x^4 the antiderivative of x^3? There is one sure way of finding out, and that is to take the derivative of x^4. The derivative of x^4 is $4x^3$. We have the correct exponent, but x^3 is multiplied by 4 instead of 1. Can you figure out what $F(x)$ should really be? There is an easy way to do this. We assume that $F(x)$ is multiplied by some unknown constant a. That is, we assume that $F(x) = ax^4$. If we differentiate $F(x)$, we know that we should get $f(x) = x^3$. Since the derivative of $F(x)$ is $4ax^3$, we know that $4a$ must equal 1. Therefore, $a = \frac{1}{4}$. An antiderivative of $f(x) = x^3$ is thus $F(x) = \frac{1}{4}x^4$. Since the derivative of a constant is 0, we find that any function of the form $F(x) = \frac{1}{4}x^4 + c$, where c is a constant, is an antiderivative of $f(x) = x^3$.   ∎

How do we know that we have found *all* possible antiderivatives of the function $f(x) = x^3$ in Example 1? That is, could there possibly be an antiderivative that does not have the form $F(x) = \frac{1}{4}x^4 + c$? The following theorems provide an answer to this question.

THEOREM

> If $F'(x) = 0$ for all x, then $F(x) = c$, where c is some constant.

It is easy to see why this theorem must be true. If $F'(x) = 0$ for all x, then the tangent line to the graph of $F(x)$ is always horizontal. Try to draw a continuous graph whose tangent line is always horizontal. Of course, you can see that the graph is itself a horizontal line, that is, a line of the form $y = c$, where c is some constant.

From this theorem we see that if $F_1(x)$ and $F_2(x)$ are both antiderivatives of the *same* function $f(x)$, then

$$\frac{d}{dx}[F_1(x) - F_2(x)] = F_1'(x) - F_2'(x)$$

$$= f(x) - f(x) = 0$$

and hence, $F_1(x) - F_2(x) = c$, a constant. We state this formally as follows:

THEOREM

> An antiderivative of a function is unique up to a constant. That is, if $F_1(x)$ and $F_2(x)$ are antiderivatives of a function $f(x)$, then
>
> $$F_1(x) - F_2(x) = \text{A constant}$$

We now see that we did a complete job in Example 1. If $F_1(x)$ is *any* antiderivative of $f(x) = x^3$, then since $F_2(x) = \frac{1}{4}x^4$ is also an antiderivative, we have

$$F_1(x) - F_2(x) = F_1(x) - \tfrac{1}{4}x^4 = c$$

for some constant. Hence,

$$F_1(x) = \tfrac{1}{4}x^4 + c$$

Example 2   Find the antiderivatives of the function $f(x) = 6x^{-3}$.

Solution   For the antiderivative, we know that the exponent must be 1 *greater*. Thus, an antiderivative must have the form $F(x) = ax^{-2} + c$, where a is some constant. Differentiating, we get $F'(x) = -2ax^{-3}$. Thus, $-2a$ must equal 6, or $a = -3$. Therefore, any antiderivative must have the form $F(x) = -3x^{-2} + c$. We may check this result by differentiation, obtaining $F'(x) = 6x^{-3} + 0 = f(x)$. It is a good habit to do this! ∎

Example 3   Find the antiderivatives of $g(x) = 2$.

Solution   We recall that the derivative of a constant times x is equal to the constant. Therefore, $G(x) = 2x + c$ is the general form of an antiderivative of $g(x)$. ∎

Example 4   Determine the antiderivatives of $h(x) = x + 2$.

Solution   We use the fact that the derivative of a sum is equal to the sum of the derivatives. We can find the antiderivative of each term separately and then add them. An antiderivative of x must have the form $ax^2 + c_1$. Since the derivative of $ax^2 + c_1$ is $2ax$, we must have $2a = 1$, or $a = \frac{1}{2}$. From Example 3, we know that the antiderivative of 2 is $2x + c_2$. Adding these together, we have that any antiderivative of $h(x) = x + 2$ has the form

$$H(x) = \tfrac{1}{2}x^2 + 2x + c$$

(Since the sum of two constants is another constant, we have simply combined c_1 and c_2 into a single constant c; that is, $c = c_1 + c_2$.) Checking the derivative, we see that $H'(x) = x + 2 = h(x)$. ∎

THE INDEFINITE INTEGRAL

The term *indefinite integral* of a function is used to refer to the general form of the antiderivative of the function. We use the symbol $\int f(x)\, dx$ to denote the indefinite integral of $f(x)$.

The **indefinite integral** of a function $f(x)$ is

$$\int f(x)\, dx = F(x) + c$$

where $F'(x) = f(x)$ and c is called a *constant of integration*. The symbol $\int$ is called an *integral sign*. The function $f(x)$ is called the *integrand*, and dx indicates that x is the *variable of integration*.

The process of determining an indefinite integral is called (indefinite) **integration.** We frequently use the simple term "integral" to mean "indefinite integral" and shall do so throughout this chapter.

Example 5 Determine the following indefinite integrals:

$$\text{a. } \int x^2 \, dx \qquad \text{b. } \int x^3 \, dx \qquad \text{c. } \int x^4 \, dx$$

Solution Using the approach we described earlier, we know that the exponents of the integrals must be 1 greater than the function we are integrating, and the correct constant may be found by assuming that $F(x) = ax^3 + c$ for part a; $F(x) = ax^4 + c$ for part b; and $F(x) = ax^5 + c$ for part c. Thus, for part a, differentiating $F(x)$ yields $F'(x) = 3ax^2$. Setting $3a = 1$, we determine the constant a to be $\frac{1}{3}$. In a similar manner, the constants for parts b and c are $\frac{1}{4}$ and $\frac{1}{5}$, respectively. Therefore, the indefinite integrals are:

$$\text{a. } \int x^2 \, dx = \tfrac{1}{3}x^3 + c$$

$$\text{b. } \int x^3 \, dx = \tfrac{1}{4}x^4 + c$$

$$\text{c. } \int x^4 \, dx = \tfrac{1}{5}x^5 + c \qquad\qquad\qquad ∎$$

Example 5 shows a definite pattern for indefinite integrals of power functions. This can be stated by the following rule:

THEOREM
Power Rule for Integration

$$\int x^p \, dx = \frac{1}{p+1}(x^{p+1}) + c = \frac{x^{p+1}}{p+1} + c \qquad p \neq -1$$

We may prove this theorem by differentiation:

$$\frac{d}{dx}\left(\frac{x^{p+1}}{p+1} + c\right) = (p+1)\frac{x^{p+1-1}}{p+1} + 0$$
$$= x^p$$

The power rule provides a simple means of finding the indefinite integrals of power functions. *One word of caution: p* cannot be equal to -1, since this would require division by 0. Therefore, this rule cannot be used to evaluate

$$\int \frac{1}{x} \, dx$$

We will see in the next chapter that this integral is related to logarithmic functions.

Example 6 Determine the indefinite integral of $f(x) = x^{12}$ using the rule for power functions given above.

Solution   We have

$$\int x^{12}\, dx = \frac{x^{12+1}}{12+1} + c$$

$$= \tfrac{1}{13}x^{13} + c \qquad \blacksquare$$

Example 7   Find the indefinite integral of $f(t) = 1/t^3$.

Solution   We first rewrite $1/t^3$ as t^{-3}. Then, using the power rule for integration, we have

$$\int t^{-3}\, dt = \frac{t^{-3+1}}{-3+1} + c$$

$$= -\tfrac{1}{2}t^{-2} + c \qquad \blacksquare$$

Example 8   Find the indefinite integral of $y = \sqrt{x}$.

Solution   Again, we must rewrite the function so that we can apply the power rule. Since $\sqrt{x} = x^{1/2}$, we have

$$\int \sqrt{x}\, dx = \int x^{1/2}\, dx$$

$$= \frac{x^{1/2+1}}{\tfrac{1}{2}+1} + c$$

$$= \tfrac{2}{3}x^{3/2} + c \qquad \blacksquare$$

A special case of the power rule occurs when $p = 0$. Since $x^0 = 1$, we have that

$$\int 1\, dx = x + c$$

That is, the indefinite integral of the function $f(x) = 1$ is simply $x + c$.

Two important properties of integrals are analogous to those we have seen for derivatives. First, the integral of the sum or difference of two functions is equal to the sum or difference of the integrals; and second, the integral of a constant times a function is equal to the constant times the integral of the function. These properties allow us to integrate many useful functions found in applied problems. Mathematically, these rules are stated below.

Properties of the Indefinite Integral

$$\int [f(x) + g(x)]\, dx = \int f(x)\, dx + \int g(x)\, dx$$

$$\int [f(x) - g(x)]\, dx = \int f(x)\, dx - \int g(x)\, dx$$

$$\int kf(x)\, dx = k \int f(x)\, dx \qquad k \text{ a constant}$$

Example 9   Determine the indefinite integral:

$$\int (x^2 + 5)\, dx$$

Solution   Noting that the integral of a sum is equal to the sum of the integrals, we have

$$\int (x^2 + 5)\, dx = \int x^2\, dx + \int 5\, dx$$
$$= (\tfrac{1}{3}x^3 + c_1) + (5x + c_2)$$
$$= \tfrac{1}{3}x^3 + 5x + c \qquad \text{\textit{Since the sum of constants is another}}$$
$$\text{\textit{constant}} \qquad\blacksquare$$

Example 10   Find the indefinite integral:

$$\int 6x^{-3}\, dx$$

Solution   In this example we use the fact that the integral of a constant times a function is equal to the constant times the integral of the function. We have

$$\int 6x^{-3}\, dx = 6 \int x^{-3}\, dx$$
$$= 6(-\tfrac{1}{2}x^{-2}) + c$$
$$= -3x^{-2} + c \qquad\blacksquare$$

Because of the constant of integration, there are infinitely many solutions for an indefinite integral. The solution to $\int f(x)\, dx = F(x) + c$ is often called the *general solution* to the indefinite integral. If we know a value of the indefinite integral $F(x) + c$ for a particular value of x, then we may solve for the constant of integration, c. The resulting antiderivative is called a *particular solution*. For instance, in Example 10, suppose we wish to find the antiderivative whose value when $x = 2$ is equal to 3. We have

$$F(2) + c = -3(2)^{-2} + c$$
$$= -\tfrac{3}{4} + c = 3$$
$$c = \tfrac{15}{4}$$

The particular solution to the indefinite integral is the antiderivative

$$F(x) = -3x^{-2} + \tfrac{15}{4}$$

The following examples illustrate this concept in practical applications.

Example 11
Stopping Distance of a Car   We know that velocity is the derivative of distance as a function of time. That is, if the distance at time t is $F(t)$ and the velocity is $f(t)$, then $f(t) = F'(t)$. To put this another way, the distance $F(t)$ is an antiderivative of the velocity $f(t)$. Suppose that the velocity of a car when braking decreases according to the function $f(t) = k - 21.5t$ feet per second, where k is the speed of the car when the brakes are first applied. Note that when $f(t) = 0$, the car comes to a stop. Therefore, setting $k - 21.5t = 0$, we see that it takes $t = k/21.5$ seconds for the car to stop. How far has it traveled during this time?

Solution   The function giving the distance while braking is represented by the integral

$$\int f(t)\, dt = \int (k - 21.5t)\, dt$$

$$= kt - 10.75t^2 + c$$

Therefore, $F(t) = kt - 10.75t^2 + c$. We can determine the constant c by realizing that at $t = 0$ (when the brakes are first applied), the distance $F(0) = 0$. Solving this equation for c gives $c = 0$. Therefore, when $t = k/21.5$, the car will have traveled

$$F\left(\frac{k}{21.5}\right) = k\left(\frac{k}{21.5}\right) - 10.75\left(\frac{k}{21.5}\right)^2$$

$$\approx \frac{k^2}{21.5} - 0.023k^2$$

$$\approx 0.024k^2 \text{ feet}$$

Notice that the braking distance is proportional to the square of the speed of the vehicle; therefore, higher speeds require much longer distances to stop.   ∎

Example 12
Production Rate and Volume   The production rate (in units per hour) on an assembly line for a new toy is increasing according to the function $f(t) = 4t + \sqrt{t} + 5$, where t is time measured in hours. What is the function $F(t)$ that gives the total number of units produced in t hours?

Solution   Since the rate of change of the total production function $F(t)$ with respect to time, that is, the function $F'(t)$, is the given production rate $f(t)$, we have

$$F'(t) = f(t)$$

In other words, $F(t)$ is an antiderivative of $f(t)$. Therefore, $F(t)$ will be represented by the indefinite integral of $f(t)$, which is

$$\int f(t)\, dt = \int (4t + \sqrt{t} + 5)\, dt$$

$$= 2t^2 + \tfrac{2}{3}t^{3/2} + 5t + c$$

Since $t = 0$ indicates the start of production, we realize that no units have been produced at this time. Therefore, $F(0) = 0$ and we have $c = 0$. Thus, the number of units produced in t hours will be

$$F(t) = 2t^2 + \tfrac{2}{3}t^{3/2} + 5t$$ ∎

PROBLEMS (Section 5-1)

Find an antiderivative of each function. Check by using differentiation.

1. $f(x) = 4$

2. $g(x) = x^5$

3. $h(x) = 2x^3$

4. $f(x) = 4x^{-2}$

5. $g(x) = 3x^{-3}$

6. $h(x) = x + 5$

7. $f(x) = 2x + 1$

8. $f(x) = \dfrac{1}{x^3} + 2x - 4$

9. $f(x) = 5\sqrt{x}$

10. $f(x) = x^{1/3}$

11. $f(x) = 2x^{1/2}$

12. $f(x) = -3x^{2/3}$

13. What is another term used for the set of all anti-derivatives of a function?

14. Antidifferentiation is the inverse of what process? What is the difference between a general and particular solution to the indefinite integral?

Find the indefinite integral of each function using the rule for power functions.

15. $\displaystyle\int x^6 \, dx$

16. $\displaystyle\int x^{15} \, dx$

17. $\displaystyle\int t^{20} \, dt$

18. $\displaystyle\int x^{-7} \, dx$

19. $\displaystyle\int x^{-4} \, dx$

20. $\displaystyle\int \dfrac{1}{\sqrt{x}} \, dx$

21. $\displaystyle\int \dfrac{dt}{\sqrt[3]{t}}$

22. $\displaystyle\int \sqrt[4]{x} \, dx$

23. $\displaystyle\int x^{0.3} \, dx$

24. $\displaystyle\int x^{-0.7} \, dx$

Compute the indefinite integral of each function using the properties of the indefinite integral.

25. $\displaystyle\int (x^2 + 9) \, dx$

26. $\displaystyle\int (x^3 - x^2 - 3) \, dx$

27. $\displaystyle\int 4x^{-6} \, dx$

28. $\displaystyle\int \tfrac{2}{3} x^{-8} \, dx$

29. $\displaystyle\int (x + 3)(x + 1) \, dx$

30. $\displaystyle\int (x - 5)^2 \, dx$

31. $\displaystyle\int (4x^3 - 5x^2 + 3x - 2) \, dx$

32. $\displaystyle\int (-2x^3 + 7x + 3) \, dx$

33. $\displaystyle\int (2x - \sqrt{x} + 6) \, dx$

34. $\displaystyle\int (5x^3 - 4x - 2\sqrt{x}) \, dx$

35. $\displaystyle\int \dfrac{x^2 + x + 1}{\sqrt{x}} \, dx$

36. Find the antiderivative $F(x)$ of $f(x) = 3x - 2$ that satisfies $F(1) = -3$.

37. Find the antiderivative $G(x)$ of $g(x) = -2x^2 + 3x + 1$ that satisfies $G(-1) = 2$.

38. Find the antiderivative $F(x)$ of $f(x) = -x^{-1/2} + x - 3$ that satisfies $F(4) = -1$.

39. An antiderivative $H(x)$ of the function $h(x) = x^2 - ax + 1$ satisfies $H(0) = 1$ and $H(2) = \tfrac{2}{3}$. What is a?

40. Suppose that the speed of a bicycle (in feet per minute) when braking is given by the function $f(t) = k - 7.3t$, where k is the speed of the bicycle when the brakes are first applied. How far does the bicycle travel before it comes to a complete stop?

41. The production rate for jeans in a clothing factory is increasing according to the function $f(t) = 2t + \sqrt{t}$, where $f(t)$ is given in jeans per hour

and t is time measured in hours. What is the function giving the total number of jeans produced in t hours?

42. Suppose that the marginal cost for producing x units of an item is given by $c'(x) = 3x - 250$ dollars per unit. Find the total cost function if there is a fixed cost of $300, that is, $c(0) = 300$.

43. Suppose that the marginal cost of producing x units of a product is given by $c'(x) = 70 + 40/\sqrt{x}$ dollars per unit. If the total cost of producing the first unit is $200, what is the cost function and what is the cost of producing the first 25 units?

44. An object is projected vertically with initial velocity of 2 feet per second, that is, $v(0) = 2$. The acceleration is given by the function $a(t) = t^2 - 5$, and the initial distance from a reference point is 4 feet [$s(0) = 4$].
    a. Find the velocity function. [*Hint:* Acceleration is the derivative of velocity.]
    b. Find the distance function.

5-2 THE AREA UNDER A CURVE

Not too long ago, before the days of free checking accounts, a Cincinnati bank charged service fees of $0.50 per month plus $0.08 per check. However, it gave a credit of $0.10 for each $100 average daily balance in the account. Therefore, if a customer wrote 12 checks in a certain month and maintained an average balance of $400, the service fee for that month would be $0.50 + $0.08(12) − 4($0.10) = $1.06. How was the average balance computed?

Let us suppose that an account had a balance of $100 at the beginning of September. Table 5-1 gives a record of the transactions (deposits and withdrawals) for this account during the month. Figure 5-2 is a graph of the checking account balance over the 30 day period.

Table 5-1
Checking account transactions during September

Day	Transaction	Balance
0	—	100
3	+50	150
10	+90	240
15	−20	220
17	−60	160
24	+140	300
30	—	300

We might think of the daily balance as a *function* of time, $f(t)$, described by the graph in Figure 5-2. (Can you write this function mathematically?) The problem of computing the average daily balance can be expressed as finding the average value of $f(t)$ over the 30 day period. This is a very different problem

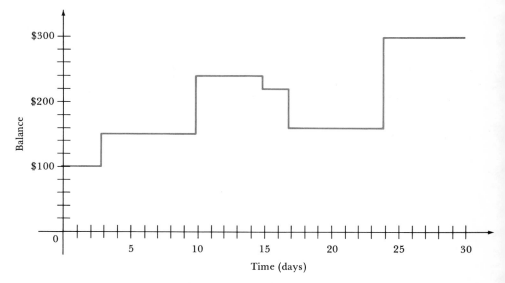

Figure 5-2
Graph of checking account balance
over time

from finding, say, the average weight of a group of individuals. For instance, if three people weigh 140, 200, and 170 pounds, respectively, then their average weight is

$$\frac{140 + 200 + 170}{3} = 170$$

We simply add their weights and divide by the number of people. Suppose we did a similar computation with the checking account balances in the last column in Table 5-1. Would this give us the true average value in the account? The sum of the balances in Table 5-1 is

$$100 + 150 + \cdots + 300 = 1470$$

Dividing by 7 gives an average value of 210. What is wrong with this method? We did not account for the length of time that the monies were held in the account. For example, suppose that you had $100 in the account for 29 days and $50 in it for 1 day. Would the average daily balance be $(100 + 50)/2 = 75$? Certainly not, since we had $100 in the account for nearly the entire month.

We must approach this in a different manner. The initial $100 is held for 3 days. Next, a balance of $150 remains for 7 days, and so on. We can think of this as shown in Table 5-2.

Now, if we add the balances on each day and divide by 30 days, we have the true average daily balance. But there is a simpler way of organizing these computations. Notice that adding $100 three times (for days 1–3) is the same as multiplying $100 by 3; similarly, adding $150 seven times (for days 4–10) is equivalent to multiplying $150 by 7. Thus, the average daily balance is computed by summing the product of the different balances times the number of days held at these values and then dividing by the number of days in the total time period:

$$\frac{(100)(3) + (150)(7) + (240)(5) + (220)(2) + (160)(7) + (300)(6)}{30} = \$197$$

Table 5-2

Day	Balance	Day	Balance
1	100	8	150
2	100	9	150
3	100	10	150
4	150	.	.
5	150	.	.
6	150	.	.
6	150	30	300
7	150		

Observe the dimensions of this quantity. The numerator is the product of dollars times days, or "dollar-days." The denominator is expressed in days. The dimensions of this ratio are thus given by dollar-days/days, or dollars. The ratio represents the (average) number of dollars held in the account over the time period.

Let us return to Figure 5-2, the graph of the daily balance. Can you see any relationship between the computation of the average balance and the graph? Each term in the numerator, (100)(3), (150)(7), and so on, is the product of the *height* of the function along the y-axis (dollars) times the *width* of the function along the x-axis (days). Geometrically, this gives the area of each of the rectangles shown in Figure 5-3. Therefore, the average daily balance is the area under the graph of $f(t)$ divided by the time interval over which the area is computed!

Figure 5-3
Area under the graph of f(t)

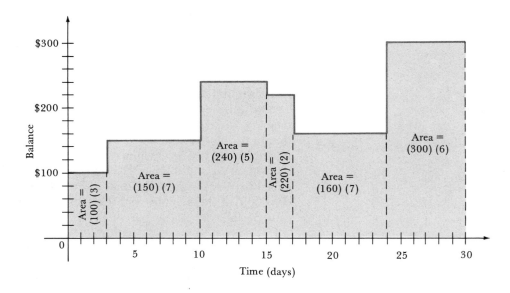

From this example we can make the following general observation:

> Let $f(x)$ be a function defined over an interval $[a, b]$, and assume that $f(x)$ is nonnegative over this interval. Then the area under the graph of $f(x)$, that is, the area between the graph of $f(x)$ and the x-axis, has dimensions equal to the dimensions of the dependent variable times the dimensions of the independent variable.

Example 13   Describe the dimensions of the quantity represented by the area under the curve for each of the following functions:

a. The production rate in units per hour is a function of time.
b. The yearly income (dollars per year) of an individual is a function of time.
c. The learning rate of an assembly worker performing a new task is given by a function $g(x)$ with dimensions of hours per unit, where x is the number of units assembled by the worker.
d. The marginal cost of an item (in dollars per unit) is a function of the number of units produced.

Solution

a. The area under the production rate curve has dimensions of

$$\left(\frac{\text{Units}}{\text{Hour}}\right)(\text{Hours}) = \text{Units}$$

b. The area under the yearly income curve has dimensions of

$$\left(\frac{\text{Dollars}}{\text{Year}}\right)(\text{Years}) = \text{Dollars}$$

c. The area under the curve $g(x)$ has dimensions

$$\left(\frac{\text{Hours}}{\text{Unit}}\right)(\text{Units}) = \text{Hours}$$

d. The area under the marginal cost curve has dimensions

$$\left(\frac{\text{Dollars}}{\text{Unit}}\right)(\text{Units}) = \text{Dollars}$$

THE DEFINITE INTEGRAL

The area under the checking account balance curve was easy to compute since the function could be broken down into nice distinct rectangles. However, suppose that we are given some arbitrary positive function and are asked to find the area under its graph. For example, consider the function

$$f(x) = 2x \qquad 0 \le x \le 5$$

which represents the demand rate (in thousands of units per month) for a new microcomputer graphics software package over the first 5 months on the market. A graph of $f(x)$ is given in Figure 5-4. The area under the curve from $x = 0$

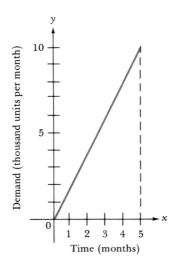

Figure 5-4
Graph of f(x) = 2x

to $x = 5$ represents the total demand (thousands of units per month $\times$ months = thousands of units) over this 5 month period. From geometry we know that the area of a triangle is $\frac{1}{2} \times$ height $\times$ base. Thus, the area under $y = 2x$ from $x = 0$ to $x = 5$ is $\frac{1}{2}(10)(5) = 25$ thousand units.

Suppose we did not know this geometric formula. One way of at least approximating the area is to "cover" the area with small rectangles as shown in Figure 5-5. The area covered by these rectangles is equal to

$$(2)(1) + (4)(1) + (6)(1) + (8)(1) + (10)(1) = 30 \text{ thousand units}$$

It is easy to see from the figure that we have overestimated the true value of the area. Notice that the height of each rectangle is equal to the value of $f(x)$ for $x = 1, 2, 3, 4$, and 5, and that the width of each rectangle is 1, or the difference between successive x values.

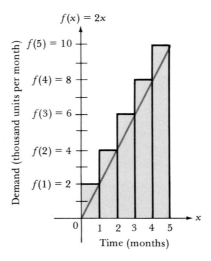

Figure 5-5
Estimating the area under the graph of f(x) = 2x with rectangles

We digress for a moment to state this process in general terms. Suppose that we wish to find the area under the graph of a function $f(x) \geq 0$ from $x = a$ to

$x = b$. Choose a set of points $a = x_0 < x_1 < x_2 < \cdots < x_n = b$. This divides the interval $[a, b]$ into n subintervals $[x_0, x_1], [x_1, x_2], \ldots, [x_{n-1}, x_n]$. The widths of these subintervals are $\Delta x_1 = x_1 - x_0$, $\Delta x_2 = x_2 - x_1$, and so on. Next, compute $f(x_j)$ for each j from 1 to n, and multiply the value of $f(x_j)$ by the width of the subinterval Δx_j. This gives the area of each rectangle, as shown in Figure 5-6. Finally, add up all the areas. (See Algebra Review 16 on page 234 for a discussion of the summation notation that is used throughout this chapter.) The result is an approximation to the area under the curve. Mathematically, such a summation is called a *Riemann sum*.

A **Riemann sum** of a function $f(x)$ is an expression of the form

$$\sum_{j=1}^{n} f(x_j)\Delta x_j = f(x_1)\Delta x_1 + f(x_2)\Delta x_2 + \cdots + f(x_n)\Delta x_n$$

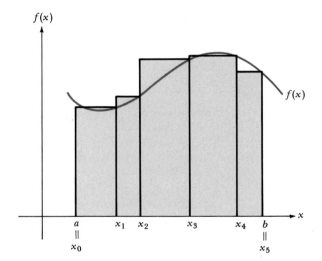

Figure 5-6
A Riemann sum with $n = 5$

Notice that the dimensions of a Riemann sum are equal to the dimensions of the dependent variable $f(x)$ times the dimensions of the independent variable x (since the length of every subinterval Δx_j has the same dimensions as x).

Example 14   Let $f(x) = \sqrt{x}$. Consider the points $x_0 = 0, x_1 = 1, x_2 = 4, x_3 = 9$, and $x_4 = 16$. We have $\Delta x_1 = x_1 - x_0 = 1$, $\Delta x_2 = x_2 - x_1 = 3$, $\Delta x_3 = x_3 - x_2 = 5$, and $\Delta x_4 = x_4 - x_3 = 7$. Also, $f(x_1) = 1, f(x_2) = 2, f(x_3) = 3$, and $f(x_4) = 4$. The Riemann sum of $f(x)$ for this set of points is thus

$$\sum_{j=1}^{4} f(x_j)\Delta x_j = f(x_1)\Delta x_1 + f(x_2)\Delta x_2 + f(x_3)\Delta x_3 + f(x_4)\Delta x_4$$
$$= (1)(1) + (2)(3) + (3)(5) + (4)(7)$$
$$= 1 + 6 + 15 + 28 = 50$$

Example 15   Estimate the area under the curve $f(x) = x^2$ from $x = 1$ to $x = 4$ using subintervals of length $\Delta x = 1$. (In the case when all subintervals have equal length, we simply use Δx to denote the common length.)

Solution   We let $x_0 = 1$, $x_1 = 2$, $x_2 = 3$, and $x_3 = 4$. (Here, $n = 3$.) Each subinterval has width equal to 1. Computing $f(x_j)$ for j equal 1 to 3, we have

$$f(2) = 2^2 = 4$$
$$f(3) = 3^2 = 9$$
$$f(4) = 4^2 = 16$$

Multiplying these by the subinterval widths $\Delta x = 1$, we estimate the area to be

$$\text{Area} \approx 4(1) + 9(1) + 16(1) = 29$$

This is illustrated in Figure 5-7.   ■

Now let us return to the problem posed at the beginning of this section. In trying to find the area under $f(x) = 2x$ from $x = 0$ to $x = 5$, we saw that using rectangles of width equal to 1 gives an estimate of 30, which is not very close to

ALGEBRA REVIEW 16: SUMMATION NOTATION

A symbol often used in mathematics to represent *summation* is the Greek letter sigma, $\sum$. If we have n numbers $x_1, x_2, \ldots, x_n$, the sum of these is denoted by

$$\sum_{j=1}^{n} x_j$$

The letter j is called the *index of summation*. The symbol $\sum_{j=1}^{n}$ means that we let j vary from 1 to n and add the quantity that follows the summation symbol for each value of j. Therefore,

$$\sum_{j=1}^{n} x_j = x_1 + x_2 + \cdots + x_n$$

For example, if $x_1 = 3$, $x_2 = 5$, and $x_3 = 1$,

$$\sum_{j=1}^{3} x_j = 3 + 5 + 1 = 9$$

As another example, consider the following:

$$\sum_{j=0}^{4} 2^j = 2^0 + 2^1 + 2^2 + 2^3 + 2^4$$
$$= 1 + 2 + 4 + 8 + 16$$
$$= 31$$

EXERCISES*

Expand the following summations:

a. $\displaystyle\sum_{i=2}^{5} 3i$ b. $\displaystyle\sum_{j=1}^{3} f(x_j)$ c. $\displaystyle\sum_{k=-1}^{3} (k-2)^2$

In exercises d and e, let $x_1 = 2$, $x_2 = 4$, $x_3 = 1$, $x_4 = 6$, and $x_5 = -1$. Find each sum.

d. $\displaystyle\sum_{j=1}^{5} x_j$ e. $\displaystyle\sum_{j=0}^{2} x_{5-2j}$

---

* Answers to algebra review exercises can be found at the end of the answer section in the back of the book.

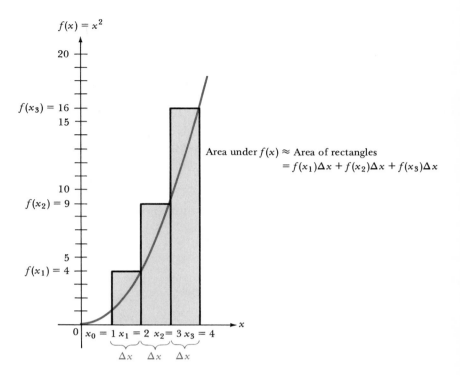

Figure 5-7
Estimating the area under
$f(x) = x^2$ from $x = 1$ to $x = 4$

Area under $f(x) \approx$ Area of rectangles
$= f(x_1)\Delta x + f(x_2)\Delta x + f(x_3)\Delta x$

the true value of 25 (see Figure 5-5). One way of getting a better approximation is to reduce the size of the subintervals. This is the same as choosing a larger number of subintervals. For example, suppose that we use 10 subintervals, each of length $\frac{1}{2}$. This is illustrated in Figure 5-8. We leave it to you as an exercise to show that the area will be estimated to be 27.5. If we further reduce the size of the subintervals by choosing more points closer together, we will continue to get better approximations to the true area.

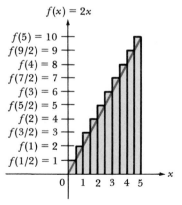

Figure 5-8
Estimating the area under
$f(x) = 2x$ from $x = 0$ to $x = 5$
using 10 subintervals

To state this idea mathematically: The area under the graph of $f(x) \geq 0$ from $x = a$ to $x = b$ can be estimated using n subintervals, having widths Δx_j, by the sum $\sum_{j=1}^{n} f(x_j) \, \Delta x_j$. We may let the widths of the subintervals get as small as we wish. (Clearly, as we do this, the *number* of subintervals increases.) This process allows us to obtain better and better approximations to the true area.

Hence, we define the area under the graph of $f(x) \geq 0$ from $x = a$ to $x = b$ to be

$$\lim_{\Delta x_j \to 0} \sum_{j=1}^{n} f(x_j) \, \Delta x_j$$

provided this limit exists. The fact that such a limit does exist for continuous functions is one of the fundamental notions of calculus. This limit is called the *definite integral* and is denoted by a special symbol:

The **definite integral of the continuous function $f(x)$ on the interval [a, b]** is denoted by

$$\int_{a}^{b} f(x) \, dx$$

and is equal to

$$\lim_{\Delta x_j \to 0} \sum_{j=1}^{n} f(x_j) \, \Delta x_j$$

The endpoints a and b of the interval $[a, b]$ are called the *lower* and *upper limits of integration,* respectively. We read this expression as "the integral of $f(x)$ from a to b."

Since the definite integral is equal to a limit of a Riemann sum, its dimensions are the same as that of the Riemann sum, that is, the dimensions of the function $f(x)$ times the dimensions of the variable x. If $f(x)$ is continuous and nonnegative over the interval $[a, b]$, then the definite integral represents the value of the area under the graph of $f(x)$ from $x = a$ to $x = b$. Notice the close relationship between the summation

$$\sum_{j=1}^{n} f(x_j) \, \Delta x_j$$

and the symbols used for integration. The integral sign was actually derived from an archaic symbol for summation, and dx is analogous to Δx_j. Hence, the integral represents a summation of function values times "infinitely small" widths (dx), which, as we have seen, corresponds to the area under the curve.

It is no coincidence that the notation for the definite integral is similar to that of the indefinite integral defined in the previous section. The major differences are that the integral sign in the indefinite integral does not have limits of integration, and the indefinite integral represents a *set of functions* while the definite integral represents a *number.*

Example 16
Telephone Cost

The cost per minute for a long-distance telephone call is given by the function

$$f(t) = 20 - 2t \text{ cents per minute} \qquad 0 < t \leq 5$$

and remains at 10¢ per minute after 5 minutes. Estimate the total cost of a 2 minute call using Riemann sums with 2, 4, and 8 subintervals. What limit do these sums appear to approach?

Solution   This function is shown in Figure 5-9, and we seek the area of the shaded region. Using 2 subintervals, each of width 1, we have $f(1) = 18$ and $f(2) = 16$. The Riemann sum is

$$\sum_{j=1}^{2} f(t_j) \, \Delta t_j = 18(1) + 16(1) = 34 \text{ cents}$$

Since $f(t)$ has dimensions of cents per minute and Δt has dimensions of minutes, the Riemann sum has dimensions of (cents/minute) (minutes) = cents. The Riemann sum therefore represents total cost.

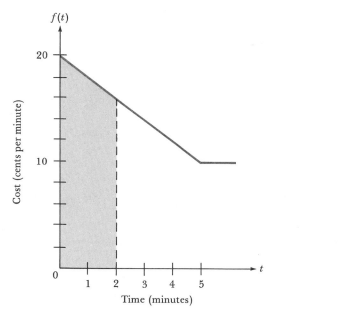

Figure 5-9
Graph of long-distance service charges

If 4 subintervals are used, we choose points located at $t = \frac{1}{2}, 1, \frac{3}{2}$, and 2 with respective function values of 19, 18, 17, and 16. The value of the Riemann sum is then

$$19(\tfrac{1}{2}) + 18(\tfrac{1}{2}) + 17(\tfrac{1}{2}) + 16(\tfrac{1}{2}) = 35 \text{ cents}$$

Finally, if 8 subintervals (of length $\frac{1}{4}$) are used, the estimate of the total cost is 35.5 cents.

By geometry, we can show that the actual area of the shaded region is 36. This is the value of the definite integral

$$\int_0^2 f(t) \, dt = \int_0^2 (20 - 2t) \, dt$$

Can you explain why the Riemann sums we computed *underestimate* this value?   ■

PROBLEMS (Section 5-2)

Describe the dimensions of the area under each curve.

1. The traveling rate in miles per hour is a function of time.

2. The weekly commission of a salesperson (in dollars) is a function of time.

3. The population growth rate of a new suburb is given by a function $p(x)$ with dimensions of people per year, where x is the number of years the suburb has been established.

In Problems 4–7, estimate the area under each curve using Riemann sums with subintervals of length 1 and length $\frac{1}{2}$. Compute the actual area by geometry.

4. $f(x) = \frac{1}{2}x$ from $x = 0$ to $x = 3$

5. $f(x) = 4 - x$ from $x = 0$ to $x = 4$

6. $f(x) = 5x + 1$ from $x = 1$ to $x = 5$

7. $f(x) = 8 - 2x$ from $x = 0$ to $x = 2$

Use Riemann sums to estimate the area under each curve. Use subintervals of length 1.

8. $f(x) = x^2 + 5$ from $x = 0$ to $x = 4$

9. $f(x) = 2x^2 + x + 3$ from $x = 0$ to $x = 6$

10. $f(x) = x^3$ from $x = 0$ to $x = 3$

11. $f(x) = 2x^3 - 1$ from $x = 1$ to $x = 6$

12. $f(x) = x^3 + 2x^2 + x$ from $x = 0$ to $x = 5$

Estimate the area under each curve using 2, 4, and 8 subintervals of equal width. What limit does each area estimate appear to approach?

13. $y = 2x + 3$ from $x = 0$ to $x = 2$

14. $f(x) = 6x + 2$ from $x = 0$ to $x = 4$

15. $f(x) = 5x - 1$ from $x = 1$ to $x = 4$

16. $y = 4x^2 + 3$ from $x = 0$ to $x = 2$

17. $f(x) = 3x^2 - 2$ from $x = 1$ to $x = 5$

18. $f(x) = \sqrt{1 - x^2}$ from $x = -1$ to $x = 1$
   [*Hint:* Sketch the graph.]

5-3 COMPUTING THE DEFINITE INTEGRAL

We have seen that the area under the graph of a function $f(x) \geq 0$ from $x = a$ to $x = b$ has many practical interpretations and is defined by the definite integral

$$\int_a^b f(x)\, dx$$

In this section we will find an easy way to compute the definite integral.

An important relationship between the definite integral and antiderivatives is given by the following:

Fundamental Theorem of Calculus

If $f(x)$ is a continuous function over the interval $[a, b]$, then there exists a function $F(x)$ such that $F'(x) = f(x)$. For any such function $F(x)$,

$$\int_a^b f(x)\, dx = F(b) - F(a)$$

The expression $F(b) - F(a)$ is commonly written $F(x)|_a^b$. This notation means we evaluate $F(x)$ for $x = b$ and then *subtract* the expression evaluated for $F(x)$ at $x = a$.

The value of a definite integral is computed using a three-step process. First, we must find an antiderivative of the function $f(x)$, which we denote by $F(x)$. Second, we substitute the upper limit of integration, b, into $F(x)$, yielding $F(b)$. Finally, we substitute the lower limit of integration, a, into $F(x)$, yielding $F(a)$, and subtract $F(a)$ from $F(b)$.

Note that *any* antiderivative $F(x)$ of $f(x)$ may be used in the fundamental theorem. If $F_1(x)$ and $F_2(x)$ are both antiderivatives of $f(x)$, then

$$F_2(x) = F_1(x) + c$$

for some constant c, and hence,

$$\begin{aligned} F_2(x)|_a^b &= F_2(b) - F_2(a) \\ &= [F_1(b) + c] - [F_1(a) + c] \\ &= F_1(b) - F_1(a) \\ &= F_1(x)|_a^b \end{aligned}$$

Therefore, when computing a definite integral, we do not need to include the constant of integration when first finding the indefinite integral.

We will not prove the fundamental theorem, but we do wish to emphasize its importance. It is, in a certain sense, the climax of calculus, and it is the culmination of centuries of mathematical labor. The derivative and the integral are the two fundamental concepts of calculus. Both of these ideas have been studied by mathematicians for many centuries. The mathematicians of classical Greece dealt with derivatives in the form of tangents to curves and with integrals in the form of areas bounded by curves, and, in fact, computed these by a limiting process. However, it was not until the 17th century that Newton and Leibniz independently established the intimate connection between the derivative $[F'(x) = f(x)]$ and the integral $[\int_a^b f(x)\, dx = F(x)|_a^b]$ expressed by the fundamental theorem.

We illustrate the use of the fundamental theorem with a few easy examples.

Example 17   Evaluate the definite integral $\int_1^4 6x^{-3}\, dx$.

Solution   The general form of an antiderivative of this function was found in Example 10 (Section 5-1). Using this result, we have

$$\int_1^4 6x^{-3}\, dx = -3x^{-2}\bigg|_1^4 = -\frac{3}{x^2}\bigg|_1^4$$
$$= \left(-\tfrac{3}{16}\right) - \left(-\tfrac{3}{1}\right) = 2\tfrac{13}{16} \qquad \blacksquare$$

Example 18   Compute the definite integral $\int_{-2}^2 (2x - 2)\, dx$.

Solution   Applying the fundamental theorem of calculus gives

$$\int_{-2}^{2} (2x - 2)\, dx = (x^2 - 2x)|_{-2}^{2} = [(2)^2 - 2(2)] - [(-2)^2 - 2(-2)]$$

$$= (4 - 4) - (4 + 4)$$
$$= -8 \qquad \blacksquare$$

Example 19   Find the area under the graph of the function $y = x + 2$ between $x = 0$ and $x = a$ for any positive value of a.

Solution   We find this by computing the integral

$$\int_{0}^{a} (x + 2)\, dx$$

This is computed using the fundamental theorem as follows:

$$\int_{0}^{a} (x + 2)\, dx = (\tfrac{1}{2}x^2 + 2x)|_{0}^{a}$$

$$= (\tfrac{1}{2}a^2 + 2a) - [\tfrac{1}{2}(0^2) + 2(0)]$$
$$= \tfrac{1}{2}a^2 + 2a$$

For instance, if $a = 1$, the area under the graph is $\tfrac{1}{2}(1)^2 + 2(1) = 2\tfrac{1}{2}$. If $a = 20$, the area is $\tfrac{1}{2}(20)^2 + 2(20) = 240$. $\qquad \blacksquare$

Be careful of the minus signs when subtracting the functions evaluated at the lower limits of integration as in Example 18. This is one of the most frequent sources of errors. Also, note that the limits of integration may be any constant, or even a variable, as shown in Example 19.

PROPERTIES OF THE DEFINITE INTEGRAL

Three important properties of the definite integral are stated below, since they are often used to simplify computations. In general terms:

Properties of the Definite Integral—I

$$\int_{a}^{a} f(x)\, dx = 0$$

$$\int_{a}^{b} f(x)\, dx = - \int_{b}^{a} f(x)\, dx$$

$$\int_{a}^{b} f(x)\, dx = \int_{a}^{c} f(x)\, dx + \int_{c}^{b} f(x)\, dx$$

The first property states that the integral of a function from a to a equals 0. The second property states that if we reverse the limits of integration, we change the sign of the value of the integral. Finally, the last property states that the integral over a range from a to b can be evaluated in two (or more) pieces.

Each of these properties may be viewed as a convention for the definite integral that is consistent with the fundamental theorem. For example,

$$\int_a^a f(x)\ dx = F(a) - F(a) = 0$$

$$\int_a^b f(x)\ dx = F(b) - F(a) = -[F(a) - F(b)] = -\int_b^a f(x)\ dx$$

where $F(x)$ is an antiderivative of $f(x)$. We leave it to you to verify the third statement by using the fundamental theorem.

Example 20   Consider the function $f(x) = 2x$.

Solution   To illustrate the first property, let both limits of integration equal 1:

$$\int_1^1 2x\ dx = x^2\big|_1^1 = 1^2 - 1^2 = 0$$

To illustrate the second property, we have

$$\int_1^3 2x\ dx = x^2\big|_1^3 = 3^2 - 1^2 = 8$$

$$\int_3^1 2x\ dx = x^2\big|_3^1 = 1^2 - 3^2 = -8$$

Finally, the third property is shown with the following:

$$\int_1^3 2x\ dx = \int_1^4 2x\ dx + \int_4^3 2x\ dx$$
$$= x^2\big|_1^4 + x^2\big|_4^3$$
$$= (16 - 1) + (9 - 16)$$
$$= 15 - 7 = 8$$

∎

Note that our illustration of the last property in Example 20 shows that the value of c does not have to lie between a and b.

The definite integral also satisfies certain arithmetical properties that correspond to properties of the indefinite integral.

Properties of the Definite Integral—II

$$\int_a^b [f(x) + g(x)]\ dx = \int_a^b f(x)\ dx + \int_a^b g(x)\ dx$$

$$\int_a^b [f(x) - g(x)]\ dx = \int_a^b f(x)\ dx - \int_a^b g(x)\ dx$$

$$\int_a^b kf(x)\ dx = k \int_a^b f(x)\ dx \qquad k \text{ a constant}$$

To verify the third property, note that if $F(x)$ is an antiderivative of $f(x)$, then $kF(x)$ is an antiderivative of $kf(x)$ since

$$\frac{d}{dx}[kF(x)] = k\frac{d}{dx}F(x) = kf(x)$$

Therefore,

$$\int_a^b kf(x)\ dx = kF(b) - kF(a) = k[F(b) - F(a)] = k \int_a^b f(x)\ dx$$

The first two properties follow from the fundamental theorem in a similar way.

Example 21   Compute:

a. $\displaystyle\int_1^3 3x\ dx$ b. $\displaystyle\int_{-1}^2 (-x^2 + 5x)\ dx$

Solution   a. We have

$$\int_1^3 3x\ dx = 3\int_1^3 x\ dx$$

$$= 3\left(\frac{x^2}{2}\bigg|_1^3\right)$$

$$= 3(\tfrac{9}{2} - \tfrac{1}{2}) = 12$$

b. Using the properties of definite integrals, we find

$$\int_{-1}^2 (-x^2 + 5x)\ dx = \int_{-1}^2 (-x^2)\ dx + \int_{-1}^2 5x\ dx$$

$$= -\int_{-1}^2 x^2\ dx + 5\int_{-1}^2 x\ dx$$

$$= -\left(\frac{x^3}{3}\bigg|_{-1}^2\right) + 5\left(\frac{x^2}{2}\bigg|_{-1}^2\right)$$

$$= -\left[\frac{8}{3} - \frac{(-1)^3}{3}\right] + 5\left[\frac{4}{2} - \frac{(-1)^2}{2}\right]$$

$$= -[\tfrac{8}{3} + \tfrac{1}{3}] + 5[2 - \tfrac{1}{2}]$$

$$= -3 + \tfrac{15}{2} = \tfrac{9}{2}$$

PROBLEMS (Section 5-3)

In Problems 1–16, evaluate each definite integral.

1. $\displaystyle\int_0^3 x^2\ dx$

2. $\displaystyle\int_0^5 3\ dx$

3. $\displaystyle\int_2^8 2x^3\ dx$

4. $\displaystyle\int_1^9 \frac{1}{\sqrt{x}}\ dx$

5. $\displaystyle\int_4^{16} 2\sqrt{x}\ dx$

6. $\displaystyle\int_5^2 \frac{3}{x^2}\ dx$

7. $\displaystyle\int_{1}^{8} \frac{5}{\sqrt[3]{x}} \, dx$

8. $\displaystyle\int_{0}^{\pi} (x^2 - 3) \, dx$

9. $\displaystyle\int_{3}^{-1} (3x^2 - 2x) \, dx$

10. $\displaystyle\int_{-1}^{4} (2x^2 + x - 3) \, dx$

11. $\displaystyle\int_{0}^{\pi} (3x^2 - 2x + 2) \, dx$

12. $\displaystyle\int_{2}^{-1} (x^2 - 2x + 5) \, dx$

13. $\displaystyle\int_{2}^{5} (x - 3)(x + 2) \, dx$

14. $\displaystyle\int_{0}^{6} (x + 5)(2x + 1) \, dx$

15. $\displaystyle\int_{0}^{4} (2x^2 + \sqrt{x} - 3) \, dx$

16. $\displaystyle\int_{-3}^{-1} (x^4 + x^2 + 1) \, dx$

17. Show that
$$\int_{1}^{4} (3x^2 + 2) \, dx = -\int_{4}^{1} (3x^2 + 2) \, dx$$

18. If $f(x) = x^2 + x - 3$, show that
$$\int_{0}^{6} f(x) \, dx = \int_{0}^{3} f(x) \, dx + \int_{3}^{6} f(x) \, dx$$

19. Suppose $F(x)$ is an antiderivative of $f(x)$ and $G(x)$ is an antiderivative of $g(x)$. Use the fundamental theorem to verify that
$$\int_{a}^{b} [f(x) + g(x)] \, dx = \int_{a}^{b} f(x) \, dx + \int_{a}^{b} g(x) \, dx$$
[*Hint:* Note that $F(x) + G(x)$ is an antiderivative of $f(x) + g(x)$.]

20. Suppose $F(x)$ is an antiderivative of $f(x)$ and $G(x)$ is an antiderivative of $g(x)$. Use the fundamental theorem to verify that
$$\int_{a}^{b} [f(x) - g(x)] \, dx = \int_{a}^{b} f(x) \, dx - \int_{a}^{b} g(x) \, dx$$

In Problems 21–24, compute the definite integral by use of the fundamental theorem. Interpret the result as an area under a curve and check your answer by using simple geometry.

21. $\displaystyle\int_{-2}^{10} 3 \, dx$

22. $\displaystyle\int_{0}^{4} 3x \, dx$

23. $\displaystyle\int_{0}^{6} f(x) \, dx$ where $f(x) = \begin{cases} 3 & \text{if } x < 2 \\ 2x - 1 & \text{if } x \ge 2 \end{cases}$

24. $\displaystyle\int_{0}^{3} f(x) \, dx$ where $f(x) = \begin{cases} 6x & \text{if } x < 1 \\ -3x + 9 & \text{if } x \ge 1 \end{cases}$

5-4 AREAS AND THE DEFINITE INTEGRAL

We interpreted the definite integral as a rather complicated limit of Riemann sums, but then showed that for positive functions it represents the area under a curve, that is, between the curve and the x-axis. What happens if the function does not lie entirely above the x-axis over the interval given by the limits of integration? What will the value of the definite integral mean in terms of area? In this section we will answer these questions and learn how to use definite integrals to solve problems involving areas located anywhere in the plane.

Example 22   Compute the area under the line $y = -2x + 10$ and above the x-axis between $x = 2$ and $x = 4$.

Solution   This area is sketched in Figure 5-10. From the figure we see that the area in question is the area of a square of side 2 plus the area of a triangle with base 2 and altitude 4. Therefore, the total area is

$$(2)(2) + \tfrac{1}{2}(2)(4) = 8$$

But we also know that the area is given by the definite integral

$$\int_2^4 (-2x + 10)\, dx$$

We can compute this integral by the fundamental theorem, as follows:

$$\int_2^4 (-2x + 10)\, dx = (-x^2 + 10x)\big|_2^4$$
$$= (-16 + 40) - (-4 + 20)$$
$$= 24 - 16 = 8$$

This verifies the result we obtained by using simple geometry.

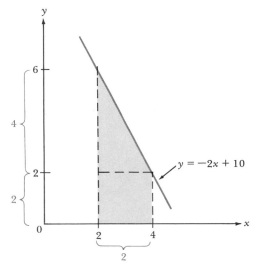

Figure 5-10
Graph of $y = -2x + 10$

Example 23   Find the area under the curve $y = -x^2 + 9$ from $x = 1$ to $x = 2$.

Solution   The first thing to do is to graph the equation to see if it lies above the x-axis over the specified interval. This is done in Figure 5-11, where the area we seek has been shaded. Since this area lies entirely above the x-axis, it is given by the integral

$$\int_1^2 (-x^2 + 9)\, dx = \left(-\frac{x^3}{3} + 9x\right)\Big|_1^2$$
$$= \left[-\frac{(2)^3}{3} + 9(2)\right] - \left[-\frac{(1)^3}{3} + 9(1)\right]$$
$$= [-\tfrac{8}{3} + 18] - [-\tfrac{1}{3} + 9]$$
$$= -\tfrac{7}{3} + 9 = 6\tfrac{2}{3}$$

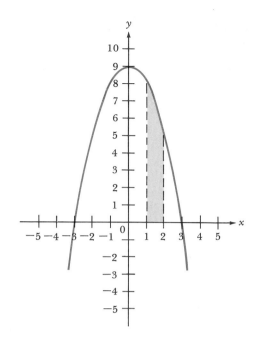

Figure 5-11
Graph of $y = -x^2 + 9$

When the curve does not lie entirely above the x-axis in the interval specified by the limits of integration, we must be careful how we interpret the integral. Consider the function $f(x) = x - 5$ for $0 \le x \le 10$. This is shown in Figure 5-12. From geometry, it can be shown that the area of each of the shaded triangles is 12.5. Now, suppose that we compute the integral of $f(x)$ from 0 to 10. This yields

$$\int_0^{10} (x - 5)\, dx = \left(\frac{x^2}{2} - 5x \right) \Big|_0^{10}$$
$$= \left[\tfrac{100}{2} - 5(10) \right] - \left[\tfrac{0}{2} - 5(0) \right] = 0$$

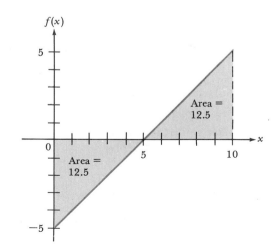

Figure 5-12
Graph of $f(x) = x - 5$

Before trying to explain this, let us see what happens if we break the integral into two parts:

$$\int_0^{10} (x-5)\,dx = \int_0^5 (x-5)\,dx + \int_5^{10} (x-5)\,dx$$

$$= \left(\frac{x^2}{2} - 5x\right)\Big|_0^5 + \left(\frac{x^2}{2} - 5x\right)\Big|_5^{10}$$

$$= -12.5 + 12.5 = 0$$

The absolute values of both integrals give the areas of the two triangles; however, the value of the integral from 0 to 5 is negative. Think back on how we computed the area of a small rectangle when we developed the notion of the definite integral. The value of the function was multiplied by the width of the interval. When $0 \le x < 5$, the value of the function is *negative*. Therefore, the value of the integral will be a negative number. When $5 < x \le 10$, the function is positive, and a positive value for the integral results. When these are added together, they algebraically cancel each other out. The only way to know this is to examine a graph of the function and determine the intervals over which the function is positive or negative. We must integrate over these intervals separately and then either add or subtract the correct values in order to determine the areas we seek.

Definite Integrals and Area

Let A be the area of the region bounded by the x-axis, the lines $x = a$ and $x = b$, and the graph of the continuous function $f(x)$.

If $f(x) \ge 0$ for $a \le x \le b$, then $A = \int_a^b f(x)\,dx.$

If $f(x) \le 0$ for $a \le x \le b$, then $A = -\int_a^b f(x)\,dx.$

Example 24   Let $y = f(x) = x^2 + 2x - 3$.

a. Find the area above this curve and below the x-axis.
b. Find the area between the curve and the x-axis from $x = 0$ to $x = 2$.
c. Using the results of part b, find the value of the integral from $x = 0$ to $x = 2$.

Solution   The graph of this function is shown in Figure 5-13.

a. To find the area above the curve and below the x-axis, we must first determine the points at which the curve crosses the x-axis. To do this, we set $y = 0$ and solve for x:

$$x^2 + 2x - 3 = 0$$

This equation can be factored as $(x + 3)(x - 1) = 0$. Therefore, the two solutions are $x = -3$ and $x = 1$. Now, since $f(x) < 0$ between $x = -3$ and $x = 1$, the value of the integral over this interval will be

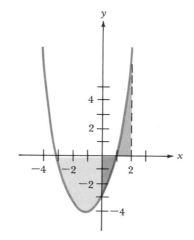

Figure 5-13
Graph of $y = x^2 + 2x - 3$

negative. Therefore, the area will be the *negative* of this integral, or

$$-\int_{-3}^{1} (x^2 + 2x - 3)\, dx = -\left(\frac{x^3}{3} + x^2 - 3x\right)\bigg|_{-3}^{1}$$

$$= -\left\{\left[\frac{(1)^3}{3} + (1)^2 - 3(1)\right] - \left[\frac{(-3)^3}{3} + (-3)^2 - 3(-3)\right]\right\}$$

$$= -\{[\tfrac{1}{3} - 2] - [-\tfrac{27}{3} + 18]\} = 10\tfrac{2}{3}$$

b. We see that from $x = 0$ to $x = 1$ the curve is below the x-axis (and will result in a negative value for the integral), and from $x = 1$ to $x = 2$ the curve is above the x-axis (and the integral will be positive). To find the area we must compute the integral in two parts as follows:

$$\text{Area} = -\int_{0}^{1} (x^2 + 2x - 3)\, dx + \int_{1}^{2} (x^2 + 2x - 3)\, dx$$

Note that we have taken the negative of the first integral since $f(x) < 0$ for $0 \le x \le 1$. Therefore,

$$\text{Area} = -\left(\frac{x^3}{3} + x^2 - 3x\right)\bigg|_{0}^{1} + \left(\frac{x^3}{3} + x^2 - 3x\right)\bigg|_{1}^{2}$$

$$= 1\tfrac{2}{3} + 2\tfrac{1}{3} = 4$$

c. Using the results from part b, we see that the value of the integral

$$\int_{0}^{2} (x^2 + 2x - 3)\, dx = \int_{0}^{1} (x^2 + 2x - 3)\, dx + \int_{1}^{2} (x^2 + 2x - 3)\, dx$$

$$= -1\tfrac{2}{3} + 2\tfrac{1}{3} = \tfrac{2}{3}$$

You might want to verify this by computing the integral directly.   ■

AREAS BETWEEN CURVES

We now address the problem of finding the area between two curves in the plane. Actually, the area between a curve and the x-axis, which we have just

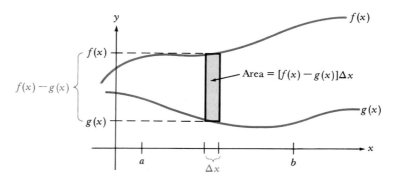

Figure 5-14

studied, is a special case of this problem, since we can think of the x-axis as the curve $y = 0$.

Suppose we are given two functions $f(x)$ and $g(x)$ such that $f(x) \geq g(x)$ over the interval from $x = a$ to $x = b$. The vertical distance between them at any point x is $f(x) - g(x)$, as shown in Figure 5-14. For a width of Δx, the area of the rectangle shown in Figure 5-14 is $[f(x) - g(x)]\Delta x$. As Δx approaches 0, the limit of the sum of all such rectangles between a and b is the integral

$$\int_a^b [f(x) - g(x)] \, dx$$

As before, we must be careful to make sure that $f(x) \geq g(x)$ over the entire range from $x = a$ to $x = b$ in order to compute the area correctly. If $f(x)$ lies below $g(x)$ over some interval, then the value of the integral will be negative and we must multiply the (negative) value of the integral by -1 to obtain the area.

Example 25   Find the areas of the regions A and B in the graph shown in Figure 5-15.

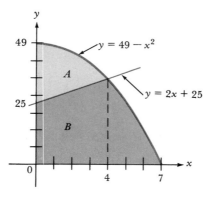

Figure 5-15

Solution   We must first find the point of intersection of the two curves. This is done by setting their y values equal to each other:

$$49 - x^2 = 2x + 25$$

This results in the equation

$$x^2 + 2x - 24 = 0$$

Factoring, we have $(x + 6)(x - 4) = 0$. Thus, the curves intersect at $x = 4$, as shown in Figure 5-15. To find the area of region A, we note that the curve $y = 49 - x^2$ lies above the curve $y = 2x + 25$ from $x = 0$ to $x = 4$. Thus, the area is given by the integral

$$\text{Area of } A = \int_0^4 [(49 - x^2) - (2x + 25)]\, dx$$

$$= \int_0^4 (24 - x^2 - 2x)\, dx$$

$$= \left(24x - \frac{x^3}{3} - x^2\right)\Big|_0^4$$

$$= \left[24(4) - \frac{(4)^3}{3} - (4)^2\right] - 0 = 58\tfrac{2}{3}$$

To find the area of region B, we must break the integral into two parts. From $x = 0$ to $x = 4$, the region is below the curve $y = 2x + 25$ and above the x-axis; from $x = 4$ to $x = 7$, the region is below the curve $y = 49 - x^2$ and above the x-axis. Thus, the total area of region B is given by

$$\text{Area of } B = \int_0^4 (2x + 25)\, dx + \int_4^7 (49 - x^2)\, dx = (x^2 + 25x)\Big|_0^4 + \left(49x - \frac{x^3}{3}\right)\Big|_4^7$$

$$= [(4)^2 + 25(4)] - [0] + \left[49(7) - \frac{(7)^3}{3}\right] - \left[49(4) - \frac{(4)^3}{3}\right]$$

$$= 170$$

Alternatively, we could find the area of region B by subtracting the area of A from the total area under the curve $y = 49 - x^2$ between $x = 0$ and $x = 7$:

$$\text{Area } B = \int_0^7 (49 - x^2)\, dx - \text{Area } A$$

$$= \left(49x - \frac{x^3}{3}\right)\Big|_0^7 - 58\tfrac{2}{3}$$

$$= 228\tfrac{2}{3} - 58\tfrac{2}{3} = 170 \qquad \blacksquare$$

Example 26
Profit and Loss

The marginal cost of producing a certain item has been found to follow the function $mc(x) = 10 + 10/\sqrt{x}$, where x is the number of units produced. The item sells for \$12 per unit. Since this is constant, the marginal revenue is 12. At what point does marginal revenue equal marginal cost, and how much loss will the firm incur if only this amount is produced and sold? What is the total profit if 100 items are produced and sold?

Solution

The total revenue is given by $R(x) = 12x$, so the marginal revenue is $R'(x) = 12$. Thus, to find the point at which marginal revenue equals marginal cost, we find the solution of the equation

$$10 + \frac{10}{\sqrt{x}} = 12$$

This is equivalent to

$$\frac{10}{\sqrt{x}} = 2$$

$$5 = \sqrt{x}$$

and hence, the solution is $x = 25$. When 25 items are produced, marginal cost equals marginal revenue. This is illustrated in Figure 5-16.

Figure 5-16

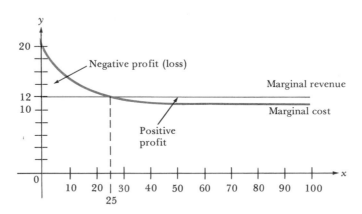

The difference $mc(x) - 12$, that is, the marginal cost minus the marginal revenue, represents the marginal loss $ml(x)$ of producing x items.

If the total loss of producing x items is $L(x)$, then by definition of the marginal loss,

$$ml(x) = L'(x)$$

that is, $L(x)$ is an antiderivative of $ml(x)$. Therefore, by the fundamental theorem,

$$L(25) - L(0) = \int_0^{25} ml(x)\, dx$$

Assuming there is no loss in producing 0 items, $L(0) = 0$, and hence, we find that the total loss in producing 25 items is

$$L(25) = \int_0^{25} ml(x)\, dx = \int_0^{25} [mc(x) - 12]\, dx$$

Therefore, the area between the curves $y = mc(x)$ and $y = 12$ from $x = 0$ to $x = 25$ represents the total loss if only 25 units are produced and sold. This is given by

$$\int_0^{25} \left(10 + \frac{10}{\sqrt{x}} - 12 \right) dx = \int_0^{25} (-2 + 10x^{-1/2})\, dx$$
$$= (-2x + 20x^{1/2})\big|_0^{25}$$
$$= -50 + 100 = 50 \text{ dollars}$$

If 100 units are produced and sold, the total profit is given by the integral of the marginal revenue minus the marginal cost function from $x = 0$ to $x = 100$, or

$$\int_0^{100} \left(12 - 10 - \frac{10}{\sqrt{x}} \right) dx = \int_0^{100} (2 - 10x^{-1/2}) \, dx$$

$$= 2x - 10x^{1/2}\big|_0^{100}$$

$$= 200 - 100 = 100 \text{ dollars}$$

Notice that the value of this integral is the sum of the negative value of the area between the two curves from $x = 0$ to $x = 25$ (representing a loss) and the area from $x = 25$ to $x = 100$ (representing a profit). We are not seeking a sum of areas in this case, but rather the (net) value of the integral. ∎

PROBLEMS (Section 5-4)

In Problems 1–5, what physical area does the definite integral of each function represent? Evaluate each integral and show its graphical interpretation.

1. $y = 2x - x^2$ from $x = 0$ to $x = 2$

2. $y = -x^2 + 2x + 3$ from $x = -1$ to $x = 3$

3. $y = \dfrac{3}{x^2}$ from $x = 1$ to $x = 3$

4. $y = x^3 + 2$ from $x = -3$ to $x = 0$

5. $y = -x^2 + 9$ from $x = 0$ to $x = 5$

6. Let $y = f(x) = x^2 + x - 6$.
   a. Find the area above this curve and below the x-axis.
   b. Find the area between the curve and the x-axis from $x = 0$ to $x = 4$.
   c. Using the results of part b, find the value of the integral from $x = 0$ to $x = 4$.

7. Let $y = f(x) = x^2 - 4$.
   a. Find the area above this curve and below the x-axis.
   b. Find the area between the curve and the x-axis from $x = -5$ to $x = 0$.
   c. Using the results of part b, find the value of the integral from $x = -5$ to $x = 0$.

In Problems 8–11, find the area between each pair of curves.

8. $y = x^2$ and $y = 2x + 8$

9. $y = 2x - x^2$ and $y = x - 2$

10. $y = x^2 - 4$ and $y = -x^2 + 4$

11. $y = x^3$ and $y = 4x$

12. In Example 26, find the total loss if only 9 units are produced and sold by evaluating the definite integral

$$\int_0^9 \left(10 + \frac{10}{\sqrt{x}} - 12 \right) dx$$

13. In Example 26, find the total profit if 81 units are produced and sold by evaluating the definite integral

$$\int_0^{81} \left(12 - 10 - \frac{10}{\sqrt{x}} \right) dx$$

14. The marginal cost of producing x items is given by the function

$$mc(x) = \frac{5400}{x^2} + 4$$

The item sells for $10 per unit.
   a. At what point does marginal revenue equal marginal cost?
   b. Suppose the item is currently selling at a rate of 25 units per month. How much additional profit will be generated if sales are increased to 50 units per month?

5-5 INTEGRATION BY SUBSTITUTION

In the applications of calculus that we have studied thus far, power functions have played a fundamental role. In Chapter 2 we learned how to differentiate simple power functions, and earlier in this chapter we saw how to find the integral of a simple power function. In Chapter 4 we considered the composition of a power function $f(x) = x^p$ with another function $g(x)$ and showed how to differentiate such functions. For instance, we showed that the derivative of $f(x) = (4x^2 - 3x)^3$ is

$$f'(x) = 3(4x^2 - 3x)^2(8x - 3)$$
$$= 3(8x - 3)(4x^2 - 3x)^2$$

Antidifferentiation, as we have seen, is the inverse of differentiation. Therefore, it must be true that

$$\int 3(8x - 3)(4x^2 - 3x)^2 \, dx = (4x^2 - 3x)^3 + c$$

In this section we show how to recognize integrals of general power functions.

The general power rule for derivatives states that if

$$f(x) = \frac{[g(x)]^{p+1}}{p + 1}$$

then

$$f'(x) = \frac{(p + 1)[g(x)]^{p+1-1}}{p + 1} g'(x) = [g(x)]^p g'(x)$$

Let us view this result in reverse. Suppose that we are given the function $[g(x)]^p g'(x)$. Then the following must be true:

THEOREM
General Power Rule for
Integration

$$\int [g(x)]^p g'(x) \, dx = \frac{[g(x)]^{p+1}}{p + 1} + c \qquad p \neq -1$$

This is very similar to the simple power rule formula that we studied earlier in this chapter. That is, we increase the exponent of $g(x)$ by 1 and divide by this new exponent. However, it is very important to realize that $[g(x)]^p$ *must* be multiplied by the derivative of $g(x)$ in order for the general power rule to apply. We need to be able to recognize functions that have this property. In the following examples, we present several different situations in which the general power rule for integration can and cannot be applied.

Example 27   Consider the integral $\int 6(6x + 2)^2 \, dx$.

Solution We first identify the function $g(x)$ as $6x + 2$. Now, the derivative of $g(x)$ is $g'(x) = 6$. Therefore, the function we wish to integrate has the form $g'(x)[g(x)]^2$. We may use the general power rule and obtain

$$\int 6(6x + 2)^2 \, dx = \frac{(6x + 2)^3}{3} + c = F(x)$$

Checking this antiderivative by differentiation, we find

$$F'(x) = \frac{3(6x + 2)^{3-1}}{3} \, (6) = 6(6x + 2)^2 \qquad \blacksquare$$

Example 28 Calculate the integral $\int (12x^2 - 2)(4x^3 - 2x)^3 \, dx$.

Solution In this case, $g(x) = 4x^3 - 2x$. Since $g'(x) = 12x^2 - 2$, the given integral satisfies the requirements for the general power rule for integration. Therefore, the integral must be

$$\int (12x^2 - 2)(4x^3 - 2x)^3 \, dx = \frac{(4x^3 - 2x)^4}{4} + c$$

We leave it as an exercise for the reader to check this by differentiation. $\qquad \blacksquare$

Example 29 Calculated the integral $\int (4x - 1)^{-2} \, dx$.

Solution In this problem, $g(x) = 4x - 1$ and $g'(x) = 4$. The function we are trying to integrate does not have the form $[g(x)]^p g'(x)$, although it is very close. In this case, the only difference is the constant factor of 4. Suppose that we multiply the function by $\frac{1}{4}(4)$. (We can always multiply any expression by 1.) Then we have

$$\int (4x - 1)^{-2} \, dx = \int \tfrac{1}{4}(4)(4x - 1)^{-2} \, dx$$

Since the integral of a constant times a function is equal to the constant times the integral of the function, we can move the constant $\frac{1}{4}$ outside the integral sign and write

$$\int \tfrac{1}{4}(4)(4x - 1)^{-2} \, dx = \tfrac{1}{4} \int \overset{g'(x)\quad\overbrace{[g(x)]^p}}{4(4x - 1)^{-2}} \, dx$$

This is now in the correct form. Applying the power rule yields

$$\frac{1}{4} \int 4(4x - 1)^{-2} \, dx = \left(\frac{1}{4}\right) \frac{(4x - 1)^{-2+1}}{-2 + 1} + c$$
$$= -\tfrac{1}{4}(4x - 1)^{-1} + c$$

You may check this by taking the derivative. $\qquad \blacksquare$

Example 30 Find the integral $\int 5x^2(x^3 + 1)^{1/2} \, dx$.

Solution Let $g(x) = x^3 + 1$; then $g'(x) = 3x^2$. Therefore, except for a constant factor, the function is in the proper form to apply the general power rule. We may multiply the function by $(\frac{1}{3})3$ to obtain

$$\int 5x^2(x^3 + 1)^{1/2} \, dx = \int (\tfrac{5}{3})3x^2(x^3 + 1)^{1/2} \, dx$$

and then move $\frac{5}{3}$ outside the integral:

$$\int 5x^2(x^3 + 1)^{1/2} \, dx = \tfrac{5}{3} \int 3x^2(x^3 + 1)^{1/2} \, dx$$

Now we may use the general power rule, which yields

$$\int 5x^2(x^3 + 1)^{1/2} \, dx = \left(\frac{5}{3}\right) \frac{(x^3 + 1)^{(1/2)+1}}{\frac{1}{2} + 1} + c$$

$$= \left(\frac{5}{3}\right) \frac{(x^3 + 1)^{3/2}}{\frac{3}{2}} + c$$

$$= \tfrac{10}{9}(x^3 + 1)^{3/2} + c$$

■

Example 31   Find the integral $\int (3x^2 + 2)^5 \, dx$.

Solution   If we let $g(x) = 3x^2 + 2$, then $g'(x) = 6x$. The function is not in the proper form. As opposed to the previous example, the difference is not simply a constant. We *can* multiply the function by $(1/6x)(6x)$, but we *cannot* then move $1/6x$ outside the integral sign because it involves the variable x. This can only be done for constants! Therefore, the general power rule for integration cannot be applied to this problem.   ■

From these examples, we can make the following observations regarding the application of the general power rule:

Guidelines for Applying the General Power Rule

1. First, identify a function $g(x)$ that is raised to a power p.
2. Find $g'(x)$, the derivative of $g(x)$.
3. Determine whether the function to be integrated has the form $[g(x)]^p g'(x)$. If the difference is only a constant factor of k appearing in $g'(x)$, then multiply the function by $(1/k)(k)$ and move all constant factors but k outside the integral sign.
4. Apply the general power rule to find the integral.
5. Check the result by differentiation.

Example 32   Integrate the following: $\int (x^3 + x)(x^4 + 2x^2)^3 \, dx$.

Solution   First, let $g(x) = x^4 + 2x^2$, since it is raised to the power 3. The derivative is $g'(x) = 4x^3 + 4x = 4(x^3 + x)$. The function differs only by the constant 4. Multiplying by $\frac{1}{4}(4)$ gives

$$\tfrac{1}{4} \int \overbrace{4(x^3 + x)}^{g'(x)} \overbrace{(x^4 + 2x^2)^3}^{[g(x)]^3} \, dx$$

and applying the general power rule, we obtain

$$\frac{1}{4} \int 4(x^3 + x)(x^4 + 2x^2)^3 \, dx = \left(\frac{1}{4}\right) \frac{(x^4 + 2x^2)^4}{4} + c$$

$$= \frac{(x^4 + 2x^2)^4}{16} + c$$

∎

The general power rule is only one method of integration by substitution. We shall discuss this topic further in Chapter 6.

PROBLEMS (Section 5-5)

In Problems 1–6, determine whether the general power rule for integration can be applied.

1. $\int (2x + 5)^4 \, dx$

2. $\int 6x(3x^2 + 1)^3 \, dx$

3. $\int 4(2x^2 - 5)^3 \, dx$

4. $\int (6x - 3)(2x^3 - 3x)^2 \, dx$

5. $\int 3x^2(5x - 4)^{-2} \, dx$

6. $\int (x^5 + x)(x^6 + 3x^2) \, dx$

Use the general power rule for integration to find each integral in Problems 7–26.

7. $\int (x + 5)^2 \, dx$

8. $\int 4(x - 7)^2 \, dx$

9. $\int \sqrt{3x - 1} \, dx$

10. $\int \sqrt{5x - 2} \, dx$

11. $\int \frac{1}{(2x + 7)^3} \, dx$

12. $\int \frac{3}{(4x + 1)^2} \, dx$

13. $\int x(x^2 - 3)^2 \, dx$

14. $\int x^2(2x^3 + 7) \, dx$

15. $\int x\sqrt{x^2 + 8} \, dx$

16. $\int x^3 \sqrt{2x^4 - 3} \, dx$

17. $\int x^2(x^3 + 5)^{3/4} \, dx$

18. $\int x^5(x^6 - 2)^{2/3} \, dx$

19. $\int (x - 1)(x^2 - 2x + 6)^{10} \, dx$

20. $\int (x^3 + x)(x^4 + 2x^2 - 3)^5 \, dx$

21. $\int x\sqrt{3 - x^2} \, dx$

22. $\int x^{-3}\sqrt{2 - x^{-2}} \, dx$

23. $\int \frac{x}{\sqrt{x^2 + 3}} \, dx$

24. $\displaystyle\int \frac{x^2}{\sqrt{x^3-2}}\,dx$

25. $\displaystyle\int \sqrt{x}\,(x^{3/2}-2)\,dx$

26. $\displaystyle\int \sqrt[3]{x}\,\frac{x^{4/3}+6}{3}\,dx$

Evaluate each definite integral.

27. $\displaystyle\int_0^3 3(3x-2)\,dx$

28. $\displaystyle\int_1^4 (5x+4)^{-2}\,dx$

29. $\displaystyle\int_0^2 x(x^2-4)^3\,dx$

30. $\displaystyle\int_0^4 x(x^2+8)^{-3}\,dx$

31. $\displaystyle\int_2^3 \frac{2x}{(x^2+5)^2}\,dx$

32. $\displaystyle\int_0^4 x\sqrt{3x^2+1}\,dx$

5-6 APPLICATIONS OF INTEGRATION

We discuss three major applications of integration in this section. These involve problems regarding *rates, average values,* and *proportions.* Remember that the key to applying integration properly is to keep in mind that the dimensions of the definite integral are the dimensions of the function times the dimensions of the independent variable. In this section we present several examples of general applications of integration; those dealing specifically with business and economics are discussed in Section 5-7.

PROBLEMS INVOLVING RATES

For problems involving rates, the integral represents a cumulative, or total, value. For instance, if the function represents marginal cost (cost/unit), then the integral represents total cost incurred by producing a specific quantity. If the function represents speed (distance/time), then the integral represents total distance traveled over time. If the function represents a learning rate (units produced/day), then the integral represents the total quantity produced over a period of time.

Example 33
Water Consumption

On a hot summer day, the demand for water in a particular township varies over time. The demand rate (in thousands of gallons per hour) is given by

$$f(t) = -t^2 + 24t + 180 \qquad 0 \le t \le 24$$

where t is the number of hours after midnight. How much water is demanded over a 24 hour period? Between noon and 5 P.M.?

Solution

We first note that the area under $f(t)$ over an interval $[a, b]$ has dimensions of thousands of gallons, since $f(t)$ is expressed in thousands of gallons per hour and t is measured in hours. Therefore, the total amount of water demanded over the interval $[a, b]$ is given by

$$F(b) - F(a) = \int_a^b f(t)\, dt \qquad \text{thousands of gallons}$$

In general, we see that if $F(t)$ denotes the total cumulative demand at time t, then the *demand rate*, $F'(t)$ is equal to $f(t)$. That is, $F(t)$ is an antiderivative of $f(t)$.

Thus, over a 24 hour period, we have

$$\int_0^{24} f(t)\, dt = \int_0^{24} (-t^2 + 24t + 180)\, dt$$

$$= \left(-\frac{t^3}{3} + 12t^2 + 180t \right)\Bigg|_0^{24}$$

$$= -\frac{(24)^3}{3} + 12(24)^2 + 180(24) - 0$$

$$= -4608 + 6912 + 4320 = 6624 \text{ thousand gallons}$$

The interval between noon and 5 P.M. corresponds to $12 \leq t \leq 17$:

$$\int_{12}^{17} f(t)\, dt = \left(-\frac{t^3}{3} + 12t^2 + 180t \right)\Bigg|_{12}^{17}$$

$$= \left[-\frac{(17)^3}{3} + 12(17)^2 + 180(17) \right] - \left[-\frac{(12)^3}{3} + 12(12)^2 + 180(12) \right]$$

$$= [-1637\tfrac{2}{3} + 3468 + 3060] - [-576 + 1728 + 2160]$$

$$= 1578\tfrac{1}{3} \text{ thousand gallons} \qquad\qquad \blacksquare$$

Example 34
Colonization of Species

In biogeography, a *colonization curve* represents the number of animals of a species present on an island, for example, as a function of time. It is determined from knowledge of the rate of immigration of a species and the rate of extinction. Suppose that the rate of immigration of a species is given by

$$I(t) = 50 + \frac{100}{t^2} \qquad t \geq 1$$

and that the rate of extinction is given by

$$E(t) = 50 - \frac{\sqrt{t} + 49}{t^2} \qquad t \geq 1$$

where t is time (in years). The difference of these rates, $I(t) - E(t)$, gives the rate at which the animal population changes. (For simplicity, we will ignore births on the island, although if we know the birth rate we can certainly take it into account.) Therefore, if $C(t)$ represents the animal count at time t, then

$$C'(t) = I(t) - E(t)$$

Assuming the island is unpopulated at $t = 1$, that is, $C(1) = 0$, by the fundamental theorem we find

$$C(T) = \int_1^T [I(t) - E(t)]\, dt$$

Therefore, the colonization curve is given by $C(T)$ and represents the number of animals present at time T. Now,

$$I(t) - E(t) = \left(50 + \frac{100}{t^2}\right) - \left(50 - \frac{\sqrt{t} + 49}{t^2}\right)$$

$$= \frac{100}{t^2} + \frac{\sqrt{t}}{t^2} + \frac{49}{t^2}$$

$$= 149t^{-2} + t^{-3/2}$$

The integral of this function from $t = 1$ to $t = T$ is

$$\int_1^T (149t^{-2} + t^{-3/2})\, dt = (-149t^{-1} - 2t^{-1/2})|_1^T$$

$$= \left[-\frac{149}{T} - \frac{2}{\sqrt{T}}\right] - [-149 - 2]$$

$$= 151 - \frac{149}{T} - \frac{2}{\sqrt{T}}$$

As T gets large, the number of animals approaches the value 151. This is illustrated in Figure 5-17.

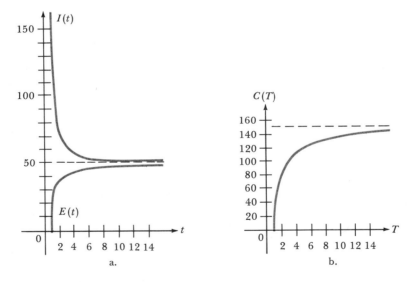

Figure 5-17
Relationships among immigration,
extinction, and colonization curves

PROBLEMS INVOLVING AVERAGE VALUES

Earlier in this chapter, we discussed the average value of a function over an interval. In general, the average value of a function $f(x)$ over the interval $[a, b]$ is given by the integral

$$\frac{1}{b - a} \int_a^b f(x)\, dx$$

Note that the denominator $b - a$ has the same dimensions as the variable x. Therefore, the dimensions of the integral above are the same as the dimensions of $f(x)$. Many situations require finding the average value of a function. We give some illustrations below.

Example 35
Average Speed

The speed of an object (in feet per second) at time t is given by the function

$$v = \sqrt{25 + 50t} \qquad 0 \le t \le 4$$

What is the average speed from $t = 0$ to $t = 4$ seconds?

Solution

We find the integral of v from $t = 0$ to $t = 4$ and divide by 4, the length of the interval. Using the general power rule for integration, we obtain

$$\int_0^4 (25 + 50t)^{1/2}\, dt = \frac{1}{50} \int_0^4 50(25 + 50t)^{1/2}\, dt$$

$$= \frac{1}{50} \left[\frac{(25 + 50t)^{3/2}}{1.5} \right]\Big|_0^4$$

$$= \frac{(25 + 50t)^{3/2}}{75}\Big|_0^4$$

$$\approx 45 - 1.67 \approx 43.3$$

The average speed is then about $\frac{1}{4}(43.3) \approx 10.8$ feet per second. ∎

Example 36
Average Temperature

The temperature in Kansas City over a 24 hour period is given by the function

$$T = 25 + 6.4t - 0.23t^2$$

where t is the number of hours since the start of the time period and T is temperature in degrees Fahrenheit. What was the average temperature during this period?

Solution

The average temperature is given by the integral

$$\int_0^{24} (25 + 6.4t - 0.23t^2)\, dt$$

divided by the length of the day (24 hours), or

$$\frac{1}{24}[(25t + 3.2t^2 - \tfrac{0.23}{3}t^3)|_0^{24}] = \frac{1}{24}[25(24) + 3.2(24)^2 - \tfrac{0.23}{3}(24)^3]$$

$$\approx \frac{1}{24}(1378.75) \approx 57.45 \text{ degrees Fahrenheit} \quad ∎$$

PROBLEMS INVOLVING PROPORTIONS

Many applications of integration involve finding the *proportion* of a total population having certain values or characteristics. To see how integration is related to proportions, let us return to the checking account example presented in Section 5-2. In that example, the account maintained a balance of $100 for 3 days, $150 for 7 days, and so on, as summarized in Table 5-3.

Suppose the following question is posed: What proportion of the time is the balance equal to $240? Since there are 30 days under consideration and the balance was $240 for 5 days, the proportion is $\frac{5}{30} = \frac{1}{6}$. Thus, if each value in the second column of Table 5-3 is divided by 30, we find the proportion of time that the balance was at the given level. Notice that it does not matter how we actually

Table 5-3

Balance	Number of Days	Proportion of time
100	3	$\frac{3}{30}$
150	7	$\frac{7}{30}$
160	7	$\frac{7}{30}$
220	2	$\frac{2}{30}$
240	5	$\frac{5}{30}$
300	6	$\frac{6}{30}$

scale these numbers; they simply represent the *relative amounts of time,* or *likelihood,* that the balance is at the given level. However, we did scale them so that the sum of the relative times is equal to 1. Let us construct a graph in which balance corresponds to the x-axis and the relative time corresponds to the y-axis, as shown in Figure 5-18. The graph clearly indicates that the relative amount of time at which the balance is \$160 is greater than that for \$220, for instance. Since the sum of all the y values is 1, we can state that the proportion of time that the balance is between \$100 and \$300 is equal to 1. Similarly, the proportion of time that the balance is between \$200 and \$280 is given by $\frac{2}{30} + \frac{5}{30} = \frac{7}{30}$ (the proportions corresponding to \$220 and \$240).

Figure 5-18
Likelihood of account balances, x

Now suppose that a very large number of transactions occurred during this 30 day period so that every balance between \$100 and \$300 occurred. A plot of such points would give a continuous curve such as the one shown in Figure 5-19.

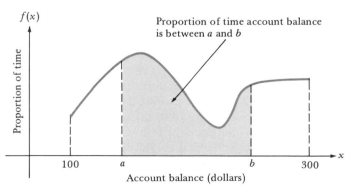

Figure 5-19
Continuous function representing likelihood of account balances, x

We cannot actually add all the *y* values, since there are now an infinite number of them. However, we can determine the area under the curve through integration. If the *y*-axis is properly scaled, the area from $x = 100$ to $x = 300$ will equal 1. In this case, it makes sense to define the area between *any* two values *a* and *b*, $100 \leq a \leq b \leq 300$, as the proportion of time that the account balance is between *a* and *b*.

The likelihood that a variable assumes a value in an interval can be described in terms of a positive function called a *density function* for the variable. If *f(x)* is a density function for the variable *x*, then the proportion of values *x* between *a* and *b* is defined to be the area under the graph of *f(x)* between $x = a$ and $x = b$, that is, $\int_a^b f(x) \, dx$.

Example 37
Social Status

Suppose that *N(s)* is a density function giving the likelihood that an individual will have a particular social status measured by the quantity *s*. For example, *s* might represent the cost of automobiles owned by individuals in a certain population. The function *N(s)* would then represent the likelihood that an individual will own a car having a value *s*. The graph in Figure 5-20 serves as an example. We see that it is more likely that a member of this population owns an automobile having value *c* than any other value. The total area under the curve must be equal to 1, since it represents the entire population of automobile owners. The area between any two points *a* and *b* will represent the proportion of the population who own automobiles having values between *a* and *b*.

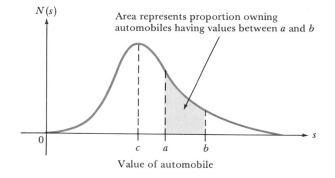

Figure 5-20
Density function representing proportions of automobile owners

Area represents proportion owning automobiles having values between *a* and *b*

Value of automobile

Example 38
Airline Flight Time

The flight time *x* of an airplane between Cincinnati and Chicago varies between 40 and 60 minutes, depending on the weather and flight traffic at O'Hare Airport. Suppose we know from historical data that any value between 40 and 60 minutes is equally likely, that is, there is the same chance that a flight will take between 40 and 41 minutes as any other 1 minute interval. Clearly, 100% of the flight times should fall between 40 and 60. We may represent the density function by

$$f(x) = \frac{1}{60 - 40} = \frac{1}{20} \qquad 40 \leq x \leq 60$$

The graph is shown in Figure 5-21.

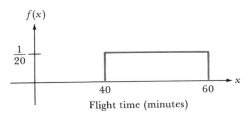

Figure 5-21
Density function for airline flight
times

$f(x)$

$\dfrac{1}{20}$

40   60 x

Flight time (minutes)

The function $f(x)$ represents the likelihood that a flight time will take x minutes. Since we have assumed that any value between 40 and 60 is equally likely, we see that $f(x)$ is constant. Also, note that the area under the graph of $f(x)$ is given by

$$\int_{40}^{60} \tfrac{1}{20}\, dx = \tfrac{1}{20}x\big|_{40}^{60}$$

$$= \tfrac{1}{20}(60 - 40) = 1$$

If we think of this as 100%, then the area under the graph of $f(x)$ between any two values of x represents the proportion of flights that have flight times between these two values. For example, the area under the graph between $x = 50$ and $x = 55$ is

$$\int_{50}^{55} \tfrac{1}{20}\, dx = \tfrac{1}{20}x\big|_{50}^{55}$$

$$= \tfrac{1}{20}(55 - 50) = 0.25$$

Thus, 25% of the flights will take between 50 and 55 minutes.   ■

Example 39
Population Density   The population density at a radius r from the central business district in a city to the city limits 15 miles away is given by

$$f(r) = -0.06r + 0.2596r^{0.3} \qquad 0 \le r \le 15$$

(The integral of this function from $r = 0$ to $r = 15$ is approximately equal to 1.) What is the proportion of the population that lives within x miles of the central business district?

Solution   This is given by the integral

$$F(x) = \int_{0}^{x} (-0.06r + 0.2596r^{0.3})\, dr$$

$$= \left(-0.03r^2 + \frac{0.2596r^{1.3}}{1.3} \right)\bigg|_{0}^{x}$$

$$= -0.03x^2 + 0.1997x^{1.3}$$

For example, if $x = 4$, then

$$F(4) = -0.03(4)^2 + 0.1997(4)^{1.3} \approx 0.73$$

That is, about 73% of the population lives within 4 miles of the central business district.   ■

PROBLEMS (Section 5-6)

1. The demand for electricity varies over time, especially during hot summer days. The demand rate (in kilowatts per hour) is given by

$$f(t) = -t^2 + 10t + 120 \qquad 0 \le t \le 24$$

where t is the number of hours after 5 A.M. What is the demand for electricity during a full day? Between 10 A.M. and 6 P.M.?

2. The preparation time (in minutes per meal) when x gourmet meals are prepared in a restaurant is given by the function

$$f(x) = \sqrt{x} + 3 \qquad 0 \le x \le 25$$

How long will it take to prepare 25 gourmet meals?

3. The rate of growth of college enrollment (in students per year) is given by the function

$$E(t) = 5000(t + 1)^{-3/2} \qquad t \ge 0$$

What is the total growth over T years?

4. It is estimated that t months from now the population of a certain city will be changing at a rate of $6 + 8t^{2/3}$ people per month. If the current population is 40,000, what will the population be 8 months from now?

5. It is estimated that t months from now the population of a small developing country will be changing at a rate of $40 + 120t^{1/2}$ people per month. If the population today is 200,000, what will the population be in 9 months?

6. A car is driven for 3 hours so that its speed after t hours have elapsed is given by the function $s(t) = 60 + 6t - t^2$ miles per hour. What is the car's average speed during the last hour of travel?

7. The number of children in a city was found to increase and decrease rather quickly because of various factors such as increases and decreases in the birth rate, movement of families in and out of the city, company transfers, and so on. If the number of children (in thousands) over a 10 year period was found to be approximately given by the function

$$N(t) = -\tfrac{1}{4}t^2 + 2t + 1 \qquad 0 \le t \le 10$$

what was the average number of children in the city over the 10 year period?

8. The air temperature during a 12 hour period is given by the function

$$T(t) = 50 + 5t - 0.4t^2 \qquad 0 \le t \le 12$$

where t is measured in hours and T in degrees Fahrenheit.
   a. What is the average temperature during the first 6 hours?
   b. What is the average temperature during the entire period?

9. The rate (in units per hour) of solar energy captured by solar panels on a California home varies according to the function

$$g(x) = -x^2 + 12x$$

where x represents the number of hours after sunrise.
   a. How many units of energy are captured in a 12 hour sunny period?
   b. What is the average number of units of energy captured per hour?

10. Suppose that the amount of oil produced by a well after it has been operating t years is $A(t)$ thousand barrels, but that we do not know the expression for $A(t)$. Upon inquiry, we learn that the *rate* of output per year was 50 thousand barrels per year at the start of pumping ($t = 0$) and is expected to decrease steadily by 2 thousand barrels per year.
   a. Write a function for the rate of production at t years.
   b. Find the total production during the first 10 years.

11. The driving time of a car between two toll booths on a tollway varies between 70 and 100 minutes, depending on the speed of the car and the amount of traffic on the tollway. Any travel time

between 70 and 100 minutes is equally likely to occur.

a. Write a function that represents the relative likelihood that travel time between the toll booths will be x minutes.

b. What proportion of cars will have travel times between 90 and 95 minutes?

12. The proportion of downtown employees who walk x miles from their car to their office is given by the density function

$$f(x) = \frac{1}{2\sqrt{x}} \qquad 0 < x \leq 1$$

a. What proportion of employees walk less than $\frac{1}{9}$ mile?

b. What proportion of employees walk between $\frac{1}{2}$ and 1 mile?

13. A wholesale vegetable distributor finds that the weights of melons (in kilograms) have the density function

$$f(x) = \frac{9}{4x^3} \qquad 1 \leq x \leq 3$$

a. What proportion of melons weighs less than 2 kilograms?

b. If the distributor sells 10,000 melons, about how many will weigh less than 2 kilograms?

14. The density function for the amount of time (in minutes) that a customer waits to use an automatic bank teller at the First National Bank is given by

$$f(t) = \tfrac{3}{2}t^2 + t \qquad 0 \leq t \leq 1$$

a. What is the proportion of customers that waits less than $\frac{3}{4}$ minute?

b. If 120 customers use the automatic teller daily, how many will wait *more* than $\frac{3}{4}$ minute?

5-7 APPLICATIONS OF INTEGRATION TO BUSINESS AND ECONOMICS

In this section we present several additional examples involving rates, average values, and proportions that specifically relate to business and economics.

Example 40
Production and Demand

A fast-food restaurant can produce burgers at a rate of $P(t) = 10$ burgers per minute. During the peak lunch period, the demand (in burgers per minute) is given by

$$D(t) = \begin{cases} \tfrac{3}{4}t & \text{if} \quad 0 \leq t \leq 20 \\ 15 & \text{if} \quad 20 < t \leq 40 \\ -\tfrac{3}{4}t + 45 & \text{if} \quad 40 < t \leq 60 \end{cases}$$

where t is the number of minutes since the start of the busy period.

a. How many burgers remain at the end of the 60 minute period if 0 are available at the start?

b. What is the supply of burgers at any point in time if 0 are available at the start of the period?

Solution   The net rate of supply is given by $P(t) - D(t)$. Since $P(t) = 10$, we have

$$P(t) - D(t) = \begin{cases} 10 - \tfrac{3}{4}t & \text{if} \quad 0 \leq t \leq 20 \\ -5 & \text{if} \quad 20 < t \leq 40 \\ \tfrac{3}{4}t - 35 & \text{if} \quad 40 < t \leq 60 \end{cases}$$

If $S(t)$ represents the supply of burgers at time t, then the rate of supply is $S'(t) = P(t) - D(t)$. That is, $S(t)$ is an antiderivative of $P(t) - D(t)$. Therefore, by the fundamental theorem (and the fact that the initial supply is 0), we have

$$S(T) = S(T) - S(0) = \int_0^T [P(t) - D(t)] \, dt$$

This integral represents the number of burgers available at time T, which is equal to the cumulative production less the cumulative demand over the interval $[0, T]$.

a. The number of burgers remaining at the end of the 60 minute period will be given by the integral of the function $P(t) - D(t)$ from $t = 0$ to $t = 60$. Since the function is defined by three different formulas over this interval, we must break the integral into three distinct parts:

$$\int_0^{60} [P(t) - D(t)] \, dt$$

$$= \int_0^{20} (10 - \tfrac{3}{4}t) \, dt + \int_{20}^{40} -5 \, dt + \int_{40}^{60} (\tfrac{3}{4}t - 35) \, dt$$

$$= (10t - \tfrac{3}{8}t^2)|_0^{20} - 5t|_{20}^{40} + (\tfrac{3}{8}t^2 - 35t)|_{40}^{60}$$

$$= [200 - \tfrac{3}{8}(400)] - [200 - 100] + [\tfrac{3}{8}(60)^2 - 35(60) - \tfrac{3}{8}(40)^2 + 35(40)]$$

$$= 200 - 150 - 200 + 100 + 1350 - 2100 - 600 + 1400$$

$$= 0$$

Therefore, at the end of the peak lunch period, no burgers remain.

b. The answer to this question depends on the interval in which T falls. Thus, if $0 \leq T \leq 20$, we have

$$S(T) = \int_0^T (10 - \tfrac{3}{4}t) \, dt$$

$$= (10t - \tfrac{3}{8}t^2)|_0^T = 10T - \tfrac{3}{8}T^2$$

If $20 < T \leq 40$,

$$S(T) = \int_0^{20} (10 - \tfrac{3}{4}t) \, dt + \int_{20}^T -5 \, dt$$

$$= (10t - \tfrac{3}{8}t^2)|_0^{20} - 5t|_{20}^T$$

$$= [10(20) - \tfrac{3}{8}(20)^2 - 0] + [-5T + 5(20)]$$

$$= 150 - 5T$$

Finally, if $40 < T \leq 60$, we have

$$S(T) = \int_0^{20} (10 - \tfrac{3}{4}t) \, dt + \int_{20}^{40} -5 \, dt + \int_{40}^T (\tfrac{3}{4}t - 35) \, dt$$

$$= 50 + (-100) + (\tfrac{3}{8}t^2 - 35t)|_{40}^T$$

$$= -50 + [\tfrac{3}{8}T^2 - 35T] - [\tfrac{3}{8}(1600) - 35(40)]$$

$$= \tfrac{3}{8}T^2 - 35T + 750$$

Note that the answer to part a is $S(60) = 0$ using this general formula.   ∎

Example 41
Average Cost

We have seen examples in this book that deal with learning curves. Understanding the nature of learning in production is important in cost estimating and budgeting, particularly for companies such as airplane manufacturers and defense contractors who often produce new products, or toy companies who often change their product lines. The production cost of each successive unit will often decrease due to more experience, new technology, and other learning factors. The general form of a production learning curve is $y = f(x) = Ax^{-b}$, where y is the cost of the xth unit produced, A is the cost of the first unit [since $f(1) = A$], and b is a learning parameter, generally between 0 and 1.

Suppose an aircraft engine contractor estimates that $b = 0.2$ and $A = \$560,000$ for a new engine. What is the average cost of the first 20 units?

Solution

The average cost of the first 20 units is given by the integral

$$\frac{1}{20} \int_0^{20} 560,000 \, x^{-0.2} \, dx = \frac{1}{20} \left[560,000 \left(\frac{x^{0.8}}{0.8} \right) \Big|_0^{20} \right]$$

$$\approx 28,000 \left(\frac{10.99}{0.8} - 0 \right)$$

$$= 384,650 \text{ dollars}$$

∎

Example 42
Consumer and Producer Surplus

In economics, both the quantity of a commodity demanded by consumers and the quantity of the commodity supplied by producers in a market are related to its price. An equation expressing price as a function of demand is called a **demand curve**, while an equation expressing price as a function of supply is called a **supply curve**. Generally, demand increases with decreasing price and supply increases with increasing price. The point of intersection of the supply and demand curves for a product is called the **equilibrium point** (see Figure 5-22).

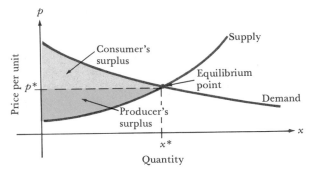

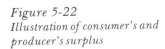

Figure 5-22
Illustration of consumer's and producer's surplus

Suppose that the price is fixed at its equilibrium value p^*, but that consumers are willing to pay the prices described by the demand curve. Then consumers have gained an amount known as the **consumer's surplus**. This is the area under the demand curve and above the line $p = p^*$, as shown in Figure 5-22. Note that the area has dimensions (dollars/unit)(units demanded) = total dollars that would have been spent if the price were not fixed at $p = p^*$. Similarly, the **producer's surplus** is defined as the area below the line $p = p^*$ and above the supply curve. It represents the amount that producers will gain by being able to sell at a price p^* higher than that described by the supply curve.

Suppose that the demand curve is $p(x) = 75 - 0.5x^2$ and the supply curve is $p(x) = 0.25x$. Then the equilibrium point is found by setting these equations equal to each other:

$$75 - 0.5x^2 = 0.25x$$
$$0.5x^2 + 0.25x - 75 = 0$$

Applying the quadratic formula yields

$$x = -12.5 \quad \text{and} \quad x = 12$$

Since only the positive root is of interest, we find the equilibrium point given as

$$x^* = 12$$
$$p^* = 0.25(12) = 3$$

Therefore, the consumer's surplus is given by

$$\int_0^{12} (75 - 0.5x^2 - 3)\, dx = \left(72x - \frac{x^3}{6} \right)\Bigg|_0^{12}$$
$$= (864 - 288) - 0 = 576 \text{ dollars}$$

The producer's surplus is

$$\int_0^{12} (3 - 0.25x)\, dx = \left(3x - \frac{x^2}{8} \right)\Bigg|_0^{12}$$
$$= (36 - 18) - 0 = 18 \text{ dollars} \qquad \blacksquare$$

Example 43
Quality Control

A density function describing the number of ounces of cola filled in a 16 ounce bottle by an automatic filling machine is given as

$$f(x) = \begin{cases} 25x - 395 & \text{if} \quad 15.8 \le x \le 16 \\ -25x + 405 & \text{if} \quad 16 < x \le 16.2 \end{cases}$$

This is shown in Figure 5-23. A new federal law requires that each bottle contain at least 15.9 ounces. What proportion of bottles will actually contain at least 15.9 ounces? What proportion will be out of compliance with the law?

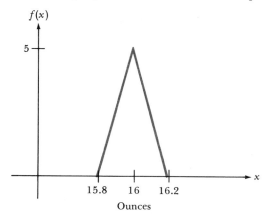

Figure 5-23
Graph of the density function f(x)

Solution

The proportion containing at least 15.9 ounces is shown by the shaded region in Figure 5-24. This area is given by the integral

$$\int_{15.9}^{16.2} f(x)\, dx = \int_{15.9}^{16} (25x - 395)\, dx + \int_{16}^{16.2} (-25x + 405)\, dx$$

$$= \left(\frac{25x^2}{2} - 395x\right)\Bigg|_{15.9}^{16} + \left(-\frac{25x^2}{2} + 405x\right)\Bigg|_{16}^{16.2}$$

$$= (3200 - 6320 - 3160\tfrac{1}{8} + 6280\tfrac{1}{2}) + (-3280\tfrac{1}{2} + 6561 + 3200 - 6480)$$

$$= \tfrac{3}{8} + \tfrac{1}{2} = 0.875$$

Thus, 87.5% of the bottles will actually contain 15.9 ounces or more. Since the total proportion is 1, the proportion out of compliance will be $1 - 0.875 = 0.125$, or 12.5%. This is equal to the value of the integral $\int_{15.8}^{15.9} f(x)\, dx$.

Figure 5-24
Proportion of bottles containing at least 15.9 ounces

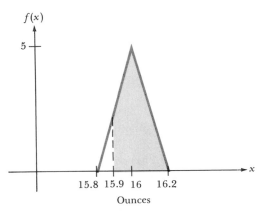

Example 44
Waiting Time in a Supermarket

Suppose that the density function describing the likelihood of waiting t minutes in a supermarket express checkout line is given by

$$f(t) = \frac{1}{4\sqrt{t}} \qquad 1 \le t \le 9$$

This function is shown in Figure 5-25. We can see, for example, that it is more likely that a customer will wait 1 minute than 9 minutes, since $f(1) = \tfrac{1}{4}$ while $f(9) = \tfrac{1}{12}$. You have probably experienced this phenomenon yourself; a high proportion of customers will not have to wait very long while a smaller percentage may have a long wait. We note that all customers will have to wait between 1

Figure 5-25
Graph of the density function for waiting time

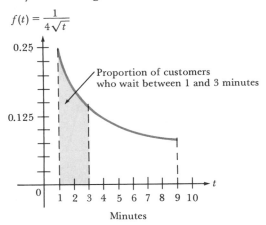

and 9 minutes since

$$\int_1^9 f(t)\, dt = \int_1^9 \tfrac{1}{4} t^{-1/2}\, dt$$
$$= \tfrac{1}{2} t^{1/2}\big|_1^9 = \tfrac{3}{2} - \tfrac{1}{2} = 1$$

We can use such a function to predict the proportion of customers that will have to wait for a specified amount of time. For example, the proportion of customers who will have to wait between 1 and 3 minutes is given by the integral

$$\int_1^3 f(t)\, dt = \int_1^3 \tfrac{1}{4} t^{-1/2}\, dt$$
$$= \tfrac{1}{2} t^{1/2}\big|_1^3$$
$$= \tfrac{1}{2}(3)^{1/2} - \tfrac{1}{2}(1)^{1/2}$$
$$\approx 0.3660, \text{ or } 36.6\%$$

PROBLEMS (Section 5-7)

1. A movie theater can sell tickets at a rate of $P(t) = 15$ tickets per minute. During the customer rush preceding a Saturday night movie, the demand is given by

$$D(t) = \begin{cases} \tfrac{1}{2}t & \text{if } 0 \le t \le 10 \\ 10 & \text{if } 10 < t \le 20 \\ -\tfrac{1}{2}t + 40 & \text{if } 20 < t \le 30 \end{cases}$$

where t is the number of minutes since the start of the pre-movie rush.
   a. What is the supply of tickets at any point in time if they are sold at the rate of 15 per minute and if 0 tickets are sold at the start of the period?
   b. How many tickets remain at the end of the 30 minute period?

2. The assembly time (in hours per unit) for a new product when x units are being produced is given by the function $f(x) = 17x^{-0.25}$ for $0 \le x \le 16$. How long does it take to produce the first 16 units? What is the average assembly time per unit?

3. Suppose that an exclusive toy company manufacturing toy robots estimates that their learning parameter is 0.4 and the cost of the first unit is $120 (see Example 41). Using the general form of the production learning curve, find the total cost of the first 10 robots and the average cost of each.

4. The rate of arrivals (in customers per minute) at a restaurant t minutes after opening is given by the function

$$r(t) = \begin{cases} 0.15 + 0.005(t - 150) & \text{if } 150 \le t \le 210 \\ 0.45 + 0.005(t - 210) & \text{if } 210 < t \le 270 \end{cases}$$

   a. How many customers arrive over this time period?
   b. What is the average rate of arrival over this time period?

5. A manufacturer purchased a machine for $5000, and its rate of change in book value in t months is given by the function $r(t) = -700(t + 2)$ dollars per month.
   a. What is the value of the machine at the end of T months?
   b. In approximately how many months will the machine have a book value of $0?

6. Suppose that the demand curve for a product is $p(x) = 40 - \tfrac{1}{4}x^2$ and the supply curve is $p(x) = 2x - 20$ (see Example 42).
   a. Find the equilibrium point.
   b. Find the consumer's surplus.
   c. Find the producer's surplus.

7. Suppose that the demand curve for a product is $p(x) = 100 - 0.2x$ and the supply curve is $p(x) = 10 + 0.1x$ (see Example 42).
   a. Find the equilibrium point.
   b. Find the consumer's surplus.
   c. Find the producer's surplus.

8. A company produces batteries. Their engineers have determined that the life, t, of a battery is from 1 to 3 years and that the density function for the life is given by

   $$f(t) = \frac{9}{4t^3} \qquad 1 \le t \le 3$$

   a. Sketch the graph of $f(t)$. Does it seem reasonable?
   b. What proportion of the batteries will last no more than 2 years?
   c. What proportion of the batteries will last from 2 to 3 years?

9. Suppose that the density function for the amount of time, x, spent on coffee breaks by a group of employees is

$$f(x) = \tfrac{3}{500}(-x^2 + 10x)$$

for coffee breaks running from 0 to 10 minutes.
   a. Show that this represents a valid density function.
   b. What proportion of the employees actually spend at least 5 minutes on coffee breaks?

10. A density function describing the actual number of ounces x of cereal dispensed into a 20 ounce box by an automated filling machine is given by

$$f(x) = \begin{cases} \tfrac{1}{9}(100x - 1970) & \text{if} \quad 19.7 \le x \le 20 \\ \tfrac{1}{9}(-100x + 2030) & \text{if} \quad 20 < x \le 20.3 \end{cases}$$

The legislature is considering passage of a new law requiring that each box contain at least 19.8 ounces.
   a. What proportion of cereal boxes will actually contain at least 19.8 ounces?
   b. What proportion will be out of compliance with the law if it is passed?
   c. What proportion will contain more than 20 ounces, thus costing the company extra money?

5-8 CONCLUSION: SOLVING SWEIGART AND ASSOCIATES' ELEVATOR PROBLEM

The problem faced by Sweigart and Associates in the Introduction to this chapter was to find the distance traveled by an elevator given the velocity as a function of time. Since distance = velocity × time, we now know that distance is the integral of the velocity with respect to time. Thus, the distance traveled during the first 10 seconds is given by

$$\int_0^{10} v(t)\, dt = \int_0^{1.75} 4t\, dt + \int_{1.75}^{8.25} 7\, dt$$
$$+ \int_{8.25}^{10} (40 - 4t)\, dt$$
$$= 2t^2|_0^{1.75} + 7t|_{1.75}^{8.25} + (40t - 2t^2)|_{8.25}^{10}$$
$$= (6.125 - 0) + (57.75 - 12.25)$$
$$+ (400 - 200 - 330 + 136.125)$$
$$= 57.75 \text{ feet}$$

Without the concept of integration, we would not have been able to determine the actual distance traveled. In a similar way, we can compute distances for any velocity function using integration. Sweigart and Associates can therefore evaluate alternative elevator designs.

Techniques of integration can easily provide answers to seemingly difficult problems involving rates.

5-9 CHAPTER REVIEW

1. The area under the graph of a function has dimensions equal to that of the dependent variable times the dimensions of the independent variable.

2. An antiderivative of a function $f(x)$ is a function $F(x)$ whose derivative is $f(x)$, that is, $F'(x) = f(x)$. The set of all antiderivatives of $f(x)$ is denoted $\int f(x)\,dx$ and is called the indefinite integral of $f(x)$. Indefinite integration is therefore the inverse of differentiation.

3. Since the derivative of a constant is 0, there are an infinite number of antiderivatives of a given function that differ only by a constant.

4. For powers of x we have

$$\int x^p\,dx = \frac{x^{p+1}}{p+1} + c \qquad p \neq -1$$

5. The area under the graph of any nonnegative function $f(x)$ from $x = a$ to $x = b$ can be approximated by breaking it up into small rectangles and adding their areas together. This concept gives rise, as the widths of the rectangles approach 0, to the definite integral

$$\int_a^b f(x)\,dx$$

Thus, the definite integral represents the area under a curve between a and b.

6. The fundamental theorem of calculus allows a simple computation of the definite integral as

$$\int_a^b f(x)\,dx = F(b) - F(a)$$

where F is an antiderivative of f. This provides a simple means of evaluating the definite integral.

7. The definite integral from a to b always gives the algebraic sum of areas between the curve and the x-axis. That is, if the curve lies *below* the x-axis, the value of the definite integral will be negative. Care must be taken when trying to evaluate areas in the plane to be sure that the correct signs are assigned to the integrals evaluated.

8. The general power rule for integration is

$$\int [g(x)]^p g'(x)\,dx = \frac{[g(x)]^{p+1}}{p+1} + c \qquad p \neq -1$$

9. Integration has many applications. Three of the most common involve rates, proportions, and average values.

REVIEW PROBLEMS (Section 5-9)

Describe the dimensions of the area under each curve when:

1. The amount of bonus in dollars is a function of the number of items sold above a quota

2. The yearly growth of a child in inches is a function of time

Use Riemann sums to estimate the area under each curve. Use subintervals of length 1.

3. $f(x) = 4x + 2$ from $x = 0$ to $x = 4$

4. $f(x) = x^2 + 3$ from $x = 1$ to $x = 4$

5. $f(x) = 3x^2 + x - 2$ from $x = 0$ to $x = 3$

6. $f(x) = 2x^3$ from $x = 1$ to $x = 5$

In Problems 7 – 18, compute each indefinite integral.

7. $\displaystyle\int x^8\,dx$

8. $\displaystyle\int x^{-5}\,dx$

9. $\displaystyle\int \sqrt[5]{x}\, dx$

10. $\displaystyle\int 5x^2\, dx$

11. $\displaystyle\int 8x^{-5}\, dx$

12. $\displaystyle\int (4x^3 - 2x)\, dx$

13. $\displaystyle\int \tfrac{3}{4}x^{-5}\, dx$

14. $\displaystyle\int (2x^2 + 3x^3 - 5)\, dx$

15. $\displaystyle\int (5x^3 - 2x^4 - 4x + 3)\, dx$

16. $\displaystyle\int (-3x^5 - 2x^{-3} + 6x)\, dx$

17. $\displaystyle\int (3x - \sqrt{x} + 2)\, dx$

18. $\displaystyle\int (x + 2)(x - 5)\, dx$

19. The marginal profit for producing x units per day is given by the function $P'(x) = 150 - 0.03x$, with $P(0) = 0$, where $P(x)$ is the profit in dollars. Find the profit function and the profit for 20 units of production per day.

20. Suppose that the marginal cost of producing x units of a product is given by the function $C'(x) = 100 + (60/\sqrt{x})$. If the total cost of producing the first unit is \$660, what is the cost function and what is the cost of producing the first 100 units?

In Problems 21–28, evaluate each definite integral.

21. $\displaystyle\int_1^4 3x^2\, dx$

22. $\displaystyle\int_4^9 4\sqrt{x}\, dx$

23. $\displaystyle\int_0^3 (5x^2 - 3x)\, dx$

24. $\displaystyle\int_{-1}^2 (4x^3 + 5x + 2)\, dx$

25. $\displaystyle\int_{-2}^3 (10x^4 + 6x + 3)\, dx$

26. $\displaystyle\int_0^\pi (9x^2 - 4x - 7)\, dx$

27. $\displaystyle\int_1^6 (x + 5)(x - 2)\, dx$

28. $\displaystyle\int_0^4 (6x^2 + \sqrt{x} - 2)\, dx$

Find the area under each curve.

29. $y = 3x - x^2$ from $x = 0$ to $x = 3$

30. $y = -x^2 + x + 6$ from $x = -2$ to $x = 3$

31. $y = \dfrac{4}{x^2}$ from $x = 1$ to $x = 5$

Find the area between each pair of curves.

32. $y = 2x^2$ and $y = 2x + 4$

33. $y = 5 - x^2$ and $y = x^2 - 3$

34. The amount of water (in thousands of gallons) flowing into a reservoir x hours after a heavy rain is given by the function

$$f(x) = \frac{10x}{(x^2 + 1)^2} \qquad 0 \le x \le 10$$

How much water accumulates during the first 10 hours after the rain?

35. A firm offers a quantity discount for an industrial cleaning product sold in 10 gallon drums. The unit price for the xth unit is given by the function $p(x) = 65 - 0.001x^2$. What is the total cost of purchasing 100 drums?

Use the general power rule for integration to find the integrals given in Problems 36–43.

36. $\displaystyle\int (x - 4)^3\, dx$

37. $\displaystyle\int \sqrt{2x + 3}\, dx$

38. $\displaystyle\int 6x(3x^2 - 3)^2\, dx$

39. $\int x(2x^2 + 6)^3 \, dx$

40. $\int x^2(3x^3 + 1) \, dx$

41. $\int x^3(x^4 + 5)^{2/3} \, dx$

42. $\int (x^2 - 1)(x^3 - 3x + 2)^6 \, dx$

43. $\int \dfrac{x}{\sqrt{x^2 - 2}} \, dx$

44. The rate of growth per month of the population in a certain town is given by the function $p(t) = 5 + 8t^{1/3}$. If the current population is 10,000, what will the population be 8 months from now?

45. Suppose that the inventory of a certain item t months after the first of the year is given approximately by the function

$$I(t) = 12 + 40t - 3t^2 \qquad 0 \le t \le 12$$

   a. What is the average inventory for the first quarter of the year?
   b. What is the average inventory for the full year?

46. Suppose that the price of an item is given by the function $p(x) = x^2 + 2x + 50$, where x is the sup-

ply. If supply and demand are in equilibrium at $x^* = 20$, find the producer's surplus.

47. The rate of production of an antibody t hours after injection into the body is given by the function

$$h(t) = \frac{5t^2}{t^3 + 10} \text{ units per hour}$$

   a. What is the rate after 1, 3, and 12 hours?
   b. How much antibody is produced in the first 12 hours?

48. Assume that the supply curve for a certain commodity is given by $p(x) = 3x + 3$, while the demand curve for the same commodity is given by $p(x) = 13 - x^2$. Determine the equilibrium point. Also, determine the consumer's surplus and the producer's surplus.

49. Suppose that the density function describing the relative likelihood of waiting x minutes in an automobile showroom before a salesperson approaches a customer is given by

$$f(x) = 0.2 - 0.02x \qquad 0 \le x \le 10$$

   a. Show that this function is a density function.
   b. What proportion of customers will have to wait 5 minutes or less before a salesperson arrives?

ADDITIONAL METHODS OF INTEGRATION

In Chapter 5 we discussed the simple power rule and the general power rule of integration. Both rules were derived as a consequence of viewing differentiation in reverse. However, there are many functions to which these rules do not apply. In this supplement we present some additional techniques of integration that extend the methods introduced in Chapter 5.

5S-1 INTEGRATION BY PARTS

Integration by parts is a technique that is based on the product rule of differentiation. Recall that the product rule states

$$\frac{d}{dx}\,[f(x)g(x)] = f(x)g'(x) + g(x)f'(x)$$

We may rearrange this as

$$g(x)f'(x) = \frac{d}{dx}\,[f(x)g(x)] - f(x)g'(x)$$

If we now integrate both sides of this equation with respect to x, we obtain

$$\int g(x)f'(x)\,dx = \int \frac{d}{dx}\,[f(x)g(x)]\,dx - \int f(x)g'(x)\,dx$$

Since an antiderivative of the derivative of a function is the function itself, we note that the first integral on the right-hand side is simply $f(x)g(x)$ plus a constant. We may ignore the constant for now since integration of $f(x)g'(x)$ will yield another constant. Therefore, we may state the following:

$$\int g(x)f'(x)\,dx = f(x)g(x) - \int f(x)g'(x)\,dx$$

We can explain this as follows: Suppose we wish to integrate a function $h(x)$. We try to write $h(x)$ as the product of two functions, which we will call $g(x)$ and $f'(x)$. In order to apply the above rule, we need to do three things. First, we must be able to integrate $f'(x)$ in order to obtain $f(x)$, which is used in both expressions on the right-hand side. Next, we need to find $g'(x)$. This is usually pretty easy to do. Finally, we must be able to integrate $f(x)g'(x)$ easily. If this is possible, then we can make the appropriate substitutions in the formula to find the integral of $h(x) = g(x)f'(x)$. This process is called *integration by parts*. There is usually some trial and error involved in attempting to use this method, and there is no guarantee that it will be successful. However, there are many types of functions for which this method is useful.

Example 1   Find the indefinite integral of $h(x) = x(x - 1)^3$.

Solution   Suppose we try to use the general power rule

$$\int g'(x)[g(x)]^p \, dx = \frac{1}{p+1} [g(x)]^{p+1} + c$$

If we let $g(x) = x - 1$, then we see that $g'(x) = 1$, which is not a multiple of x. Thus, the general power rule cannot be used. As an alternative, let us choose

$$f'(x) = x$$
$$g(x) = (x - 1)^3$$

and apply the integration by parts formula. We first integrate $f'(x)$ to obtain $f(x)$:

$$f(x) = \int f'(x) \, dx$$
$$= \int x \, dx = \frac{x^2}{2}$$

Next, find $g'(x)$:

$$g'(x) = \frac{d}{dx} [(x - 1)^3] = 3(x - 1)^2$$

Finally, try to integrate $f(x)g'(x)$:

$$\int \left(\frac{x^2}{2}\right) 3(x - 1)^2 \, dx$$

This is not an easily recognizable integral. Perhaps we made the wrong choices for $f'(x)$ and $g(x)$ initially. Let us try it the other way. Define

$$f'(x) = (x - 1)^3$$
$$g(x) = x$$

Then

$$f(x) = \int (x - 1)^3 \, dx = \frac{(x - 1)^4}{4}$$
$$g'(x) = 1$$

and

$$\int f(x)g'(x) \, dx = \int \frac{(x - 1)^4}{4} \, dx$$
$$= \frac{(x - 1)^5}{20} + c$$

Using the integration by parts formula, we have

$$\int x(x-1)^3 \, dx = \frac{(x-1)^4 x}{4} - \frac{(x-1)^5}{20} + c$$

To check this, differentiate:

$$\frac{d}{dx}\left[\frac{(x-1)^4 x}{4} - \frac{(x-1)^5}{20}\right] = \frac{(x-1)^4(1)}{4} + x(x-1)^3 - \frac{5(x-1)^4}{20}$$

$$= x(x-1)^3 \qquad \blacksquare$$

Example 2   Compute:

$$\int \frac{3x}{\sqrt{1-x}} \, dx$$

Solution   Let $f'(x) = (1-x)^{-1/2}$ and $g(x) = 3x$. Then

$$f(x) = \int (1-x)^{-1/2} \, dx = -\int (-1)(1-x)^{-1/2} \, dx = -2(1-x)^{1/2}$$

$$g'(x) = 3$$

Then we evaluate

$$\int f(x)g'(x) \, dx = \int -6(1-x)^{1/2} \, dx$$

$$= 6 \int -(1-x)^{1/2} \, dx$$

$$= 6(\tfrac{2}{3})(1-x)^{3/2} + c$$

$$= 4(1-x)^{3/2} + c$$

Therefore,

$$\int \frac{3x}{\sqrt{1-x}} \, dx = -2(1-x)^{1/2}(3x) - 4(1-x)^{3/2} + c$$

$$= -6x\sqrt{1-x} - 4\sqrt{(1-x)^3} + c \qquad \blacksquare$$

A useful device, which is often preferred for integration by parts, is the following:

$$\int u \, dv = uv - \int v \, du$$

The integration by parts formula follows from this by setting $u = g(x)$, $v = f(x)$, $dv = f'(x) \, dx$, and $du = g'(x) \, dx$:

$$\int \overbrace{g(x)}^{u}\overbrace{f'(x) \, dx}^{dv} = \overbrace{g(x)}^{u}\overbrace{f(x)}^{v} - \int \overbrace{f(x)}^{v} \overbrace{g'(x) \, dx}^{du}$$

You may find this easier to remember. However, the key to applying it is to identify u and dv properly. We shall use this formula in the next example.

Example 3   Find the integral

$$\int \frac{x^3}{\sqrt{2 + x^2}} \, dx$$

Solution   If we let $u = x^3$ and $dv = 1/\sqrt{2 + x^2} \, dx$, then we cannot find v using the techniques we currently know. If we let $u = 1/\sqrt{2 + x^2}$ and $dv = x^3 \, dx$, we still have a problem evaluating the integral of $v \, du$. We must recognize that in order to apply the general power rule to the function $(2 + x^2)^{-1/2}$, it must be multiplied by x. Therefore, if we let

$$dv = \frac{x}{\sqrt{2 + x^2}} \, dx$$

$$u = x^2$$

there is a better chance of successfully integrating by parts. We have the following:

$$v = \int \frac{x}{\sqrt{2 + x^2}} \, dx = \int x(2 + x^2)^{-1/2} \, dx$$

$$= \tfrac{1}{2} \int 2x(2 + x^2)^{-1/2} \, dx$$

$$= (2 + x^2)^{1/2}$$

$$du = 2x \, dx$$

$$\int v \, du = \int 2x(2 + x^2)^{1/2} \, dx = \tfrac{2}{3}(2 + x^2)^{3/2} + c$$

Therefore,

$$\int \frac{x^3}{\sqrt{2 + x^2}} \, dx = (2 + x^2)^{1/2}(x^2) - \tfrac{2}{3}(2 + x^2)^{3/2} + c \qquad \blacksquare$$

We will see further examples of integration by parts in Chapter 6.

To evaluate a definite integral using integration by parts we may state the following:

$$\int_a^b g(x) f'(x) \, dx = f(x)g(x) \Big|_a^b - \int_a^b f(x)g'(x) \, dx$$

or

$$\int_a^b u \, dv = uv \Big|_a^b - \int_a^b v \, du$$

Example 4   Evaluate the definite integral $\int_0^6 x\sqrt{2x + 4} \, dx$.

Solution   Let

$$dv = \sqrt{2x + 4}\ dx$$

$$u = x$$

Then

$$v = \int (2x + 4)^{1/2}\ dx = \tfrac{1}{2} \int 2(2x + 4)^{1/2}\ dx = \tfrac{1}{3}(2x + 4)^{3/2}$$

$$du = 1\ dx$$

The definite integral is

$$\int_0^6 x\sqrt{2x + 4}\ dx = \tfrac{1}{3}(2x + 4)^{3/2}(x)\ \Big|_0^6 - \int_0^6 \tfrac{1}{3}(2x + 4)^{3/2}\ dx$$

$$= \tfrac{1}{3}(2x + 4)^{3/2}(x)\ |_0^6 - \tfrac{1}{15}(2x + 4)^{5/2}\ |_0^6$$
$$= [\tfrac{1}{3}(64)(6) - 0] - [\tfrac{1}{15}(16)^{5/2} - \tfrac{1}{15}(4)^{5/2}]$$
$$\approx 128 - 66.133 = 61.867$$

■

5S-2 IMPROPER INTEGRALS

In Chapter 5 the definite integral was defined as the limit of a Riemann sum over an interval $[a, b]$. Consider the function $f(x) = 1/x^3$. The graph of this function is shown in Figure 5S-1. The graph approaches the x-axis asymptotically as x approaches infinity. Consider the area under the curve to the right of $x = 1$. Two questions arise. First, does the area have a finite value? Second, if it does, how can we compute it? We certainly can compute the area from $x = 1$ to $x = b$ for some finite value of b. This is given by the integral

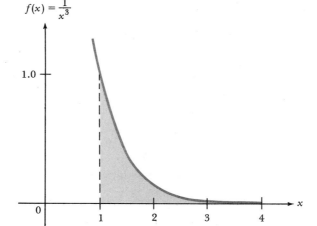

Figure 5S-1

Area under $f(x) = \dfrac{1}{x^3}$ from 1 to ∞

$$\int_1^b \frac{1}{x^3}\, dx = \int_1^b x^{-3}\, dx = -\left.\frac{x^{-2}}{2}\right|_1^b$$

$$= \left.\frac{-1}{2x^2}\right|_1^b$$

$$= \frac{-1}{2b^2} - \left[\frac{-1}{2(1)^2}\right]$$

$$= \frac{1}{2} - \frac{1}{2b^2}$$

Now suppose that we let b get arbitrarily large. Then the value of $-\frac{1}{2}b^2$ approaches 0. We can write this as

$$\lim_{b\to\infty} \left(\frac{1}{2} - \frac{1}{2b^2}\right) = \frac{1}{2}$$

Using this notion, we may define the area under the curve to the right of $x = 1$ as

$$\int_1^\infty \frac{1}{x^3}\, dx = \lim_{b\to\infty} \int_1^b \frac{1}{x^3}\, dx = \frac{1}{2}$$

Such an integral is called an *improper integral*, since one (or more) of the limits of integration is plus or minus infinity. We define three types of improper integrals for continuous functions $f(x)$:

Improper Integrals

$$\int_a^\infty f(x)\, dx = \lim_{b\to\infty} \int_a^b f(x)\, dx$$

$$\int_{-\infty}^b f(x)\, dx = \lim_{a\to-\infty} \int_a^b f(x)\, dx$$

$$\int_{-\infty}^\infty f(x)\, dx = \int_{-\infty}^c f(x)\, dx + \int_c^\infty f(x)\, dx$$

$$= \lim_{a\to-\infty} \int_a^c f(x)\, dx + \lim_{b\to\infty} \int_c^b f(x)\, dx$$

These definitions of improper integrals also give us a method for evaluating them. If the limit of the integral is finite (that is, not plus or minus infinity), then the improper integral is said to be **convergent.** Otherwise, it is said to be **divergent.** Thus, we would say that

$$\int_1^\infty \frac{1}{x^3}\, dx$$

is convergent, since it has a value of $\frac{1}{2}$.

The third type of improper integral above is convergent only if *both* of the limits on the right are finite numbers.

Example 5   Evaluate the integral

$$\int_{-\infty}^{-1} \frac{1}{x^2}\, dx$$

Solution This is equal to

$$\lim_{a\to-\infty}\int_a^{-1}\frac{1}{x^2}\,dx = \lim_{a\to-\infty}\left.\frac{-1}{x}\right|_a^{-1}$$

$$= \lim_{a\to-\infty}\left(1+\frac{1}{a}\right)$$

$$= 1$$

Therefore, the integral converges to 1.

Example 6 Evaluate the integral

$$\int_4^\infty \frac{1}{\sqrt{x}}\,dx$$

Solution We have

$$\int_4^\infty \frac{1}{\sqrt{x}}\,dx = \lim_{b\to\infty}\int_4^b x^{-1/2}\,dx$$

$$= \lim_{b\to\infty} 2x^{1/2}\,|_4^b$$

$$= \lim_{b\to\infty}(2\sqrt{b}-4)$$

As $b\to\infty$, the limit of $2\sqrt{b}-4$ approaches infinity also. Therefore, the integral is divergent and has no finite value.

Example 7 Find the integral

$$\int_{-\infty}^\infty \frac{10x}{\sqrt{x^2+1}}\,dx$$

Solution This is the third form of improper integral given above. We may select any value for c, so let us choose $c=0$:

$$\int_{-\infty}^\infty \frac{10x}{\sqrt{x^2+1}}\,dx = \int_{-\infty}^0 \frac{10x}{\sqrt{x^2+1}}\,dx + \int_0^\infty \frac{10x}{\sqrt{x^2+1}}\,dx$$

$$= \lim_{a\to-\infty}\int_a^0 \frac{10x}{\sqrt{x^2+1}}\,dx + \lim_{b\to\infty}\int_0^b \frac{10x}{\sqrt{x^2+1}}\,dx$$

We may evaluate each of these integrals using the general power rule:

$$\int \frac{10x}{\sqrt{x^2+1}}\,dx = \int 10x(x^2+1)^{-1/2}\,dx$$

$$= 5\int 2x(x^2+1)^{-1/2}\,dx$$

$$= 5\int g'(x)[g(x)]^{-1/2}\,dx$$

$$= 10\sqrt{g(x)} + c$$

where $g(x) = x^2 + 1$. Therefore,

$$\lim_{a \to -\infty} \int_a^0 \frac{10x}{\sqrt{x^2 + 1}} \, dx = \lim_{a \to -\infty} 10\sqrt{x^2 + 1} \,|_a^0$$

$$= \lim_{a \to -\infty} (10 - 10\sqrt{a^2 + 1}) = -\infty$$

Also,

$$\lim_{b \to \infty} \int_0^b \frac{10x}{\sqrt{x^2 + 1}} \, dx = \lim_{b \to \infty} 10\sqrt{x^2 + 1} \,|_0^b$$

$$= \lim_{b \to \infty} (10\sqrt{b^2 + 1} - 10) = \infty$$

Therefore, since each of these integrals diverges, the original integral is also divergent. (Note that ∞ is *not* a number; we cannot say that $-\infty + \infty = 0$.) ■

Example 8
Total Sales

The rate of sales of a new product (in thousands of units per month) is estimated to be

$$f(t) = \frac{1000t}{(t^2 + 50)^{3/2}}$$

Find the total estimated sales over the life of the product, assuming an indefinite life. (Remember, models are only approximations to reality.)

Solution

The total sales is given by the integral

$$\int_0^\infty \frac{1000t}{(t^2 + 50)^{3/2}} \, dt = \int_0^\infty 1000t(t^2 + 50)^{-3/2} \, dt$$

$$= \lim_{b \to \infty} \int_0^b 1000t(t^2 + 50)^{-3/2} \, dt$$

$$= \lim_{b \to \infty} -1000(t^2 + 50)^{-1/2} \,|_0^b$$

$$= \lim_{b \to \infty} \left(-\frac{1000}{\sqrt{b^2 + 50}} + \frac{1000}{\sqrt{50}} \right)$$

$$= 0 + \frac{1000}{\sqrt{50}}$$

$$\approx 141 \text{ thousand units}$$

■

Example 9
Waiting Time at a Bank

In Example 44 of Chapter 5 we discussed how areas under certain curves can be used to find proportions of customers who wait for service. Suppose that the distribution of the proportion of customers who wait x minutes at an automatic bank teller during a busy period has been found to be $f(x) = 1/x^2$, $x \geq 1$. Show that the area under the curve equals 1. What proportion of customers will wait more than 5 minutes?

Solution

The total area under the curve is

$$\int_1^\infty \frac{1}{x^2}\,dx = \lim_{b\to\infty} \int_1^b x^{-2}\,dx$$

$$= \lim_{b\to\infty} \left. \frac{-1}{x}\right|_1^b$$

$$= \lim_{b\to\infty} \left(\frac{-1}{b} - \frac{-1}{1}\right)$$

$$= 1$$

The proportion of customers that will wait more than 5 minutes is

$$\int_5^\infty \frac{1}{x^2}\,dx = \lim_{b\to\infty} \left.\frac{-1}{x}\right|_5^b$$

$$= \lim_{b\to\infty} \left(\frac{-1}{b} - \frac{-1}{5}\right)$$

$$= \tfrac{1}{5}, \text{ or } 20\%$$

5S-3 USING INTEGRAL TABLES

Although we have seen methods for finding the integrals of a variety of different functions, these methods are by no means adequate to solve every problem. Over the years, mathematicians have found the integrals of many standard functional forms, and tables of integrals have been published. The table inside the back cover of this text lists the integrals of several different functions that are more difficult to derive than the examples we have studied in this book. (Many of these involve exponential and logarithmic functions, which are the subject of the next chapter.) Such tables of integrals can be used to solve more difficult problems. More extensive tables can be found in reference books in your library.

Example 10   Evaluate the integral below using the table of integrals inside the back cover of this book.

$$\int_0^3 \frac{1}{(x^2 + 9)^{3/2}}\,dx$$

Solution   We first look for a function that has this form. In the table, the a's and b's represent arbitrary constants. Scanning the list, we see that integral 10 has the same form as the integral we are trying to evaluate with $a = 3$. Therefore, the indefinite integral of this function is

$$\frac{+x}{9\sqrt{x^2 + 9}} + c$$

We can apply the fundamental theorem of calculus to evaluate the definite integral as follows:

$$\int_0^3 \frac{1}{(x^2+9)^{3/2}}\,dx = \frac{x}{9\sqrt{x^2+9}}\Big|_0^3$$

$$= \frac{3}{9(3\sqrt{2})} - 0$$

$$= \frac{1}{9\sqrt{2}}$$

■

PROBLEMS (Supplement to Chapter 5)

In Problems 1–16, integrate each function using the technique of integration by parts.

1. $\displaystyle\int x(x-2)^2\,dx$

2. $\displaystyle\int 2x(x+3)^3\,dx$

3. $\displaystyle\int x(3x+1)^4\,dx$

4. $\displaystyle\int x^3(x^2+6)^3\,dx$

5. $\displaystyle\int \frac{5x}{\sqrt{2+x}}\,dx$

6. $\displaystyle\int \frac{3x}{\sqrt{4-2x}}\,dx$

7. $\displaystyle\int \frac{2x}{\sqrt{3x+4}}\,dx$

8. $\displaystyle\int \frac{2x^3}{\sqrt{x^2+1}}\,dx$

9. $\displaystyle\int \frac{3x^3}{\sqrt{4+x^2}}\,dx$

10. $\displaystyle\int \frac{x^5}{\sqrt{3-x^3}}\,dx$

11. $\displaystyle\int_0^3 x\sqrt{x+1}\,dx$

12. $\displaystyle\int_2^3 \frac{2x}{(x-1)^3}\,dx$

13. $\displaystyle\int_0^1 \frac{x^3}{(2+x^2)^4}\,dx$

14. $\displaystyle\int_1^4 x\sqrt{3x-3}\,dx$

15. $\displaystyle\int_0^3 \frac{x}{\sqrt{4-x}}\,dx$

16. $\displaystyle\int_1^2 \frac{x^3}{\sqrt{5-x^2}}\,dx$

17. Rework Example 1 using the notation

$$\int u\,dv = uv - \int v\,du$$

18. Rework Example 2 using the notation

$$\int u\,dv = uv - \int v\,du$$

In Problems 19–34, determine whether each improper integral converges or diverges.

19. $\displaystyle\int_1^\infty \frac{1}{x^2}\,dx$

20. $\displaystyle\int_1^\infty \frac{1}{x^5}\,dx$

21. $\displaystyle\int_{-\infty}^1 x^{-3}\,dx$

22. $\displaystyle\int_{-\infty}^{1} x^{-2/3}\, dx$

23. $\displaystyle\int_{9}^{\infty} \frac{1}{\sqrt{x}}\, dx$

24. $\displaystyle\int_{-\infty}^{0} \frac{1}{\sqrt[3]{x}}\, dx$

25. $\displaystyle\int_{3}^{\infty} \frac{2}{x^3}\, dx$

26. $\displaystyle\int_{-\infty}^{-1} \frac{3}{x^4}\, dx$

27. $\displaystyle\int_{0}^{\infty} \frac{1}{(x+1)^{1/2}}\, dx$

28. $\displaystyle\int_{2}^{\infty} \frac{1}{(x+2)^{3/2}}\, dx$

29. $\displaystyle\int_{-\infty}^{\infty} \frac{8x}{(x^2+1)^2}\, dx$

30. $\displaystyle\int_{-\infty}^{\infty} \frac{20x}{(x^2+4)^3}\, dx$

31. $\displaystyle\int_{-\infty}^{\infty} \frac{x}{\sqrt{x^2+9}}\, dx$

32. $\displaystyle\int_{-\infty}^{\infty} \frac{2x}{(x^2+4)^{3/2}}\, dx$

33. $\displaystyle\int_{0}^{\infty} \frac{250x}{(x^2+16)^{1.5}}\, dx$

34. $\displaystyle\int_{-\infty}^{0} \frac{800x}{(x^2-50)^2}\, dx$

35. Suppose that the density of the proportion of customers who wait x minutes at an automatic car wash during a busy period has been found to be $f(x) = 2/x^3, x \geq 1$. What proportion of customers wait more than 5 minutes? More than 10 minutes?

36. On a particular take-home test, the length of time each student spent on a problem was found to follow the density

$$f(x) = \frac{2}{(x+2)^2} \qquad x \geq 0$$

where x is the time spent in hours. What proportion of students will spend at least 3 hours on the problem?

Compute the integrals given in Problems 37 – 40 using the table of integrals inside the back cover of this book.

37. $\displaystyle\int_{4}^{6} \frac{1}{x^2\sqrt{x^2-16}}\, dx$

38. $\displaystyle\int_{1}^{5} \frac{1}{x^2\sqrt{x^2+25}}\, dx$

39. $\displaystyle\int_{0}^{5} \frac{1}{(x^2-9)^{3/2}}\, dx$

40. $\displaystyle\int_{0}^{8} \frac{1}{(x^2-36)^{3/2}}\, dx$

6

EXPONENTIAL AND LOGARITHMIC FUNCTIONS

INTRODUCTION

Since computers were initially developed in the 1940's, their capabilities have grown at an incredibly rapid pace. Table 6-1 lists the changes in two significant aspects of computer hardware: size, measured in the number of circuits per cubic foot; and instruction time, the time to execute an instruction in the central processing unit (CPU). Figure 6-1 shows the graphs of the data from Table 6-1. In Figure 6-1a, we see a very rapid increase in the number of circuits per cubic foot. This drastic reduction in size has made today's microcomputers a reality. In Figure 6-1b, we see a similar decrease in instruction time.

Can we model these functions mathematically? Consider the size factor. In 1950, the number of circuits per cubic foot was 1000. Ten years later (1960), this value increased to 100,000, or 100 times the value in 1950. In 1970, the value again increased 100 times the level in 1960. Thus, it appears that every 10 years, the value increased 100 times. Let N represent the number of circuits per cubic foot and x represent

Table 6-1
Approximate computer size and time factors

Year	Size (circuits/ft³)	Time (nanoseconds*/instruction)
1950	1,000	300,000
1960	100,000	5,000
1970	10,000,000	80
1980	1,000,000,000	1.5

* 1 nanosecond is 10^{-9} second.

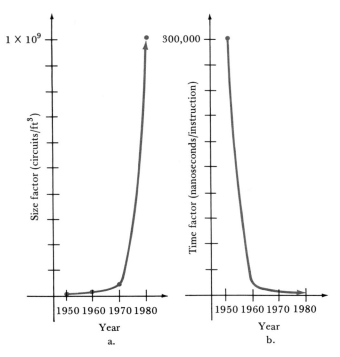

Figure 6-1
Graphs of computer size and time factors

time in 10 year increments since 1950 ($x = 0$ corresponds to 1950). Then we have

1950: $N = 1000$
       $= 1000(100)^0$ *Since $100^0 = 1$*

1960: $N = 100,000$
       $= 1000(100)^1$

1970: $N = 10,000,000$
       $= 100,000(100)$
       $= [1000(100)](100)$
       $= 1000(100)^2$

1980: $N = 1,000,000,000$
       $= 1000(100)^2(100)$
       $= 1000(100)^3$

From the definition of x we may write

$$N = 1000(100)^x$$

A function for the time factor can be developed in a similar fashion. From Table 6-1, we see that the value in 1960 is $\frac{1}{60}$ that of the 1950 value; the value in 1970 is approximately $\frac{1}{60}$ that of the 1960 value; and so on. (Remember that models are only *approximations* of reality. For example, the graph of the typing model in Chapter 2 did not pass through all data points but nevertheless described the behavior of the data quite well. Here, we should not expect the 1970 value to be *exactly* $\frac{1}{60}$ that of the 1960 value!) From this observation, we might conjecture that the time factor (T) follows the relation

$$T = 300,000(\tfrac{1}{60})^x$$

Since $(\tfrac{1}{60})^x = (60)^{-x}$, we can also write this as

$$T = 300,000(60)^{-x}$$

Both functions we have developed have the independent variable x as an *exponent*. Thus, we call them *exponential functions*. The graph in Figure 6-1a is called an **exponential growth curve**, since the function is constantly growing; the one in Figure 6-1b is called an **exponential decay curve**, since the value of the function continually decreases. Suppose we are interested in the rate of growth of N or the rate of decay of T. This would be given by the derivatives of the respective functions. At first you might suspect that the power rule for derivatives applies to these functions. Unfortunately, this is not the case; the power rule can be used only when the independent variable is raised to a constant power — *not* when the variable is an exponent. Therefore, we need new methods for finding the derivatives of such functions.

In this chapter we discuss the properties of exponential functions and a closely related class of functions called *logarithmic functions*. We also show how to find derivatives and integrals of such functions and how they are used in many diverse applications.

6-1 EXPONENTIAL FUNCTIONS

> If b is a positive real number not equal to 1, then the function
>
> $$f(x) = b^x$$
>
> is called an **exponential function** and b is called the **base**.

The domain of an exponential function is the set of all real numbers. This might seem a little odd at first, but from the properties of exponents (see Algebra Review 8 in Chapter 2) we know that if x is a positive integer, then $b^x = bb \cdot \cdots \cdot b$ (x factors). We also know that if x is a negative integer, then $b^x = 1/b^{-x} = 1/(bb \cdot \cdots \cdot b)$. Further, if x is a rational number (that is, $x = p/q$ where p and q are integers), then $b^x = b^{p/q} = \sqrt[q]{b^p}$. However, what does b^x mean when x is an irrational number such as π or $\sqrt{2}$? Any irrational number can be approximated by decimal fractions to any precision (for instance, to five decimal places). For example,

$$\sqrt{2} \approx 1.41421$$

The number 1.41421 can be written as

$$1 + \tfrac{4}{10} + \tfrac{1}{100} + \tfrac{4}{1000} + \tfrac{2}{10,000} + \tfrac{1}{100,000}$$

Thus, a number such as $2^{\sqrt{2}}$ may be approximated as $2^{1+4/10+\cdots+1/100,000}$. Then, since $2^{x+y} = 2^x 2^y$, we have

$$2^{\sqrt{2}} \approx (2)(2^{4/10}) \cdot \cdots \cdot (2^{1/100,000})$$

Each exponent is a rational number and we may therefore approximate $2^{\sqrt{2}}$ to five decimal places in this fashion. This can be extended to an arbitrary number of decimal places. By taking limits of such approximations, we see that b^x can be defined for all real numbers x.

Exponential functions are useful in many applications in business, economics, the social sciences, and the physical sciences.

Example 1
Return on Investment

An investor places $10,000 into 3 month Treasury bills earning an annual rate of 12%. Therefore, after 3 months, the investor will receive the initial investment of $10,000 plus the interest of

$$(\tfrac{1}{4})(0.12)(10,000) = 0.03(10,000) = \$300$$

Assuming that the interest rate remains stable, how much will this initial investment be worth if all earnings are reinvested every 3 months?

Solution

Every 3 months, the investor will receive $1 + 0.03 = 1.03$ times the amount invested. Thus, after the first 3 months, the investor will have

$$10,000(1.03) = 10,300$$

If this amount is reinvested,

$$10,300(1.03) = 10,609$$

will be received after the second 3 month period. This can be written as

$$10,300(1.03) = [10,000(1.03)](1.03)$$
$$= 10,000(1.03)^2$$

In general, after x periods of 3 months, the total amount (principal plus interest) will be $10,000(1.03)^x$. This is an exponential function of the form $y = ab^x$, where $a = 10,000$ and $b = 1.03$. Notice that a is the value when the exponent equals 0.

Example 2
Half-Life of Natural Elements

The half-life of the isotope tritium is 12.3 years. This means that any quantity of tritium is reduced by half after 12.3 years. Suppose we have a grams of tritium. After 12.3 years, we will have $a/2$ grams. After another 12.3 years, this amount will be reduced by half, or to $\frac{1}{2}(a/2) = a/4$. Letting x represent the number of 12.3 year periods, the amount remaining will be

$$A = a(\tfrac{1}{2})^x$$

We can compute the amount remaining after t years by noting that $x = t/12.3$, since x is measured in units of 12.3 years. Thus, if we want to find the amount left after 10 years, we have

$$x = \tfrac{10}{12.3} \approx 0.813$$

and

$$A \approx a(\tfrac{1}{2})^{0.813} \approx 0.569a$$

This calculation requires a calculator with the capability of computing exponents, or a computer. We can express A in terms of t by substituting $x = t/12.3 \approx 0.0813t$ into the general equation. We obtain $A = a(\tfrac{1}{2})^{0.0813t}$. This model is easier to work with since the independent variable is expressed in years.

One fact we can observe through the above examples is the following:

For an exponential function, the percentage change in the dependent variable is constant whenever the independent variable is increased by 1.

For instance, in Example 1, the amount of money increased by 3% every 3 month time period; in Example 2, the amount of tritium decreased by 50% every 12.3 year time period. We can easily prove this fact mathematically. (See Algebra Review 17 for a brief review of some of the fundamental laws of exponents.) Let $y = b^x$. The percentage change in y when x is increased by 1 is 100 times the ratio of the change in y to its original value, or

$$100 \frac{b^{x+1} - b^x}{b^x} = 100 \frac{b^x b - b^x}{b^x}$$

$$= \frac{100 b^x (b - 1)}{b^x}$$

$$= 100(b - 1)$$

which is a constant. Therefore, if the percentage change of some quantity is p, the base of an exponential function that represents growth or decay can be computed as

$$100(b - 1) = p$$

$$b - 1 = \frac{p}{100}$$

$$b = 1 + \frac{p}{100}$$

ALGEBRA REVIEW 17: LAWS OF EXPONENTS

There are several basic laws of exponents that are useful when dealing with exponential functions. These were introduced in Algebra Review 8 in Chapter 2. At that time, however, we discussed only exponents that were constants. The same basic laws also apply when the exponents are variables. We present them here for additional review:

  1. *Product law:* $a^x a^y = a^{x+y}$

  2. *Quotient law:* $\dfrac{a^x}{a^y} = a^{x-y}$

  3. *Power law:* $(a^x)^y = a^{xy}$

These laws hold whether a, x, and y are constants *or* variables. For example:

  a. $3^x 3^y = 3^{x+y}$

  $x^4 x^6 = x^{10}$

  b. $\dfrac{2^x}{2^y} = 2^{x-y}$

  $\dfrac{x^4}{x^6} = x^{-2}$

  c. $(2^x)^y = 2^{xy}$

  $(x^3)^5 = x^{15}$

These rules will be used throughout this chapter in simplifying and solving equations involving exponential functions.

EXERCISES*

Write the following expressions using a single base to a power:

  a. $y^4 y^{-2}$ b. $\dfrac{6^y}{6^{3y}}$ c. $(4^x)^{2x}$

Write each of the following expressions as *both* a product and a quotient with the base b:

  d. b^{3x-y} e. b^{-x+2}

* Answers to algebra review exercises can be found at the end of the answer section in the back of the book.

In Example 1, $p = 3$ and therefore $b = 1 + 0.03 = 1.03$. Similarly, in Example 2, $p = -50$ (the negative sign indicates a decrease) and $b = 1 - \frac{50}{100} = \frac{1}{2}$. This is a very useful fact in recognizing when an exponential function is appropriate in applications.

Example 3
New Product Sales

The sales of a new product are initially 6000 units and are expected to increase by 7% per year over the next 5 years. Develop a function that relates sales to time.

Solution

Let $S =$ sales and $t =$ years. We have $b = 1 + \frac{7}{100} = 1.07$. The appropriate function is therefore the exponential function

$$S = 6000(1.07)^t \qquad 0 \le t \le 5 \qquad \blacksquare$$

A second observation that can be made from these examples is that when $b > 1$, the function $y = b^x$ is increasing (as in Examples 1 and 3); and when $b < 1$, the function is decreasing (as in Example 2). Therefore, the graphs of these functions will have shapes dependent on the value of b.

GRAPHS OF EXPONENTIAL FUNCTIONS

Consider the function $y = 3^x$. We construct a table of values for x and y as follows:

x	$y = 3^x$
-2	$3^{-2} = \dfrac{1}{3^2} = \dfrac{1}{9}$
-1	$3^{-1} = \dfrac{1}{3}$
0	$3^0 = 1$
1	$3^1 = 3$
2	$3^2 = 9$

Figure 6-2
Graph of $y = 3^x$

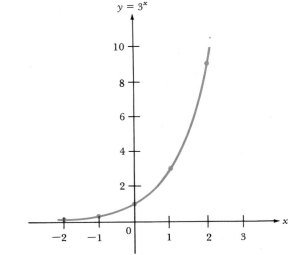

The graph of $y = 3^x$ is shown in Figure 6-2. The graph is increasing for all values of x. Also, as x approaches $-\infty$, $y = 3^x$ gets closer to 0 and the graph approaches the x-axis. Therefore, the line $y = 0$ is a horizontal asymptote of the function as $x \to -\infty$. This will be true for any base $b > 1$.

Next, suppose that $0 < b < 1$, say $b = \frac{1}{3}$. A table of function values is given below, and the graph of $y = (\frac{1}{3})^x$ is shown in Figure 6-3.

x	$y = (\frac{1}{3})^x$
-2	$(\frac{1}{3})^{-2} = \dfrac{1}{(\frac{1}{3})^2} = \dfrac{1}{\frac{1}{9}} = 9$
-1	$(\frac{1}{3})^{-1} = 3$
0	$(\frac{1}{3})^0 = 1$
1	$(\frac{1}{3})^1 = \frac{1}{3}$
2	$(\frac{1}{3})^2 = \frac{1}{9}$

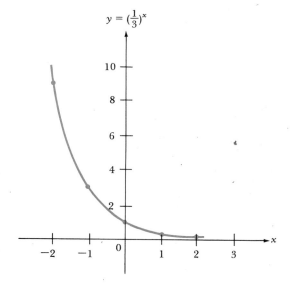

Figure 6-3
Graph of $y = (\frac{1}{3})^x$

In this case, the function is decreasing, and as x approaches $+\infty$, $y = (\frac{1}{3})^x$ approaches 0. The line $y = 0$ is a horizontal asymptote as $x \to +\infty$.

No matter what value b has, the y-intercept of the curve $y = b^x$ is always equal to 1 since $b^0 = 1$. We can change the y-intercept and the shape of an exponential function by multiplying b^x by the constant a to obtain $y = ab^x$. If $a > 0$, the general shape of the graph is similar to that in Figure 6-2. The y-intercept is a, and for large values of a the graph is steeper than it is for small values of a. Figure 6-4 on page 294 illustrates this.

If $a < 0$, then the graph takes on a very different shape. When $a < 0$, all values of $y = ab^x$ will be negative. The graph is thus the *mirror image* across the

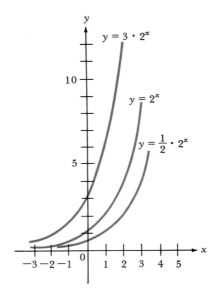

x-axis of $y = ab^x$ when $a > 0$. Consider, for example, the functions $y = 2^x$ and $y = (-1)2^x$:

x	$y = 2^x$	$y = (-1)2^x$
-2	$\frac{1}{4}$	$-\frac{1}{4}$
-1	$\frac{1}{2}$	$-\frac{1}{2}$
0	1	-1
1	2	-2
2	4	-4

Also, consider the functions $y = (\frac{1}{2})^x$ and $y = (-1)(\frac{1}{2})^x$:

x	$y = (\frac{1}{2})^x$	$y = (-1)(\frac{1}{2})^x$
-2	4	-4
-1	2	-2
0	1	-1
1	$\frac{1}{2}$	$-\frac{1}{2}$
2	$\frac{1}{4}$	$-\frac{1}{4}$

Figure 6-5 shows the graphs of $y = (-1)2^x$ and $y = (-1)(\frac{1}{2})^x$; the functions $y = 2^x$ and $y = (\frac{1}{2})^x$ are shown by dashed lines.

You can readily see that the exponential function $y = ab^x$ assumes a variety of shapes depending on the values of a and b. Because of this flexibility and for

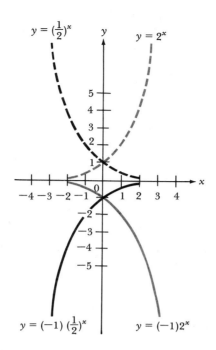

Figure 6-5
Graphs of $y = ab^x$

other reasons we shall later discover, exponential functions are used to model many practical problems.

Example 4
Physical / Psychological
Sensation

The psychological sensation to a physical stimulus does not necessarily change at the same rate as the magnitude of the stimulus. For example, the sensation of an electric shock grows rapidly and at an increasing rate as electric current is increased (we do not suggest that you try this experiment!). On the other hand, the perceived brightness of a light bulb grows at a decreasing rate as the wattage increases. Typical curves are shown in Figure 6-6. As a result of experimentation, psychologists have proposed an exponential model relating physical intensity x to the psychological sensation y:

$$y = ab^x + c$$

Figure 6-6

*Graphs of sensation as a function
of intensity of stimuli*

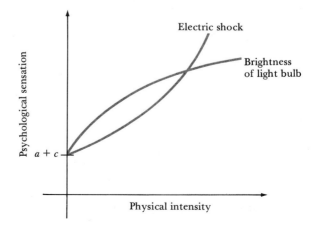

The constants a and b are determined by the nature of the physical stimulus. For a given value of a, the constant c determines the y-intercept. For electric shock, $a > 0$ and $b > 1$; for light brightness, $a < 0$ and $0 < b < 1$. Thus, we see that the same exponential model can represent very different response behavior by changing the value of the constants as determined by observation and data analysis.   ■

Figure 6-7 summarizes the four basic shapes that an exponential function $y = ab^x$ may assume. From these shapes, we can determine the nature of the first two derivatives. For instance, when $a > 0$ and $0 < b < 1$, the graph is decreasing; thus, $y' < 0$. Also, the curve is concave up, and therefore $y'' > 0$. Similar reasoning applies to the other cases shown in Figure 6-7. Later in this chapter we will discuss how to find the derivative of an exponential function.

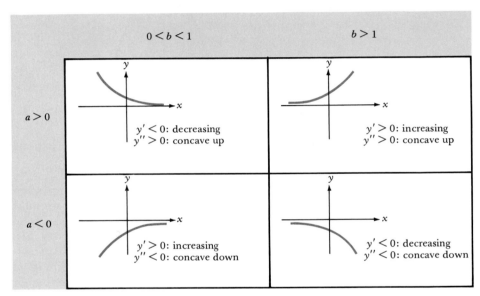

Figure 6-7
Graphs of exponential functions
$y = ab^x$ *and their properties*

PROBLEMS (Section 6-1)

1. Mr. Brown places $5000 into 3 month certificates of deposit earning an annual rate of 12%. Find the value of his initial investment after the first 3 months. After the second 3 months. After 1 year.

2. Sue Campbell places $15,000 into 6 month certificates of deposit earning an annual rate of 13%. Express what her initial investment will be worth after the first 6 months. After 1 year. After 2 years.

3. The half-life of a certain element is 2.6 years. Write an expression for the amount left after 2 years (express A in terms of t, as in Example 2).

4. The half-life of an element is 54.5 days. Write an expression for the amount left after 600 days (express A in terms of t, as in Example 2).

5. The half-life of an element is 2.6 months. Write an expression for the amount left after 6 months (express A in terms of t, as in Example 2).

6. The sales of a new convenience food item are initially 10,000 units and are expected to increase by 5% per year over the next 4 years. Develop a function that relates sales to time.

7. The television audience of a new game show was initially 500,000 viewers and is expected to increase by 8% per month over the next 3 months. Develop a function that relates viewers to time.

8. A local bank investor places $100,000 into 3 month Treasury bills earning an annual rate of 12% and $50,000 into 6 month Treasury bills earning an annual rate of 14%. Express what the dual investment will be worth after the first 6 months. After 1 year.

9. A chain letter is sent to 5 different people and they in turn must send it to 5 more different people. After x mailings, how many will have received the letter?

10. It is projected that t years from now, the population of a certain country will be $P(t) = 40(4^{0.02t})$ million. What is the current population of the country? What will be the population 25 years from now?

11. The population density x miles from the center of a certain city is $D(x) = 10(3^{-0.05x})$ thousand people per square mile. What is the population density 10 miles from the center of the city?

12. In March 1985, an official of the National Institute of Allergy and Infectious Diseases estimated that the number of cases of the deadly AIDS (Acquired Immune Deficiency Syndrome) disease doubles every 9 months. In March 1981, there were 245 reported cases of the disease. How many cases may be expected by January 1987? By January 1989? A newspaper report in March 1985 reported that 8697 Americans were diagnosed as having AIDS. Does this correspond to the model?

13. In 1939, the median starting salary of Harvard Business School graduates was $1800 per year.

Since that time, it has approximately doubled every 10 years. Develop a model for predicting the median starting salary for any year since 1939. If the median salary was actually $6600 in 1959, $14,000 in 1969, and $27,500 in 1979, how well would the model have estimated these values? What is the prediction for 1989?

Sketch the graph of each function.

14. $y = 4^x$

15. $y = \frac{1}{100}(4^x)$

16. $y = -(3^x)$

17. $y = -50(3^x)$

18. $y = (\frac{1}{5})^x$

19. $y = -(\frac{1}{5})^x$

20. $y = 0.01(\frac{1}{5})^x$

Without plotting any points, which general shape will the graphs of the following exponential functions have?

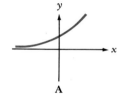

A

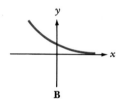

B

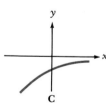

C

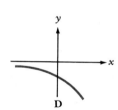

D

21. $y = (10)4^x$

22. $y = -\frac{1}{50}(3^x)$

23. $y = 100(\frac{1}{5})^x$

24. $y = -100(\frac{1}{5})^x$

25. $y = -0.01(\frac{1}{5})^x$

6-2 LOGARITHMIC FUNCTIONS

In Figure 6-2 we presented a graph of the function $y = 3^x$. From this graph we see that y can take on any positive real value; that is, the range of the function is the set of all positive real numbers. We have also noted that the domain of an exponential function is the set of all real numbers. Suppose that we are first given a value of y and asked to find the corresponding value of x. If the graph is accurate, we can "read across and down," as shown in Figure 6-8 to do this. Thus, from Figure 6-8 we see that if $y = 9$, then $x = 2$; if $y = 3$, then $x = 1$; and so on. We may think of this process as letting y represent the independent variable and x represent the dependent variable. If for every positive value of y there is associated only one value of x, then we can say that x is a function of y. This is clearly evident from the graph in Figure 6-8.

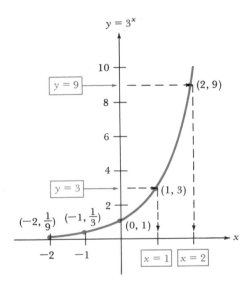

Figure 6-8
Graph of y = 3ˣ

Let us construct a table of values of y and corresponding values of x using the points shown on the graph in Figure 6-8:

y	x
9	2
3	1
1	0
$\frac{1}{3}$	-1
$\frac{1}{9}$	-2

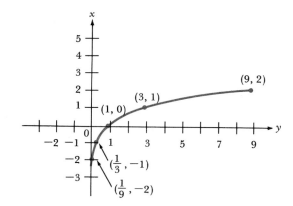

Figure 6-9

Graph of x as a function of y for the function y = 3ˣ

We can now draw a graph where y is the independent variable (on the horizontal axis) and x is the dependent variable (on the vertical axis). This is shown in Figure 6-9.

The function given by the graph in Figure 6-9 (note the relabeling of the axes!) is the *inverse function* of the function $y = 3^x$. What we have done is to "solve" the equation $y = 3^x$ for x in order to express x as a function of y. The inverse function of an exponential function has a special name, a *logarithmic function*. The logarithmic function is therefore characterized by the following relationship:

$$y = b^x \qquad \text{if and only if} \qquad x = \log_b y$$

The expression $\log_b y$ is read "the logarithm of y to the base b." Thus, the inverse of the function $y = f(x) = 3^x$ is the function $x = g(y) = \log_3 y$. Since our usual convention is to write $y = f(x)$, where y is the dependent variable and x is the independent variable, we define the function $f(x) = \log_b x$ as follows:

The Logarithmic Function

> Given $b > 0$ and $b \neq 1$, then for each $x > 0$, the **logarithmic function** is defined as
>
> $$f(x) = \log_b x \qquad \text{if and only if} \qquad b^{f(x)} = x$$

In other words, $\log_b x$ is the *power* to which the base b is raised in order to yield x. Note that since $\log_b x$ is defined only for numbers x of the form $x = b^y > 0$, the domain of the function $f(x) = \log_b x$ is the set of positive numbers.

Example 5 Write the following expressions in corresponding exponential or logarithmic form:

a. $y = 6^x$ b. $x = 4^y$
c. $81 = 3^4$ d. $\log_2 8 = 3$
e. $\log_b a = c$

Solution   Using the definition above we have

a. $\log_6 y = x$ b. $\log_4 x = y$
c. $\log_3 81 = 4$ d. $2^3 = 8$
e. $b^c = a$

We list the important properties of logarithms in the box. These properties can be used to simplify many numerical calculations such as multiplication and division, and formerly, they were used for this reason. They are also useful in solving equations involving logarithmic functions, as we shall see later.

Properties of Logarithms

a. $\log_b xy = \log_b x + \log_b y$

b. $\log_b \dfrac{x}{y} = \log_b x - \log_b y$

c. $\log_b x^y = y \log_b x$
d. $\log_b b = 1$
e. $\log_b 1 = 0$

Why do these properties hold? Consider, for example, property a. According to our definition of the logarithm, $z = \log_b xy$ is the unique number that satisfies

$$b^z = xy$$

However, by the product law of exponents (see Algebra Review 17) and the definition of the logarithmic function:

$$b^{\log_b x + \log_b y} = b^{\log_b x} \, b^{\log_b y} = xy = b^z$$

Hence,

$$\log_b xy = z = \log_b x + \log_b y$$

In a similar way, the other properties of logarithms follow from corresponding properties of exponents.

Example 6   Our number system is constructed using *base 10*. That is, any number can be written as a sum of a digit from 0 through 9 times a power of 10. For instance, $204 = (2 \times 10^2) + (0 \times 10^1) + (4 \times 10^0)$. Since $10^2 = 100, \log_{10} 100 = 2$; similarly, $\log_{10} 1000 = 3$. The logarithmic function is continuous, so there must be a value between 2 and 3 that equals $\log_{10} 204$. This number is approximately 2.31. Thus, $204 = 10^{2.31}$. If you have a calculator that can compute this, check it out.

Tables of logarithms (particularly to the base 10) were developed many years ago. Suppose one wishes to multiply 204 by 3.16 ($\approx \sqrt{10}$). From the properties of logarithms (and the use of tables or a calculator), we find that

$$\log_{10}[(204)(3.16)] = \log_{10} 204 + \log_{10} 3.16$$
$$\approx 2.31 + 0.5 = 2.81$$

Thus, $(204)(3.16) = 10^{2.81} \approx 645$. The number $10^{2.81}$ is called the *antilog* of 2.81, and tables of these values are also readily available. Thus, to multiply 204 by 3.16, all one needs to do is to find the logarithms of 204 and 3.16, add them together, and find the antilog of the result. The process of multiplication is reduced by the use of logarithms to the simpler process of addition. Although this may seem a strange way to multiply two numbers today, it was the preferred method before the use of calculators became widespread. Multiplication of large numbers could be quite tedious, and logarithms simplified the calculations considerably. In fact, logarithms formed the basis for slide rule calculations until the electronic calculator displaced the slide rule as a scientific aid. ∎

Logarithmic functions have many other useful contemporary applications. The following two examples illustrate some of their uses.

Example 7
Binary Numbers and
Computer Memory

"Computer arithmetic" is based on the *binary number system*. A binary number consists only of the digits 0 and 1. This is useful in electronic computers, since electric components can only be *off* (corresponding to 0) or *on* (corresponding to 1). A binary number is a string of 0's and 1's, for example, 1101_2, where the subscript 2 indicates that the number is to be interpreted in *base 2*. To convert from binary to our ordinary base 10 number system, we multiply the kth digit from the right by 2^{k-1} and add. Thus, 1101_2 equals

$$1 \cdot 2^3 + 1 \cdot 2^2 + 0 \cdot 2^1 + 1 \cdot 2^0 = 8 + 4 + 0 + 1$$
$$= 13$$

An important consideration to computer scientists is the amount of storage needed to represent data. The largest number that can be represented by k binary digits (called "bits") is $2^k - 1$. For example, the largest 3 digit binary number is

$$111_2 = 1 \cdot 2^2 + 1 \cdot 2^1 + 1 \cdot 2^0 = 7 = 2^3 - 1$$

This, of course, is essentially an exponential function. But we are interested in the inverse question. Namely, how many bits are required to represent a given positive decimal integer x in binary? It is not surprising that the answer to this question will involve the inverse of the base 2 exponential function, that is, $\log_2 x$. In fact, it can be shown that the number of bits, y, needed to represent the positive integer x in binary is given by

$$y = 1 + \lfloor \log_2 x \rfloor$$

where the symbol $\lfloor a \rfloor$ represents the greatest integer in a. For example, $\lfloor 2.356 \rfloor = 2$. Therefore, if $x = 8$, $\log_2 8 = 3$ and $y = 4$. If $x = 4$, $\log_2 4 = 2$ and $y = 3$. This means that we need four bits to represent the number 8 and 3 bits to represent the number 4 using binary numbers. For $4 < x < 8$, $2 < \log_2 x < 3$ and $\lfloor \log_2 x \rfloor = 2$. Hence, $y = 3$. So to represent the number 6 in binary notation we need to use 3 bits: 110_2. ∎

Example 8
The Decibel Scale for
Sound Intensity

Sound intensity, I, is the physical measurement of sound in watts per square meter. *Loudness,* the ear's perception of sound intensity, is measured in decibels. A 10 decibel increase in loudness is equivalent to multiplying the sound intensity by 10. The measurement of noise levels has taken on increased importance in industry since the Occupational Safety and Health Act of 1970 set maximum limits on noise levels in a work environment. For example, a worker may sustain a 90 decibel noise level for 9 hours. The decibel level is related to sound intensity (which can be measured by instruments) by the equation

$$\text{Loudness} = 10 \log_{10}\left(\frac{I}{I_0}\right)$$

where $I_0 = 10^{-12}$ watt per square meter. Thus, a sound of intensity $I = 0.1$ watt per square meter has a loudness of

$$10 \log_{10}\left(\frac{0.1}{10^{-12}}\right) = 10 \log_{10}\left(\frac{10^{-1}}{10^{-12}}\right)$$
$$= 10 \log_{10}(10^{11})$$
$$= 10(11) = 110 \text{ decibels}$$

GRAPHS OF LOGARITHMIC FUNCTIONS

As with exponential functions, graphs of logarithmic functions can take on different shapes and can be modified by multiplying the function by a constant. When the base $b > 1$, the graph of $y = \log_b x$ appears as shown in Figure 6-10. That is, it is increasing and concave down. Note that the line $x = 0$ is a vertical asymptote.

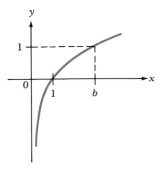

Figure 6-10
Graph of $y = \log_b x$, $b > 1$

When $0 < b < 1$, the graph is significantly different. Since $y = \log_b x$ is the same as $x = b^y$, we can use the corresponding exponential function to compute some points on the graph. Suppose $b = \frac{1}{3}$. Then we have the following values:

x	y
9	-2
3	-1
1	0
$\frac{1}{3}$	1
$\frac{1}{9}$	2

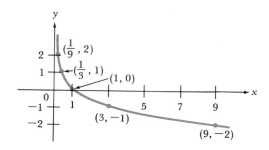

Figure 6-11
Graph of $y = \log_{1/3} x$

The graph of $y = \log_{1/3} x$ is shown in Figure 6-11. This function is decreasing and concave up. The line $x = 0$ is a vertical asymptote of this function also.

In a similar fashion to our analysis of the exponential function, we can multiply a logarithmic function by a constant and modify its shape. Consider the general form $y = a \log_b x$. For different values of a, the graph takes the shapes shown in Figure 6-12. Note that when $a < 0$, the graph of $y = a \log_b x$ is simply the mirror image across the x-axis of the graph of $y = a \log_b x$ when $a > 0$.

Figure 6-12
Graphs of $y = a \log_3 x$

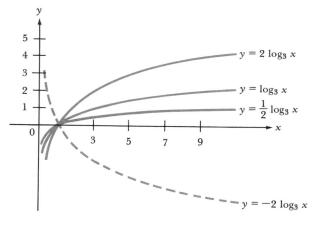

We summarize the general shapes of the logarithmic function $y = a \log_b x$ in Figure 6-13 (page 304). It may seem odd that the graph corresponding to $a > 0$ and $0 < b < 1$ is similar to that corresponding to $a < 0$ and $b > 1$. The same can be said for $a > 0$, $b > 1$ and $a < 0$ and $0 < b < 1$. We can, in fact, write a logarithmic function in either of two ways. Thus, there are only two basic forms of the logarithmic function. We illustrate this in the next example.

Example 9   Write $y = -3 \log_8 x$ in the form $y = a \log_b x$ with $a > 0$ and $0 < b < 1$.

Solution
$$y = -3 \log_8 x$$

$$\frac{-y}{3} = \log_8 x$$

Thus, $x = (8)^{-y/3} = (8^{-1/3})^y = (\tfrac{1}{2})^y$. We therefore have

$$y = \log_{1/2} x$$

that is, $a = 1$ and $b = \tfrac{1}{2}$.

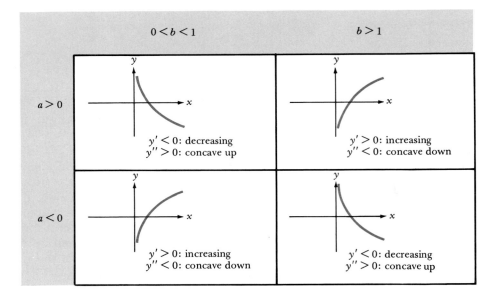

Figure 6-13
Graphs of logarithmic functions
$y = a \log_b x$ *and their properties*

EQUATIONS INVOLVING EXPONENTS AND LOGARITHMS

We have seen that the basic relationship between exponential and logarithmic functions is given by

$$y = b^x \quad \text{if and only if} \quad x = \log_b y$$

By making the appropriate substitutions, it easily follows that

> $\log_b b^x = x \qquad$ for all x
>
> and
>
> $b^{\log_b x} = x \qquad$ for $x > 0$

These relationships are useful in solving equations involving exponents and logarithms. If we have an equation $A = B$, where A and B are algebraic expressions, we may perform either of the following operations:

1. Take logarithms of both sides; that is,

   $$\log_b A = \log_b B$$

2. Raise a base b to the powers A and B on both sides of the equation; that is,

   $$b^A = b^B$$

To see how these operations are used, suppose we wish to solve the equation

$$10^{2x-4} = 3$$

for x. Taking logarithms to the base 10 on both sides yields

$$\log_{10}(10^{2x-4}) = \log_{10} 3$$
$$2x - 4 = \log_{10} 3$$
$$x = \frac{4 + \log_{10} 3}{2}$$

As a second example, consider the equation

$$\log_2(x - 1) = 3$$

Applying the second operation gives

$$2^{\log_2(x-1)} = 2^3$$
$$x - 1 = 8$$
$$x = 9$$

Example 10
Sound Intensity (Continued)

Referring to the discussion in Example 8, suppose that loudness is known to be 50 decibels. What is the sound intensity?

Solution

We have the equation

$$50 = 10 \log_{10}\left(\frac{I}{I_0}\right)$$
$$5 = \log_{10}\left(\frac{I}{I_0}\right)$$

Using the definition of the logarithmic function, this is equivalent to

$$10^5 = 10^{\log_{10}(I/I_0)} = \frac{I}{I_0}$$

Since $I_0 = 10^{-12}$, we have

$$I = 10^5(10^{-12}) = 10^{5-12} = 10^{-7} \text{ watt per square meter}$$

PROBLEMS (Section 6-2)

In Problems 1–8, write each expression in logarithmic form.

1. $16 = 2^4$

2. $y = 3^x$

3. $\frac{1}{8} = 2^{-3}$

4. $1 = 5^0$

5. $5^{-1} = 0.2$

6. $(\frac{1}{2})^{-2} = 4$

7. $(\frac{1}{3})^{-3} = 27$

8. $(\sqrt{2})^4 = 4$

In Problems 9–17, write each expression in exponential form.

9. $\log_4 64 = 3$

10. $\log_9 81 = 2$

11. $\log_{25} 625 = 2$

12. $\log_{1/2} 8 = -3$

13. $\log_5 0.2 = -1$

14. $\log_{10} 0.01 = -2$

15. $\log_{1/2}(\frac{1}{4}) = 2$

16. $\log_{1/16}(\frac{1}{4}) = \frac{1}{2}$

17. $\log_{\sqrt{3}} 9 = 4$

In Problems 18–23, simplify each expression.

18. $\log_2 4$

19. $\log_5 125$

20. $\log_2(\frac{1}{2})$

21. $\log_{10} 0.1$

22. $\log_{1/2}(\frac{1}{16})$

23. $\log_{1/3}(\frac{1}{81})$

In Problems 24–29, convert each binary number to the base 10 number system.

24. 101_2

25. 1010_2

26. 1110_2

27. 101101_2

28. 110011_2

29. 1101111_2

In Problems 30–33, convert the given sound intensity to decibels (use the equation in Example 8).

30. 0.01 watt per square meter

31. 0.001 watt per square meter

32. 1 watt per square meter

33. 10 watts per square meter

What is the sound intensity of each of the following decibel levels?

34. 100 decibels

35. 90 decibels

36. 1 decibel

In Problems 37–43, write each logarithmic equation in the form $y = a \log_b x$ with $a > 0$ and $0 < b < 1$.

37. $y = -3 \log_{27} x$

38. $y = -2 \log_{16} x$

39. $y = -4 \log_{16} x$

40. $y = -3 \log_{125} x$

41. $y = -\frac{1}{2} \log_5 x$

42. $y = -\frac{1}{3} \log_3 x$

43. $y = -\frac{1}{5} \log_2 x$

44. Seismologists use the Richter scale to measure the intensity of earthquakes. The energy of a quake, E (in watts per cubic meter), is related to its Richter value R by

$$1.5R = \log_{10} E - 11.4$$

a. Find E as a function of R.

b. The Alaskan earthquake of 1964 measured 8.5 on the Richter scale. How does the energy E of this quake compare with that of the 1985 Santiago, Chile quake, which measured 7.4 on the Richter scale?

Sketch the graphs of the logarithmic functions given in Problems 45–51.

45. $y = \log_2 x$

46. $y = -\log_2 x$

47. $y = \frac{1}{5} \log_2 x$

48. $y = 50 \log_2 x$

49. $y = \log_{1/3} x$

50. $y = 20 \log_{1/3} x$

51. $y = -100 \log_{1/3} x$

In Problems 52–59, solve each equation for x (you do not have to evaluate the logarithm of a number).

52. $3^{2x} = 4$

53. $20(2^{-x/2}) = 4$

54. $2^{x^2 - 2x + 1} = 1$

55. $4(10^x)(10^{-5x}) = 3$

56. $\log_2 x^2 = 1$

57. $\log_{10}(x - 3) = 2$

58. $\log_9(x^2 + 2x) = \frac{1}{2}$

59. $\log_2(\log_2 x) = 0$

6-3 THE DERIVATIVE OF AN EXPONENTIAL FUNCTION

Let us consider the exponential function $f(x) = b^x$. Recall the limit definition of the derivative:

$$f'(x) = \lim_{h \to 0} \frac{f(x + h) - f(x)}{h}$$

Applying this to the function $f(x) = b^x$, we have

$$f'(x) = \lim_{h \to 0} \frac{b^{x+h} - b^x}{h}$$

$$= \lim_{h \to 0} \frac{b^x b^h - b^x}{h}$$

$$= \lim_{h \to 0} \frac{b^x (b^h - 1)}{h}$$

$$= b^x \lim_{h \to 0} \frac{b^h - 1}{h}$$

Note that we can factor b^x out of the limit since it is independent of h. The question is now reduced to evaluating this limit. We first do this numerically for $b = 2$ and $b = 3$ in the following table:

h	$(2^h - 1)/h$	$(3^h - 1)/h$
0.1	0.7177	1.1612
0.01	0.6956	1.1047
0.001	0.6934	1.0992
0.0001	0.6932	1.0987
0.00001	0.6931	1.0986
0.000001	0.6931	1.0986

It appears that

$$\lim_{h \to 0} \frac{2^h - 1}{h} \approx 0.6931$$

and

$$\lim_{h \to 0} \frac{3^h - 1}{h} \approx 1.0986$$

Hence, we may state that

$$\frac{d}{dx}(2^x) \approx 2^x(0.6931)$$

and

$$\frac{d}{dx}(3^x) \approx 3^x(1.0986)$$

Figure 6-14 shows the graphs of $y = 2^x$ and $y = 3^x$. Since both 2^0 and 3^0 equal 1, when $x = 0$, the derivative of $y = 2^x$ is about 0.6931 and the derivative of $y = 3^x$ is about 1.0986. Also, since these values represent the slopes of the tangent lines to the respective curves at $x = 0$, it is reasonable to conclude that there is an exponential function with a base somewhere between 2 and 3 whose tangent line to the graph at $x = 0$ has a slope of exactly 1. That is,

$$\lim_{h \to 0} \frac{b^h - 1}{h} = 1 \qquad \text{for some } b \text{ with } 2 < b < 3$$

There is indeed a unique such number, and it is called e. The number e is an irrational number which, approximated to 5 decimal places, is

$$e \approx 2.71828$$

We can therefore state the following:

$$\lim_{h \to 0} \frac{e^h - 1}{h} = 1$$

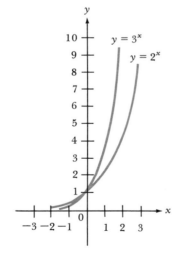

Figure 6-14
Graphs of $y = 2^x$ and $y = 3^x$

Further,

$$\frac{d}{dx}e^x = e^x \lim_{h \to 0} \frac{e^h - 1}{h} = e^x$$

The function e^x has a derivative equal to itself!

Derivative of e^x

$$\frac{d}{dx}e^x = e^x$$

Like any exponential function, $y = e^x$ has all real numbers as a domain and all positive real numbers as a range. Also, it is a continuous function. Suppose we are given a certain value of y, say $y = b$. Can we find a value of x for which $b = e^x$? If the graph is accurate, we can "read across and down" to find the corresponding value of x, as illustrated in Figure 6-15. This means that for any positive number b, we can always find a unique number c such that $b = e^c$.

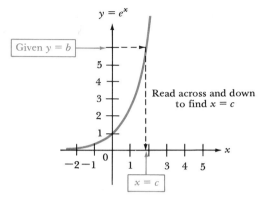

Figure 6-15
Graph of $y = e^x$

While this is an interesting result, how useful is it? Suppose we have an exponential function $y = b^x$. Expressing b in terms of c, we can write

$$y = b^x = (e^c)^x = e^{cx}$$

That is, *we can write any exponential function using the base e* (provided, of course, that we can find the value of c). We can now apply the chain rule to find the derivative of $y = e^{cx}$.

Recall from Chapter 4 that the chain rule states that if we have the composition of functions f and g, $y = f(g(x))$, then $y' = f'(g(x))g'(x)$. Define $f(x) = e^x$ and $g(x) = cx$. Then $y = e^{cx}$ is the composition $f(g(x))$. Now, $f'(x) = e^x$ and hence, $f'(g(x)) = e^{cx}$. Also, $g'(x) = c$. Applying the chain rule, we obtain

$$\frac{d}{dx} e^{cx} = f'(g(x))g'(x) = e^{cx}(c) = ce^{cx}$$

We may specialize the chain rule to an exponential function of the form $y = e^{g(x)}$ as follows:

$$y = f(g(x)) \qquad \text{where} \qquad f(x) = e^x$$

Therefore,

$$y' = f'(g(x))g'(x) = e^{g(x)}g'(x) = g'(x)e^{g(x)}$$

If $y = e^{g(x)}$, then $y' = g'(x)e^{g(x)}$.

Thus, to find the derivative of $y = e^{x^2}$, we have $g(x) = x^2$ and

$$y' = \overset{g'(x)}{(2x)}\overset{e^{g(x)}}{e^{x^2}}$$

Example 11   Find the derivative of each of the following functions:

a. $y = e^{3x}$ b. $f(x) = e^{6x^3 + 2x}$

c. $y = 4xe^x$ d. $f(x) = \dfrac{e^{4x}}{x^2 + 9}$

Solution   For parts a and b, we apply the chain rule directly.

a. Letting $g(x) = 3x$, the derivative of $y = e^{3x}$ is $y' = 3e^{3x}$.
b. Here, $g(x) = 6x^3 + 2x$ and $g'(x) = 18x^2 + 2$. Thus,

$$f'(x) = (18x^2 + 2)e^{6x^3 + 2x}$$

c. We use the product rule:

$$
\underbrace{y = 4x}_{\substack{First \\ function}}\underbrace{(e^x)}_{\substack{Derivative \\ of\ second \\ function}} \quad + \quad \underbrace{e^x}_{\substack{Second \\ function}}\underbrace{(4)}_{\substack{Derivative \\ of\ first \\ function}}
$$

$$= e^x(4x + 4)$$
$$= 4e^x(x + 1)$$

d. The quotient rule applies. We also use the chain rule for e^{4x}:

$$f'(x) = \frac{(x^2 + 9)4e^{4x} - e^{4x}(2x)}{(x^2 + 9)^2}$$

$$= \frac{(4x^2 + 36)e^{4x} - 2xe^{4x}}{(x^2 + 9)^2}$$

$$= \frac{(4x^2 + 36 - 2x)e^{4x}}{(x^2 + 9)^2}$$

■

Example 12
Respiration   The volume of nitrogen (in milliliters) eliminated t minutes after beginning to breathe pure oxygen is given by the function $y = 850(1 - e^{-0.023t})$. What is the rate of change of y with respect to time?

Solution   The derivative of this function is

$$y' = 850(0.023e^{-0.023t}) = 19.55e^{-0.023t}$$

Figures 6-16 and 6-17 show both the graph of y and that of its derivative. From Figure 6-16, we see that the graph approaches 850 as t gets large; thus, the line $y = 850$ is a horizontal asymptote. That is, 850 milliliters is the limiting value of the amount of nitrogen that can be eliminated. The derivative in Figure 6-17 is continuously decreasing and approaches 0 as t gets large. This should be expected since the function itself approaches a constant. You can see that the slope of y starts out large in Figure 6-16 when $t = 0$ and gradually gets smaller. This is what the graph of the derivative shows.   ■

Example 13
Reproduction Rate   The reproduction rate of flies has been found to be related to the number of flies kept in an enclosed area (such as in a bottle). If x represents the number of flies, the daily reproduction rate is given by

$$y = 34.5e^{-0.02x}x^{-0.66} \qquad \text{flies per day}$$

What is the derivative of y with respect to x?

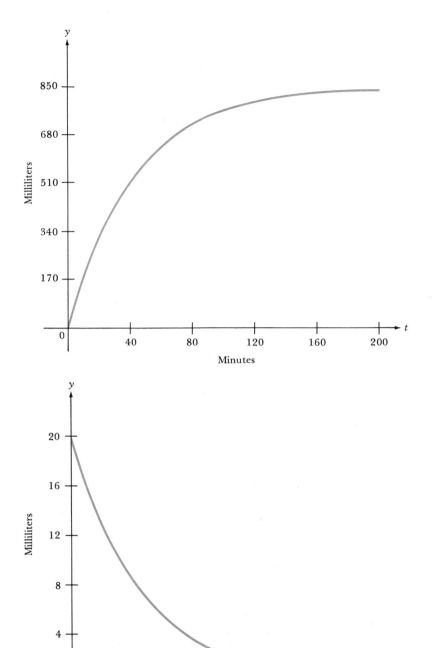

Figure 6-16
Graph of $y = 850(1 - e^{-0.023t})$

Figure 6-17
Graph of $y' = 19.55e^{-0.023t}$

Solution   **Using the product rule, we obtain**

$$y' = 34.5e^{-0.02x}(-0.66x^{-1.66}) + (x^{-0.66})(34.5)(-0.02e^{-0.02x})$$
$$= -e^{-0.02x}(22.77x^{-1.66} + 0.69x^{-0.66})$$

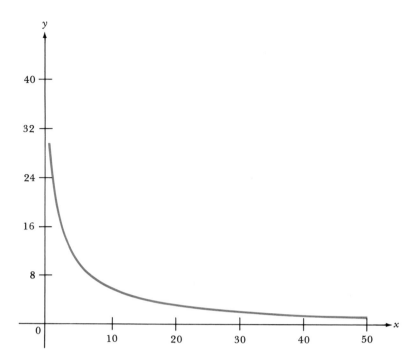

Figure 6-18
Graph of $y = 34.5e^{-0.02x}x^{-0.66}$

Figure 6-18 shows the graph of the reproduction rate. This bears out common sense, for as the population in the bottle gets congested, the reproductive rate gets smaller and eventually approaches 0.   ∎

Although e is commonly used in modeling with exponential functions, in some cases it is more appropriate to use a base other than e. For instance, in Example 1 (Section 6-1), a base of 1.03 has more meaning in the context of the application than if we were to express the function in terms of e. Thus, it is useful to know how to differentiate exponential functions having arbitrary bases. To do this we need the concept of the natural logarithm.

THE NATURAL LOGARITHM

Let us return to our original question of finding the derivative of $y = b^x$. We noted that we can find a number c such that $b = e^c$. Therefore, $y = b^x = e^{cx}$. Since the derivative of $y = e^{cx}$ is $y' = ce^{cx} = c(e^c)^x$, we have $y' = cb^x$. Using the definition of logarithm, we see that $c = \log_e b$, since $e^c = e^{\log_e b} = b$. Therefore, if $y = b^x$, then $y' = (\log_e b)b^x$.

A logarithm to the base e is called a **natural logarithm** and is given a special symbol:

$$\ln x = \log_e x$$

The symbol $\ln x$ is read "the natural logarithm of x."

We may therefore state the following:

> If $y = b^x$, then $y' = (\ln b)b^x$.

For example, if $y = 10^x$, then $y' = (\ln 10)10^x$. While if $y = x^2 3^x$, then by the product rule,

$$y' = x^2(\ln 3)3^x + (2x)3^x$$

Example 14
Return on Investment

Determine and interpret the rate of change of the interest function developed in Example 1 (Section 6-1).

Solution

The function is $y = 10,000(1.03)^x$, and the derivative is given by

$$y' = 10,000(\ln 1.03)(1.03)^x \approx 295.6(1.03)^x \qquad \textit{Since from Appendix B,*}$$
$$\textit{ln 1.03} \approx \textit{0.02956}$$

This function gives the rate of change of money after x 3 month periods. For example, if $x = 4$, then $1.03^4 \approx 1.1255$ and $y' \approx 332.70$. After 1 year, the amount of money is changing at a rate of about \$332.70 every 3 months (remember that x is measured in 3 month periods). ∎

The differentiation rule given above can be easily extended to functions of the form $y = b^{g(x)}$. We can express such a function as a composite:

$$y = b^{g(x)} = f(g(x)) \qquad \text{where } f(x) = b^x$$

By the chain rule and the differentiation rule above, we then have

$$y' = f'(g(x))g'(x) = (\ln b)b^{g(x)}g'(x)$$

> If $y = b^{g(x)}$, then $y' = (\ln b)b^{g(x)}g'(x)$.

Example 15
Half-Life of Natural Elements

The function giving the amount of tritium after t years was shown to be $A = a(\frac{1}{2})^{0.0813t}$ in Example 2 (Section 6-1). Explain the meaning of the derivative when $t = 10$.

Solution

$$A' = a(\ln \tfrac{1}{2})(\tfrac{1}{2})^{0.0813t}(0.0813) \qquad \textit{Use a calculator with an "ln" key or Appendix B.†}$$
$$\approx a(-0.693)(\tfrac{1}{2})^{0.0813t}(0.0813)$$
$$\approx -0.056a(\tfrac{1}{2})^{0.0813t}$$

When $t = 10$, $A' \approx -0.056a(0.569) \approx -0.032a$. After 10 years, the amount of tritium is decreasing at a rate of about 3.2% of the original amount per year. ∎

---

* Appendixes A and B give values of e^x, e^{-x}, and $\ln x$ for various values of x. However, we recommend that you use a calculator to compute these when necessary.
† Note that to use Appendix B to find $\ln \frac{1}{2}$, we write $\ln \frac{1}{2} = \ln 1 - \ln 2 = 0 - 0.69315 \approx -0.693$.

Although there are times when bases other than e are used, we see that taking derivatives is a little messy. In the majority of calculus applications involving exponential functions, the base e is used. We stress that it does not matter since any exponential function can be written in terms of e.

Example 16
Radioactive Decay

Radium has a half-life of 1620 years. Write exponential functions representing the amount of radium left after t years using both the base $\frac{1}{2}$ and the base e.

Solution

Let $x =$ the number of 1620 year increments. Then it is easy to see that $y = (\frac{1}{2})^x$ represents the amount of radium left after x time periods. We can express x in terms of years by the relation $x = t/1620$. Therefore, we have

$$y = (\tfrac{1}{2})^{t/1620}$$

To express this relation using the base e as an exponential function of the form $y = e^{kt}$, we need to find a constant k that satisfies the following relation:

$$y = (\tfrac{1}{2})^{t/1620} = e^{kt}$$

Raising both sides to the power $1620/t$, we obtain

$$\tfrac{1}{2} = e^{k(1620)}$$

Taking the natural logarithm of both sides yields

$$\ln \tfrac{1}{2} = 1620k$$

Thus, $k = \ln(\frac{1}{2})/1620 \approx -0.000428$. The exponential function can be written as

$$y = e^{-0.000428t}$$

PROBLEMS (Section 6-3)

Find the derivative of each function.

1. $y = e^{7x}$
2. $y = e^{20x}$
3. $f(x) = e^{-9x}$
4. $f(x) = e^{-0.04x}$
5. $y = 5e^{8x}$
6. $y = -0.16e^{5x}$
7. $f(x) = e^{2x+3}$
8. $y = e^{-5x+2}$
9. $y = e^{x^2-4x+3}$
10. $f(x) = -12e^{x+3}$

11. $y = 3e^{2x^3-4}$
12. $y = 6xe^x$
13. $f(x) = 3x^2e^x$
14. $y = 3xe^{2x}$
15. $f(x) = x^2e^{-3x}$
16. $f(x) = \dfrac{e^{2x}}{x^2+12}$
17. $y = \dfrac{e^x}{6x-1}$
18. $f(x) = \dfrac{e^{x+1}}{x^2+1}$

19. $y = \dfrac{x^2 - 1}{e^x}$

20. $f(x) = (x^2 - 3)e^{-x}$

21. $f(x) = (1 + e^{-x})x^3$

22. $y = 50 - 80e^{-0.4t}$

23. $y = 80 - 100e^{-0.02t}$

24. $f(x) = \dfrac{e^t - e^{-t}}{4}$

25. The average typing student's performance is given by the function

$$N(t) = 80 - 80e^{-0.05t}$$

where $N(t)$ is the number of words typed per minute after t weeks of instruction. What is the rate of change of $N(t)$ with respect to time?

26. Determine the rate of change of the investment function in Problem 1 of Section 6-1. What is the rate of change after 1 year?

27. Determine the rate of change of the investment function in Problem 2 in Section 6-1. What is the rate of change after 1 year?

28. Find the derivative of the function developed in Problem 3 of Section 6-1. Explain the meaning of the derivative when $t = 2$.

29. Find the derivative of the function developed in Problem 4 of Section 6-1. Explain the meaning of the derivative when $t = 50$.

30. Carbon-14 has a half-life of 5568 years. Write exponential functions representing the amount of carbon-14 left after t years using both the base $\frac{1}{2}$ and the base e.

Compute the first and second derivatives and verify that each function in Problems 31–36 has the properties given in Figure 6-7.

31. $y = 3^x$

32. $y = (\frac{1}{2})^x$

33. $y = 3(\frac{1}{2})^x$

34. $y = 2(3)^x$

35. $y = -\frac{1}{2}(2)^x$

36. $y = -\frac{1}{3}(\frac{1}{2})^x$

37. The size of a population grows according to the function

$$P = \frac{100}{1 + 25e^{-0.12t}}$$

Does this function achieve a maximum? If so, what is it?

Sketch the graph of each function in Problems 38–40 using appropriate curve sketching techniques.

38. $y = xe^x$

39. $y = xe^{-x}$

40. $y = x^2 e^x$

6-4 THE DERIVATIVE OF A LOGARITHMIC FUNCTION

In this section we address the problem of differentiating the function $f(x) = \log_b x$. Let us first consider the natural logarithm function $y = \ln x$. If $y = \ln x$, then $x = e^y$, and by substitution we can write $x = e^{\ln x}$. Let us differentiate both sides of this equation with respect to x:

$$\frac{d}{dx} x = \frac{d}{dx} e^{\ln x}$$

The derivative of the left-hand side is 1. To differentiate the right-hand side, we use the chain rule:

$$1 = e^{\ln x} \left(\frac{d}{dx} \ln x \right)$$

$$1 = x \left(\frac{d}{dx} \ln x \right)$$

This implies the following:

$$\frac{d}{dx} \ln x = \frac{1}{x}$$

Thus, the derivative of the natural logarithm function is simply $1/x$. We can easily generalize this rule to compositions of functions $f(g(x))$ where $f(x) = \ln x$ by using the chain rule from Chapter 4. This is stated below.

If $y = \ln g(x)$, then

$$y' = \left[\frac{1}{g(x)} \right] g'(x) = \frac{g'(x)}{g(x)}$$

For example, to find the derivative of $y = \ln x^2$, we have $g(x) = x^2$ and

$$\overbrace{y' = \left(\frac{1}{x^2} \right)}^{1/g(x)} \overbrace{(2x)}^{g'(x)}$$

$$= \frac{2x}{x^2} = \frac{2}{x}$$

Example 17   Differentiate the following functions:

  a. $y = \ln(2x^2 + 4x)$ b. $f(x) = (\ln x^4)^2$
  c. $y = \ln[(x + 4)(x - 2)^2]$ d. $y = x^2 \ln(x^4 + 6)$

Solution   a. We have

$$y' = \frac{4x + 4}{2x^2 + 4x}$$

$$= \frac{4(x + 1)}{2x(x + 2)}$$

$$= \frac{2(x + 1)}{x(x + 2)}$$

  b. Note that $(\ln x^4)^2$ *does not equal* $\ln x^6$! It is equal to $(4 \ln x)^2 = 16(\ln x)^2$. Therefore, we must use the general power rule:

$$f'(x) = 2(16) \ln x \frac{d}{dx} \ln x$$

$$= (32 \ln x) \left(\frac{1}{x} \right)$$

$$= \frac{32 \ln x}{x}$$

c. We can apply the chain rule and the product rule. By the chain rule,

$$y' = \frac{1}{(x+4)(x-2)^2} \frac{d}{dx} [(x+4)(x-2)^2]$$

The product rule then gives

$$y' = \frac{1}{(x+4)(x-2)^2} [(x+4)(2)(x-2)(1) + (x-2)^2(1)]$$

$$= \frac{2(x+4)(x-2) + (x-2)^2}{(x+4)(x-2)^2}$$

$$= \frac{(x-2)[2(x+4) + (x-2)]}{(x+4)(x-2)^2}$$

$$= \frac{2(x+4) + (x-2)}{(x+4)(x-2)}$$

$$= \frac{3x+6}{x^2 + 2x - 8}$$

Alternatively, we could have simplified y first as $y = \ln(x+4) + 2 \ln(x-2)$ before differentiating. Show that this yields the same result.

d. We first apply the product rule:

$$y' = x^2 \frac{d}{dx} \ln(x^4 + 6) + 2x \ln(x^4 + 6)$$

However, by the chain rule,

$$\frac{d}{dx} \ln(x^4 + 6) = \left(\frac{1}{x^4 + 6} \right) \frac{d}{dx} (x^4 + 6)$$

$$= \left(\frac{1}{x^4 + 6} \right) 4x^3$$

$$= \frac{4x^3}{x^4 + 6}$$

Substituting this last expression in the equation for y', we have

$$y' = x^2 \left(\frac{4x^3}{x^4 + 6} \right) + 2x \ln(x^4 + 6)$$

$$= \frac{4x^5}{x^4 + 6} + 2x \ln(x^4 + 6) \qquad \blacksquare$$

Example 18
Neurological Reaction Time   The reaction time t in the central nervous system in response to a stimulus decreases as the intensity of the stimulus S increases. A model that has been

proposed relating t and S is

$$t = a + \frac{1}{b} \ln \left(\frac{cS - k}{cS - h} \right)$$

The values of $a, b, c, k,$ and h are determined by observation and measurement. For instance, suppose that

$$t = 0.01 + \frac{1}{2} \ln \left(\frac{3S - 1}{3S - 2} \right)$$

Determine a function that gives the rate of change of t with respect to S.

Solution   Since

$$t = 0.01 + \tfrac{1}{2} \ln(3S - 1) - \tfrac{1}{2} \ln(3S - 2)$$

we have

$$\frac{dt}{dS} = \frac{1}{2} \left(\frac{3}{3S - 1} \right) - \frac{1}{2} \left(\frac{3}{3S - 2} \right)$$

$$= \frac{3[(3S - 2) - (3S - 1)]}{2(3S - 1)(3S - 2)}$$

$$= \frac{-3}{18S^2 - 18S + 4}$$

Example 19
Social Satisfaction   A model that has been proposed by sociologists to represent the satisfaction resulting from having a level x of some material good (money, for example) is

$$S = \ln(1 + x) - bx \qquad 0 < b < 1$$

Suppose that b is estimated to be 0.20 for some good. At what level is satisfaction a maximum?

Solution   The derivative is given by

$$S' = \frac{1}{1 + x} - 0.2$$

Setting this equal to 0 and solving for x yields

$$\frac{1}{1 + x} = 0.2$$

$$1 = 0.2(1 + x)$$

$$5 = 1 + x$$

$$x = 4$$

The second derivative is $S'' = -1/(1 + x)^2$ and is less than 0 for all x. Therefore, we have a relative maximum at $x = 4$. The graph of this function is shown in Figure 6-19.

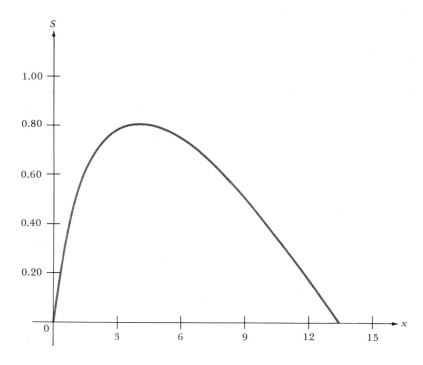

Figure 6-19
Graph of S = ln(1 + x) − 0.2x

We may now use the derivative of the natural logarithm function to help us find the derivative of a logarithmic function to any base b, that is, $y = \log_b x$. First, note that $y = \log_b x$ is equivalent to stating $x = b^y$. Earlier in this chapter we saw that we can always find a constant c such that $b = e^c$. We therefore have $x = (e^c)^y = e^{cy}$. Now let us take the natural logarithm of both sides of this equation:

$$\ln x = \ln e^{cy}$$

Since $\ln e^{cy} = cy$, we have

$$\ln x = cy$$

$$y = \frac{1}{c} \ln x$$

Now we can differentiate with respect to x and obtain

$$y' = \left(\frac{1}{c}\right)\left(\frac{1}{x}\right)$$

But, since $b = e^c$, then $c = \ln b$, and so we can finally state the following:

> If $y = \log_b x$, then
>
> $$y' = \frac{1}{x \ln b}$$

The chain rule can be applied to give an extended version of this result:

If $y = \log_b g(x)$, then

$$y' = \frac{g'(x)}{g(x) \ln b}$$

Example 20
Loudness and Sound Intensity

In Example 8 (Section 6-2), we stated that loudness L (in decibels) is given by the function

$$L(I) = 10 \log_{10} \left(\frac{I}{I_0} \right)$$

where I is the sound intensity and $I_0 = 10^{-12}$ watt per square meter. The derivative $L'(I)$ represents the rate of change of loudness with respect to sound intensity I. This derivative is found by applying the chain rule to $L(I)$:

$$L'(I) = 10 \left(\frac{1}{I_0} \right) \left[\frac{1}{(I/I_0) \ln 10} \right]$$

$$= \frac{10}{I \ln 10}$$

Since $\ln 10 \approx 2.3026$, then $L'(I) \approx 4.3429/I$. As I increases by 1 watt per square meter, the loudness will increase by approximately $4.3429/I$ decibels.   ∎

LOGARITHMIC DIFFERENTIATION

A useful application of logarithmic functions is in simplifying the ordinary differentiation of products of functions. In Chapter 4 (Example 3), we saw how to use the product rule in a "nested" fashion in order to differentiate the function

$$f(x) = (x + 3)(x^2 - 2x + 1)(x^3 - 3x^2)$$

We now discuss an alternative method of computing derivatives of products that uses properties of logarithms. Suppose that we first take the natural logarithm of both sides of the equation. This yields

$$\ln f(x) = \ln[(x + 3)(x^2 - 2x + 1)(x^3 - 3x^2)]$$

From the properties of logarithms, we know that the logarithm of a product of factors is the sum of the logarithms of the factors. Therefore, the above equation is equivalent to

$$\ln f(x) = \ln(x + 3) + \ln(x^2 - 2x + 1) + \ln(x^3 - 3x^2)$$

We *cannot* simplify this any further, since the logarithm of a sum or difference *does not equal* the sum or difference of the logarithms. This is a common misconception.

Next, let us differentiate both sides of the equation. Applying the chain rule to each term, we obtain

$$\frac{f'(x)}{f(x)} = \frac{1}{x+3} + \frac{2x-2}{x^2-2x+1} + \frac{3x^2-6x}{x^3-3x^2}$$

Finally, solving for $f'(x)$ gives the following result:

$$f'(x) = f(x)\left(\frac{1}{x+3} + \frac{2x-2}{x^2-2x+1} + \frac{3x^2-6x}{x^3-3x^2}\right)$$

$$= (x+3)(x^2-2x+1)(x^3-3x^2)\left(\frac{1}{x+3} + \frac{2x-2}{x^2-2x+1} + \frac{3x^2-6x}{x^3-3x^2}\right)$$

Upon simplifying, we find that $f'(x) = 6x^5 + 2x^4 - 20x^3 - 18x^2 - 18x$, which is the same result found using the product rule in Example 3 of Chapter 4. When a function consists of the product of several factors, it is usually easier to find the derivative in this way than to apply the product rule. This process is called *logarithmic differentiation* and we summarize it as follows:

Logarithmic Differentiation

---

1. Take the natural logarithm of both sides of the equation.
2. Express the logarithm of a product of factors as the sum of the logarithms of the factors.
3. Differentiate each term using the chain rule for logarithmic functions.
4. Solve for $f'(x)$ and replace $f(x)$ by its original definition.

Logarithmic differentiation also applies to quotients, as the next example illustrates.

Example 21   Differentiate the function

$$y = \frac{(x^2-2)(e^x+3x)}{2x-1}$$

Solution   When we take the logarithm of the right-hand side, we use the fact that $\ln(a/b) = \ln a - \ln b$:

$$\ln y = \ln(x^2-2) + \ln(e^x+3x) - \ln(2x-1)$$

Differentiating both sides yields

$$\frac{y'}{y} = \frac{2x}{x^2-2} + \frac{e^x+3}{e^x+3x} - \frac{2}{2x-1}$$

Therefore,

$$y' = \frac{(x^2-2)(e^x+3x)}{2x-1}\left(\frac{2x}{x^2-2} + \frac{e^x+3}{e^x+3x} - \frac{2}{2x-1}\right)$$

∎

We may also use logarithmic differentiation to provide a simple proof of the product rule for differentiation. Consider the function $y = g(x)f(x)$. Taking logarithms, we have

$$\ln y = \ln g(x) + \ln f(x)$$

Differentiating both sides of this equation yields

$$\frac{y'}{y} = \frac{g'(x)}{g(x)} + \frac{f'(x)}{f(x)}$$

Therefore,

$$y' = \frac{f(x)g(x)g'(x)}{g(x)} + \frac{f(x)g(x)f'(x)}{f(x)}$$
$$= f(x)g'(x) + g(x)f'(x)$$

which is the statement of the product rule in Chapter 4.

RELATIVE RATES OF CHANGE

The derivative of $\ln f(x)$ was shown to be equal to $f'(x)/f(x)$. This is interpreted as the rate of change of the function divided by the value of the function for any x.

> The ratio $f'(x)/f(x)$ is often called the **relative rate of change of $f(x)$** with respect to x.

Relative rates of change are useful when comparing the growth rates of different quantities. Consider Table 6-2, which gives the prices of certain automotive operating expenses in 1979 and 1983.

Table 6-2
Selected automotive operating expenses

Item	Unit Price ($)		Change	Change/Year
	1979	*1983*		
Gasoline (gallon)	0.93	1.15	0.22	0.055
Oil (quart)	0.95	1.75	0.80	0.20
Insurance (standard annual policy)	230	350	120	30

The average change per year from 1979 to 1983 is given in the last column of the table. Gasoline increased an average of $0.055 per year over this period, oil increased by $0.20, and insurance increased by $30 per year. Clearly, it is not correct to say that insurance has increased faster than gasoline or oil simply because the average increase per year is larger. A more meaningful comparison is on a percentage basis. The percentage change is the actual change divided by the 1979 value times 100. Thus, the average percentage change per year for gasoline is $0.055/0.93 \times 100 = 5.91\%$. For oil we have $0.20/0.95 \times 100 = 21.05\%$, and for insurance the figure is $30/230 \times 100 = 13.04\%$. Oil therefore has increased the most *relative to its base price in 1979*.

For any function $f(x)$, the relative rate of change is given by $f'(x)/f(x)$, which also happens to be the derivative of the natural logarithm of $f(x)$. Thus, to find the relative rate of change, we can either compute $f'(x)$ directly and divide by $f(x)$ or use logarithmic differentiation when it is applicable.

Example 22   Find the relative rate of change for the following:

   a. $y = 6x^2 + 2x$ at $x = 1$
   b. $g(t) = t^2 e^t$ at $t = 1$

Solution   a. $y' = 12x + 2$. When $x = 1$, $y = 6(1)^2 + 2(1) = 8$ and $y' = 12(1) + 2 = 14$. Thus, the relative rate of change is $\frac{14}{8} = 1.75$, or 175%.

   b. Logarithmic differentiation is helpful here. Taking the logarithms of both sides, we have

$$\ln g(t) = \ln t^2 + \ln e^t$$
$$= 2 \ln t + t$$

Differentiating yields

$$\frac{g'(t)}{g(t)} = \frac{2}{t} + 1$$

When $t = 1$, $g'(t)/g(t) = \frac{2}{1} + 1 = 3$, or 300%. Note that when logarithmic differentiation is used, we do not have to evaluate both $g(t)$ and $g'(t)$ explicitly.   ∎

Example 23
Real Wages   The hourly wages of factory workers can be modeled by the function $f(t) = 0.50t + 5$, where t is the number of years since 1976. However, the consumer price index has changed approximately according to the function

$$g(t) = 1.5t^2 + 3.5t + 100 \qquad 0 \le t \le 4$$

The real value of wages in year t adjusted for inflation is given by $h(t) = 100f(t)/g(t)$. What was the actual change in hourly wages in 1978 and 1980 and the relative rate of change of real dollars in these years as compared to 1976?

Solution   The year 1978 corresponds to $t = 2$ and 1980 to $t = 4$. We see that $f(2) = 6.00$ and $f(4) = 7.00$. Since $f(0) = 5$, the hourly wage increased by 20% in 1978 and by 40% in 1980 relative to 1976. The real wages adjusted for inflation is given by the function

$$h(t) = 100 \frac{0.50t + 5}{1.5t^2 + 3.5t + 100}$$

Taking logarithms, we have

$$\ln h(t) = \ln 100(0.50t + 5) - \ln(1.5t^2 + 3.5t + 100)$$

Differentiating, we obtain

$$\frac{h'(t)}{h(t)} = \frac{100(0.50)}{100(0.50t + 5)} - \frac{3t + 3.5}{1.5t^2 + 3.5t + 100}$$

At $t = 2$, we get

$$\frac{h'(t)}{h(t)} = \frac{0.5}{1.5} - \frac{9.5}{113}$$

$$\approx 0.333 - 0.084 = 0.249, \text{ or } 24.9\%$$

When $t = 4$,

$$\frac{h'(t)}{h(t)} = \frac{0.5}{2.5} - \frac{15.5}{138}$$

$$\approx 0.200 - 0.112 = 0.088, \text{ or } 8.8\%$$

Thus, in 1978, the average factory worker's purchasing power was increasing at a rate of 24.9% relative to 1976, but due to inflation, by 1980, it was increasing at a rate of only 8.8% relative to 1976 even though wages increased by 40%. ■

PROBLEMS (Section 6-4)

In Problems 1–18, differentiate each function.

1. $y = \ln(2x^3 + 3x)$

2. $y = \ln(3x^2 - 5x + 6)$

3. $f(x) = 3 \ln x$

4. $y = -8 \ln 2x^2$

5. $f(x) = (\ln x^3)^2$

6. $y = (\ln x^4)^{-2}$

7. $y = (\ln 2x^2)^3$

8. $f(x) = (\ln 3x^3)^2$

9. $f(x) = \ln[(x + 3)(x - 7)]$

10. $y = \ln[(2x + 1)(3x - 6)]$

11. $y = \ln[(x - 4)^4(x - 3)^6]$

12. $y = \ln[(2x + 5)(x + 4)^2]$

13. $f(x) = \ln\left(\dfrac{3x - 1}{x}\right)$

14. $f(x) = \ln\left(\dfrac{x^2 + 4}{x}\right)$

15. $y = \ln(x^2 + 4)^3$

16. $f(x) = \ln(2x^3 + 3x + 6)^2$

17. $f(x) = x^2 \ln(x + 3)$

18. $f(x) = x^3 \ln(2x - 4)$

Find the relative extrema for each function.

19. $y = \ln(1 + x) - 4x$

20. $y = \ln(2 - x) + 3x$

21. $y = x - \ln x$

22. $y = \ln x - x^2$

In Problems 23–34, differentiate each logarithmic function.

23. $y = \log_3 x$

24. $y = \log_5 x$

25. $y = \log_3 4x$

26. $f(x) = \log_7 8x$

27. $f(x) = \log_{10} x^3$

28. $y = \log_3 x^2$

29. $y = \log_{10}(3x + 7)$

30. $y = \log_6(x + 1)$

31. $f(x) = \log_3(x^2 + 4x)^5$

32. $f(x) = \log_{10}\left(\dfrac{x}{x^2 + 1}\right)$

33. $y = \log_7\left(\dfrac{2x}{x - 1}\right)$

34. $y = \log_{10}(x - 3)^2$

35. The pH of a substance is a measure of its acidity. Chemists define pH by

$$pH = \log_{10}(1/H)$$

where H is the concentration (in moles per liter) of hydrogen ions in a solution of the substance. What is the rate of change of pH with respect to H when $H = 10^{-4}$?

Find the derivative of each function using logarithmic differentiation.

36. $f(x) = (x + 2)(x^2 + 3x + 2)(x^3 - 2x^2)$

37. $f(x) = (2x + 6)(x^2 - 5x - 7)(2x^3 + 6)$

38. $f(x) = (x^2 + x)(3x^2 + 2x + 1)(4x^3 - 2x)$

39. $y = (x^3 + 2x)(5x^2 - x - 3)(2x^4 - 6x^2 + x)$

40. $y = \dfrac{(x^2 - 4)(e^x + 4x)}{3x + 2}$

41. $f(x) = \dfrac{(2x^2 + 3)^2(e^x - 6x)^3}{(2x^2 + 1)^4}$

42. $y = \dfrac{(e^x + 2x)(3x^2 - 4)}{e^{3x} - 4x}$

43. $f(x) = \dfrac{(e^{5x} - 7x)(e^x + 3x^2)}{e^{0.5x} - x^2}$

44. $y = x^x$

45. $y = x^{x^2}$

Find the relative rate of change for each function.

46. $y = x^2 + 3x + 6$ at $x = 8$

47. $y = x^3 - 8x^2 + 4x$ at $x = 2$

48. $g(t) = t^3 e^t$ at $t = 2$

49. $g(t) = 3t^2 e^t$ at $t = 1$

50. $g(t) = (2t^2 + 1)e^t$ at $t = 3$

51. $g(t) = 3e^t(t^2 - 7)$ at $t = 1$

52. $g(t) = (t^2 - 4)(2t + 6)(2e^t)$ at $t = 2$

Sketch the graph of each function.

53. $y = \dfrac{\ln x}{x}$

54. $y = \dfrac{1 + \ln x}{x}$

6-5 INTEGRATION OF EXPONENTIAL AND LOGARITHMIC FUNCTIONS

In Chapter 5 we discussed the concept of integration and its relationship to the derivative. In summary, if the derivative of $F(x)$ is $f(x)$, then $F(x)$ is an antiderivative of $f(x)$, and $\int f(x)\, dx = F(x) + c$. It is easy to see how to integrate exponential and logarithmic functions now that we know how to differentiate them. In Table 6-3 (page 326) we list some of the results concerning derivatives that we have developed in this chapter.

We can view Table 6-3 in reverse to find antiderivatives of the functions in the right-hand column, being sure to add a constant of integration. Thus, we have the results listed in Table 6-4.

Table 6-3

Function	Derivative
e^x	e^x
$e^{g(x)}$	$g'(x)e^{g(x)}$
$\ln x$	$1/x, \quad x > 0$
$\ln g(x)$	$g'(x)/g(x), \quad g(x) > 0$
b^x	$(\ln b)b^x$

Table 6-4
Integrals involving exponential and
logarithmic functions

$$\int e^x \, dx = e^x + c$$

$$\int g'(x)e^{g(x)} \, dx = e^{g(x)} + c$$

$$\int \frac{1}{x} \, dx = \ln|x| + c \qquad x \neq 0$$

$$\int \frac{g'(x)}{g(x)} \, dx = \ln|g(x)| + c \qquad g(x) \neq 0$$

$$\int b^x \, dx = \frac{b^x}{\ln b} + c$$

A few words are in order concerning the absolute values appearing in the formulas in Table 6-4. If $x > 0$, then since $|x| = x$,

$$\frac{d}{dx} \ln|x| = \frac{d}{dx} \ln x = \frac{1}{x}$$

While, if $x < 0$, then $|x| = -x$, and hence, by the chain rule,

$$\frac{d}{dx} \ln|x| = \frac{d}{dx} \ln(-x) = \frac{1}{-x}(-1) = \frac{1}{x}$$

Therefore, we see that for any $x \neq 0$, the function $F(x) = \ln|x|$ is an antiderivative of the function $f(x) = 1/x$, that is,

$$\int \frac{1}{x} \, dx = \ln|x| + c$$

In a similar way we find that

$$\int \frac{g'(x)}{g(x)} \, dx = \ln|g(x)| + c$$

We have now expanded our capabilities of integration to include a wide variety of useful functions. The tricky part, as we saw in Chapter 5, is identifying functions that are the result of the chain rule and recognizing the simple form of their antiderivative.

Example 24   Find the following integral:

$$\int 2e^{6x}\, dx$$

Solution   Whenever you see $e^{g(x)}$ in an integral, always check to see if it is multiplied by $g'(x)$ to within a constant. For e^{6x}, we see that we require a factor of the derivative of $6x$, that is, 6. Since constant factors may be moved in and out of the integral sign, obtaining the required factor is a simple matter:

$$\int 2e^{6x}\, dx = 2 \int e^{6x}\, dx$$

$$= \tfrac{2}{6} \int 6e^{6x}\, dx$$

$$= \tfrac{1}{3} \int 6e^{6x}\, dx$$

$$= \tfrac{1}{3}e^{6x} + c$$

∎

Example 25   Determine

$$\int xe^{3x^2}\, dx$$

Solution   The derivative of $3x^2$ is $6x$. This differs from x by only a constant, namely 6. Therefore,

$$\int xe^{3x^2}\, dx = \tfrac{1}{6} \int 6xe^{3x^2}\, dx$$

$$= \tfrac{1}{6}e^{3x^2} + c$$

∎

Example 26   Find the following integral:

$$\int \frac{6}{x}\, dx$$

Solution   We can factor the constant 6 out of the integral, yielding

$$\int \frac{6}{x}\, dx = 6 \int \frac{1}{x}\, dx$$

$$= 6 \ln|x| + c$$

∎

Example 27   Determine

$$\int \frac{x}{6x^2 + 1}\, dx$$

Solution   Whenever the integrand is a ratio of functions, consider the following result from Table 6-4:

$$\int \frac{g'(x)}{g(x)}\, dx = \ln|g(x)| + c$$

If we let $g(x) = 6x^2 + 1$, then

$$\frac{g'(x)}{g(x)} = \frac{12x}{6x^2 + 1}$$

which differs from our integrand by a factor of 12. We can easily compensate for this constant factor in the following way:

$$\int \frac{x}{6x^2 + 1}\, dx = \frac{1}{12} \int \frac{12x}{6x^2 + 1}\, dx$$
$$= \tfrac{1}{12} \ln|6x^2 + 1| + c$$
$$= \tfrac{1}{12} \ln(6x^2 + 1) + c$$

Note that we have dropped the absolute value bars, since $6x^2 + 1$ is always positive. ∎

Example 28
New Product Sales   When a new product is introduced, early sales often increase in an exponential fashion as more and more customers become aware of the product. Suppose that the weekly sales (in units per week) of a new product follows the exponential model

$$S = 500e^{0.12t}$$

How many units are sold in the first 40 weeks?

Solution   This is given by the integral

$$\int_0^{40} 500e^{0.12t}\, dt = 500 \int_0^{40} e^{0.12t}\, dt$$
$$= \frac{500}{0.12} e^{0.12t} \Big|_0^{40}$$
$$= 4116.67(e^{0.12(40)} - e^0)$$
$$\approx 4116.67(121.51 - 1)$$
$$\approx 502{,}125 \text{ units}$$ ∎

Example 29
Value of an Annuity   Individual Retirement Accounts (IRA's) have become very popular in recent years. If an individual invests A dollars at a rate of r, the value of the investment in t years will be approximately Ae^{rt}. This is called *continuous compounding* of interest. (A more complete discussion of this concept can be found in the supplement to Chapter 2.) If the same investment is made each year for T years, the value at the end of T years will be approximately

$$\int_0^T Ae^{rt}\, dt$$

This is called an *annuity*. The maximum IRA contribution is $2000. Suppose that an interest rate of $r = 0.10$ is assumed. If a person invests $2000 yearly in an IRA beginning at age 29, how much money will have accumulated by the time the investor reaches age 59?

Solution This is given by the integral

$$\int_0^{30} 2000e^{0.10t}\, dt = \frac{2000}{0.10} e^{0.10t} \Big|_0^{30}$$

$$= 20{,}000(e^3 - e^0)$$
$$\approx 20{,}000(20.086 - 1)$$
$$= \$381{,}720$$

∎

Example 30
Exponential Population Models

Example 29 showed that an exponential function represents the growth of money over time. There are many other applications where exponential growth models are appropriate. These occur in biology, sociology, pharmacology, and many other disciplines. Consider, for instance, a bacteria culture that initially contains 500 bacteria. If left undisturbed, the population will double every 5 days. Find the number of bacteria as a function of time. What is the average number of bacteria over an 8 day period?

Solution An exponential model for this situation is given by

$$y = 500e^{kt}$$

Since the population doubles in 5 days, we have

$$1000 = 500e^{k(5)}$$
$$2 = e^{5k}$$

Taking logarithms of both sides, we obtain

$$\ln 2 = 5k$$
$$k = \frac{\ln 2}{5}$$
$$\approx 0.1386$$

Thus, the function relating the number of bacteria to time is given by $y = 500e^{0.1386t}$.

The average number of bacteria over an 8 day period is given by

$$\frac{1}{8}\int_0^8 500e^{0.1386t}\, dt = \frac{1}{8}\left(\frac{500}{0.1386} e^{0.1386t}\right)\Big|_0^8$$

$$= \tfrac{1}{8}(3607.5)(e^{0.1386(8)} - e^0)$$
$$= 450.94(3.031 - 1)$$
$$\approx 916 \text{ bacteria}$$

∎

INTEGRATION BY PARTS*

Recall that the formula for applying integration by parts is

$$\int g(x)f'(x)\, dx = f(x)g(x) - \int f(x)g'(x)\, dx$$

---

* This section requires previous coverage of the supplement to Chapter 5; it may be omitted without loss of continuity.

or in the notation $u = g(x)$, $du = g'(x)\,dx$, $v = f(x)$, $dv = f'(x)\,dx$:

$$\int u\,dv = uv - \int v\,du$$

Integration by parts is useful in many applications involving exponential and logarithmic functions. We present two examples below.

Example 31   Find the integral $\int xe^x\,dx$.

Solution   Let $g(x) = x$ and $f'(x) = e^x$. Then

$$g'(x) = 1$$

$$f(x) = \int e^x\,dx = e^x$$

$$\int f(x)g'(x)\,dx = \int e^x\,dx = e^x + c$$

Hence,

$$\int xe^x\,dx = xe^x - e^x + c$$

Example 32   Integrate $\int x \ln x\,dx$.

Solution   In this example, let $u = \ln x$ and $dv = x\,dx$. Then $v = x^2/2$ and $du = 1/x\,dx$:

$$\int x \ln x\,dx = \overbrace{(\ln x)}^{u}\overbrace{\left(\frac{x^2}{2}\right)}^{v} - \int \overbrace{\left(\frac{x^2}{2}\right)}^{v}\overbrace{\left(\frac{1}{x}\right)}^{du}\,dx$$

$$= \left(\frac{x^2}{2}\right)\ln x - \int \frac{x}{2}\,dx$$

$$= \left(\frac{x^2}{2}\right)\ln x - \frac{x^2}{4} + c$$

PROBLEMS (Section 6-5)

In Problems 1–22, find the integral of each function.

1. $\int 3e^{6x}\,dx$

2. $\int 2e^{10x}\,dx$

3. $\int e^{-4x}\,dx$

4. $\int 5e^{5x-3}\,dx$

5. $\int xe^{5x^2}\,dx$

6. $\int 2xe^{4x^2}\,dx$

7. $\int 3x^3 e^{2x^4} \, dx$

8. $\int 3x^2 e^{3x^3} \, dx$

9. $\int \dfrac{9}{x} \, dx$

10. $\int \dfrac{15}{x} \, dx$

11. $\int \dfrac{6x^2}{x^3} \, dx$

12. $\int -\dfrac{8x^4}{x^5} \, dx$

13. $\int \dfrac{4}{4x + 2} \, dx$

14. $\int \dfrac{1}{x - 7} \, dx$

15. $\int \dfrac{2x}{x^2 + 5} \, dx$

16. $\int \dfrac{x}{3x^2 + 2} \, dx$

17. $\int \dfrac{x}{5x^2 - 3} \, dx$

18. $\int \dfrac{3x^2}{x^3 + 2} \, dx$

19. $\int \dfrac{4x^3}{x^4 - 2} \, dx$

20. $\int \dfrac{2x + 4}{x^2 + 4x + 3} \, dx$

21. $\int \dfrac{x^2 - 1}{x^3 - 3x} \, dx$

22. $\int \dfrac{5 - 5x}{6x - 3x^2} \, dx$

Find each definite integral.

23. $\int_0^4 e^x \, dx$

24. $\int_0^2 5e^{5x} \, dx$

25. $\int_0^3 4e^{4x-2} \, dx$

26. $\int_1^4 \dfrac{3}{x} \, dx$

27. $\int_0^5 \dfrac{2x}{x^2 + 5} \, dx$

28. $\int_0^2 \dfrac{x}{5x^2 - 3} \, dx$

29. When a new television program is introduced, viewers often increase in an exponential fashion as more and more people become aware of the program. Suppose that the number of weekly viewers of a new program follows the exponential model

$$V = 1000e^{0.15t}$$

where t is given in weeks. How many viewers will there be in the first 10 weeks?

30. Mary Campbell contributed $2000 a year to an IRA paying an interest rate of 12%. If she began to invest at age 35, how much money will she have accumulated by the time she reaches age 55?

31. The rate at which a substance grows is given by $G(t) = 300e^t$, where t is the time in days. What is the total accumulated growth after 3 days?

32. An oil field is estimated to produce $P(t)$ thousand barrels of oil per month t months from now, as given by the function

$$P(t) = 10te^{-0.2t}$$

Estimate the total production in the first 6 months of operation.

33. The population density D at a distance from the center of a city often exhibits exponential decay. Suppose the population density is approximated by $D = 0.14e^{-0.14r}$, where r is the distance from the center. What percentage of the population lives within 5 miles?

34. The rate of population change in a certain community has been estimated to be $f(t) = 900e^{0.018t}$,

where t equals the number of years since 1984. What is the net population change by 1987?

35. Standard Oil has about 30 billion barrels of oil in reserve. Solve the equation

$$\int_0^T 1.2e^{0.03t}\, dt = 30$$

to find the number of years, T, that this amount will last if nothing is added to it and it is withdrawn at a rate of $1.2e^{0.03t}$ billion barrels per year in year t.

36. The rate of production (in units per hour) of an antibody t hours after injection into the body is given by the function

$$h(t) = \frac{5t^2}{t^3 + 10}$$

a. After how many hours is the rate a maximum?
b. What is the rate after 1, 3, and 12 hours?

c. How much antibody is produced in the first 12 hours?

In Problems 37 – 42, use integration by parts to determine each integral.

37. $\displaystyle\int xe^{4x}\, dx$

38. $\displaystyle\int xe^{-x}\, dx$

39. $\displaystyle\int (x + 2)e^x\, dx$

40. $\displaystyle\int 3x^3 e^{x^2}\, dx$

41. $\displaystyle\int x^{-1/2} \ln x\, dx$

42. $\displaystyle\int \frac{(\ln x)^4}{x}\, dx$

6-6 CONCLUSION: IMPROVEMENT RATES OF COMPUTER HARDWARE

In the Introduction to this chapter, we developed two functions that represented the change in computer size and speed characteristics:

$$N = 1000(100)^x$$
$$T = 300,000(60)^{-x}$$

Both of these functions are exponential functions. In Section 6-3 we showed how to differentiate an exponential function with an arbitrary base b. Thus, if we are interested in the rate of change of these characteristics, we have

$$N' = 1000(\ln 100)(100)^x$$
$$= 4605.17(100)^x \text{ circuits per cubic foot per year}$$

and

$$T' = -300,000(\ln 60)(60)^{-x}$$
$$= -1,228,303(60)^{-x} \text{ nanoseconds per instruction per year}$$

The negative sign in T' indicates a decrease in instruction time, as the graph in Figure 6-1b shows.

Suppose that we wish to determine when T will be reduced to 0.1, assuming that the instruction time function is the exponential function T given above. We can use logarithms to solve the equation

$$0.1 = 300,000(60)^{-x}$$

Taking logarithms of both sides and using the properties we discussed in this chapter, we have

$$\ln 0.1 = \ln 300,000 - x \ln 60$$
$$-2.3026 = 12.61 - x(4.09)$$

Solving for x, we find $x \approx 3.65$. Since x represents 10 year increments, we would predict that this advance in technology would occur in 1986. How good a prediction was this? Check with your library or computer center!

6-7 CHAPTER REVIEW

1. An exponential function is a function of the form $f(x) = b^x$, where b is called the base. The domain of an exponential function is the set of all real numbers.

2. The percentage change in the dependent variable of an exponential function is constant if the value of the independent variable is increased by 1.

3. Graphs of a general exponential function $y = ab^x$ take on different shapes, depending on the values of a and b. This makes the exponential function useful in modeling a wide variety of practical problems.

4. A logarithmic function is the inverse of an exponential function. That is, $y = \log_b x$ if and only if $b^y = x$.

5. Like exponential functions, graphs of logarithmic functions assume a variety of different shapes, depending on the values of the parameters.

6. A special exponential function is $y = e^x$. This has the property that its derivative is itself, that is, $y' = e^x$.

7. We can express any exponential function in terms of the base e. Specifically, $b^x = e^{cx}$, where $c = \ln b$.

8. If $y = e^{g(x)}$, then $y' = g'(x)e^{g(x)}$.

9. The natural logarithm is the logarithm to the base e: $\ln x = \log_e x$.

10. If $y = b^x$, then $y' = (\ln b)b^x$.

11. If $y = \ln g(x)$, then $y' = g'(x)/g(x)$. This expression is often called the relative rate of change of $g(x)$.

12. If $y = \log_b x$, then $y' = 1/(x \ln b)$.

REVIEW PROBLEMS (Section 6-7)

1. Mike Smith places $8000 into 3 month certificates of deposit earning an annual rate of 10%. What will his initial investment be worth after the first 3 months? After the second 3 months? After 1 year?

2. The half-life of an element is 25.6 years. Write an expression for the amount left after 10 years.

3. It is estimated that the population of a certain country grows exponentially. If the population was 60 million in 1978 and 90 million in 1982, what is the anticipated population in 1990?

4. The amount of a certain radioactive substance remaining after t years is given by a function of the form $Q(t) = Q_0 e^{-0.004t}$. Find the half-life of the substance.

5. Radium decays exponentially. Its half-life is 1690 years. How long will it take for a 60 gram sample of radium to be reduced to 5 grams?

Write the following expressions in logarithmic form:

6. $y = 4^x$

7. $\frac{1}{27} = 3^{-3}$

8. $(\frac{1}{2})^{-4} = 16$

9. $(\sqrt{3})^4 = 9$

Write the following expressions in exponential form:

10. $\log_5 125 = 3$

11. $\log_{1/2} 32 = -5$

12. $\log_5 0.04 = -2$

Convert the following binary numbers to the base 10 number system:

13. 110_2

14. 1011_2

15. 11011_2

16. 101010_2

Write the following logarithmic equations in the form $y = a \log_b x$ with $a > 0$ and $0 < b < 1$.

17. $y = -3 \log_{64} x$

18. $y = -2 \log_{81} x$

Find the derivatives of the following functions:

19. $y = e^{5x}$

20. $f(x) = e^{-0.3x}$

21. $y = -4e^{7x}$

22. $y = xe^{-x^2}$

23. $f(x) = \dfrac{e^{2x}}{x^2 + 8}$

24. $f(x) = \dfrac{x^2}{e^x}$

25. The sales of a new product usually grow rapidly and then level off with time. For a certain product, sales are represented by an equation of the form

$$S(t) = 200 - 120e^{-0.3t}$$

where t represents time in years and $S(t)$ represents the level of sales. Find the rate of change of sales when $t = 1$; when $t = 10$.

26. Suppose that the population of a certain insect colony is given by $P(t) = 1000e^{0.3t}$, where t represents time in days. Find the rate of change of the population when $t = 2$; when $t = 10$.

Differentiate the following functions:

27. $y = \ln(x^2 + 6x - 2)$

28. $f(x) = -6 \ln 2x^2$

29. $f(x) = (\ln x^3)^{-2}$

30. $y = (\ln 2x^3)^3$

31. $y = x \ln x$

32. $y = \dfrac{\ln x}{x^2}$

Find the relative extrema for each function.

33. $y = \ln(2 + x) + 5x$

34. $y = x + \ln x$

Differentiate the following:

35. $y = \log_{10}(4x - 3)$

36. $y = \log_5 9x$

37. $f(x) = \log_6 \left(\dfrac{3x}{x + 1} \right)$

Find the derivative of each function using logarithmic differentiation.

38. $y = (x + 3)(x^2 - 2x + 5)(x^3 + 3x^2)$

39. $f(x) = \dfrac{(x^2 + 2)(e^x - 5x)}{2x + 6}$

40. $f(x) = \dfrac{(e^x + 5x)(2x^2 + 3)}{e^x - 3x}$

Find the relative rate of change for each function.

41. $y = x^3 + 2x^2 + 6x$ at $x = 3$

42. $g(t) = 2t^2 e^t$ at $t = 2$

43. $g(t) = (3t^2 + 6)e^t$ at $t = 2$

Compute the following indefinite integrals:

44. $\displaystyle\int 5e^{8x}\, dx$

45. $\displaystyle\int 4x^2 e^{2x^3}\, dx$

46. $\displaystyle\int \dfrac{10}{x}\, dx$

47. $\displaystyle\int \dfrac{2x}{x^2 - 4}\, dx$

48. $\displaystyle\int \dfrac{2x + 2}{x^2 + 2x + 1}\, dx$

49. $\displaystyle\int \dfrac{1}{x \ln 8}\, dx$

50. The amount of a sample of a radioactive substance remaining after t years is given by a function $A(t) = A_0 e^{-0.0002t}$. At the end of 6000 years, 3000 grams of the substance remain. How many grams were present initially?

51. A radioactive substance decays exponentially. If 600 grams of the substance were present initially and 400 grams are present 100 years later, how many grams will be present after 200 years?

52. The amount of a certain radioactive substance remaining after t years is given by a function of the form $Q(t) = Q_0 e^{-0.002t}$. Find the half-life of the substance.

53. It is known that 20% of a radioactive substance disintegrates in 100 years. What is the half-life of the substance?

54. How quickly will money double if it is invested at an annual interest rate of 8% compounded continuously?

55. Money deposited in a certain bank doubles every 12 years. The bank compounds interest continuously. What annual interest rate does the bank offer?

56. Tim Walker contributed $2000 a year to an IRA paying an interest rate of 11%. If he began to invest at age 30, how much money will he have accumulated by the time he reaches age 60?

57. The marginal cost of an item is given by $C'(x) = 200 + e^{0.004x}$, where x is the number of units of a product in hundreds. If x increases from 100 to 500, find the total increase in cost.

58. If a bacteria culture is growing at a rate given by $P'(t) = 3000e^{0.2t}$, where t is the time in hours and $P(0) = 5000$, find $P(t)$ and the number of bacteria after 5 hours.

DIFFERENTIAL EQUATIONS

OUTLINE

INTRODUCTION

Our environment has become an important source of concern over the last several decades. Environmental groups, government agencies, and industry have combined their efforts to clean up the land and water. Some questions that environmental agencies might wish to answer are the following: Suppose that a river is polluted to an extent of 5% and flows at a constant rate of 10 cubic kilometers per year (km^3/year) into a lake having a volume of 50 cubic kilometers. The river water is mixed uniformly throughout the lake and the resulting polluted water leaves the lake (through runoff and absorption) at the same rate of 10 km^3/year. (Hence, the volume of the lake remains at 50 cubic kilometers.) If the river is immediately and completely cleaned up so that no more pollution enters the lake, how long will it take for the concentration (that is, quantity/volume) of pollutants in the

lake to reach one-tenth of its present level? Or, suppose the present concentration of pollutants in the lake is 3% and it is decided to reach the more modest goal of reducing pollution to half its present value (that is, to 1.5%) over a 5 year period by a partial cleanup of the river. To what level must the pollution in the river be cut in order to achieve this goal?

To answer such questions, we might try to develop a mathematical model that represents the pollution level in the lake as a function of time, so that predictions can be made and planning can take place. To construct a model, the information from which we draw consists of observations or hypotheses about the nature of the pollution process. For instance, measurements indicate that water flows into the lake at a rate of 10 km^3/year. This implies that the concentration of pollutants in the lake is also changing at a certain rate. Of course, such rates of change are expressed mathematically as *derivatives*, and the problem translates into the mathematical problem of determining a function from given information about its derivative(s).

In the first problem posed, notice that since 5% of the river water is polluted, pollutants flow into the lake at a rate of

$$10 \text{ km}^3/\text{year} \times 0.05 = 0.5 \text{ km}^3/\text{year}$$

If we denote the quantity of pollutants (measured in cubic kilometers) in the lake at time t by $q(t)$, then (assuming the water in the lake is uniformly mixed) the concentration of pollutants in the lake at time t is $q(t)/50$. Since the water also flows out of the lake at a rate of $10 \text{ km}^3/\text{year}$, the rate at which pollutants leave the lake is

$$10 \text{ km}^3/\text{year} \times \frac{q(t)}{50} = \frac{q(t)}{5} \text{ km}^3/\text{year}$$

The rate at which the quantity of pollutants in the lake changes with respect to time, dq/dt, is equal to the rate at which pollutants flow into the lake minus the rate at which they flow out:

$$\frac{dq}{dt} = 0.5 - \frac{q(t)}{5} \text{ km}^3/\text{year}$$

If we let $c(t)$ represent the concentration of pollutants in the lake at time t, then $c(t) = q(t)/50$. Therefore, dividing the equation above by 50, we obtain the following equation for the concentration of pollutants in the lake:

$$\frac{dc}{dt} = \frac{d}{dt}\left(\frac{q(t)}{50}\right) = 0.01 - \frac{c(t)}{5}$$

Many pressing problems in today's technological society, such as that sketched above, involve *rates of change* of certain quantities. Inflation rates, the heating rate of the atmosphere due to the greenhouse effect, the rate at which a cancer grows, the nuclear proliferation rate, and many other examples immediately come to mind.

An equation that relates an unknown function with certain of its derivatives is called a *differential equation*. The equation relating $c(t)$ with dc/dt above is an example of a differential equation. The striking feature of differential equations is their wide applicability. We will see that diverse phenomena in economics, medicine, physical science, biology, and social science can be modeled by the same type of differential equation. While working through this chapter we strongly urge you to seek out the mathematical analogies between the seemingly different problems presented.

We wish to point out once more, as we did in Chapter 1, that differential equations, like all mathematical models, are only imperfect representations of the truth. Mathematical models based on differential equations are by nature *simplifications* that take into consideration only the most important factors. Therefore, the results of a mathematical analysis of a differential equation provide important insights into a realistic problem — not the last word on its solution.

7-1 DIFFERENTIAL EQUATIONS AND THEIR SOLUTIONS

By a **differential equation** we mean an equation involving an unknown function as well as one or more of its derivatives. For example, the differential equation

$$t^2 y'' + t y' - y = 0$$

relates an unknown function $y = f(t)$ and its first two derivatives. Because of the appearance of the second derivative in this equation we say that it is of *second order*. In general, the **order** of a differential equation refers to the highest-order derivative of the unknown function that appears in the equation. By a **solution** of a differential equation we mean any function $y = f(t)$ that satisfies the equation, that is, a function which when substituted into the differential equation results in a true statement.

For example, the function $y = t$ is a solution of the differential equation $t^2 y'' + t y' - y = 0$, since

$$y' = 1, \qquad y'' = 0$$

and hence,

$$t^2 y'' + t y' - y = t^2 \cdot 0 + t \cdot 1 - t = 0$$

In fact, it is quite easy to see that any function of the form $y = at$, where a is a constant, is a solution of this equation. Indeed, for this function we have

$$y' = a, \qquad y'' = 0$$

and therefore,

$$t^2 y'' + t y' - y = t^2 \cdot 0 + ta - at = 0$$

Similarly, any function of the form $y = bt^{-1}$, where b is a constant, is also a solution of $t^2 y'' + t y' - y = 0$. To see this, notice that if $y = bt^{-1}$, then

$$y' = -bt^{-2}, \qquad y'' = 2bt^{-3}$$

and hence,

$$\begin{aligned} t^2 y'' + t y' - y &= t^2 \cdot 2bt^{-3} + t \cdot (-bt^{-2}) - bt^{-1} \\ &= 2bt^{-1} - bt^{-1} - bt^{-1} = 0 \end{aligned}$$

Example 1   Show that $f(t) = t^2 + 2t + 1$ and $g(t) = t^2 + 2t + 5$ are both solutions of the differential equation

$$y' - 2t - 2 = 0$$

Solution   If $y = f(t)$, then

$$y' = f'(t) = 2t + 2$$

and hence,

$$y' - 2t - 2 = (2t + 2) - 2t - 2 = 0$$

Thus, $y = f(t)$ satisfies the differential equation.

Similarly, if $y = g(t)$, then

$$y' = g'(t) = 2t + 2$$

and again

$$y' - 2t - 2 = (2t + 2) - 2t - 2 = 0$$

In fact, it should be clear that *any* function of the form

$$y = t^2 + 2t + c$$

where c is a constant, is also a solution. Therefore, this differential equation has *infinitely many* solutions which differ only by the choice of the constant c (see Figure 7-1).

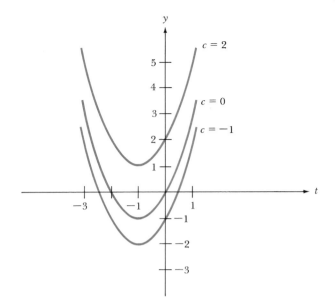

Figure 7-1
The family of functions
$y = t^2 + 2t + c$

What is behind this multiplicity of solutions of differential equations? Basically, the culprit is the differentiation process itself. In Chapter 2 we learned that differentiation destroys some of the information contained in a function. The reason for this is that the derivative of a constant, *any* constant, is 0. For example, the functions

$$f(t) = 4t + 7 \quad \text{and} \quad g(t) = 4t - 10$$

both have the same derivative, even though as functions they are quite distinct.

Another way to view this question of constants is to note that in solving a first-order differential equation we must recover a function from knowledge of its derivative. Therefore, a part of this recovery process must involve integration. In Example 1, we wish to find a function y whose derivative is $y' = 2t + 2$. In Chapter 5 we saw that this can be accomplished by finding any antiderivative of $2t + 2$. As always, this will introduce a constant of integration that will appear as an arbitrary constant in the solution. If an additional condition (usually called an **initial condition**) is imposed upon the solution, then this constant is uniquely determined.

Example 2   Find the solution of the differential equation $y' = 2t^3 + t + 1$ that satisfies the initial condition $y(1) = 1$. [By $y(1)$ we mean the value of the function $y = f(t)$ at $t = 1$.]

Solution   From the differential equation, we find

$$y = \int (2t^3 + t + 1) \, dt$$

$$\int (2t^3 + t + 1)\, dt = 2\frac{t^4}{4} + \frac{t^2}{2} + t + c$$
$$= \tfrac{1}{2}t^4 + \tfrac{1}{2}t^2 + t + c$$

where c is a constant. In order to satisfy the initial condition $y(1) = 1$, we must have

$$1 = y(1) = \tfrac{1}{2}(1)^4 + \tfrac{1}{2}(1)^2 + 1 + c$$
$$= 2 + c$$

and hence, $c = -1$. The solution of the differential equation that satisfies the initial condition is therefore

$$y = \tfrac{1}{2}t^4 + \tfrac{1}{2}t^2 + t - 1$$
∎

You should recall that the exponential function enjoys a very special property with respect to differentiation. Namely, if r and a are constants and $f(t) = ae^{rt}$, then we saw in Chapter 6 that

$$f'(t) = rae^{rt}$$
$$= rf(t)$$

This equation says quite simply that differentiating $f(t) = ae^{rt}$ is exactly equivalent to multiplying $f(t)$ by r, a very simple operation. For this reason, the exponential function plays an important role in the solution of differential equations.

Example 3   Solve the differential equation $y' = 3y$. Also, find the solution of this differential equation that satisfies the initial condition $y(0) = 7$.

Solution   From the discussion above we see that a solution has the form $y = ae^{3t}$. In order to find the solution that satisfies the initial condition, the constant a must be chosen properly. On substituting $t = 0$ into the formula for y, we obtain

$$y(0) = ae^0 = a$$

Since the initial condition is $y(0) = 7$, we have

$$7 = y(0) = a$$

Hence, $y = 7e^{3t}$ is the solution of the differential equation that satisfies the initial condition. ∎

In a similar manner, solving a second-order differential equation involves integrating twice, with the consequent introduction of two arbitrary constants. For example, if we wish to solve the differential equation

$$y'' = 1$$

a single integration produces

$$y' = \int y''\, dt = \int 1\, dt = t + c$$

and introduces a constant of integration c. In order to find y we must integrate once more:

$$y = \int y' \, dt = \int (t + c) \, dt$$

$$= \frac{t^2}{2} + ct + d$$

which introduces a second constant of integration, d.

Example 4   Find the solution of the differential equation $y'' = e^t$ that satisfies the initial conditions $y(0) = 3$ and $y'(0) = 2$.

Solution   Since $y'' = e^t$, we find

$$y' = \int e^t \, dt = e^t + c$$

The initial condition $y'(0) = 2$ gives

$$2 = y'(0) = e^0 + c$$
$$= 1 + c$$

Therefore, $c = 1$ and $y' = e^t + 1$.
   A further integration yields

$$y = \int (e^t + 1) \, dt = e^t + t + d$$

From the initial condition $y(0) = 3$, we then obtain

$$3 = y(0) = e^0 + 0 + d$$
$$= 1 + d$$

That is, $d = 2$. The solution of the differential equation that satisfies the initial conditions is therefore

$$y = e^t + t + 2$$

It is important to keep in mind that in solving a differential equation we are seeking a *function*, not a number. This function will be determined only to within certain arbitrary constants agreeing in number with the order of the equation. The constants can be uniquely determined if the appropriate initial conditions are specified.

PROBLEMS (Section 7-1)

In Problems 1 – 5, show that the given function is a solution of the differential equation.

1. $2t^2y' - 4ty = 0$, $y = 2t^2$

2. $ty' + y = 2t$, $y = t - t^{-1}$

3. $3y' + 2y = 0$, $y = 6e^{-(2/3)t}$

4. $(y')^2 - 8y = -16$, $y = 2t^2 + 2$

5. $y'' - y' - 2y = 0$, $y = e^{2t} - e^{-t}$

6. Show that each of the functions $y = t^{1/2}$ and $y = t^4$ is a solution of

$$2t^2 y'' - 7ty' + 4y = 0$$

7. Show that each of the functions $y = e^t$ and $y = e^{-t}$ is a solution of $y'' - y = 0$. Show that $y = ae^t + be^{-t}$ is also a solution for any constants a and b.

8. Solve the differential equation $y' = t^2 + t$.

9. Solve $y' = t + 1$, with $y(0) = 1$.

10. Solve $y' = -y$, with $y(0) = 6$.

11. Solve $y' = 6y$, with $y(0) = -1$.

12. Solve $y' = e^{-2t}$, with $y(0) = 2$.

13. Solve $y'' = 2e^{2t}$, with $y(0) = 1$ and $y'(0) = 2$.

14. Solve $y'' = -e^{2t}$, with $y(0) = 0$ and $y'(0) = 1$.

15. Solve $y' = \dfrac{1}{t+1}$, with $y(0) = 7$.

16. Solve $y' = \dfrac{t}{t^2 + 1}$, with $y(0) = 2$.

17. Solve $y' = (2t - 1)^5$, with $y(1) = 0$.

18. Solve $y' = e^{2t} + 2t$, with $y(0) = 2$.

19. Solve $y' = \dfrac{1}{t^2} + e^t - 1$, with $y(1) = e$.

20. Solve $y' = te^t - t$, with $y(0) = -1$.

7-2 MODELING WITH DIFFERENTIAL EQUATIONS

In the introduction to this chapter we indicated that many of the quantitative phenomena of physical science, biology, and economics are most easily described in terms of the *rates of change* of certain quantities. For example, one might hear statements such as "The rate of return on investments is double what it was. . . ," "The bacteria colony increased at a rate proportional to its size. . . ," or "The cooling rate is proportional to the temperature difference. . . ." In each case, instead of specifying the quantity itself, information is given on the rate at which a quantity changes (usually with respect to time).

In Chapter 2 we pointed out that the instantaneous rate of change of a function $f(t)$ at the point t is its derivative $f'(t)$. Indeed, derivatives were invented specifically to give easy expression to the notion of instantaneous rate of change. Therefore, each of the statements above concerns the derivative of some function. Such statements lead naturally, as we shall see in the examples below, to differential equations for an unknown quantity.

Example 5
Bacterial Growth   Biologists have observed that, within certain limits, the rate at which a colony of bacteria grows is proportional to the amount of bacteria present. If we let $Q(t)$ stand for the quantity of bacteria present at time t (we consider the colony to be continuous so that Q can be measured conveniently as a mass or volume), then the differential equation governing the growth of the colony is

$$\frac{dQ}{dt} = kQ$$

Here, k is a positive constant of proportionality called the **growth constant.** Clearly, the larger k is, the faster the colony will grow.   ∎

Example 6
A Learning Model

In many learning situations, for example, in the learning of typewriting which we discussed briefly in the Introduction to Chapter 2, direct observations of the rate of learning can be made. Suppose that $y(t)$ represents the typing speed (in words per minute) of a student at time t days. Then dy/dt, the rate at which typing speed changes with time, can be taken as a measure of the learning rate. Suppose it is found that the learning rate is proportional to $y - d$, the amount by which y differs from a certain fixed typing speed of d words per minute. Then the differential equation governing the typing speed is

$$\frac{dy}{dt} = a(y - d)$$

where a is a constant of proportionality.   ∎

Example 7
Spread of Diseases

A simple model for the spread of a communicable disease can be developed in terms of the fraction of a population that has the illness. Suppose the fraction of a given population that has the disease is denoted by s ($0 \le s \le 1$). Of course, s is a function of time. If s increases with time, then the disease is spreading, while if s is decreasing with time, the disease is in remission. If a fraction s of a population has the disease, then the fraction of the population that does not have the disease is $1 - s$. The basic premise of this model is that sick individuals interact freely with healthy individuals, and the disease spreads as a result of contacts between the sick and healthy. The number of such contacts is proportional to $s(1 - s)$. For example, if at a given time in a population of 100 individuals, 20 individuals have the disease, then $s = 0.2$, $1 - s = 0.8$, and there are $20 \times 80 = 10{,}000 \times 0.2 \times 0.8$ possible contacts between sick and well individuals. The rate of spread of the disease is then measured by the quantity ds/dt, the rate at which the fraction of diseased individuals changes, and the differential equation that models the spread of the disease is

$$\frac{ds}{dt} = ks(1 - s)$$

where k is a positive constant of proportionality.   ∎

Example 8
Price Dynamics

It is common knowledge that the supply S and demand D of a commodity are related to its price P. Suppose that the supply and demand are linear functions of the price, that is,

$$S = a + bP \qquad \text{and} \qquad D = c + dP$$

where a, b, c, and d are constants. Prices tend to change in direct proportion to the *excess demand*, or more precisely, the rate of change of price is proportional to the amount by which demand exceeds supply:

$$\frac{dP}{dt} = k(D - S)$$

where k is a positive constant of proportionality. Substituting the expressions for S and D above, we obtain

$$\frac{dP}{dt} = k[c + dP - (a + bP)]$$
$$= k[(c - a) + (d - b)P]$$

∎

Example 9
Cooling Rates

It is known that the surface temperature of a hot object will decrease at a rate proportional to the difference between the surface temperature of the object and the temperature of the surrounding medium (this so-called *ambient temperature* is assumed to be constant). This principle is known as **Newton's law of cooling.** If we denote the surface temperature of the object by T and the ambient temperature by A, then the law may be written as

$$\frac{dT}{dt} = -k(T - A)$$

where k is a positive constant (called the **cooling rate**). Note that the constant of proportionality, $-k$, is *negative*. The reason for this is that for $T > A$ the object cools down, that is, T is a *decreasing* function of time and therefore has a negative derivative. ∎

Example 10
Gravity

Isaac Newton invented calculus in order to give precise mathematical formulation to the laws of dynamics. The fundamental concepts in this science are velocity (rate of change of position with respect to time), acceleration (rate of change of velocity with respect to time), and force, which according to **Newton's second law of motion** is equal to the product of mass and acceleration.

Near the surface of the earth the gravitational force on an object can be taken to be essentially constant, and therefore, the acceleration due to gravity is essentially constant. This constant is typically denoted by g and in metric units has a value of about 9.8 meters per second per second (m/sec²). Consider an object in free-fall near the surface of the earth. Let the vertical distance from the surface of the earth to the object at time t be $y(t)$ (see Figure 7-2). The acceleration of the object is then y''. Neglecting all effects except gravity, this acceleration must be entirely due to gravity and hence the motion of the object can be described by the very simple second-order differential equation

$$y'' = -g$$

(Do you see the reason for the negative sign? If not, consider whether y' increases or decreases with respect to time.)

Figure 7-2
An object in free-fall

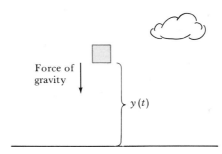

Force of gravity

$y(t)$

∎

Example 11
Advertising

Businesspeople know that advertising costs are related to sales. If we denote the profit level corresponding to an advertising expenditure x by $P(x)$, then dP/dx represents the rate at which profit changes with respect to advertising expenditure. A common model assumes that there is an equilibrium profit level P_0 and that the rate at which P changes with respect to advertising expenditure is directly proportional to the difference $P_0 - P$:

$$\frac{dP}{dx} = k(P_0 - P)$$

Compare this with the model in Example 9. Can you describe the analogy? ▪

Example 12
Mechanical Vibrations

Consider an object of mass m resting on a frictionless horizontal surface and connected to a vertical wall by a stiff spring (see Figure 7-3). If the spring is stretched (or compressed) by s units, then by **Hooke's law** the object will experience a force proportional to s. We can use s to indicate the position of the mass (with respect to its equilibrium position), and hence its acceleration is s''. Newton's second law (see Example 10) then says that the total force on the object is ms''. Since we are ignoring friction, the only force on the object is due to the spring; therefore, by Hooke's law we find that the motion of the object is modeled by the second-order differential equation

$$ms'' = -ks$$

where k is a positive constant of proportionality called the *stiffness* of the spring.

Figure 7-3
Mechanical vibrations

Restoring force ▪

Example 13
Continuous Investing

Suppose a bank compounds interest continuously at an annual interest rate of r. A sum of money S in such a bank account then grows at a rate

$$\frac{dS}{dt} = rS$$

(see Chapter 6). Suppose, however, that an additional amount of k dollars per year is deposited continuously into the account (say, through an automatic transfer account). Then the growth of the sum in the account is affected by two factors, the compounding of interest and the continuous investing. If no interest were paid, then the sum would increase only due to the continuous investment at the rate

$$\frac{dS}{dt} = k$$

Taking both factors into account, we find that S grows so as to satisfy the differential equation

$$\frac{dS}{dt} = rS + k$$

  ▪

Example 14
A Mixing Problem

Suppose water containing 3 pounds of salt per gallon flows into a vat, which initially contains 20 gallons of fresh water, at a rate of 5 gallons per minute. The

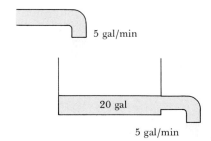

Figure 7-4
One-compartment mixing problem

liquid in the vat is uniformly mixed and allowed to flow out at the same rate of 5 gallons per minute (see Figure 7-4). We want to develop a model for $q(t)$, the quantity of salt in the tank at time t. Clearly, dq/dt, the rate at which the quantity of salt changes with respect to time, is equal to the rate at which salt flows into the vat minus the rate at which salt flows out of the vat. Since salt water with a concentration of 3 pounds per gallon flows into the vat at a rate of 5 gallons per minute, salt enters the vat at a rate of

$$3 \; \frac{\text{lb}}{\text{gal}} \times 5 \; \frac{\text{gal}}{\text{min}} = 15 \; \frac{\text{lb}}{\text{min}}$$

At what rate does salt leave the vat? Note that the volume of liquid in the vat is constantly 20 gallons. Since the quantity of salt in the vat at time t is $q(t)$, the concentration of salt in the vat at time t is $q(t)/20$ pounds per gallon. This liquid leaves at a rate of 5 gallons per minute, so the salt leaves the vat at a rate of

$$\frac{q(t)}{20} \frac{\text{lb}}{\text{gal}} \times 5 \; \frac{\text{gal}}{\text{min}} = \frac{q(t)}{4} \frac{\text{lb}}{\text{min}}$$

Therefore,

$$\frac{dq}{dt} = \text{Rate in} - \text{Rate out}$$

$$= 15 - \frac{q(t)}{4} \; \text{lb/min}$$

models the quantity of salt in the vat at time t. Do you see the analogy between this example and the introduction to this chapter? ∎

PROBLEMS (Section 7-2)

1. Show that $Q = Q_0 e^{kt}$ is the solution of the differential equation in Example 5 that satisfies the initial condition $Q(0) = Q_0$.

2. It is known that a radioactive isotope decays at a rate proportional to the amount of isotope present. Write a differential equation that describes this situation. (Compare with Example 5.)

3. A simple model for investment of capital goes like this: Suppose $C(t)$ is the capital invested at time t and A is a preassigned base line for the capital

invested. Whenever the capital rises above A, it is withdrawn at a rate proportional to the difference between $C(t)$ and A. On the other hand, whenever the capital decreases below A, it is invested at a rate proportional to the difference. Write a differential equation for this policy. (Compare with Examples 9 and 11.)

4. The spread of rumors has been modeled in the following way: It is assumed that a certain fraction x of a population knows a rumor and spreads it by contact with the fraction $1 - x$ of the population that does not know the rumor. Write a differential equation that models the rate of spread in terms of the quantity dx/dt. (Compare with Example 7.)

5. Suppose an ice cube melts (that is, loses volume) at a rate proportional to its surface area. Write a differential equation describing the rate at which a *side* of the cube shrinks.

6. Suppose an object falls near the surface of the earth and experiences a force due to air resistance that is proportional to the square of its velocity. Write a differential equation describing the object's velocity.

7. Let $P(t)$ stand for the population of a species of animals at time t. Suppose the population increases at a rate that is a sum of two terms. The first term is proportional to P (compare with Example 5) and the second term, which models the pernicious effects of crowding, is proportional (with a negative constant of proportionality) to P^2. Write a differential equation that models the evolution of this population. (This differential equation is sometimes called **Verhulst's law.**)

8. Show that for any constant C, $T = A + Ce^{-kt}$ is a solution of the differential equation in Example 9.

9. Develop a differential equation for the following simple model for drug infusion: Suppose the drug is infused at a constant rate k but is removed from the bloodstream and absorbed into the body tissues at a rate proportional to the amount of drug in the bloodstream. (Compare with Example 13.)

10. Write a differential equation for the following simple pricing policy: An equilibrium price p_0 is

established. Whenever the actual market price p differs from p_0, it is changed in such a way that the rate of change of p with respect to time is proportional to the difference between p_0 and p. (Compare with Example 7 and Problem 3.) Compare your model with Example 8. Identify p_0 in terms of the constants in Example 8.

11. Suppose that the surface in Example 12 is not frictionless and the force of friction is assumed to be proportional to the velocity of the object. Find a differential equation that describes the motion of the object.

12. Consider a population of animals, say catfish on a catfish farm, which if undisturbed would tend to increase their number at a rate proportional to the number present. Suppose, however, that the fish are subjected to continuous harvesting of k fish per day. Write a differential equation that describes the evolution of the population. (Compare with Example 13.)

13. A vat initially contains 100 gallons of pure water. A 10% salt solution enters the vat through a pipe at a rate of 2 gallons per minute while the mixed solution is allowed to escape through a tap at the same rate. Write a differential equation, with initial condition, that describes the concentration of the salt solution in the vat. (Compare with Example 14.)

14. Suppose a 100 gallon vat initially contains 20 gallons of pure water. A 10% salt solution enters the vat at a rate of 2 gallons per minute while the mixed solution escapes through a tap at the rate of 1 gallon per minute. Write a differential equation, with initial condition, to describe the quantity of salt in the vat. Is the equation valid for all times? Explain.

15. Consider a chemical reaction in which a certain substance X is transformed into another substance Y. Let $x(t)$ and $y(t)$ represent the concentrations of X and Y, respectively, at time t. Let c be the initial concentration of X; then $x(t) = c - y(t)$. The reaction is called *autocatalytic* if it "feeds on itself" in the sense that the rate at which substance Y is produced is proportional to the product of $y(t)$ and $x(t)$. Find a differential equation for the quantity $y(t)$.

7-3 LINEAR DIFFERENTIAL EQUATIONS

Many of the differential equations of the previous section have (or can be put into) the form

$$y' + ay = g$$

where a is a given constant and g is a given function of the independent variable. For instance, in Example 9, $a = k$ and g is the constant function $g(t) = kA$. An equation of this type is called a **linear first-order differential equation** with constant coefficients. The term *linear* refers to a special mathematical feature of solutions to such equations which need not concern us here. Our interest in such equations comes from the fact that they occur frequently and they are very easy to solve. The key to solving them is the rule for differentiating the exponential function,

$$\frac{d}{dt} e^{at} = ae^{at}$$

along with the product rule for differentiation,

$$\frac{d}{dt} e^{at}y = e^{at}y' + ae^{at}y$$
$$= e^{at}(y' + ay)$$

That is, the right-hand side of this equation is e^{at} times the left-hand side of the original differential equation. To put it another way, multiplying $y' + ay = g$ by e^{at} transforms it into an equation in which the left-hand side is the derivative of $e^{at}y$:

$$e^{at}(y' + ay) = e^{at}g$$

or

$$\frac{d}{dt} e^{at}y = e^{at}g$$

To solve this differential equation, we integrate both sides:

$$\int \frac{d}{dt} e^{at}y \, dt = \int e^{at}g \, dt$$
$$e^{at}y = \int e^{at}g \, dt$$

To solve for y, we simply divide by e^{at} (equivalently, multiply by e^{-at}), keeping in mind that the indefinite integral on the right-hand side will introduce an arbitrary constant of integration.

Solving Linear Differential Equations

> The solution of the linear differential equation
> $$y' + ay = g$$
> satisfies
> $$e^{at}y = \int e^{at}g \, dt$$

We now illustrate the technique of solution with several examples.

Example 15   Solve $y' + 2y = 0$.

Solution   Here, $a = 2$ and $g(t) = 0$. Therefore,

$$e^{2t}y = \int 0 \, dt$$
$$= c$$
$$y = ce^{-2t}$$

where c is an arbitrary constant.   ∎

Example 16   Solve $y' = 3y + e^t$.

Solution   First put the equation in the standard form:

$$y' - 3y = e^t$$

We then see that $a = -3$ and $g(t) = e^t$. Therefore, we have

$$e^{-3t}y = \int e^{-3t}e^t \, dt$$
$$= \int e^{-2t} \, dt$$
$$= -\tfrac{1}{2}e^{-2t} + c$$
$$y = -\tfrac{1}{2}e^t + ce^{3t}$$

where c is an arbitrary constant.   ∎

As always, the constant in the solution can be determined if an initial condition is imposed.

Example 17   Solve $y' = 7y - 4$ with the initial condition $y(0) = 10$.

Solution   In this problem, $a = -7$ and $g(t) = -4$. Therefore,

$$e^{-7t}y = \int -4e^{-7t} \, dt$$
$$= \tfrac{4}{7}e^{-7t} + c$$
$$y = \tfrac{4}{7} + ce^{7t}$$

The initial condition then gives

$$10 = y(0) = \tfrac{4}{7} + ce^0$$
$$= \tfrac{4}{7} + c$$

Therefore,

$$c = 10 - \tfrac{4}{7} = \tfrac{66}{7}$$

and

$$y = \tfrac{4}{7} + \tfrac{66}{7}e^{7t}$$
$$= \tfrac{1}{7}(4 + 66e^{7t}) \qquad \blacksquare$$

Example 18
An Investment Problem

A sum of money is invested at an annual percentage rate of 14% compounded continuously. How long does it take the money to double?

Solution

Let S represent the sum of money at time t. Then under continuous compounding we know that the sum grows at a rate of

$$\frac{dS}{dt} = 0.14S$$

or in standard form,

$$\frac{dS}{dt} - 0.14S = 0$$

We therefore have $a = -0.14$ and $g(t) = 0$. Hence,

$$e^{-0.14t}S = \int 0 \, dt$$
$$= c$$
$$S = ce^{0.14t}$$

Note that $S(0) = ce^0 = c$ is the initial investment. We would then like to find the time T (the doubling time) so that $S(T) = 2c$. Substituting these expressions into the formula for S above, we have

$$2c = ce^{0.14T}$$
$$2 = e^{0.14T}$$

Therefore,

$$\ln 2 = 0.14T$$

and

$$T = \frac{\ln 2}{0.14} \approx 4.95 \text{ years} \qquad \blacksquare$$

Example 19
A Cooling Problem

A tub of hot water at 150°F is placed in an air-conditioned room which is kept at 70°F. A quarter hour later, the temperature of the water is 120°F. How long does it take for the temperature of the water to reach 90°F?

Solution   Let $T(t)$ represent the temperature of the water t hours after it has been brought into the room. According to Newton's law of cooling (see Example 9 in Section 7-2), T satisfies

$$\frac{dT}{dt} = -k(T - 70) \qquad T(0) = 150$$

Therefore,

$$\frac{dT}{dt} + kT = 70k$$

We then have

$$e^{kt}T = \int 70ke^{kt}\, dt$$
$$= 70e^{kt} + c$$
$$T = 70 + ce^{-kt}$$

However, the initial condition $T(0) = 150$ gives

$$150 = 70 + c$$
$$c = 80$$

and hence,

$$T(t) = 70 + 80e^{-kt}$$

In order to proceed further we must determine a value for k. Fortunately, we have been supplied with a piece of information that allows us to do this, namely, we are told that after a quarter hour the temperature is $120°$F. Therefore,

$$120 = T\left(\frac{1}{4}\right) = 70 + 80e^{-k/4}$$

$$e^{-k/4} = \frac{120 - 70}{80} = \frac{5}{8}$$

Therefore,

$$-\frac{k}{4} = \ln\left(\frac{5}{8}\right)$$

$$k = -4\ln\left(\frac{5}{8}\right) \approx 1.88$$

Substituting this value for k into the formula for T above gives

$$T(t) = 70 + 80e^{-1.88t}$$

When will the temperature reach $90°$F? Call this time τ. Then

$$90 = T(\tau) = 70 + 80e^{-1.88\tau}$$

We then have

$$e^{-1.88\tau} = \frac{90 - 70}{80} = \frac{1}{4}$$

and hence,

$$-1.88\tau = \ln\left(\frac{1}{4}\right) \approx -1.386$$

That is, $\tau \approx 0.74$, or about three-quarters of an hour. ∎

Example 20
Manufacturing Costs

The manufacturing cost K of a certain item is found to be related to the number of units produced, x, by the differential equation

$$\frac{dK}{dx} + K = 3 + 2x$$

It follows that

$$e^x K = \int (3 + 2x)e^x \, dx$$
$$= 3e^x + 2(xe^x - e^x) + c$$
$$= e^x + 2xe^x + c$$

(You can check the integration formula $\int xe^x \, dx = xe^x - e^x + c$ for yourself; simply differentiate the right-hand side! Or you can compute the integral directly by using integration by parts, as described in the supplement to Chapter 5.) Therefore,

$$K(x) = 2x + 1 + ce^{-x}$$

The term $2x$ is directly proportional to the number of units manufactured and is called the *direct cost*. The term $1 + ce^{-x}$ decreases with the number of units manufactured and is called the *learning cost*, because it represents the effect of experience in the manufacturing process. ∎

Example 21
Carbon-14 Dating

The radioactive isotope carbon-14 decays at a rate proportional to the amount present (see Problem 2 of Section 7-2). If Q represents the amount of carbon-14 present at time t, then it follows that

$$\frac{dQ}{dt} = -kQ$$

where k is a positive constant. We now have

$$\frac{dQ}{dt} + kQ = 0$$

and hence,

$$e^{kt}Q = \int 0 \, dt$$
$$= c$$
$$Q = ce^{-kt}$$

Note that $Q(0) = ce^0 = c$, so the constant c represents the initial quantity of carbon-14.

It has been determined experimentally that the **half-life** of carbon-14 is 5568 years; that is, a given quantity of carbon-14 will decay to one-half its original mass in 5568 years. Since the initial quantity of carbon-14 is the constant $c = Q(0)$, this is the same as saying $Q(5568) = Q(0)/2$ (where t is measured in years). Substituting into the formula for Q above, we have

$$\frac{Q(0)}{2} = Q(0)e^{-k(5568)}$$

$$\frac{1}{2} = e^{-k(5568)}$$

Therefore,

$$-k(5568) = \ln\left(\frac{1}{2}\right) = -\ln 2$$

We then find that the decay constant k for carbon-14 is

$$k = \frac{\ln 2}{5568} \approx 1.245 \times 10^{-4}$$

W. Libby won the 1960 Nobel prize in chemistry for developing the radio-carbon technique for dating archaeological remains. Organic matter, such as wood, contains a certain amount of carbon-14 which begins to decay when the tree dies. Libby's technique involves measuring the ratio $Q/Q(0)$ for carbon-14. Since

$$\frac{Q}{Q(0)} = e^{-kt}$$

and k is known, this allows the determination of the time t since the death of the remains. Suppose, for example, that $Q/Q(0)$ is measured to be 0.15 for an oak ax handle. Then

$$0.15 = e^{-kt}$$

where t is the time that has elapsed since the oak was felled and the ax handle was made. We then find that

$$\ln 0.15 = -kt$$

and

$$t = \frac{\ln 0.15}{-k} = \frac{-\ln 0.15}{1.245 \times 10^{-4}} \approx 15,238 \text{ years}$$

That is, the ax handle is about 15,000 years old. ∎

Example 22
Advertising

Consider the model for advertising and profit from Example 11 (Section 7-2):

$$\frac{dP}{dx} = k(P_0 - P)$$

Suppose that profit and advertising expenditures are both measured in units of $10,000. Furthermore, suppose that if no advertising is done, the profit level of the firm is $100,000 [that is, $P(0) = 10$]. If $10,000 is spent on advertising, the firm's profit increases to $120,000 [that is, $P(1) = 12$], and if $20,000 is spent on advertising, the profit is $136,000 [that is, $P(2) = 13.6$]. What is the equilibrium profit level P_0?

Solution   To answer this question we must solve the differential equation

$$\frac{dP}{dx} = k(P_0 - P)$$

or

$$\frac{dP}{dx} + kP = kP_0$$

as well as satisfy the conditions $P(0) = 10$, $P(1) = 12$, and $P(2) = 13.6$. To solve the differential equation, notice that

$$e^{kx}P = \int kP_0 e^{kx} \, dx$$

$$= P_0 e^{kx} + c$$

$$P = P_0 + ce^{-kx}$$

Since $P(0) = 10$, we find

$$10 = P(0) = P_0 + ce^{-k0}$$

$$= P_0 + c$$

$$c = 10 - P_0$$

Therefore,

$$P(x) = P_0 + (10 - P_0)e^{-kx}$$

Notice that, since $k > 0$,

$$\lim_{x \to \infty} P(x) = P_0 + (10 - P_0) \lim_{x \to \infty} e^{-kx}$$

$$= P_0 + (10 - P_0)0 = P_0$$

Therefore, P_0 represents the highest profit level attainable if there are unlimited advertising expenditures ($x \to \infty$). To find P_0 we use the two pieces of information $P(1) = 12$ and $P(2) = 13.6$:

$$12 = P(1) = P_0 + (10 - P_0)e^{-k}$$

$$13.6 = P(2) = P_0 + (10 - P_0)e^{-2k}$$

From the first equation we have

$$\frac{12 - P_0}{10 - P_0} = e^{-k}$$

and from the second equation,

$$\frac{13.6 - P_0}{10 - P_0} = e^{-2k}$$

Now, if we square both sides of the first equation, we obtain

$$\left(\frac{12 - P_0}{10 - P_0}\right)^2 = e^{-2k}$$

and then, since the left-hand side of each equation is equal to e^{-2k}, we have

$$\frac{13.6 - P_0}{10 - P_0} = \left(\frac{12 - P_0}{10 - P_0}\right)^2$$

Multiplying both sides of this equation by $(10 - P_0)^2$, we have

$$\frac{13.6 - P_0}{10 - P_0}(10 - P_0)^2 = \frac{(12 - P_0)^2}{(10 - P_0)^2}(10 - P_0)^2$$

$$(13.6 - P_0)(10 - P_0) = (12 - P_0)^2$$

That is,

$$136 - 23.6P_0 + P_0^2 = 144 - 24P_0 + P_0^2$$

$$136 - 23.6P_0 = 144 - 24P_0$$

$$24P_0 - 23.6P_0 = 144 - 136$$

$$0.4P_0 = 8$$

$$P_0 = \frac{8}{0.4} = 20$$

Therefore, allowing unlimited expenditures for advertising, the maximum profit level is $200,000. ∎

PROBLEMS (Section 7-3)

1. Solve $y' + 3y = 2$.

2. Solve $y' = 2y - 5$.

3. Solve $y' = -y + 4$, with $y(0) = 1$.

4. Solve $y' = 2y + e^t$, with $y(0) = 2$.

5. Solve $y' + 3y = te^{-3t}$, with $y(0) = -7$.

6. Solve $y' = -4y + t^2 e^{-4t}$, with $y(0) = 6$.

7. Solve $y' - 2y = 2e^{2t}$, with $y(1) = 4$.

8. Suppose a population increases at a rate proportional to its size.
   a. Write a differential equation describing this population.
   b. Solve the differential equation found in part a.
   c. Find a formula, in terms of the proportionality constant, for the time it takes for the population to double. (This formula is called the *law of Malthus;* compare with Example 5 in Section 7-2.)
   d. Suppose that in 50 years the population increased sixfold. How long did it take it to double?

9. Consider the differential equation for the bacteria colony in Example 5 in Section 7-2.
   a. Solve this differential equation.
   b. Suppose the colony is observed to double in 16 hours. How long will it take for the colony to increase threefold?

10. a. Solve the differential equation for Newton's law of cooling given in Example 9 in Section 7-2.
    b. What value does the surface temperature ultimately approach as time goes on indefinitely?

11. a. Solve the differential equation for continuous investing in Example 13 in Section 7-2.

b. Suppose the annual interest rate is 11% and $5000 is invested continuously each year. If the initial sum is $50,000, find a formula for $S(t)$, the total sum invested after t years.

c. Referring to part b, how long will it take for the sum invested to reach $100,000?

12. Consider the drug infusion model of Problem 9 in Section 7-2. Let $q(t)$ stand for the amount of the drug in the bloodstream at time t. Since the drug is infused at a constant rate k and removed at a rate proportional to $q(t)$, the differential equation for q is

$$\frac{dq}{dt} = k - aq$$

where a is a positive constant of proportionality.

a. Suppose that the infusion rate is 250 milligrams per hour and initially there is no drug in the bloodstream. Solve the differential equation.

b. What is the limiting value of the amount of the drug in the bloodstream as t becomes arbitrarily large?

c. Suppose that after 1 hour there is 200 milligrams of the drug in the bloodstream. Can you find the constant a?

13. Suppose the population of catfish on a catfish farm is subjected to continuous harvesting so that it satisfies a differential equation of the form

$$\frac{dP}{dt} = aP - k$$

(See Problem 12 of Section 7-2.)

a. Suppose $a = 0.02$, the initial population is 50,000, and the catfish are harvested at a rate of 500 per day. Solve the differential equation.

b. Is the model realistic? What important effect is not being taken into account?

14. When light enters a translucent substance, its intensity I decreases at a rate proportional to the intensity $I(x)$ at depth x in the medium. (This is known as *Bouguer's law*.)

a. Write a differential equation for $I(x)$.

b. Suppose that after traversing 2 feet of a certain medium a beam of light has lost 30% of its intensity. How much of the intensity is lost in traversing 5 feet of the medium?

15. The concentration of a solute on the outside of a cell membrane is assumed to be a constant c_0. Inside the cell the concentration of the solute is a function $c(t)$. The concentration $c(t)$ changes at a rate proportional to $c_0 - c(t)$, the difference of the concentrations.

a. Write a differential equation for $c(t)$.

b. Solve the differential equation.

c. Sketch a graph of $c(t)$ considering both the case $c(0) > c_0$ and the case $c(0) < c_0$.

16. A 20% dye solution flows at a rate of 3 gallons per minute into a vat that initially contains 50 gallons of fresh water. The uniformly mixed solution then flows out of the vat at the same rate of 3 gallons per minute.

a. Write a differential equation for the concentration of dye in the vat.

b. Solve the equation.

c. How long does it take for the concentration of dye in the vat to reach 10%?

17. Suppose the discharge from the vat in Problem 16 flows into a tank that initially holds 30 gallons of fresh water. The uniformly mixed solution in this second tank then flows out at a rate of 3 gallons per minute (see the figure). Find a formula for the concentration of dye in the second tank as a function of time.

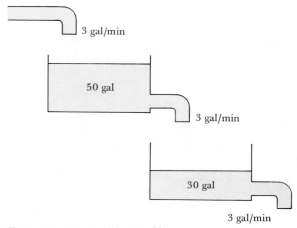

Two-compartment mixing problem

18. Water containing an unknown concentration of pollutants seeps at an unknown rate into a 1000 gallon underground cistern. The water leaks out of the cistern at the same rate. Measurements

show that the initial concentration of pollutants in the cistern is 1%. After 1 day, the concentration of pollutants is 1.12%, and after 2 days, it is 1.23%. What is the concentration of pollutants in the groundwater and at what rate is it seeping into the cistern?

19. A fossilized bone is found to contain 0.5% of the original amount of carbon-14. What is the age of the bone?

20. The half-life of plutonium-239 is about 24,180 years. An accident at a military installation spreads plutonium-239 uniformly over a certain area. It is determined that in order for this area to be safe again the concentration of plutonium-239 must be decreased by 72%. How long will it be before the area is again habitable?

21. Suppose a dense object of mass m falling through a liquid experiences a resistive force that is proportional to its velocity (see the figure). The net force on the object will then be the difference

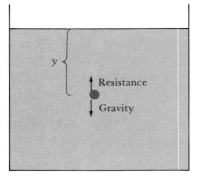

Falling with resistance

between the force of gravity, mg, and the force kv, where k is a constant and v is the velocity. According to Newton's second law, this net force must equal my'', where y is the depth of the object.
   a. Write a differential equation for the velocity of the object. (Remember that $v = y'$.)
   b. What is the limiting velocity of the object (that is, the velocity that is approached as time grows indefinitely)?

22. Suppose sales volume S is related to advertising effort x by the differential equation

$$\frac{dS}{dx} = a(b - S)$$

where a and b are positive constants. Solve this differential equation. Sketch the graph of the solution assuming that the initial sales volume $S(0)$ is less than b.

23. Consider the model for the melting ice cube given in Problem 5 in Section 7-2. Suppose the ice cube is initially 10 centimeters on a side, and 5 minutes later it is 8 centimeters on a side. Solve the differential equation for the length of the side of the cube. How long will it take for 75% of the cube to melt?

24. Suppose that (in appropriate units) the sales and demand for a certain commodity are

$$S = 66 + 0.2P \qquad \text{and} \qquad D = 80 + 0.1P$$

where P is the price. Suppose also that $P(0) = 1$ and $P(2) = 2$. Solve the differential equation for P given in Example 8 in Section 7-2.

7-4 SEPARABLE DIFFERENTIAL EQUATIONS

A certain type of differential equation, called the *separable differential equation,* can be solved simply by integration. We say that a differential equation is **separable** if it has, or can be put into, the form

$$g(y) \frac{dy}{dt} = f(t)$$

where $g(y)$ is a function of y alone and $f(t)$ is a function of t alone. For example, the equation

$$(1 + y)\frac{dy}{dt} = t^2 + t$$

is separable as can be seen by setting $g(y) = 1 + y$ and $f(t) = t^2 + t$. The differential equation

$$\frac{dy}{dt} = \frac{y^2 + 2}{t^2 + 1}$$

is also separable since it can be rearranged to yield

$$\frac{1}{y^2 + 2}\frac{dy}{dt} = \frac{1}{t^2 + 1}$$

in which

$$g(y) = \frac{1}{y^2 + 2} \quad \text{and} \quad f(t) = \frac{1}{t^2 + 1}$$

A differential equation of the form above can be solved simply by integrating both sides with respect to t:

$$\int g(y)\frac{dy}{dt}\,dt = \int f(t)\,dt$$

We will show below that the dt in the integral on the left can be "canceled" to give

$$\int g(y)\,dy = \int f(t)\,dt$$

It follows that solving a separable differential equation can be reduced to two main procedures:

Step 1. Perform the necessary algebra to reduce the equation to the form

$$g(y)\frac{dy}{dt} = f(t)$$

Step 2. Integrate both sides with respect to t, "canceling" the dt in the left-hand integral.

We will refine this program for solving separable differential equations after we have seen a few simple examples.

Example 23 Solve

$$y^4\frac{dy}{dt} = t^2 + 1$$

Solution   This equation is already in the form

$$g(y)\frac{dy}{dt} = f(t)$$

with $g(y) = y^4$ and $f(t) = t^2 + 1$. Integrating with respect to t and "canceling" dt gives

$$\int y^4 \, dy = \int (t^2 + 1) \, dt$$

and therefore,

$$\tfrac{1}{5}y^5 = \tfrac{1}{3}t^3 + t + c$$

Note that we include only one constant of integration in this equation. If a constant of integration were included on both sides of the equation, we would have

$$\tfrac{1}{5}y^5 + c_1 = \tfrac{1}{3}t^3 + t + c_2$$

But then

$$\tfrac{1}{5}y^5 = \tfrac{1}{3}t^3 + t + c_2 - c_1$$

and since we may set $c = c_2 - c_1$, we have

$$y^5 = \tfrac{5}{3}t^3 + 5t + c$$

$$y = \sqrt[5]{\tfrac{5}{3}t^3 + 5t + c}$$

where c is an arbitrary constant.   ∎

We have followed the common practice here of letting the symbol c stand for a general arbitrary constant. That is, in multiplying the equation in Example 23 by 5, we wrote c rather than $5c$ on the right-hand side. The reason for this is that c is an *arbitrary* constant and therefore $5c$ is also an arbitrary constant, so we may as well replace the symbol $5c$ by the symbol c.

Example 24   Solve the equation $y' = y^2 + y^2 e^t$, with $y(0) = 1$.

Solution   Note that $y' = y^2(1 + e^t)$ and hence the equation can be put in the standard form

$$y^{-2}\frac{dy}{dt} = 1 + e^t$$

In this equation, $g(y) = y^{-2}$ and $f(t) = 1 + e^t$. Applying the two-step procedure outlined on page 359, we then have

$$\int y^{-2} \, dy = \int (1 + e^t) \, dt$$

Therefore,

$$-y^{-1} = t + e^t + c$$

We find that

$$y = \frac{-1}{t + e^t + c}$$

The initial condition then gives

$$1 = y(0) = \frac{-1}{0 + 1 + c}$$

and hence, $c = -2$. Therefore,

$$y = \frac{-1}{t + e^t - 2}$$ ∎

It turns out that "canceling" the dt in the method used above is perfectly justified. In fact, it is a simple consequence of the general chain rule and the fundamental theorem of calculus. Let $G(y)$ be an antiderivative of $g(y)$. Then by the fundamental theorem of calculus,

$$\int g(y)\, dy = G(y) + c$$

Suppose now that y is some function of t, say $y = h(t)$. Then

$$\frac{dy}{dt} = h'(t)$$

and hence,

$$\int g(h(t))h'(t)\, dt = \int g(y)\, \frac{dy}{dt}\, dt$$

However, by the general chain rule,

$$\frac{d}{dt}\, G(h(t)) = G'(h(t))h'(t)$$

$$= g(y)\, \frac{dy}{dt}$$

Therefore, by the fundamental theorem,

$$\int g(y)\, \frac{dy}{dt}\, dt = \int \frac{d}{dt}\, G(h(t))\, dt$$
$$= G(h(t)) + c$$
$$= G(y) + c$$
$$= \int g(y)\, dy$$

and the "canceling" of dt is shown to be justified.

We would like to warn the reader that separating the variables involves more than just getting all the y's on one side of the equation. For example, it is tempting to try to solve

$$y' = y + 2$$

by adding $-y$ to both sides to obtain

$$-y + y' = 2$$

However, note that dy/dt in the equation

$$g(y)\, \frac{dy}{dt} = f(t)$$

is a *factor* of the left-hand side, which is not the case in the equation $-y + y' = 2$.

Example 25   Solve $y' = y + 2$, with $y(0) = 3$.

Solution   We write y' as dy/dt and divide by $y + 2$:

$$\frac{1}{y+2}\frac{dy}{dt} = 1$$

Integrating with respect to t and "canceling dt," this gives

$$\int \frac{dy}{y+2} = \int dt$$

$$\ln|y+2| = t + c$$

Therefore,

$$|y+2| = e^{t+c} = e^c e^t$$

That is, either $y + 2 = e^c e^t$ or $y + 2 = -e^c e^t$ for some constant c (recall the definition of absolute value).

However, since $y(0) = 3$, the possibility

$$y + 2 = -e^c e^t$$

would imply the contradiction

$$5 = -e^c e^0 = -e^c < 0$$

Hence, we may conclude that $y + 2 = e^c e^t$. The initial condition $y(0) = 3$ then gives

$$3 + 2 = e^c e^0$$

$$5 = e^c$$

and we finally find that

$$y + 2 = 5e^t$$

$$y = 5e^t - 2 \qquad \blacksquare$$

Note that the equation in Example 25 is also linear and therefore can be solved by the method of Section 7-3. This would, of course, give exactly the same solution. (Check it for yourself.)

Before proceeding to additional examples, we outline the steps to be taken in solving a separable differential equation:

Solving a Separable Differential Equation

Step 1. Write the equation in the form

$$g(y)\frac{dy}{dt} = f(t)$$

Step 2. Integrate both sides, "canceling the dt" on the left-hand side:

$$\int g(y)\frac{dy}{dt}\, dt = \int g(y)\, dy = \int f(t)\, dt$$

Step 3. Find antiderivatives $G(y)$ for $g(y)$ and $F(t)$ for $f(t)$:

$$G(y) = F(t) + c$$

Step 4. Solve for y (if possible).

Example 26   A biologist observes that a bizarre population of microbes tends to increase at a rate proportional to the square of its size. If we let $y(t)$ represent the size of the population at time t, then we have

$$\frac{dy}{dt} = ky^2$$

for a suitable positive constant k. Therefore,

$$y^{-2} \frac{dy}{dt} = k \qquad \text{Step 1}$$

and hence,

$$\int y^{-2} \, dy = \int k \, dt \qquad \text{Step 2}$$

That is,

$$-y^{-1} = kt + c \qquad \text{Step 3}$$

Finally,

$$y = \frac{-1}{kt + c} \qquad \text{Step 4}$$

Note that if $y(0) = y_0$ is the initial population, then

$$y_0 = \frac{-1}{0 + c}$$

$$c = -y_0$$

We then find

$$y = \frac{-1}{kt - y_0}$$

From this we see that the model cannot be realistic for all times, for once t grows large enough so that $kt > y_0$, y becomes negative — very strange behavior indeed!   ∎

Example 27
Pricing and Sales   It is found that the sales volume S of a commodity is related to its unit price p in such a way that S decreases with respect to p at a rate proportional to S and inversely proportional to the price plus a constant. The sales volume then satisfies the differential equation

$$\frac{dS}{dp} = \frac{-aS}{b + p}$$

where a and b are positive constants. With obvious changes in notation (S for y and p for t), we may follow our step-by-step program for solving separable differential equations. First,

$$\frac{1}{S} \frac{dS}{dp} = \frac{-a}{b + p} \qquad \text{Step 1}$$

Then,

$$\int \frac{1}{S}\, dS = -a \int \frac{dp}{b+p} \qquad \text{Step 2}$$

$$\ln S = -a \ln(b+p) + c \qquad \text{Step 3}$$

Therefore,

$$e^{\ln S} = e^{-a \ln(b+p)+c}$$

$$S = e^{c} e^{\ln(b+p)^{-a}}$$

$$= A(b+p)^{-a} \qquad \text{Step 4}$$

where $A = e^{c}$ is a positive constant. ∎

Example 28
Spread of Diseases

We will now solve the equation

$$\frac{ds}{dt} = ks(1-s)$$

of Example 7 in Section 7-2, which models the spread of a communicable disease. Writing it in standard form, we have

$$\frac{1}{s(1-s)} \frac{ds}{dt} = k \qquad \text{Step 1}$$

and hence,

$$\int \frac{ds}{s(1-s)} = \int k\, dt \qquad \text{Step 2}$$

In order to compute the integral on the left, note that

$$\frac{1}{s(1-s)} = \frac{1}{s} + \frac{1}{1-s}$$

and that both of these terms are positive since $0 < s < 1$. Therefore,

$$\int \frac{1}{s}\, ds + \int \frac{1}{1-s}\, ds = \int k\, dt$$

or

$$\ln s - \ln(1-s) = kt + c \qquad \text{Step 3}$$

However, by one of the laws of logarithms (see Chapter 6),

$$\ln s - \ln(1-s) = \ln\left(\frac{s}{1-s}\right)$$

Therefore,

$$\ln\left(\frac{s}{1-s}\right) = kt + c$$

and hence,

$$e^{\ln[s/(1-s)]} = e^{kt+c} = e^{c} e^{kt}$$

That is,

$$\frac{s}{1-s} = Ae^{kt}$$

where $A = e^c$ is a positive constant. But this gives

$$s = Ae^{kt} - sAe^{kt}$$

$$s(1 + Ae^{kt}) = Ae^{kt}$$

and finally,

$$s = \frac{Ae^{kt}}{1 + Ae^{kt}} \qquad Step \; 4$$

It is instructive to rewrite this as

$$s = \frac{1 + Ae^{kt} - 1}{1 + Ae^{kt}} = 1 - \frac{1}{1 + Ae^{kt}}$$

From this it is clear that s is always less than 1 but approaches 1 as t becomes arbitrarily large. Also note from the original equation that

$$\frac{d^2s}{dt^2} = \frac{d}{dt}\,ks(1-s) = -ks\frac{ds}{dt} + k\frac{ds}{dt}(1-s)$$

$$= k\frac{ds}{dt}(1 - 2s)$$

Since $0 < s < 1$, $ds/dt = ks(1-s) > 0$, and hence s has a unique point of inflection at $s = \frac{1}{2}$. (Check that $d^2s/dt^2 < 0$ for $s < \frac{1}{2}$ and $d^2s/dt^2 > 0$ for $s > \frac{1}{2}$.) The solution s is sketched in Figure 7-5; it is called a *logistic curve* (see also Problems 28–30 at the end of this section).

Figure 7-5

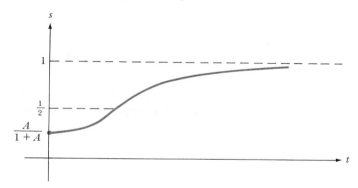

Before going any further, we would like to give a warning concerning the algebra needed to transform an equation to the standard form

$$g(y)\frac{dy}{dt} = f(t)$$

Often, this requires division, and the most important algebraic rule is: *Never divide by zero!* In Example 24 we solved the equation

$$y' = y^2(1 + e^t)$$

by first dividing by y^2 and then integrating to obtain

$$y = \frac{-1}{t + e^t + c}$$

In dividing by y^2 we assumed that $y \neq 0$, but note that $y = 0$ is itself a solution of the differential equation, since

$$0' = 0^2(1 + e^t)$$

However, this identically zero solution is not represented by

$$y = \frac{-1}{t + e^t + c}$$

since no choice of c in this formula will produce $y = 0$. In Example 24 this is of no concern, because the initial condition $y(0) = 1$ precludes the zero solution. However, if the goal is to find all solutions of the equation $y' = y^2(1 + e^t)$, with no initial condition specified, we have to include the solution $y = 0$. Any functions that are eliminated to prevent division by zero must be checked separately to see if they are also solutions of the original differential equation.

Example 29   Find all solutions of the equation

$$\frac{dy}{dt} = (y - 3)^2(t^2 + 1)$$

Solution   The first step is to separate the variables by dividing by $(y - 3)^2$. Therefore, for $y \neq 3$ the given equation is equivalent to

$$(y - 3)^{-2}\frac{dy}{dt} = t^2 + 1$$

Integrating this equation, we find

$$\int (y - 3)^{-2}\, dy = \int (t^2 + 1)\, dt$$

$$-(y - 3)^{-1} = \tfrac{1}{3}t^3 + t + c$$

$$y = 3 - \frac{1}{\tfrac{1}{3}t^3 + t + c}$$

where c is an arbitrary constant. However, not all solutions are of this form since in deriving this solution we precluded the function $y = 3$, which is also a solution since

$$\frac{d}{dt}(3) = 0 = (3 - 3)^2(t^2 + 1)$$

Therefore, the complete list of solutions is

$$y = 3 \qquad \text{and} \qquad y = 3 - \frac{1}{\tfrac{1}{3}t^3 + t + c}$$

Our final remark concerns the final step in the solution process, namely "solve for y (if possible)." If the antiderivative $G(y)$ is at all complicated, it may be virtually impossible to solve explicitly for y and we may be forced to settle for the equation

$$G(y) = F(t) + c$$

which only gives y *implicitly* as a function of t. Furthermore, even if this equation can be solved, it may have more than one solution.

Example 30 Solve

$$\frac{dy}{dt} = \frac{t}{y}$$

Solution Multiplying by y and integrating we have

$$\int y \, dy = \int t \, dt$$

$$\frac{y^2}{2} = \frac{t^2}{2} + c$$

$$y^2 = t^2 + c$$

Therefore, each function of the form

$$y = \sqrt{t^2 + c} \qquad \text{or} \qquad y = -\sqrt{t^2 + c}$$

is a solution of the differential equation. ∎

PROBLEMS (Section 7-4)

Use the technique of separation of variables to solve the differential equations in Problems 1–25.

1. $6y^2 \dfrac{dy}{dt} = 5$

2. $(2 + t) \dfrac{dy}{dt} = 2y$

3. $e^y \dfrac{dy}{dt} - 3t^2 = 2$

4. $y' = \dfrac{t^2}{y^2}$

5. $\dfrac{dy}{dt} = y^2 t$

6. $y' = y^{2/3} t$

7. $1 - t^2 \dfrac{dy}{dt} = 0$

8. $(t + 2) \dfrac{dy}{dt} - t = 0$

9. $\dfrac{dP}{dx} = xP \quad (P > 0)$

10. $u \ln x = x \dfrac{du}{dx}$

11. $y' = -2(y - 1), \quad y(0) = 2$

12. $y \dfrac{dy}{dt} = e^t, \quad y(0) = 3$

13. $-y(t + 1) = y', \quad y(0) = 1$

14. $\dfrac{dy}{dt} = 2y + 5, \quad y(0) = -1$

15. $y\dfrac{dy}{dt} = t^3, \quad y(0) = 1$

16. $y' - y^{-2}t = 0, \quad y(0) = -1$

17. $(t^2 + 1)y' = 2ty^2, \quad y(0) = 2$

18. $y' = e^{2t-y}, \quad y(0) = 0$

19. $y' = \dfrac{e^{3t}}{y}, \quad y(0) = 4$

20. $\dfrac{dw}{dx} = xw^2 + w^2, \quad w(0) = -1$

21. $u' = 3t^2u^4, \quad u(0) = 1$

22. $\dfrac{dy}{dt} = (t + 2)e^{-y}, \quad y(0) = 2$

23. $\dfrac{dy}{dx} = \dfrac{\ln x}{x}y^{-2}, \quad y(1) = 1$

24. $\dfrac{dy}{dx} = \left(\dfrac{x}{1+y}\right)^2, \quad y(0) = 1$

25. $3y' = 2ty - y, \quad y(0) = 1$

26. Use the technique of separation of variables to solve the temperature equation of Newton's law of cooling (see Example 9 in Section 7-2).

27. Suppose a total amount of information M is to be learned and the rate at which learning takes place at time t is proportional to the difference between M and $l(t)$, the amount learned by time t.
    a. Write a differential equation for $l(t)$ (see Example 6 in Section 7-2).
    b. Solve the differential equation from part a using the technique of separation of variables. Take $l(0) = 0$.

28. The general logistic equation is a differential equation of the form

$$\dfrac{dy}{dt} = ay(b - y)$$

where a and b are positive constants.
    a. Find all solutions of this differential equation. [Hint: $1/y(b - y) = (1/b)/y + (1/b)/(b - y)$]
    b. Sketch the solutions of the differential equation with initial condition $y(0) = y_0$ considering the cases $y_0 > b$, $y_0 = b$, and $0 < y_0 < b$ separately.

29. Suppose a population of animals evolves in such a way that

$$\dfrac{dP}{dt} = 0.5P - 0.01P^2 \qquad P(0) = 10$$

(See Problem 7 in Section 7-2.)
    a. Find a formula for P as a function of time.
    b. Sketch your solution.
    c. At what value of the population does the rate of growth "turn around," that is, at what population level does d^2P/dt^2 change sign?

30. Solve the differential equation

$$\dfrac{dy}{dt} = 0.2y(0.8 - y) \qquad y(0) = 0.2$$

which arises from an autocatalytic reaction $X \rightarrow Y$, where the initial concentration of the chemical X is 80% (see Problem 15 in Section 7-2). Sketch your solution.

31. A population y is said to grow according to *Gompertz's law* if

$$\dfrac{dy}{dt} = -ay\dfrac{\ln y}{b} \qquad y(0) = e$$

where a and b are positive constants. Solve this differential equation.

7-5 CONCLUSION: EVALUATING POLLUTION CONTROL STRATEGIES

We introduced this chapter with a problem in water pollution. Recall that the model gave rise to the differential equation $dc/dt = 0.01 - \tfrac{1}{5}c(t)$ for the concentration of pollutants in the lake if the concentration of pollutants in the river water is 0.05 (that is, 5%). Suppose, however, that the concentration of pollutants in the river is a. Then the rate of flow of pollutants into the lake is

$$10 \text{ km}^3/\text{year} \times a = 10a \text{ km}^3/\text{year}$$

If, as in the Introduction, we convert from quantity of pollutants to concentration of pollutants by dividing by the volume of the lake (50 km³), we obtain

$$\frac{10a}{50} = \frac{1}{5}a$$

as the term that will tend to increase the concentration of pollutants. The differential equation will then be

$$\frac{dc}{dt} = \frac{1}{5}a - \frac{1}{5}c$$

Notice that when $a = 0.05$, this agrees with the differential equation above. To solve this equation, we rewrite it as

$$\frac{dc}{dt} + \frac{1}{5}c = \frac{1}{5}a$$

and hence,

$$e^{t/5}c = \int \frac{1}{5}ae^{t/5}\,dt = ae^{t/5} + d$$

or

$$c(t) = a + de^{-t/5}$$

Notice that

$$c(0) = a + de^0$$
$$= a + d$$
$$d = c(0) - a$$

and therefore,

$$c(t) = a + [c(0) - a]e^{-t/5}$$

We can now answer the first question posed in the Introduction, namely, if the river is completely cleaned up, how long will it take for the concentration of pollutants in the lake to reach 10% of its present level at the time of cleanup? If the river is cleaned up, then $a = 0$ and the formula for the concentration of pollutants in the lake t years after the cleanup is

$$c(t) = c(0)e^{-t/5}$$

where $c(0)$ is the concentration at the time of cleanup. Let T represent the time it takes for the concentration to reach 10% of this value, that is,

$$c(T) = \frac{c(0)}{10}$$

Substituting this expression into the preceding formula, we find

$$\frac{c(0)}{10} = c(0)e^{-T/5}$$

$$\frac{1}{10} = e^{-T/5}$$

Therefore,

$$\frac{-T}{5} = \ln\left(\frac{1}{10}\right) = -\ln 10$$

$$T = 5 \ln 10 \approx 11.5$$

Therefore, it will take about 11.5 years after a complete cleanup for the level of pollution in the lake to drop by 90%.

Suppose that the current level of pollution in the lake is 3%, that is, $c(0) = 0.03$ and a partial cleanup is made with the aim of reducing the pollution level to 1.5% in 5 years, that is, $c(5) = 0.015$. If the concentration of pollutants in the river is immediately reduced to $(0.05)r$, where r is some fraction $(0 \le r \le 1)$, then the equation for the concentration of pollutants in the lake is (setting $a = 0.05r$ above)

$$c(t) = 0.05r + (0.03 - 0.05r)e^{-t/5}$$

In 5 years we then find that

$$0.015 = 0.05r + (0.03 - 0.05r)e^{-1}$$
$$= 0.05r(1 - e^{-1}) + 0.03e^{-1}$$

Therefore,

$$r = \frac{0.015 - 0.03e^{-1}}{0.05(1 - e^{-1})} \approx 0.1254$$

That is, in order to achieve the 5 year plan, pollution in the river must be immediately cut by $100 - 12.54 \approx 87.5\%$.

7-6 CHAPTER REVIEW

1. Many problems in business and the physical, natural, and social sciences can be modeled in terms of rates of change of certain quantities. These rates of change may be interpreted as derivatives, and the resulting equations are differential equations.

2. A differential equation of the form

$$\frac{dy}{dt} + ay = g$$

where g is a given function of t, is called a linear differential equation. The solution of such a linear differential equation is

$$y = e^{-at}\left[\int e^{at}g(t)\,dt + c\right]$$

The constant c can be determined if an initial condition is specified.

3. A differential equation of the form

$$g(y)\frac{dy}{dt} = f(t)$$

is called a separable differential equation. This type of equation may be solved by "multiplying by dt" and integrating:

$$\int g(y)\,dy = \int f(t)\,dt$$

REVIEW PROBLEMS (Section 7-6)

1. Show that $y(t) = t^2$ is a solution of the differential equation $ty' - 2y = 0$.

2. Show that $y(t) = t^3 + 2t - 1$ is a solution of the differential equation $ty' - 3y = 3 - 4t$.

3. Show that $y(t) = te^t$ solves the problem

$$ty' - (1 + t)y = 0 \qquad y(1) = e$$

4. Show that $y(t) = t \ln t$ solves the problem

$$ty' - y = t \qquad y(1) = 0$$

5. Solve the differential equation $y' + 2y = 1$.

6. Solve the differential equation $2y' - 4y = 8$.

7. Solve the differential equation $3y' = 4y - 1$.

8. Solve the differential equation $-2y' = -6y + t$.

9. Solve $y' - y = 0$, with $y(0) = 1$.

10. Solve $2y' = 3y + e^{4t}$, with $y(0) = 2$.

11. Solve $y' - y = e^t$, with $y(0) = 1$.

12. Solve $-3y' = 6y - e^{2t}$, with $y(0) = 1$.

13. A certain radioactive isotope has a half-life of 2420 years. What percentage of a given portion of the isotope remains after 8000 years?

14. Suppose \$2000 is invested continuously each year into an account that earns 12% per annum, compounded continuously. How long will it take for the value of the account to be \$50,000?

15. Suppose an investment account grows at a rate proportional to its square root.
    a. Write a differential equation for the sum $S(t)$ of money in the account at time t.
    b. Solve the differential equation assuming the initial sum in the account is \$1000 and the sum in the account after 2 years is \$1600.

16. Suppose a vat contains 100 gallons of a 2% dye solution. If fresh water flows into the vat at a rate of 1 gallon per minute and the well-mixed dye solution flows out of the vat at the same rate, how

long will it take the dye in the vat to reach a concentration of 0.5%?

17. One morning a body was found in an office which is kept at 70°F. At 9:00 A.M. the coroner arrived and found that the temperature of the body was 86°F. An hour later the temperature of the body was 80°F. Can you establish the time of death? (Assume that at the time of death the temperature was normal body temperature, 98.6°F, and that Newton's law of cooling applies.)

18. Solve $e^y y' - 2t = 2$.

19. Solve $y' = \dfrac{(y-1)^2}{t}$.

20. Solve $t^2 y' = (t^2 + 1)y$, with $y(1) = 1$.

21. Solve $(ty)y' = 1$, with $y(1) = 1$.

22. Solve $y' = te^{-y}$, with $y(0) = 1$.

23. Solve $t^2 y' = \dfrac{t^2 + 1}{3y^2 + 1}$, with $y(1) = 2$.

APPROXIMATIONS FOR DIFFERENTIAL EQUATIONS

Many (in fact, most) of the differential equations that occur in real life cannot be solved by the methods we have studied so far. Nevertheless, the topics we have covered until now are important because they have, we hope, given the reader some important insights into the nature of differential equations and their solutions as well as some specific techniques for certain important classes of differential equations.

But what do you do when you are faced with a differential equation for which the usual methods of solution fail? Fortunately, it happens that an exact explicit solution is seldom required. The manager, engineer, or scientist is usually quite satisfied with a good approximation on which predictions or courses of action can be confidently based. In fact, it is often the case that only *qualitative* information on a solution is needed. Such questions as: is the solution positive, or increasing, or bounded, or concave up on a given interval might be asked. Clearly, a general picture of the solution would be just as good as (or even better than!) an explicit exact solution to answer these questions.

7S-1 GRAPHICAL APPROXIMATIONS

There is a simple, although somewhat tedious, way to get a good picture of the possible solutions of a differential equation. Suppose the differential equation has the form

$$\frac{dy}{dt} = f(t, y)$$

where $f(t, y)$ is some given function of t and y. For example, the differential equation

$$\frac{dy}{dt} = t^2 + y^2$$

which is neither linear nor separable, has this form with $f(t, y) = t^2 + y^2$. What this says, quite simply, is that given a value for t and for y, that is, a point in the ty-coordinate plane, the numerical value of $f(t, y)$ is equal to dy/dt, the slope of the tangent line to the solution which passes through this point. We can then indicate this tangent line by drawing a short line segment through the point with slope equal to $f(t, y)$. Consider, for example, the following five points in the ty-coordinate plane and the slopes corresponding to them:

t	y	$f(t, y) = t^2 + y^2$
0	1	1
-1	1	2
-0.5	-1	1.25
1	1.5	3.25
2	3	13

For each of these five points (t, y) a small line segment with slope $f(t, y)$ has been drawn in Figure 7S-1. Of course, with such a small number of segments it is quite impossible to visualize the solution curves. In Figure 7S-2 more segments are plotted and we get a much better idea of the "flow" which the differential equation sets up in the ty-coordinate plane. Since no initial condition is specified, there are infinitely many solutions and hence there are infinitely many curves in the ty-coordinate plane whose tangents could be represented by the **direction field.** We have sketched one representative solution in Figure 7S-2, namely, the one that satisfies $y(0) = 0$.

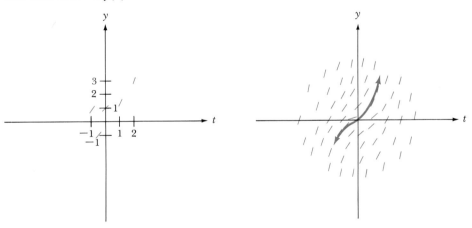

Figure 7S-1   Figure 7S-2

Example 1   Sketch the direction field of the equation

$$\frac{dy}{dt} = y(1 - y)(1 + y)$$

Solution   Note that in this example $f(t, y) = y(1 - y)(1 + y)$ is independent of t. Therefore, for constant y (that is, along a horizontal line in the ty-coordinate plane), the value of $f(t, y)$ does not change with t. This makes all the small line segments on a given horizontal line parallel. Note also that the constants $y = 0$, $y = 1$, and $y = -1$ are solutions of the differential equation. We have sketched these constant solutions as well as a portion of the direction field and the particular solution satisfying $y(0) = \frac{1}{2}$ in Figure 7S-3.

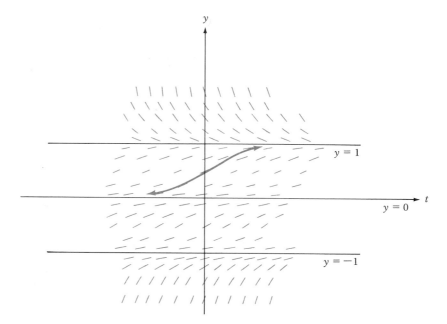

Figure 7S-3

7S-2 NUMERICAL APPROXIMATIONS

While a direction field gives a good qualitative description of the solutions of a differential equation, it does not give any *quantitative* information. There are a number of methods of obtaining numerical approximations to the solution of a differential equation with initial condition, all of which are based to some extent on the tangent line approximation to the graph of a function. We will describe the simplest of such methods, **Euler's method.** Although more sophisticated numerical methods are typically used in realistic applications, Euler's method serves as an excellent introduction to the field of numerical methods for differential equations.

In Figure 7S-4 (page 376) we recall the tangent line approximation to the graph of a function $y(t)$ (see Chapter 2). If $y(t_0) = y_0$ and $t_1 = t_0 + h$, then

$$y(t_1) \approx y_0 + h y'(t_0)$$

Now suppose that y is the solution of the following differential equation with the given initial condition:

$$\frac{dy}{dt} = f(t, y) \qquad y(t_0) = y_0$$

where $f(t, y)$ is some given function of two variables. For a given number h (called the **step size**), Euler's method will produce numerical approximations to the solution at the points $t_1 = t_0 + h$, $t_2 = t_0 + 2h$, $t_3 = t_0 + 3h$, etc. For example, to

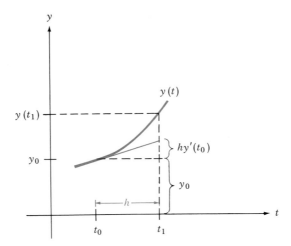

Figure 7S-4

get a numerical approximation to $y(t_1)$, the value of the solution at $t_1 = t_0 + h$, we simply use $y_0 + hy'(t_0)$. We will call the approximation that we get in this way y_1:

$$y_1 = y_0 + hy'(t_0)$$

However, from the differential equation we see that $y'(t_0) = f(t_0, y_0)$. Therefore,

$$y_1 = y_0 + hf(t_0, y_0)$$

This is the first step in Euler's method.

Example 2   Approximate $y(1.1)$ if

$$\frac{dy}{dt} = -ty^2 \qquad y(1) = \frac{2}{3}$$

Solution   Let $t_0 = 1$ and $y_0 = \frac{2}{3}$. We wish to approximate y at $t = 1.1 = 1 + 0.1 = t_0 + h$, where $h = 0.1$. Here, $f(t, y) = -ty^2$; therefore, our approximation is

$$\begin{aligned} y_1 &= y_0 + hf(t_0, y_0) \\ &= \tfrac{2}{3} + (0.1)(-1)(\tfrac{2}{3})^2 \approx 0.62222 \end{aligned}$$

The true solution of the differential equation is

$$y(t) = \frac{2}{t^2 + 2}$$

and hence the actual value of y at $t_1 = 1.1$ is

$$y(1.1) = \frac{2}{(1.1)^2 + 2} \approx 0.62305 \qquad \blacksquare$$

The next step is to approximate $y(t_2)$ where $t_2 = t_1 + h$. By using the tangent line approximation and the differential equation, we have

$$\begin{aligned} y(t_2) &\approx y(t_1) + hy'(t_1) \\ &= y(t_1) + hf(t_1, y(t_1)) \end{aligned}$$

However, we have at our disposal only y_1, which is an approximation to $y(t_1)$ and we use this to obtain an approximation y_2 to $y(t_2)$:

$$y_2 = y_1 + hf(t_1, y_1)$$

Note that this has exactly the same form as before and corresponds to following the line with slope $f(t_1, y_1)$ from the point (t_1, y_1) to obtain the next approximation y_2 (see Figure 7S-5). Further approximations are computed by using this same idea over and over again:

$$y_3 = y_2 + hf(t_2, y_2)$$
$$y_4 = y_3 + hf(t_3, y_3)$$
$$\vdots$$
$$y_{n+1} = y_n + hf(t_n, y_n)$$

Figure 7S-5

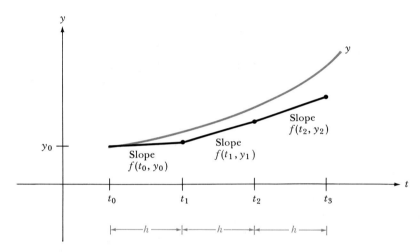

In order to use Euler's method, we need an initial value $y(t_0) = y_0$, a step size h, a number N indicating how many approximations are to be calculated, and, of course, the function $f(t, y)$ specified in the differential equation. A flow-chart for Euler's method is given in Figure 7S-6 on page 378.

Example 3   Use Euler's method to approximate the solution of

$$\frac{dy}{dt} = -ty^2 \qquad y(1) = \frac{2}{3}$$

at the points $t = 1.1, 1.2, 1.3, 1.4$, and 1.5.

Solution   Here we have $t_0 = 1$, $y_0 = \frac{2}{3}$, and $h = 0.1$. In Example 2 we found $y_1 \approx 0.62222$. Continuing with Euler's method, we have

$$\begin{aligned} y_2 &= y_1 + hf(t_1, y_1) \\ &\approx 0.62222 + (0.1)(-1.1)(0.62222)^2 \\ &\approx 0.57963 \end{aligned}$$

$$\begin{aligned} y_3 &= y_2 + hf(t_2, y_2) \\ &\approx 0.57963 + (0.1)(-1.2)(0.57963)^2 \\ &\approx 0.53931 \end{aligned}$$

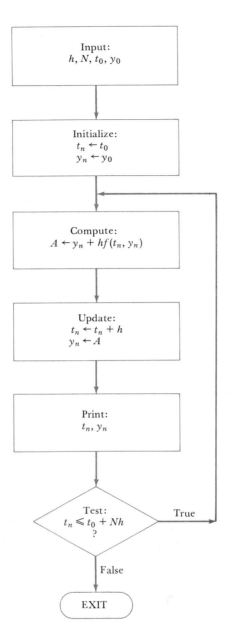

Figure 7S-6

$$y_4 = y_3 + hf(t_3, y_3)$$
$$\approx 0.53931 + (0.1)(-1.3)(0.53931)^2$$
$$\approx 0.50150$$

$$y_5 = y_4 + hf(t_4, y_4)$$
$$\approx 0.50150 + (0.1)(-1.4)(0.50150)^2$$
$$\approx 0.46629$$

Since at each step of Euler's method an error is made in approximating $f(t_n, y(t_n))$ in the tangent line approximation formula by $f(t_n, y_n)$, one should expect that generally these errors will accumulate as more and more steps are taken. In Table 7S-1 we compare the true solution of the differential equation in Example 3 with the Euler approximations. The error propagation is apparent in this table.

Table 7S-1

t_n	y_n	$y(t_n)$	Error: $y(t_n) - y_n$
1.1	0.62222	0.62305	0.00083
1.2	0.57963	0.58140	0.00177
1.3	0.53931	0.54201	0.00270
1.4	0.50150	0.50505	0.00355
1.5	0.46629	0.47059	0.00430

PROBLEMS (Supplement to Chapter 7)

1. What is the slope of the graph of the solution of the differential equation $y' = t^2 y + t$ that passes through the point $(1, 3)$?

2. Consider the solution of the differential equation $y' + t^2 y = 2 + t$ that satisfies the condition $y(2) = 4$. What is the slope of this solution at the point $(2, 4)$?

3. Draw a direction field for the differential equation $y' = y$ in the rectangle $-4 \le t \le 4, 2 \le y \le 2$ in the ty-plane. Plot at least 40 tangent segments.

4. Sketch a direction field for the differential equation $dy/dt = -ty^2$ for $1 \le t \le 2, -1 \le y \le 1$ in the ty-plane which is sufficiently fine to indicate the general nature of the solutions.

5. Sketch a direction field for the logistic equation $dy/dt = 2y(1 - y)$ for $-2 \le t \le 2, -2 \le y \le 2$ in the ty-plane.

6. Use Euler's method with $h = 0.5$ to approximate the solution of $y' = ty^2$, with $y(1) = -1$, on the interval $1 \le t \le 2$.

7. Use Euler's method with $h = 0.25$ to estimate $y(1)$ if $y' = (t - 1)y$, with $y(0) = 1$.

8. Estimate $y(1)$ if $y' = y + 1$, with $y(0) = 1$, using Euler's method with $h = 1$. Repeat the problem using $h = 0.5$, $h = 0.2$, and $h = 0.1$. In each case, compare the estimate with the true value $y(1)$.

9. Use Euler's method with $h = 0.25$ to approximate the solution of the logistic equation $dy/dt = y(1 - y)$, with $y(0) = 0.25$, on the interval $0 \le t \le 3$. Compare with the true solution.

10. The *modified Euler method* for the equation $dy/dt = f(t, y)$, with $y(t_0) = y_0$, is given by

$$y_1 = y_0 + hf\left(t_0 + \frac{h}{2}, y_0 + \frac{1}{2}hf(t_0, y_0)\right)$$

$$y_2 = y_1 + hf\left(t_1 + \frac{h}{2}, y_1 + \frac{1}{2}hf(t_1, y_1)\right)$$

$$\vdots$$

$$y_{n+1} = y_n + hf\left(t_n + \frac{h}{2}, y_n + \frac{1}{2}hf(t_n, y_n)\right)$$

a. Draw a flowchart for the modified Euler method.

b. Use $h = 0.1$ and the modified Euler method to approximate the solution of the equation in Example 3 on the interval $1 \le t \le 1.5$. Compare with the true solution and the results in Table 7S-1.

FUNCTIONS OF SEVERAL VARIABLES

INTRODUCTION

The Phone Store sells standard rotary and push-button desk telephones as well as many special and one-of-a-kind styles. The desk phones, however, are the biggest sellers, and the company wishes to establish a pricing policy that will maximize the contribution to revenue from this product line. The sales of rotary and push-button desk phones are not independent of each other. In economics, these are called *substitutable products.* That is, if the price of the rotary telephone is increased, then sales will usually decrease, but there will be a corresponding increase in sales of the push-button phone. Similarly, if the price of the push-button phone increases, then consumers will substitute the rotary model; that is, sales of the push-button phone will decrease while the sales of the rotary phone will increase. Because of such dependencies, The Phone Store must consider the prices of both models together in trying to develop a revenue-maximizing strategy.

A study of price and sales data has resulted in the establishment of the following relationships between the quantity of each model sold and the prices of both models:

$$n_1 = 19.5 - 0.6p_1 + 0.25p_2$$
$$n_2 = 30.1 + 0.08p_1 - 0.4p_2$$

where

- n_1 = Number of rotary desk phones sold per month
- n_2 = Number of push-button desk phones sold per month
- p_1 = Price of a rotary phone
- p_2 = Price of a push-button phone

Both n_1 and n_2 can be considered to be functions of the two variables p_1 and p_2. Using standard function notation, we can write

$$n_1 = f_1(p_1, p_2)$$
$$n_2 = f_2(p_1, p_2)$$

Since revenue equals price times quantity sold, the total revenue received from sales of the two models of desk phones will be

$$\text{Revenue} = p_1 n_1 + p_2 n_2$$
$$= p_1(19.5 - 0.6p_1 + 0.25p_2)$$
$$+ p_2(30.1 + 0.08p_1 - 0.4p_2)$$

$$= 19.5p_1 - 0.6p_1^2 + 0.25p_1p_2$$
$$+ 30.1p_2 + 0.08p_1p_2 - 0.4p_2^2$$
$$= -0.6p_1^2 - 0.4p_2^2 + 0.33p_1p_2$$
$$+ 19.5p_1 + 30.1p_2$$

Since this is also a function of two variables, we can denote revenue by $r(p_1, p_2)$. The problem facing The Phone Store is to determine values for p_1 and p_2 that will maximize the revenue function.

Functions of more than one variable are often called **multivariable functions.** Multivariable functions have widespread use, since many quantities naturally depend on more than one variable. For example, a person's income may depend on education, age, and sex (though not always legal); the stopping distance of a car is a function of its weight and the force applied on the brakes; and sales of a product may be a function of the dollars spent on advertising and the price of the competing product. In previous chapters, we studied methods and applications of calculus for functions of a single variable. In this chapter, we extend these ideas to functions of more than one variable.

8-1 MULTIVARIABLE FUNCTIONS

In Chapter 1 we developed a model for the monthly driving cost of an automobile in terms of the number of miles driven per day, x, and the number of miles per gallon (fuel economy) of the vehicle, y. This was determined to be

$$MC = \frac{31.5x}{y} + 0.564375x + 42$$

(in dollars). Monthly cost is a function of two independent variables, x and y. As with any function, there can be only one dependent variable — in this case, cost.

The notation that is commonly used to denote a function of two variables is $z = f(x, y)$, where z represents the dependent variable. [This is analogous to the notation "$y = f(x)$."] A function of two variables is thus a rule that associates with each *pair* of values of the independent variables a unique value of the dependent variable. We can visualize such a rule as follows:

$$(x,y) \xrightarrow{\quad f \quad} z$$

Independent variables   *Function*   *Dependent variable*

Thus, we can write the monthly driving cost function as

$$MC = f(x, y) = \frac{31.5x}{y} + 0.564375x + 42$$

If we substitute specific values for x and y, say $x = a$ and $y = b$, then $f(a, b)$ is the value of the function when $x = a$ and $y = b$. Thus, for this example, if $x = 20$ and $y = 30$, we have

$$f(20, 30) = \frac{31.5(20)}{30} + 0.564375(20) + 42$$

$$= 74.2875 \approx \$74.29$$

Example 1 Let $z = f(x, y) = 3x^2 + 2y^2 - 6xy + 25$. Evaluate the following:

  a. $f(2, 0)$ b. $f(0, -1)$
  c. $f(a, a^2)$ d. $f(x + h, y)$

Solution Substituting for x and y in the function $f(x, y)$, we obtain:

  a. $f(2, 0) = 3(2)^2 + 2(0)^2 - 6(2)(0) + 25 = 37$
  b. $f(0, -1) = 3(0)^2 + 2(-1)^2 - 6(0)(-1) + 25 = 27$
  c. $f(a, a^2) = 3(a)^2 + 2(a^2)^2 - 6(a)(a^2) + 25$
      $= 3a^2 + 2a^4 - 6a^3 + 25$
  d. $f(x + h, y) = 3(x + h)^2 + 2y^2 - 6(x + h)y + 25$
      $= 3x^2 + 6xh + 3h^2 + 2y^2 - 6xy - 6hy + 25$ ∎

We can easily extend these concepts to functions involving more than two independent variables. For example, we may write $w = f(x, y, z)$. Here, w is a function of three independent variables, x, y, and z. Another example would be $y = f(x_1, x_2, x_3, x_4)$, where y is the dependent variable and x_1 through x_4 are the independent variables. In all cases, we evaluate a multivariable function by substituting the values of each independent variable into the rule defining the function.

Example 2
Electronic Display Design In designing electronic display screens such as for radar, reconnaissance, and so on, a formula for the minimum screen size for target identification is

$$D = \frac{0.0058VR}{T}$$

where

  D = Screen size (in inches)
  V = Viewing distance from the screen (in inches)
  R = Range on ground being displayed (in feet)
  T = Smallest target dimension (in feet)

Here, D is a function of three variables, V, R, and T. Thus, we write

$$D = f(V, R, T) = \frac{0.0058VR}{T}$$

If $V = 12$ inches, $T = 1000$ feet, and $R = 243{,}000$ feet (about 40 nautical miles), we have

$$D = f(12, 243{,}000, 1000) = \frac{0.0058(12)(243{,}000)}{1000}$$

$$\approx 17 \text{ inches} \qquad \blacksquare$$

Example 3
Container Design

A closed metal box is to be constructed with a volume of 3 cubic feet. The cost of sheet metal is 8 cents per square foot. Express the cost of the box as a function of the base dimensions x and y (see Figure 8-1).

Figure 8-1

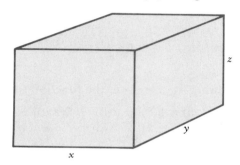

Solution

The surface area is given by the multivariable function

$$S = 2xy + 2xz + 2yz$$

and the volume is

$$V = xyz = 3$$

Solving the volume equation for the height z, we have $z = 3/xy$. Since the cost is $8S$, that is, 8 times the surface area, we have

$$\text{Cost} = f(x, y) = 8(2xy + 2xz + 2yz)$$

$$= 16xy + 16x\left(\frac{3}{xy}\right) + 16y\left(\frac{3}{xy}\right)$$

$$= 16xy + \frac{48}{y} + \frac{48}{x} \qquad \blacksquare$$

GRAPHS OF MULTIVARIABLE FUNCTIONS

We know that the graph of a single-variable function can be drawn on a two-dimensional coordinate system, where one dimension corresponds to the independent variable and the second dimension to the dependent variable. Graphs of two-variable functions require three dimensions: two for the independent variables and a third for the dependent variable.

> The **three-dimensional rectangular coordinate system** consists of three
> mutually perpendicular axes, usually denoted x, y, and z. Any two of these
> axes define a **coordinate plane.**

Figure 8-2 illustrates the three-dimensional rectangular coordinate system
and the coordinate planes defined by pairs of axes. Any point in this coordinate
system is denoted by an ordered triple of numbers (x, y, z). Notice that the
yz-plane consists of all points where $x = 0$, that is, points of the form $(0, y, z)$.
Similarly, the xy-plane consists of all points of the form $(x, y, 0)$, and the xz-plane
consists of all points of the form $(x, 0, z)$.

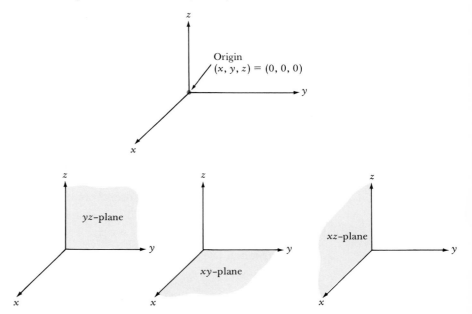

Figure 8-2
Three-dimensional rectangular
coordinate system

To plot a point (a, b, c), we first locate the point $(a, b, 0)$ on the xy-plane and
then "move up" a distance c (or down if c is negative). To obtain proper visual
perspective, the line from the origin to $(a, b, 0)$ on the xy-plane should be parallel
to the line from the point $(0, 0, c)$ to (a, b, c). This is illustrated in Example 4.

Example 4 Plot the point $(4, 3, 5)$ on a three-dimensional coordinate system.

Solution The point is shown in Figure 8-3 (page 386). The perspective is easier to see
if a box is drawn for which the plotted point is the farthest point from the
origin. (Perhaps a useful analogy is to think of the coordinate system as the
corner of a room.) ∎

To deal with multivariable functions effectively, it is useful to be able to
draw the entire function in three dimensions rather than simply plotting individual points. We admit that this is not always easy to do, since most people are

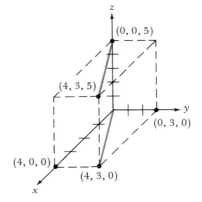

Figure 8-3
Plotting the point (4, 3, 5) in three
dimensions

unaccustomed to drawing in three dimensions. We shall only present some basic ideas using simple functions; more complicated functions are better drawn using computer graphic techniques.

Consider the function $z = f(x, y)$. The domain of this function is the set of all points (x, y) for which the function value $f(x, y)$ is defined. For each pair of values (x, y) in the domain we can compute a value for z. The collection of all points (x, y, z), where (x, y) is in the domain of the function, forms a *surface* in the three-dimensional coordinate system, called the *graph* of the function. In sketching such surfaces it is often convenient to first restrict our attention to the individual coordinate planes and sketch the *traces* of the surface. These traces are the curves that result when one of the variables is set equal to 0 and hence are analogous to the intercepts that are used when sketching the graph of a function of one variable.

Example 5   Sketch the function $z = f(x, y) = 6 - 2x - 3y$.

Solution   Suppose we set $z = 0$. Then we have the equation $2x + 3y = 6$. This is simply a straight line in the xy-plane, which we can graph easily (refer to Figure 8-4). Next, suppose we set $x = 0$. This yields the equation $z = 6 - 3y$ or $3y + z = 6$, which is also a straight line. We graph this line in the yz-plane. Finally, if we set $y = 0$, we get $2x + z = 6$, a straight line in the xz-plane. These three lines outline a plane passing through the points $(3, 0, 0)$, $(0, 2, 0)$, and $(0, 0, 6)$. Any point on this plane will satisfy the original equation. For example, when $x = \frac{1}{2}$ and $y = 1$, we have $z = 6 - 2(\frac{1}{2}) - 3(1) = 2$. The point $(\frac{1}{2}, 1, 2)$ is also shown in Figure 8-4.   ∎

Example 6   Sketch a graph of the function $z = f(x, y) = 9 - x^2 - y^2$.

Solution   If we set $z = 0$, we get the equation $x^2 + y^2 = 9$. This is the equation of a circle in the xy-plane with center $(0, 0, 0)$ and radius 3. When $x = 0$, we have $z = 9 - y^2$. This is the equation of a parabola in the yz-plane. Finally, if $y = 0$, then $z = 9 - x^2$, which is also the equation of a parabola but positioned in the xz-plane. The surface $z = f(x, y)$ is sketched in Figure 8-5.   ∎

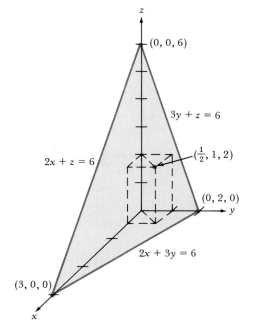

Figure 8-4
Graph of $z = f(x, y) = 6 - 2x - 3y$

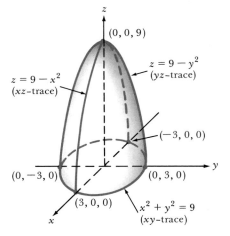

Figure 8-5
Graph of $z = f(x, y) = 9 - x^2 - y^2$

In this last example, suppose we fix z at various different values, say $-1, 0, 6$, and 9. This yields the equations

$$x^2 + y^2 = 10 \qquad x^2 + y^2 = 3$$
$$x^2 + y^2 = 9 \qquad x^2 + y^2 = 0$$

Each of these is the equation of a circle in the xy-plane. (The equation $x^2 + y^2 = 0$ represents a single point at the origin.) If we graph these in the xy-coordinate system, we obtain the set of curves shown in Figure 8-6. These curves are called *level curves* and are a way to represent three-dimensional properties in two dimensions. Think of looking at the function straight down over the xy-plane. You have undoubtedly seen similar representations on contour maps that show elevation and weather maps that show curves of constant temperature (isotherms) or curves of constant barometric pressure (isobars). A level curve shows

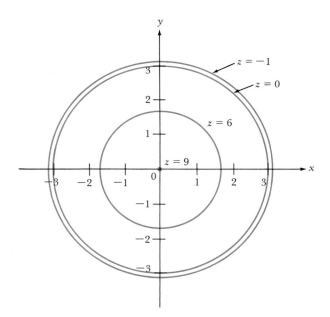

Figure 8-6
Level curves of the surface
$z = f(x, y) = 9 - x^2 - y^2$

curves of constant function value, z. Geometrically, this can be thought of as "slicing" through the surface in Figure 8-5 with a plane parallel to the xy-plane and projecting the curve of intersection onto the xy-plane. (Try this with an old tennis ball cut in half.)

Figures 8-7 and 8-8 are examples of computer plots of several more complicated functions. One thing you will probably notice from these plots are "peaks" and "valleys" that represent points where relative maxima and minima occur. In the following sections, we will develop the calculus techniques necessary to identify relative optima of multivariable functions.

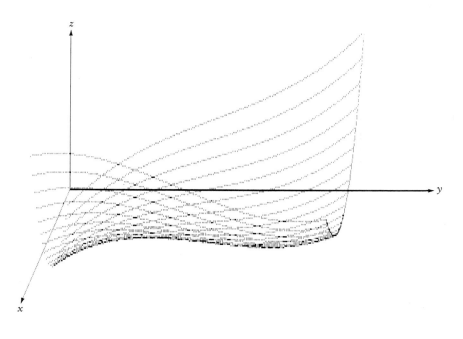

Figure 8-7
Computer plot of
$z = 3x^2 - 2xy + y^3 - 4x + 5y - 3y^2$

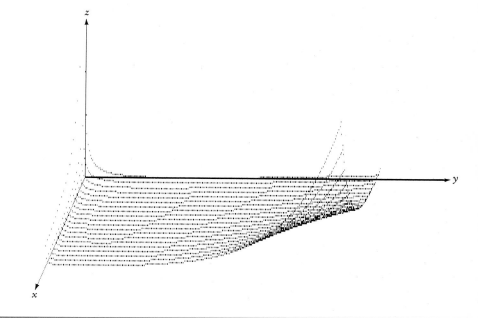

Figure 8-8
Computer plot of

$$z = ye^{xy} - 3xy + y^2 + \frac{1}{xy}$$

PROBLEMS (Section 8-1)

Let $z = f(x, y) = 4x^2 + 3y^2 - 5xy + 20$. *Evaluate:*

1. $f(3, 0)$
2. $f(0, -2)$
3. $f(2, 2)$
4. $f(1, -3)$
5. $f(a, a^2)$
6. $f(x + h, y)$

Let $z = f(x, y) = 2x^2 - 4y^2 + 3xy - 15$. *Evaluate:*

7. $f(2, 0)$
8. $f(0, -1)$
9. $f(3, 3)$
10. $f(-2, 4)$
11. $f(b, b^2)$
12. $f(x, y + h)$

Let $z = f(x, y) = 2x^3 + y^3 - 2xy + 50$. *Evaluate:*

13. $f(3, 0)$
14. $f(0, -2)$

15. $f(2, 3)$
16. $f(-2, -3)$
17. $f(c^2, c)$
18. $f(x + h, y + h)$

19. A rubber company produces two types of tires, bias-ply and steel-belted radials. The annual revenue and cost functions (in millions of dollars) for producing x thousand bias-ply tires and y thousand steel-belted radials are

$$R(x, y) = 3x + 4y$$
$$C(x, y) = x^2 - xy + 2y^2 + 5x - 6y + 10$$

  a. Find $R(2, 4)$.
  b. Find $C(3, 2)$.
  c. If profit is denoted by

$$P(x, y) = R(x, y) - C(x, y),$$

    find $P(2, 3)$.

20. A closed plastic box is to be constructed with a volume of 2 cubic feet. The cost of plastic is 3 cents per square foot. Express the cost of the box

as a function of the base dimensions x and y. [*Hint:* Use the volume requirement to eliminate the height dimension z.]

21. A closed rectangular cardboard box is to be constructed with a volume of 240 cubic inches. The price of cardboard is 0.01 cent per square inch. Express the cost of the box as a function of the width x and the height z. [*Hint:* Use the required volume to eliminate the variable length y.]

Plot the following points on a three-dimensional coordinate system:

22. (2, 1, 0)

23. (3, 0, 4)

24. (0, 2, 3)

25. (0, 0, 5)

26. (1, 2, −3)

27. (4, 2, 3)

28. (3, −5, 2)

29. (2, 1, 6)

In Problems 30–35, sketch each function in a three-dimensional coordinate system.

30. $z = f(x, y) = 4 - 2x - y$

31. $z = f(x, y) = 8 + 4x - 2y$

32. $z = f(x, y) = 3 + x + 3y$

33. $z = f(x, y) = 4 - x^2 - y^2$

34. $z = f(x, y) = x^2 + y^2 - 25$

35. $z = f(x, y) = x^2 + y - 9$

8-2 DERIVATIVES OF MULTIVARIABLE FUNCTIONS

How do we differentiate a function of several variables? Obviously, if there is more than one independent variable, there is some freedom in choosing a variable with respect to which to differentiate. With multivariable functions, we can differentiate with respect to *any* of the independent variables.

> The derivative of a multivariable function with respect to one independent variable, treating all other independent variables as constants, is called a **partial derivative.**

The most common notation used to denote a partial derivative is analogous to the symbol df/dx for functions of a single variable.

> If $z = f(x, y)$, then the partial derivative of z with respect to x is denoted by $\partial z / \partial x$ or $\partial f / \partial x$, and the partial derivative with respect to y is denoted by $\partial z / \partial y$ or $\partial f / \partial y$.

As mentioned in the box above, the partial derivative with respect to a particular independent variable is the derivative of the function with respect to that variable, *treating all other independent variables as constants.*

Let us find the partial derivatives $\partial f/\partial x$ and $\partial f/\partial y$ of the function $f(x, y) = 3x^2 + 2xy - y^2$. To find $\partial f/\partial x$, we treat y as a constant. The derivative of the first term, $3x^2$, is simply $6x$. The second term can be written as $(2y)x$. Since we treat $2y$ as constant, the partial derivative of this term with respect to x is $2y$. Finally, the third term, $-y^2$, is considered to be a constant, so its derivative is 0. We therefore have

$$\frac{\partial f}{\partial x} = 6x + 2y$$

Using similar reasoning and treating x as a constant, we find

$$\frac{\partial f}{\partial y} = 2x - 2y$$

Example 7   Find the partial derivatives $\partial f/\partial x$ and $\partial f/\partial y$ of the function

$$f(x, y) = 4x^2y + 2x + e^{y^2}$$

Solution   First, treat y as a constant. We have

$$\frac{\partial f}{\partial x} = 8xy + 2 + 0 = 8xy + 2$$

Next, assume that x is constant. Differentiating with respect to y yields

$$\frac{\partial f}{\partial y} = 4x^2 + 0 + 2ye^{y^2} = 4x^2 + 2ye^{y^2}$$ ∎

Example 8   Find the partial derivatives $\partial f/\partial x$ and $\partial f/\partial y$ of the function

$$f(x, y) = (3x^2 + 2yx^3)^4$$

Solution   We use the general power rule for this function. When y is assumed constant, we have

$$\frac{\partial f}{\partial x} = 4(3x^2 + 2yx^3)^3(6x + 6yx^2)$$

Treating x as constant and differentiating with respect to y yields

$$\frac{\partial f}{\partial y} = 4(3x^2 + 2yx^3)^3(2x^3)$$ ∎

A partial derivative of a multivariable function is also a multivariable function and can therefore be evaluated at specific values of the variables. The symbols

$$\frac{\partial f}{\partial x}(a, b) \quad \text{and} \quad \frac{\partial f}{\partial y}(a, b)$$

are used to denote the values of the partial derivatives $\partial f/\partial x$ and $\partial f/\partial y$ at the point $x = a$, $y = b$.

Example 9   Evaluate the partial derivatives $(\partial f/\partial x)(2, 1)$ and $(\partial f/\partial y)(1, 2)$ of the function

$$f(x, y) = \ln xy + xe^y$$

Solution   When y is assumed to be constant, we have

$$\frac{\partial f}{\partial x} = \left(\frac{1}{xy}\right)(y) + e^y = \frac{1}{x} + e^y$$

Therefore,

$$\frac{\partial f}{\partial x}(2, 1) = \frac{1}{2} + e^1 \approx 3.21828$$

Treating x as a constant and differentiating with respect to y gives

$$\frac{\partial f}{\partial y} = \left(\frac{1}{xy}\right)(x) + xe^y = \frac{1}{y} + xe^y$$

Hence,

$$\frac{\partial f}{\partial y}(1, 2) = \frac{1}{2} + 1(e^2) \approx 7.88905$$

■

For a function of one variable, we have seen that the derivative represents the rate of change of the dependent variable with respect to the independent variable and also represents the slope of the tangent line to the curve. We now discuss how these notions apply to partial derivatives of multivariable functions.

INTERPRETING PARTIAL DERIVATIVES AS RATES OF CHANGE

Let us consider the monthly driving cost function discussed earlier:

$$MC = f(x, y) = \frac{31.5x}{y} + 0.564375x + 42$$

The partial derivatives are

$$\frac{\partial f}{\partial x} = \frac{31.5}{y} + 0.564375 \qquad \text{and} \qquad \frac{\partial f}{\partial y} = \frac{-31.5x}{y^2}$$

Now suppose that $x = 20$ and $y = 30$. We saw earlier in this chapter that $f(20, 30) = 74.2875$. That is, a vehicle driven 20 miles per day that gets 30 miles per gallon will cost approximately \$74.29 per month to operate. If y is fixed at 30, we have

$$f(x, 30) = \frac{31.5x}{30} + 0.564375x + 42$$

Since this is a function of x alone, we can define $f(x, 30)$ to be a function of a single variable, $g(x)$. The derivative of $g(x)$ is

$$\frac{dg}{dx} = \frac{31.5}{30} + 0.564375 = 1.614375$$

Note, however, that this is the same as $\partial f/\partial x$ when $y = 30$. Therefore, at $y = 30$, we can interpret $\partial f/\partial x$ in the same fashion as we would interpret dg/dx. That is, $\partial f/\partial x$ represents the instantaneous rate of change of the function f with respect to x when y is held constant. For this example, $\partial f / \partial x$ is the rate of change of monthly driving cost with respect to the number of miles driven per day. When $x = 20$ and $y = 30$, this rate of change is approximately $1.61 per mile per day. If x is increased by 1, we would expect the monthly cost to increase by approximately $1.61, assuming the fuel economy *remains fixed at y = 30*.

In a similar fashion, $\partial f / \partial y$ represents the rate of change of cost with respect to fuel economy. Thus, when $x = 20$ and $y = 30$,

$$\frac{\partial f}{\partial y} = \frac{-31.5(20)}{(30)^2} = -0.70$$

This means that when the driving distance *remains fixed at x = 20*, the cost will decrease by approximately $0.70 if the fuel economy can be increased by 1 mile per gallon from $y = 30$ to $y = 31$. Since $\partial f / \partial y$ is a function of x and y, we cannot say that this will be true for all other values of x and y. We must reevaluate the partial derivative for different values of x and y.

In general, we may state the following:

A partial derivative is the rate of change of the dependent variable with respect to a single independent variable when all other independent variables are held constant.

Example 10
Mass Transit Capacity

In the analysis of mass transit operations such as subway trains, a measure of track capacity, Q, is the number of passengers per hour that can be served. This is given by the formula

$$Q = \frac{60kL}{H}$$

where

k = Number of passengers per foot of train length
L = Total train length (in feet)
H = Headway (distance) between trains (in minutes)

Suppose for a certain city during rush hour, k is estimated to be 3.1. Then

$$Q = \frac{60(3.1)L}{H} = \frac{186L}{H}$$

Interpret the partial derivatives of Q when $L = 500$ feet and $H = 1.4$ minutes.

Solution

The partial derivatives are

$$\frac{\partial Q}{\partial L} = \frac{186}{H} \quad \text{and} \quad \frac{\partial Q}{\partial H} = \frac{-186L}{H^2}$$

When $L = 500$ and $H = 1.4$, we have

$$\frac{\partial Q}{\partial L} = \frac{186}{1.4} \approx 132.86 \quad \text{and} \quad \frac{\partial Q}{\partial H} = \frac{-186(500)}{(1.4)^2} \approx -47,448.98$$

The rate of change of passengers per hour with respect to train length is 132.86. That is, increasing train length from 500 to 501 feet will increase passenger flow by about 133 passengers per hour when headway is fixed at 1.4 minutes. The interpretation of $\partial Q/\partial H$ is that the rate of change of passengers per hour with respect to headway is approximately $-47,449$; that is, a 1 minute *increase* in headway from 1.4 to 2.4 minutes will *decrease* passenger flow by over 47,000 passengers per hour when the train length is fixed at 500 feet.   ∎

Example 11
Production Function

A function that specifies the amount of production that can be achieved with x units of labor and y units of capital (investment in equipment, material, and so on) is called a *production function*. A particular form of a production function that is often used in economics is the *Cobb–Douglas production function*:

$$p(x, y) = ax^b y^{1-b} \qquad 0 < b < 1$$

where $p(x, y)$ is the number of units produced. The partial derivatives are

$$\frac{\partial p}{\partial x} = abx^{b-1}y^{1-b} \qquad \text{and} \qquad \frac{\partial p}{\partial y} = a(1-b)x^b y^{-b}$$

The partial derivative $\partial p/\partial x$ is called the *marginal productivity of labor* and represents the rate of change in production per unit of labor input (for fixed capital investment). Similarly, $\partial p/\partial y$ is called the *marginal productivity of capital* and represents the rate of change in production per unit of capital investment (for fixed labor input).

Suppose that a firm's production function is estimated to be

$$p(x, y) = 4x^{0.6}y^{0.4}$$

(that is, $b = 0.6$). When $x = 2000$ and $y = 600,000$, we have

$$\frac{\partial p}{\partial x} = 4(0.6)(2000)^{-0.4}(600,000)^{0.4} \approx 23.5$$

$$\frac{\partial p}{\partial y} = 4(0.4)(2000)^{0.6}(600,000)^{-0.6} \approx 0.0522$$

When $x = 2000$, a 1 unit increase in labor will increase production by approximately 23.5 units when capital investment is held at $y = 600,000$. When $y = 600,000$, a 1 unit increase in capital will increase production by approximately 0.0522 unit when the level of labor input is held at $x = 2000$.   ∎

Example 12
Metabolism

The rate at which a chemical substance is absorbed into a bacterium and distributed throughout its volume is called the *rate of metabolism, M*. This is proportional to the surface area S and inversely proportional to the volume V of the bacterium. That is, $M = aS/V$, where a is a constant. Suppose that a bacterium is cylindrical in shape with length l and radius r. Then

$$S = 2\pi rl + 2\pi r^2$$
$$V = \pi r^2 l$$

How does a change in length or radius affect the rate of metabolism?

Solution We write the rate of metabolism as

$$M = \frac{aS}{V} = \frac{2a\pi(rl + r^2)}{\pi r^2 l}$$

$$= \frac{2a(rl + r^2)}{r^2 l}$$

$$= \frac{2a(l + r)}{rl}$$

$$= \frac{2a}{r} + \frac{2a}{l}$$

The rate of change of metabolism with respect to the length l (for fixed radius) is

$$\frac{\partial M}{\partial l} = \frac{-2a}{l^2}$$

For fixed radius, as the length increases, metabolism decreases at a rate inversely proportional to the square of the length.

The rate of change of metabolism with respect to the radius r (for fixed length) is

$$\frac{\partial M}{\partial r} = \frac{-2a}{r^2}$$

Thus, for fixed length, as the radius increases, the rate of metabolism decreases at a rate inversely proportional to the square of the radius. ■

THE TOTAL DIFFERENTIAL*

In the supplement to Chapter 3 we saw that the derivative of a function is useful in estimating the approximate change in the function when the independent variable is changed by a small amount, Δx. We called this the differential, and it was defined as

$$dy = \frac{dy}{dx} \Delta x$$

We can extend this concept to multivariable functions by using the fact that each partial derivative represents the rate of change of the function with respect to a single independent variable. If $z = f(x, y)$ and x and y are both changed by small amounts Δx and Δy, respectively, then we define the total differential as follows:

If $z = f(x, y)$, the **total differential** is

$$dz = \frac{\partial z}{\partial x} \Delta x + \frac{\partial z}{\partial y} \Delta y$$

---

* This section can be skipped without loss of continuity.

This is an approximation of the change in z at a given point if x and y are both changed by a small amount. Thus, for the monthly driving cost example, when $x = 20$ and $y = 30$, we have $\partial f/\partial x = 1.614375$ and $\partial f/\partial y = -0.70$, and the total differential is

$$dz = \frac{\partial f}{\partial x}\,\Delta x + \frac{\partial f}{\partial y}\,\Delta y$$
$$= 1.614375\,\Delta x - 0.70\,\Delta y$$

So, if the vehicle is driven an extra 1.5 miles to pick up a passenger ($\Delta x = 1.5$) and the fuel economy decreases by 2 miles per gallon due to a need for a tuneup ($\Delta y = -2$), then the approximate change in monthly cost will be

$$dz = 1.614375(1.5) - 0.70(-2)$$
$$\approx \$3.82 \text{ per month}$$

Note that $\Delta y = -2$ since there is a *decrease* in the value of y.

Example 13
Mass Transit Capacity

In Example 10, estimate the change in the number of passengers per hour when the total train length is increased by 50 feet (from 500 to 550 feet) and the headway is decreased by 0.1 minute (from 1.4 to 1.3 minutes). What is the actual change and why is it different?

Solution   The total differential is

$$dQ = \frac{\partial Q}{\partial L}\,\Delta L + \frac{\partial Q}{\partial H}\,\Delta H$$
$$\approx 132.86(50) - 47{,}448.98(-0.1)$$
$$\approx 11{,}388 \text{ passengers per hour}$$

The actual change is found by computing the difference in Q for the new and old values:

$$\text{Actual change} = Q(550, 1.3) - Q(500, 1.4)$$
$$= \frac{186(550)}{1.3} - \frac{186(500)}{1.4}$$
$$\approx 12{,}264 \text{ passengers per hour}$$

Since derivatives represent instantaneous rates of change at one particular point, the total differential is only an approximation. This accounts for the discrepancy from the actual change. ∎

GEOMETRIC INTERPRETATION OF PARTIAL DERIVATIVES

The derivative of a single-variable function $f(x)$ represents the slope of the tangent line to the curve at a specific point $(x, f(x))$. Consider the two-variable function $z = f(x, y)$. At a specific point (a, b, c) there are an infinite number of tangent lines, as Figure 8-9 illustrates.

Suppose that we fix the value of $y = b$ and "slice" through the surface with a plane parallel to the xz-plane. This is shown in Figure 8-10a. The function $z = f(x, b)$ is now a function of x alone. If we look straight down the y-axis toward the xz-plane, we see the function $z = f(x, b)$, as shown in Figure 8-10b.

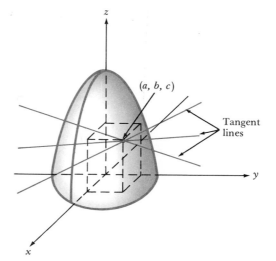

Figure 8-9
Tangent lines to a point on a surface

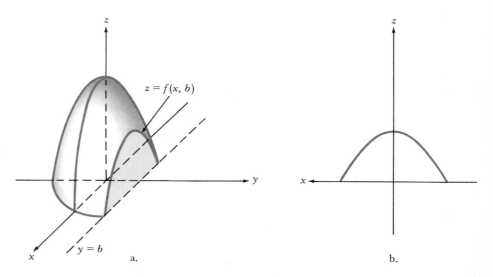

Figure 8-10
a. If $y = b$, $z = f(x, b)$ is a curve
b. The curve $z = f(x, b)$ as seen
from the end of the y-axis (the y-axis
comes out from the page)

Now, for a given value of x, say $x = a$, there is a unique tangent line to the curve $z = f(x, b)$. We show this from two perspectives in Figure 8-11. The tangent line is parallel to the xz-plane and is therefore unique. What is the value of the slope of this line? You may have guessed that the slope has the value of the partial derivative $\partial f / \partial x$ at $x = a$ and $y = b$. This is indeed correct. The value of a partial derivative $\partial z / \partial x$ is the slope of the tangent line to the surface that is parallel to the xz-plane; the value of $\partial z / \partial y$ is the slope of the tangent line to the surface that is parallel to the yz-plane. This interpretation is useful in developing concepts of relative maxima and minima for multivariable functions.

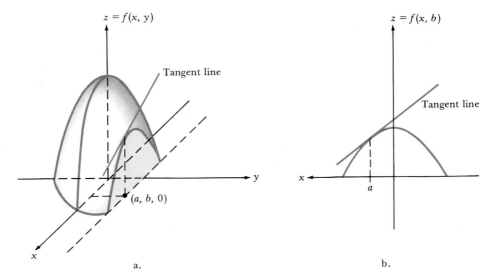

Figure 8-11
Tangent line to z = f(x, y) at x = a and y = b

a.

b.

Example 14   Find the slopes of the tangent lines parallel to the xz-plane and the yz-plane for the function $z = f(x, y) = 3x^2 - 2xy + y^3$ at the point $(1, -2, -1)$.

Solution   The partial derivatives of z are

$$\frac{\partial z}{\partial x} = 6x - 2y \quad \text{and} \quad \frac{\partial z}{\partial y} = -2x + 3y^2$$

If we substitute $x = 1$ and $y = -2$, we obtain

$$\frac{\partial z}{\partial x} = 10 \quad \text{and} \quad \frac{\partial z}{\partial y} = 10$$

Therefore, the slopes of both lines are equal to 10.

PROBLEMS (Section 8-2)

In Problems 1–11, find $\partial f / \partial x$ and $\partial f / \partial y$ for each function.

1. $f(x, y) = x^2 + 2xy^2 - 2x$

2. $f(x, y) = 2x^2 - 4xy + 3x$

3. $f(x, y) = x^3y^2 + x^2y^3 + 2$

4. $f(x, y) = x^2y^4 + 5xy - 4$

5. $f(x, y) = x^3 + y^3 - 2x^2y + 2xy^2 + 3$

6. $f(x, y) = x^3 - y^3 + 3x^2y - 5xy^2 - 2$

7. $f(x, y) = x^4y^3 - 2x^3y^2 + 5x^2 + 3y - 2xy + 5$

8. $f(x, y) = x^6 + 3x^5y^4 + 2x^4y^4 - x^3y^3 + 2x - 4$

9. $f(x, y) = \sqrt{x + y}$

10. $f(x, y) = \sqrt{x^2 - y}$

11. $f(x, y) = \dfrac{x^2 + 3}{y - 2}$

For each function in Problems 12–22 below, compute $(\partial f/\partial x)(a, b)$ and $(\partial f/\partial y)(a, b)$ at the point indicated.

12. $f(x, y) = \dfrac{2xy^2}{x^2 + 3y^2}$ at $(1, 1)$

13. $f(x, y) = \ln xy + 2xe^y$ at $(2, 1)$

14. $f(x, y) = \ln(x^2 + y^2)$ at $(-1, 0)$

15. $f(x, y) = 3x \ln(x^2 + y)$ at $(-1, 1)$

16. $f(x, y) = \ln(2x^2 + 3y^3)$ at $(2, -1)$

17. $f(x, y) = e^{4xy}$ at $(0, 1)$

18. $f(x, y) = 5x^3 - 3x + 2e^{y^2}$ at $(1, 0)$

19. $f(x, y) = (2x^2 + 3xy)^3$ at $(-2, -1)$

20. $f(x, y) = (4x^2 - 3x^2y)^3$ at $(2, 2)$

21. $f(x, y) = (3x^3 + 4x^3y)^4$ at $(-1, 1)$

22. $f(x, y) = (2x^4 - 5x^3y^2)^4$ at $(1, 2)$

23. In Example 10, estimate the change in the number of passengers per hour when the total train length in increased by 30 feet and the headway is decreased by 0.2 minute. What is the actual change and why is your answer different?

24. The cost of a product is given by
$$C(x, y) = z = 120 + 10x + 20y^2 + 2xy$$
where

$z =$ Unit cost of producing a product (in dollars)

$x =$ Level of materials (in pounds per unit)

$y =$ Amount of labor (in hours per unit)

a. Find the rate of change in cost with respect to the level of raw materials if $y = 2$ and $x = 3$. Interpret your result.

b. Find the rate of change in cost with respect to the level of labor if $x = 3$ and $y = 3$. Interpret your result.

25. A snack food company has found that its weekly sales S are a function of the number of people, x, viewing their television commercials and the number of cities, y, showing the commercials. If $S = 48xy + 2x + 800$, then:

a. Find $\partial f/\partial x$ and interpret your result.

b. Find $\partial f/\partial y$ and interpret your result.

26. The total number of matings per day between mosquitoes is approximated by
$$M(x, y) = 2xy + 10xy^2 + 20y^2 + 10$$
where x represents the temperature (°C) and y represents the number of days since the last rain.

a. Find the rate of change in mating related to the temperature when $x = 20°C$ and $y = 3$ days.

b. Find the rate of change in mating related to the number of dry days when $x = 20°C$ and $y = 5$ days.

27. In economics, if the utility function of an individual takes the form
$$U = U(x, y) = (x + 2)^2(y + 3)^3$$
where U is total utility and x and y are the quantities of two commodities consumed, then:

a. Find the marginal utility function of each of the two commodities.

b. Find the value of the marginal utility of the first commodity when 4 units of each commodity are consumed.

28. Find the slopes of the tangent lines parallel to the xz-plane and the yz-plane for the function
$$z = f(x, y) = 3x^2 - xy + y^2$$
at the point $(2, 2, -1)$.

29. Find the slopes of the tangent lines parallel to the xz-plane and the yz-plane for the function
$$z = f(x, y) = 4x^2 + 2xy + 2y^2$$
at the point $(1, 2, 2)$.

30. Given
$$f(x, y) = x^2 - y^2 + xy \ln\left(\frac{y}{x}\right)$$
show that
$$x\frac{\partial f}{\partial x} + y\frac{\partial f}{\partial y} = 2f(x, y)$$

8-3 SECOND-ORDER PARTIAL DERIVATIVES

We find the second derivative of a function of one variable by differentiating the first derivative. For multivariable functions, we have seen that the first partial derivative can be found with respect to each independent variable. Therefore, we can also differentiate each first partial derivative with respect to any of the independent variables. For instance, consider the function

$$f(x, y) = 6x^2 - 3x^2y + 2y^3$$

The first partial derivatives are

$$\frac{\partial f}{\partial x} = 12x - 6xy \qquad \text{and} \qquad \frac{\partial f}{\partial y} = -3x^2 + 6y^2$$

If we now differentiate $\partial f / \partial x$ with respect to x (remembering to treat y as a constant), we obtain the *second partial derivative with respect to* x, denoted by $\partial^2 f / \partial x^2$:

$$\frac{\partial^2 f}{\partial x^2} = \frac{\partial}{\partial x}(12x - 6xy)$$

$$= 12 - 6y$$

Similarly, if we differentiate $\partial f / \partial y$ with respect to y, we obtain

$$\frac{\partial^2 f}{\partial y^2} = \frac{\partial}{\partial y}(-3x^2 + 6y^2)$$

$$= 12y$$

There are also two other choices. We may differentiate $\partial f / \partial x$ with respect to y, or we may differentiate $\partial f / \partial y$ with respect to x. Such partial derivatives are called *mixed partial derivatives* and are denoted by $\partial^2 f / \partial y \partial x$ and $\partial^2 f / \partial x \partial y$, respectively. For the example above, we obtain

$$\frac{\partial^2 f}{\partial y \partial x} = \frac{\partial}{\partial y}(12x - 6xy)$$

$$= -6x$$

$$\frac{\partial^2 f}{\partial x \partial y} = \frac{\partial}{\partial x}(-3x^2 + 6y^2)$$

$$= -6x$$

It is no coincidence that these are equal. In fact, for all functions that we shall encounter (and except for certain rare cases), $\partial^2 f / \partial y \partial x$ will always be equal to $\partial^2 f / \partial x \partial y$. Therefore, we need only compute one of them, although the other is a useful check of our mathematics. We can also find higher-order derivatives of multivariable functions, but these will not be considered in this text.

Example 15   Find all second-order partial derivatives for the following:

$$f(x, y) = 4x^2y^2 + e^{xy}$$

Solution   We must first find the first partial derivatives $\partial f/\partial x$ and $\partial f/\partial y$:

$$\frac{\partial f}{\partial x} = 8xy^2 + ye^{xy} \qquad \text{and} \qquad \frac{\partial f}{\partial y} = 8x^2y + xe^{xy}$$

Then the second-order partial derivatives are computed as

$$\frac{\partial^2 f}{\partial x^2} = 8y^2 + y^2 e^{xy}$$

$$\frac{\partial^2 f}{\partial y^2} = 8x^2 + x^2 e^{xy}$$

$$\frac{\partial^2 f}{\partial x \partial y} = 16xy + y(e^{xy}x) + e^{xy}(1)$$

$$= 16xy + xye^{xy} + e^{xy} = \frac{\partial^2 f}{\partial y \partial x}$$

In finding the mixed partial derivative, we could have used either $\partial f/\partial x$ or $\partial f/\partial y$ and differentiated with respect to the other variable. In either case, we needed to use the product rule for the second term. ∎

Example 16   Find all second-order partial derivatives for the function

$$z = f(x, y) = x \ln y + (x + y)^3$$

Solution   We have the following (applying the general power rule to the second term):

$$\frac{\partial z}{\partial x} = \ln y + 3(x + y)^2$$

$$\frac{\partial z}{\partial y} = \frac{x}{y} + 3(x + y)^2$$

$$\frac{\partial^2 z}{\partial x^2} = 6(x + y)$$

$$\frac{\partial^2 z}{\partial y^2} = \frac{-x}{y^2} + 6(x + y)$$

$$\frac{\partial^2 z}{\partial x \partial y} = \frac{1}{y} + 6(x + y)$$

∎

PROBLEMS (Section 8-3)

Find all second-order partial derivatives for each function. Verify that $\partial^2 f/\partial y \partial x = \partial^2 f/\partial x \partial y$.

1. $f(x, y) = x^2 + 2xy - 3y^2$

2. $f(x, y) = 3x^2 + 4xy + 2y^2 - 4x + 3y - 5$

3. $f(x, y) = x^3 - 2x^2y + 3xy^2$

4. $f(x, y) = x^4 + 3x^3y - 2x^2y^2 + 6xy^3 + 8y^4$

5. $f(x, y) = 3x^2y^2 - e^{xy}$

6. $f(x, y) = e^{2xy} - 4x^3y$

7. $f(x, y) = e^{xy} + x \ln y$

8. $f(x, y) = e^{xy} - x \ln y$

9. $f(x, y) = 4x^2y + (x + y)^2$

10. $f(x, y) = (x^2 + y)^2 + (x + y)^3$

11. $f(x, y) = \ln(x + y) + 3$

12. $f(x, y) = \ln(x^2 + y^2) - 2$

13. $f(x, y) = (x^2 + xy + 2)(x^2 - xy + 1)$

14. $f(x, y) = (x^2 + xy + y^2)(x^2 + xy + 3)$

15. $f(x, y) = (x + y)^2(xy)$

16. $f(x, y) = (x + y)^2(x - y)^3$

8-4 OPTIMIZATION OF MULTIVARIABLE FUNCTIONS

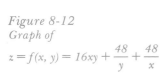

In Example 3 we developed a function representing the cost of constructing a metal box with a volume of 3 cubic feet. This was given as

$$f(x, y) = 16xy + \frac{48}{y} + \frac{48}{x}$$

Suppose a company that manufactures these metal boxes wants to choose the dimensions of the box in order to minimize costs. Figure 8-12 shows the surface which is the graph of this function. From this figure, we can see that as x and y approach 0, the value of z becomes large. Also, as x and y grow large, z again increases. Later in this section, we will find that the minimum value of z is located at the point $x = \sqrt[3]{3}$, $y = \sqrt[3]{3}$, $z \approx 196$. Around this point we see that the surface is relatively flat. In this section we show how calculus can be used to find relative minimum and maximum points of multivariable functions.

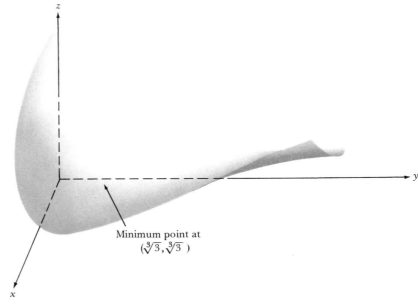

Figure 8-12
Graph of
$z = f(x, y) = 16xy + \dfrac{48}{y} + \dfrac{48}{x}$

Minimum point at
$(\sqrt[3]{3}, \sqrt[3]{3})$

For a differentiable single-variable function, the first derivative is 0 at a relative maximum or minimum; that is, the tangent line to the curve at such a point is parallel to the x-axis and hence has 0 slope (see Figure 8-13a). You might suspect that relative optima of multivariable functions have similar properties,

Figure 8-13
Illustration of tangents in two and
three dimensions

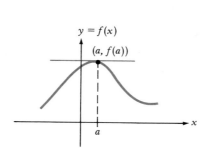

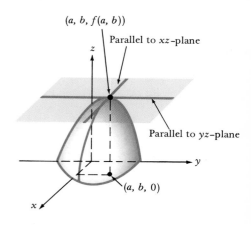

a. Tangent line to a curve at
    a relative maximum point

b. Tangent plane to a surface at
    a relative maximum point

namely that the first partial derivatives are 0. This is indeed the case, but let us first examine why this must be true.

A three-dimensional analogy of a tangent line to a curve is a tangent plane to a surface, as illustrated in Figure 8-13b. At a point that is a relative maximum or minimum, the tangent plane must be parallel to the *xy*-plane. Suppose that, in Figure 8-13b, the function has a relative maximum at $x = a$ and $y = b$. Then the function $g(x) = f(x, b)$ has a relative maximum at $x = a$, and hence,

$$0 = g'(a) = \frac{\partial f}{\partial x}(a, b)$$

In geometrical terms this says that the line in the tangent plane that passes through the point $(a, b, f(a, b))$ and is parallel to the *xz*-plane has slope equal to 0 (see Figure 8-13b).

Similarly, since $f(x, y)$ has a relative maximum at $x = a$, $y = b$, the function $h(y) = f(a, y)$ has a relative maximum at $y = b$, and hence,

$$0 = h'(b) = \frac{\partial f}{\partial y}(a, b)$$

indicating that the line in the tangent plane that passes through the point $(a, b, f(a, b))$ and is parallel to the *yz*-plane has slope equal to 0 (see Figure 8-13b). In the same way, we can show that if $f(x, y)$ has a relative minimum at a point, then both first partials must vanish at that point. Thus, we may state the following theorem:

THEOREM

If a function $z = f(x, y)$ has a relative maximum or minimum at a point $(a, b, f(a, b))$, then both first partial derivatives must be 0 at that point (assuming they exist). That is,

$$\frac{\partial f}{\partial x}(a, b) = \frac{\partial f}{\partial y}(a, b) = 0$$

As with single-variable functions, points at which both first partial derivatives are 0 are called *critical points*. We can identify critical points by setting the first partial derivatives equal to 0 and solving the resulting equations for x and y (see Algebra Review 18). Thus, for the minimum cost container design example mentioned at the beginning of this section, we have

$$\frac{\partial f}{\partial x} = 16y - \frac{48}{x^2} = 0$$

$$\frac{\partial f}{\partial y} = 16x - \frac{48}{y^2} = 0$$

Solving the first equation for y, we obtain

$$y = \frac{3}{x^2}$$

Substituting this into the second equation yields

$$16x - \frac{48}{(3/x^2)^2} = 0$$

$$16x = \frac{48}{9/x^4}$$

ALGEBRA REVIEW 18: SOLVING SYSTEMS OF EQUATIONS

We often have to solve two equations in two unknowns. For example, consider the system of equations

$$6x + 2y = 5$$
$$3x - y = 2$$

A general method for solving such a system of equations is to solve one equation for one of the variables and then substitute the result into the other equation. For instance, if we solve the second equation above for y, we get

$$-y = 2 - 3x$$
$$y = -2 + 3x$$

Substituting this into the first equation yields

$$6x + 2y = 5$$
$$6x + 2(-2 + 3x) = 5$$
$$6x - 4 + 6x = 5$$

$$12x = 9$$
$$x = \tfrac{3}{4}$$

Using this value for x, we may find y from either equation, although it is usually easiest to use the expression for y found in the first step. Thus, we have

$$y = -2 + 3x$$
$$= -2 + 3(\tfrac{3}{4}) = \tfrac{1}{4}$$

Therefore, the point $(\tfrac{3}{4}, \tfrac{1}{4})$ is the only solution to the original system of equations. You should verify this by substitution.

As another example, consider the system of equations

$$x + 2y = 2$$
$$x^2 + y^2 = 4$$

If we solve the first equation for x, we get

$$x = 2 - 2y$$

$$x = \frac{3}{9/x^4}$$

$$x = \tfrac{1}{3}x^4$$

$$3x = x^4$$

$$x(3 - x^3) = 0$$

The solutions to this equation are $x = 0$ (clearly, we disregard this root since our container must have positive dimensions), and $x = \sqrt[3]{3}$. Substituting $x = \sqrt[3]{3}$ into $y = 3/x^2$ yields $y = 3/3^{2/3} = \sqrt[3]{3}$, and hence, we have a critical point at $x = y = \sqrt[3]{3}$. However, we do not yet know whether the function has a relative maximum or minimum at this critical point. We shall consider this shortly.

Example 17   Find all critical points of the function

$$z = x^2 + 4xy + 6y^2 - 2x + 4y$$

Solution   Setting the first partial derivatives equal to 0, we have

$$\frac{\partial z}{\partial x} = 2x + 4y - 2 = 0 \qquad \frac{\partial z}{\partial y} = 4x + 12y + 4 = 0$$

Substituting this into the second equation yields

$$(2 - 2y)^2 + y^2 = 4$$

$$4 - 8y + 4y^2 + y^2 = 4$$

$$5y^2 - 8y = 0$$

Factoring this equation, we have two roots for y:

$$y(5y - 8) = 0$$

$$y = 0 \qquad \text{and} \qquad y = \tfrac{8}{5}$$

Using the expression we found for x in terms of y, $x = 2 - 2y$, we see that when $y = 0$, $x = 2$, and when $y = \tfrac{8}{5}$, $x = -\tfrac{6}{5}$. Therefore, this system of equations has two solutions $(2, 0)$ and $(-\tfrac{6}{5}, \tfrac{8}{5})$. Again, you should verify that each of these points is a solution to the original system of equations.

This method of substitution can also be used to solve three equations in three unknowns. First, solve one equation for one of the variables. Then, substitute the result in the other two equations, yielding a system of two equations in two unknowns that can be solved by the method outlined above. (This process can be extended to larger systems of equations.)

EXERCISES*

Solve the following systems of equations:

a. $x - y = 4$
    $2x + 3y = 3$

b. $x + y - 2 = 0$
    $-4x + y + 3 = 0$

c. $2x^2 + y = 0$
    $x^2 - y = 3$

d. $w + 2x + 4y = 6$
    $x + 2y = 3$
    $w + x - 2y = 1$

e. $-3x + 2y - z = -1$
    $4x - 2y + z = 3$
    $6x - 3y + z = 3$

---

* Answers to algebra review exercises can be found at the end of the answer section in the back of the book.

Solving the first equation for x, we have

$$2x = -4y + 2$$
$$x = -2y + 1$$

Substituting this into the second equation yields

$$4(-2y + 1) + 12y + 4 = 0$$
$$-8y + 4 + 12y + 4 = 0$$
$$4y + 8 = 0$$
$$y = -2$$

Substituting this value into the equation for x, we find $x = -2(-2) + 1 = 5$. Substituting these values into the original equation for z yields

$$z = (5)^2 + 4(5)(-2) + 6(-2)^2 - 2(5) + 4(-2) = -9$$

The point $(5, -2, -9)$ is the only critical point of this function.   ∎

Example 18   Find the critical points of the function

$$z = y^2 - x^2$$

Solution   Setting the first partial derivatives of this function equal to 0, we have

$$\frac{\partial z}{\partial x} = -2x = 0$$

$$\frac{\partial z}{\partial y} = -2y = 0$$

This system of equations has solution $x = 0$ and $y = 0$. Thus, $(0, 0, 0)$ is the only critical point.   ∎

Example 19   Find the critical points of the function

$$z = x^2 + \tfrac{1}{2}y^4 + 6xy$$

Solution   Setting the first partial derivatives equal to 0 yields

$$\frac{\partial z}{\partial x} = 2x + 6y = 0$$

$$\frac{\partial z}{\partial y} = 2y^3 + 6x = 0$$

Solving the first equation for x, we obtain $x = -3y$. Substituting this into the second equation yields

$$2y^3 + 6(-3y) = 0$$
$$y(2y^2 - 18) = 0$$
$$2y(y - 3)(y + 3) = 0$$

The roots of this equation are $y = 0$, $y = 3$, and $y = -3$. Since $x = -3y$, we have critical points at $(x, y) = (0, 0)$, $(-9, 3)$, and $(9, -3)$.   ∎

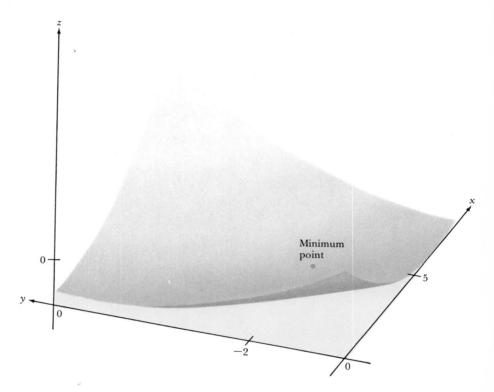

Figure 8-14
Graph of
$z = x^2 + 4xy + 6y^2 - 2x + 4y$

Graphs of the functions discussed in Examples 17, 18, and 19 are shown in Figures 8-14, 8-15, and 8-16, respectively. For the function $z = x^2 + 4xy + 6y^2 - 2x + 4y$ in Figure 8-14, we see that there is a shallow valley around the relative minimum at $x = 5$ and $y = -2$. For the function $z = y^2 - x^2$, the only critical point shown in Figure 8-15 is $(0, 0, 0)$. Such a critical point is called a *saddle point:* as we move in the x direction (keeping the y value fixed at 0), the function appears to have a relative maximum at the critical point, but as we move in the y direction (keeping the x value fixed at 0), it seems to have a relative minimum at the critical point. Thus, $z = y^2 - x^2$ has neither a relative maximum

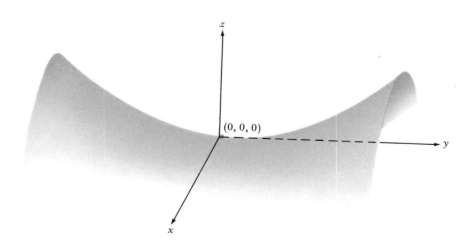

Figure 8-15
Graph of $z = y^2 - x^2$

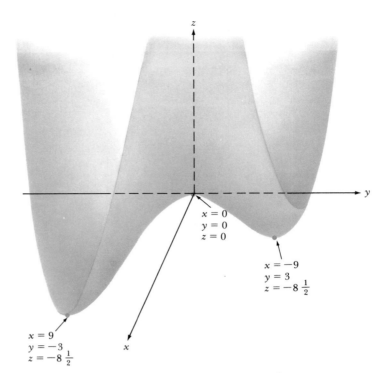

Figure 8-16
Graph of $z = x^2 + \frac{1}{2}y^4 + 6xy$

nor a relative minimum at $(0, 0, 0)$. You may recall that we found a similar phenomenon for single-variable functions, namely, a point of inflection.

Finally, for the function $z = x^2 + \frac{1}{2}y^4 + 6xy$ in Example 19, relative minima occur at $x = 9$, $y = -3$ and at $x = -9$, $y = 3$, as shown in Figure 8-16. This function has a saddle point when $x = 0$ and $y = 0$.

SECOND DERIVATIVE TEST

Once we have located critical points we need to determine whether they are relative maximum, minimum, or saddle points. For functions of one variable, we derived a simple second derivative test based on concavity properties of functions. For multivariable functions, there is a similar test, although it is more complicated. We state it below without proof.

Second Derivative Test for Two-Variable Functions

Let $(a, b, f(a, b))$ be a critical point of the function $z = f(x, y)$. Let

$$K = \left(\frac{\partial^2 f}{\partial x^2}\right)\left(\frac{\partial^2 f}{\partial y^2}\right) - \left(\frac{\partial^2 f}{\partial x \partial y}\right)^2$$

If at $x = a$ and $y = b$:

a. $K > 0$ and $\partial^2 f / \partial x^2 > 0$ (or $\partial^2 f / \partial y^2 > 0$), then $(a, b, f(a, b))$ is a relative minimum point.

b. $K > 0$ and $\partial^2 f / \partial x^2 < 0$ (or $\partial^2 f / \partial y^2 < 0$), then $(a, b, f(a, b))$ is a relative maximum point.

c. $K < 0$, then $(a, b, f(a, b))$ is a saddle point.

d. $K = 0$, then the test is inconclusive.

Some comments are in order. First, note that K can never be positive if $\partial^2 f/\partial x^2$ and $\partial^2 f/\partial y^2$ have opposite signs. Therefore, whenever $K > 0$, they must have the same sign, so it does not matter which one we check in steps a and b. Second, if $\partial^2 f/\partial x^2$ and $\partial^2 f/\partial y^2$ have opposite signs, then we can interpret this to mean that the rate of change of the slope of the tangent lines in one direction is increasing (implying a relative minimum), but decreasing in the other direction (implying a relative maximum). Thus, when $K < 0$, a saddle point results.

Let us reconsider the container design example in which $f(x, y) = 16xy + (48/y) + (48/x)$. We found that a critical point occurs at $x = \sqrt[3]{3}$ and $y = \sqrt[3]{3}$. The second partial derivatives are

$$\frac{\partial^2 f}{\partial x^2} = \frac{96}{x^3}$$

$$\frac{\partial^2 f}{\partial y^2} = \frac{96}{y^3}$$

$$\frac{\partial^2 f}{\partial x \partial y} = 16$$

Therefore, at $x = \sqrt[3]{3}$, $y = \sqrt[3]{3}$,

$$K = (\tfrac{96}{3})(\tfrac{96}{3}) - (16)^2 = 768 > 0$$

Since $\partial^2 f/\partial x^2 = 32 > 0$, we find that at $x = y = \sqrt[3]{3}$, the function has a relative minimum value of $f(\sqrt[3]{3}, \sqrt[3]{3}) \approx 196$.

Example 20   Determine the type of critical point for each of the functions in Examples 17–19.

Solution   We compute all second partial derivatives and apply the second derivative test.

Example 17:   For the function $z = x^2 + 4xy + 6y^2 - 2x + 4y$, we have

$$\frac{\partial^2 z}{\partial x^2} = 2$$

$$\frac{\partial^2 z}{\partial y^2} = 12$$

$$\frac{\partial^2 z}{\partial x \partial y} = 4$$

Thus, $K = (2)(12) - (4)^2 = 8 > 0$. Since $\partial^2 z/\partial x^2 > 0$, the point $(5, -2, -9)$ is a relative minimum point (see Figure 8-14).

Example 18:   For the function $z = y^2 - x^2$, we obtain

$$\frac{\partial^2 z}{\partial x^2} = -2$$

$$\frac{\partial^2 z}{\partial y^2} = 2$$

$$\frac{\partial^2 z}{\partial x \partial y} = 0$$

Thus, $K = (-2)(2) - 0 = -4 < 0$. Hence, the point $(0, 0, 0)$ is a saddle point (see Figure 8-15).

Example 19:   The second partial derivatives for $z = x^2 + \frac{1}{2}y^4 + 6xy$ are

$$\frac{\partial^2 z}{\partial x^2} = 2$$

$$\frac{\partial^2 z}{\partial y^2} = 6y^2$$

$$\frac{\partial^2 z}{\partial x \partial y} = 6$$

We must test each point individually: At $x = 0$ and $y = 0$, we have $K = 2(0) - (6)^2 = -36 < 0$. Therefore, $(0, 0, 0)$ is a saddle point. For $x = 9$ and $y = -3$, $K = (2)(54) - (6)^2 = 72 > 0$. Since $\partial^2 z/\partial x^2 > 0$, we find that $(9, -3, -8\frac{1}{2})$ is a relative minimum point. At $x = -9$ and $y = 3$, $K = 2(54) - (6)^2 = 72 > 0$ and $\partial^2 z/\partial x^2 > 0$. Therefore, $(-9, 3, -8\frac{1}{2})$ is also a relative minimum point (see Figure 8-16).   ∎

We may now use these results to solve a variety of practical optimization problems involving two independent variables.

Example 21
Process Control

The cost per batch of a certain chemical production process depends on the setting of two control variables, pressure and temperature. Statistically designed experiments have enabled process engineers to determine a function relating cost per batch (in dollars) to the values of temperature (x, in degrees Celsius) and pressure (y, in atmospheres) as follows:

$$z = f(x, y) = 4x^2 + 2y^2 + 4xy - 8x - 6y + 50$$

Determine the temperature and pressure settings that will minimize cost.

Solution   We set the first partial derivatives of z equal to 0:

$$\frac{\partial z}{\partial x} = 8x + 4y - 8 = 0$$

$$\frac{\partial z}{\partial y} = 4y + 4x - 6 = 0$$

Solving this system of equations yields $x = \frac{1}{2}$ and $y = 1$. The value of z at this point is 45. We next apply the second derivative test to this point:

$$\frac{\partial^2 z}{\partial x^2} = 8$$

$$\frac{\partial^2 z}{\partial y^2} = 4$$

$$\frac{\partial^2 z}{\partial x \partial y} = 4$$

$$K = 8(4) - (4)^2 = 16 > 0$$

Since $\partial^2 z/\partial x^2 > 0$, the point $(\frac{1}{2}, 1, 45)$ is a relative minimum. Since this is also the only critical point and the function values z grow arbitrarily large as x and y increase without bound, this point is also an absolute minimum point. Therefore, a minimum cost of $45 results when the temperature is set at $\frac{1}{2}°$C and the pressure is set at 1 atmosphere. ∎

Example 22
International Pricing

A Silicon Valley computer chip manufacturer has two major markets, the United States and Japan. A new high-speed microprocessor will soon be introduced in these markets. Based on accounting records, the total cost (in dollars) of producing and marketing z chips is estimated to be

$$C(z) = 800,000 + 12.5z + 0.00025z^2$$

The demand in the United States as a function of price is estimated to be

$$x = 140,000 - 570p(x)$$

where x is the demand and $p(x)$ is the price. In Japan, the demand function is estimated to be

$$y = 136,000 - 450p(y)$$

How many chips should be produced for each market in order to maximize profit?

Solution

Since $z = x + y$, the profit function is given by

$$p(x, y) = xp(x) + yp(y) - C(x + y)$$
$$= \frac{x(140,000 - x)}{570} + \frac{y(136,000 - y)}{450} - 800,000 - 12.5(x + y) - 0.00025(x + y)^2$$

To maximize $p(x, y)$, we set $\partial p/\partial x$ and $\partial p/\partial y$ equal to 0:

$$\frac{\partial p}{\partial x} = \frac{140,000}{570} - \frac{2x}{570} - 12.5 - 2(0.00025)(x + y)$$
$$\approx 233.11 - 0.004x - 0.0005y = 0$$

$$\frac{\partial p}{\partial y} = \frac{136,000}{450} - \frac{2y}{450} - 12.5 - 2(0.00025)(x + y)$$
$$\approx 289.72 - 0.0005x - 0.0049y = 0$$

Solving these equations yields $x = 51,544$ and $y = 53,867$. The second partial derivatives are

$$\frac{\partial^2 p}{\partial x^2} = -\frac{2}{570} - 0.0005 \approx -0.004$$

$$\frac{\partial^2 p}{\partial y^2} = -\frac{2}{450} - 0.0005 \approx -0.005$$

$$\frac{\partial^2 p}{\partial x \partial y} = -0.0005$$

Therefore, $K = 0.000015 > 0$ and since $\partial^2 p / \partial x^2 < 0$, we have a relative maximum at $x = 51,554$ and $y = 53,867$. Since there is only one critical point and $p(x, y)$ approaches $-\infty$ as x and y become arbitrarily large, this relative maximum is also an absolute maximum. Thus, to maximize profit, 51,554 chips should be produced for the United States market and 53,867 for the Japanese market. The profit is $p(51,554, 53,867) \approx \$12,935,079$.   ∎

PROBLEMS (Section 8-4)

In Problems 1–8, find all critical points for each function.

1. $z = 5 - x^2 - 4x - y^2$

2. $z = 4 - y^2 - x^2 + 6y$

3. $z = x^2 + y^2 - 5x + 4y + xy$

4. $z = x^2 - 6x + 4y^2 + 16y$

5. $z = x^2 + y^2 + 2x - 6y + 10$

6. $z = 2x - 3y + xy - 4$

7. $z = 2x^3 + y^3 - 3x^2 + 1.5y^2 - 12x - 90y$

8. $z = xy - \dfrac{1}{x} - \dfrac{1}{y}$

In Problems 9–20, locate all critical points and use the second derivative test to identify each as a relative maximum, relative minimum, or saddle point.

9. $f(x, y) = x^2 + y^2 - 3x$

10. $f(x, y) = x^2 + 2y^2 - 4x - 4y + 4$

11. $f(x, y) = x^2 + 2x - y^2 - 4y - 3$

12. $f(x, y) = x^2 + 2xy + 3y^2 + 2x + 10y + 7$

13. $f(x, y) = x^2 + y^2 + 13x - 12y - 3xy + 10$

14. $f(x, y) = y^3 + 6y + x^2 + 3x - 6xy - 5$

15. $f(x, y) = x^3 - 4x + 2xy + y^2 - 3y - 1$

16. $f(x, y) = 3x^2 y + x^2 - 6x - 3y - 4$

17. $f(x, y) = xy + \dfrac{4}{x} + \dfrac{2}{y}$

18. $f(x, y) = xy + \dfrac{8}{y} + \dfrac{8}{x}$

19. $f(x, y) = x^3 - 6xy + y^3$

20. $f(x, y) = (x - 1)^2 + y^3 - 3y^2 - 9y + 6$

21. Suppose $Q = 1.08x^2 - 0.03x^3 - 1.68y^2 - 0.08y^3$ is a production function for some company. Find the quantities of input, x and y, that maximize the output Q.

22. Does $f(x, y) = 10xy - 5x^2 - 6y^2 + 10x$ have any maximum or minimum points? If so, what are they?

23. A firm produces two products, x and y, selling for $\$25$ and $\$70$, respectively. The cost function is determined to be

$$C(x, y) = 2x^2 + y^2 - 2xy + 8x - 5y$$

How many of each product should be made to maximize profit?

24. The demand and cost functions for two products are

$$p_1(q_1, q_2) = 100 - 0.2q_1 - 0.1q_2$$
$$p_2(q_1, q_2) = 80 - 0.1q_1 - 0.2q_2$$
$$C(q_1, q_2) = 25q_1 + 38q_2$$

where p_1 is the unit price of product 1 and p_2 is the unit price of product 2 when the demand for products 1 and 2 are q_1 and q_2, respectively.
a. Find the quantities and unit prices that maximize profit.
b. Find the value of the maximum profit.

25. A consumer products company wishes to design a rectangular box having a capacity of 64 cubic inches. What should the dimensions be in order to use the minimum amount of material?

26. Turtle Express Overnite Delivery has a restriction that the length plus girth of any package cannot exceed 100 inches (see the figure). If a package is to have the maximum volume, what should its dimensions be?

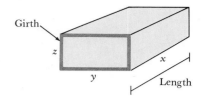

8-5 *LAGRANGE MULTIPLIERS*

In Chapter 3 we saw examples of optimization problems involving two variables and subject to a constraint. The forest fire model in the Introduction to Chapter 3 was one such example. We were able to solve that problem in Section 3-8 by using techniques of single-variable calculus, that is, by solving for one variable in the constraint and substituting into the function to be optimized. For instance, consider the problem

Minimize $f(x, y) = x^2 - 14x + y^2 - 16y$
Subject to $2x + 3y = 12$

Solving for x in the constraint yields

$$x = 6 - \tfrac{3}{2}y$$

If we substitute this expression into the function $f(x, y)$, we obtain a function of one variable, y:

$$h(y) = (6 - \tfrac{3}{2}y)^2 - 14(6 - \tfrac{3}{2}y) + y^2 - 16y$$
$$= \tfrac{13}{4}y^2 - 13y - 48$$

Setting the derivative of $h(y)$ equal to 0 yields

$$\frac{dh}{dy} = \frac{13}{2}y - 13 = 0$$

$$y = 2$$

The second derivative is $d^2h/dy^2 = 13/2 > 0$; hence, $y = 2$ minimizes $h(y)$. The value for x is determined from the constraint:

$$x = 6 - \tfrac{3}{2}y$$
$$= 6 - \tfrac{3}{2}(2) = 3$$

Sometimes, the nature of a constraint is such that we cannot solve explicitly for one variable. This is the case with the equation

$$xy^3 + 3y^2x + yx^3 - 1 = 0$$

Even if we can solve the constraint for one variable, the substitution may result in a very complicated function.

In this section we introduce the **method of Lagrange multipliers,** a general procedure for optimizing a function $f(x, y)$ subject to a constraint $g(x, y) = 0$.

This method was developed by Joseph Louis Lagrange, an 18th century mathematician. The idea behind the method is to eliminate the constraint, not by substitution, but by introducing a *new* variable, λ (the Greek letter lambda), called the *Lagrange multiplier*. We create a new function $F(x, y, \lambda)$, called the *Lagrangian function*, which is defined as

$$F(x, y, \lambda) = f(x, y) + \lambda g(x, y)$$

Notice that F is an unconstrained function of three variables and is equal to the original function to be optimized plus λ times the constraint function. The relationship between the original problem and the Lagrangian function is stated in the following theorem:

THEOREM

Consider the problem of minimizing or maximizing $f(x, y)$ subject to a constraint $g(x, y) = 0$. Suppose that (x^*, y^*) is an optimal solution. Then there is a value λ^* such that the partial derivatives of $F(x, y, \lambda)$, evaluated at (x^*, y^*, λ^*), are all equal to 0. That is,

$$\frac{\partial F}{\partial x} = \frac{\partial F}{\partial y} = \frac{\partial F}{\partial \lambda} = 0 \qquad \text{at } (x^*, y^*, \lambda^*)$$

The theorem suggests that if we set the partial derivatives $\partial F/\partial x$, $\partial F/\partial y$, and $\partial F/\partial \lambda$ equal to 0 and solve for x, y, and λ, we will obtain points that are candidates for the optimal solution to the problem of minimizing or maximizing $f(x, y)$, subject to $g(x, y) = 0$. First, we rewrite the constraint $2x + 3y = 12$ as $2x + 3y - 12 = 0$ in order to get it into the form $g(x, y) = 0$. The Lagrangian function is then

$$F(x, y, \lambda) = f(x, y) + \lambda g(x, y)$$
$$= x^2 - 14x + y^2 - 16y + \lambda(2x + 3y - 12)$$

The partial derivatives of F are

$$\frac{\partial F}{\partial x} = 2x - 14 + 2\lambda$$

$$\frac{\partial F}{\partial y} = 2y - 16 + 3\lambda$$

$$\frac{\partial F}{\partial \lambda} = 2x + 3y - 12$$

Thus, we have to solve a system of three equations in three unknowns:

$$\frac{\partial F}{\partial x} = 2x - 14 + 2\lambda = 0$$

$$\frac{\partial F}{\partial y} = 2y - 16 + 3\lambda = 0$$

$$\frac{\partial F}{\partial \lambda} = 2x + 3y - 12 = 0$$

Notice that $\partial F/\partial \lambda = 0$ is simply the same as the constraint $g(x, y) = 0$. Therefore, any solution to these equations will satisfy the constraint in the original problem. Such a system of equations is usually solved by substitution (refer back to Algebra Review 18). For example, let us first solve the third equation for x:

$$2x + 3y - 12 = 0$$
$$2x = -3y + 12$$
$$x = -\tfrac{3}{2}y + 6$$

Next, solve the second equation for λ:

$$'2y - 16 + 3\lambda = 0$$
$$3\lambda = -2y + 16$$
$$\lambda = -\tfrac{2}{3}y + \tfrac{16}{3}$$

Finally, substitute both of these expressions into the first equation:

$$2x - 14 + 2\lambda = 0$$
$$2(-\tfrac{3}{2}y + 6) - 14 + 2(-\tfrac{2}{3}y + \tfrac{16}{3}) = 0$$

Solving for y yields $y = 2$. Using the equations for x and λ above, we have

$$x = -\tfrac{3}{2}(2) + 6 = 3$$
$$\lambda = -\tfrac{2}{3}(2) + \tfrac{16}{3} = 4$$

Thus, $x = 3$, $y = 2$, and $\lambda = 4$ is a candidate for the optimal solution. We have not actually proven that this point yields a minimum. There is a version of the second derivative test that does this, but this is a topic for a more advanced text. Let it suffice to say that for the problems in this chapter, any solution obtained by this method will be optimal.

We may summarize the method of Lagrange multipliers as follows:

Method of Lagrange Multipliers

To minimize or maximize $f(x, y)$ subject to $g(x, y) = 0$:

Step 1 Form the Lagrangian function:

$$F(x, y, \lambda) = f(x, y) + \lambda g(x, y)$$

Step 2 Set the first partial derivatives of F equal to 0:

$$\frac{\partial F}{\partial x} = 0 \qquad \frac{\partial F}{\partial y} = 0 \qquad \frac{\partial F}{\partial \lambda} = 0$$

Step 3 Solve this system of equations for x, y, and λ.

Step 4 Check the function values at the points obtained in step 3.

Example 23 Solve the problem:

Minimize $f(x, y) = x^2 + 2y^2 - 8x - 12y$
Subject to $x^2 + 2y^2 = 5$

Solution We form the Lagrangian function:

$$F(x, y, \lambda) = x^2 + 2y^2 - 8x - 12y + \lambda(x^2 + 2y^2 - 5)$$

Next, set the first partial derivatives equal to 0:

$$\frac{\partial F}{\partial x} = 2x - 8 + 2\lambda x = 0$$

$$\frac{\partial F}{\partial y} = 4y - 12 + 4\lambda y = 0$$

$$\frac{\partial F}{\partial \lambda} = x^2 + 2y^2 - 5 = 0$$

Solve this system of equations. Although this can be done in several ways, we will solve the first two equations for λ. The first equation yields

$$\lambda = \frac{-2x + 8}{2x} = \frac{4 - x}{x}$$

From the second equation, we obtain

$$\lambda = \frac{-4y + 12}{4y} = \frac{3 - y}{y}$$

Therefore,

$$\frac{4 - x}{x} = \frac{3 - y}{y}$$
$$(4 - x)y = (3 - y)x$$
$$4y - xy = 3x - yx$$
$$y = \tfrac{3}{4}x$$

Substituting this expression into the last equation, we find

$$x^2 + 2(\tfrac{3}{4}x)^2 - 5 = 0$$
$$\tfrac{17}{8}x^2 = 5$$
$$x^2 = \tfrac{40}{17}$$
$$x = \pm\sqrt{\tfrac{40}{17}} \approx \pm 1.534$$

Substitution into the equation for y yields

$$y = \tfrac{3}{4}x = \pm\tfrac{3}{4}\sqrt{\tfrac{40}{17}} \approx \pm 1.150$$

Therefore, the two points $(\sqrt{\tfrac{40}{17}}, \tfrac{3}{4}\sqrt{\tfrac{40}{17}})$ and $(-\sqrt{\tfrac{40}{17}}, -\tfrac{3}{4}\sqrt{\tfrac{40}{17}})$ are candidates for points at which a minimum occurs. Checking the function values at these points, we find

$$f(\sqrt{\tfrac{40}{17}}, \tfrac{3}{4}\sqrt{\tfrac{40}{17}}) \approx -20.072$$
$$f(-\sqrt{\tfrac{40}{17}}, -\tfrac{3}{4}\sqrt{\tfrac{40}{17}}) \approx 30.072$$

Hence, the minimum value of -20.072 occurs at $x \approx 1.534$ and $y \approx 1.150$. ∎

Example 24
Pest Control   A farm is sprayed with a mixture of x pounds of one pesticide and y pounds of another. Agricultural scientists have determined that the fraction of pests killed

will be

$$p(x, y) = 1 - \tfrac{1}{2}e^{-x/40} - \tfrac{1}{2}e^{-y/80}$$

The airplane used to spray the farm has a 100 pound limit. How much of each pesticide should be used in order to maximize the fraction of pests killed?

Solution   We have the problem:

Maximize $p(x, y) = 1 - \tfrac{1}{2}e^{-x/40} - \tfrac{1}{2}e^{-y/80}$
Subject to $x + y = 100$

We form the Lagrangian function:

$$F(x, y, \lambda) = 1 - \tfrac{1}{2}e^{-x/40} - \tfrac{1}{2}e^{-y/80} + \lambda(x + y - 100)$$

Taking partial derivatives of F, we have

$$\frac{\partial F}{\partial x} = \left(\frac{1}{2}\right)\left(\frac{1}{40}\right)e^{-x/40} + \lambda = 0$$

$$\frac{\partial F}{\partial y} = \left(\frac{1}{2}\right)\left(\frac{1}{80}\right)e^{-y/80} + \lambda = 0$$

$$\frac{\partial F}{\partial \lambda} = x + y - 100 = 0$$

Solve the third equation for x, the second for λ, and then substitute the expressions found into the first equation:

$$x = 100 - y$$
$$\lambda = -\tfrac{1}{160}e^{-y/80}$$
$$\tfrac{1}{80}e^{-(100-y)/40} - \tfrac{1}{160}e^{-y/80} = 0$$

We simplify this last equation and take logarithms to solve:

$$2e^{(-100+y)/40} = e^{-y/80}$$

$$\ln 2 + \left(\frac{-100}{40} + \frac{y}{40}\right) = \frac{-y}{80}$$

$$\ln 2 - \frac{100}{40} = \frac{-3y}{80}$$

$$80 \ln 2 - 200 = -3y$$

$$y = \frac{200 - 80 \ln 2}{3}$$

$$\approx 48.182$$

Since $x = 100 - y$, $x \approx 100 - 48.182 = 51.818$. The value of λ is computed as

$$\lambda = -\tfrac{1}{160}e^{-y/80} \approx -0.003$$

The desired maximum value is $p(x, y) \approx 0.862$. Therefore, if about 48 pounds of the first pesticide is mixed with 52 pounds of the second pesticide, then about 86% of the pests will be eradicated.   ∎

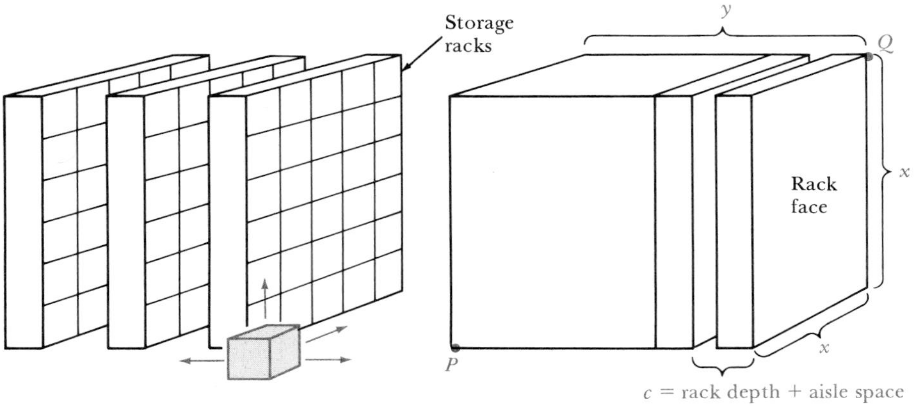

Figure 8-17
Illustration of automated storage
and retrieval system (AS/RS)

a. Storage and retrieval unit   b. Simplified diagram of AS/RS
   can move in three dimensions

Example 25
Automated Storage and
Retrieval

Automation has become an important aspect of industrial productivity. In a warehouse, for example, automated storage and retrieval systems (AS/RS) are commonly used to store and remove bulk products such as bags of dog food or cases of canned goods. An AS/RS typically consists of racks of storage locations and a computer-controlled storage and retrieval unit that moves horizontally and vertically within an aisle to store or retrieve goods (see Figure 8-17a). Suppose that a company wishes to design such a system consisting of square storage racks having dimension x feet and total length of y feet, as shown in Figure 8-17b. The depth of each storage rack is fixed. A measure of storage capacity is therefore given by the total amount of rack face area. The farthest that the storage and retrieval unit has to travel to a location is from point P to point Q in Figure 8-17b. Assume that the horizontal speed of the retrieval unit is 2 feet per minute and the vertical speed is 4 feet per minute. The company wishes to minimize the total time required to go from point P to point Q. Since the unit must travel a total horizontal distance of $x + y$ feet and a total vertical distance of x feet, the total time required is

$$\frac{x + y}{2} + \frac{y}{4}$$

Therefore, the problem of minimizing the time is

$$\text{Minimize } f(x, y) = \frac{x + y}{2} + \frac{x}{4}$$

If n storage racks are used, the amount of usable rack face is nx^2. From Figure 8-17b, we see that $y = nc$, or $n = y/c$, where c is the rack depth plus aisle width. Therefore, if an amount of A square feet of usable rack face is desired, we have the constraint

$$nx^2 = A$$

or

$$yx^2 = cA, \text{ since } n = y/c.$$

Thus, the problem can be modeled as

Minimize $f(x, y) = \dfrac{x + y}{2} + \dfrac{x}{4}$

Subject to $yx^2 = cA$

To solve this using the method of Lagrange multipliers, we write the Lagrangian function:

$$F(x, y, \lambda) = \frac{x + y}{2} + \frac{x}{4} + \lambda(yx^2 - cA)$$

Taking partial derivatives, we have

$$\frac{\partial F}{\partial x} = \frac{1}{2} + \frac{1}{4} + 2\lambda yx = 0$$

$$\frac{\partial F}{\partial y} = \frac{1}{2} + \lambda x^2 = 0$$

$$\frac{\partial F}{\partial \lambda} = yx^2 - cA = 0$$

From the second equation, we have

$$\lambda = \frac{-1}{2x^2}$$

Substituting this expression into the first equation, we find

$$\frac{3}{4} + 2\left(\frac{-1}{2x^2}\right)yx = 0$$

$$\frac{3}{4} = \frac{y}{x}$$

$$y = \tfrac{3}{4}x$$

Substituting this expression into the third equation gives

$$(\tfrac{3}{4}x)x^2 - cA = 0$$

$$x^3 = \tfrac{4}{3}cA$$

$$x = \sqrt[3]{4cA/3}$$

The equation for y then gives

$$y = \tfrac{3}{4}x = \tfrac{3}{4}\sqrt[3]{4cA/3}$$

So, for example, if $A = 2000$ square feet and $c = 12$ feet then the dimensions

$$x = \sqrt[3]{(4)(12)(2000)/3} = \sqrt[3]{32{,}000} \approx 31.748 \text{ feet}$$

$$y = \tfrac{3}{4}x \approx 23.811 \text{ feet}$$

will result in quickest access from point P to point Q.   ∎

INTERPRETATION OF LAGRANGE MULTIPLIERS

The constraint $g(x, y) = 0$ is often derived from a constraint of the form $h(x, y) = b$. For instance, in Example 24, the weight limit of the airplane was expressed as $x + y = 100$. Therefore, if we have a constraint of the form $h(x, y) = b$, then we take $g(x, y) = h(x, y) - b = 0$ in the method of Lagrange multipliers. The Lagrangian function becomes

$$F(x, y, \lambda) = f(x, y) + \lambda[h(x, y) - b]$$
$$= f(x, y) + \lambda h(x, y) - \lambda b$$

Suppose we treat b as a variable. Then the partial derivative of F with respect to b is

$$\frac{\partial F}{\partial b} = -\lambda$$

For a given value of b, the optimal values x^*, y^*, λ^* of x, y, λ are functions of b. Hence, since $h(x^*, y^*) = b$, we may write

$$F(b) = F(x^*, y^*, \lambda^*) = f(x^*, y^*)$$

to express the optimal value of F as a function of the variable b. We are interested in how the optimal value of the objective function $f(x^*, y^*)$ changes with respect to b. From the two equations above, we see that

$$\frac{d}{db} f(x^*, y^*) = \frac{d}{db} F(b) \approx \frac{\partial F}{\partial b} = -\lambda^*$$

This states that at the optimal solution, the rate of change of the function f with respect to b is equal to the negative value of the Lagrange multiplier. We can interpret this to mean that if b is increased by 1 unit, then f will change by approximately $-\lambda^*$ units. In Example 24, for instance, $\lambda^* = -0.003$. This means that a 1 pound increase in the airplane's capacity will result in approximately $-(-0.003) = 0.3\%$ increase in the fraction of pests killed. Such an interpretation is extremely useful in business and economics and other applications where changes in resource availabilities are of interest.

Example 26
Optimal Product Mix

A firm has determined that total profit (in thousands of dollars) as a function of the number of units of two products manufactured is given by

$$p(x, y) = -x^2 - 4y^2 + 6x + 16y + 1000$$

where x and y are the number of units (in thousands) of the two products. In the next planning period, only 8000 pounds of a certain ingredient is available to produce these products. Both products require 2 pounds of this ingredient per unit. What is the optimal product mix, and how much is 1 additional pound of the ingredient worth?

Solution

We write the constraint as

$$2000x + 2000y = 8000$$

(Remember that x and y are in thousands of units.) The Lagrangian function is therefore

$$F(x, y, \lambda) = -x^2 - 4y^2 + 6x + 16y + 1000 + \lambda(2000x + 2000y - 8000)$$

Taking partial derivatives and setting them equal to 0, we obtain

$$\frac{\partial F}{\partial x} = -2x + 6 + 2000\lambda = 0$$

$$\frac{\partial F}{\partial y} = -8y + 16 + 2000\lambda = 0$$

$$\frac{\partial F}{\partial \lambda} = 2000x + 2000y - 8000 = 0$$

The solution to these equations is $x^* = 2.2$, $y^* = 1.8$, and $\lambda^* = -0.0008$. The company should produce 2200 units of the first product and 1800 units of the second product in the next planning period. An additional pound of the ingredient would change the value of profit by approximately $-\lambda^* = -(-0.0008) = 0.0008$ thousand dollars or $0.80. ∎

PROBLEMS (Section 8-5)

In Problems 1 – 10, use the method of Lagrange multipliers to find critical points of each function subject to the given constraint.

1. $f(x, y) = 4xy$; $x + y = 8$

2. $f(x, y) = x^2y$; $2x + y = 6$

3. $f(x, y) = x^2 + 2y^2 - xy$; $x + y = 8$

4. $f(x, y) = 25 - x^2 - y^2$; $2x + y = 10$

5. $f(x, y) = xy$; $x + y = 100$

6. $f(x, y) = x^2 + y^2$; $x + 3y = 4$

7. $f(x, y) = x^2 + y^2 - xy$; $x - y = 10$

8. $f(x, y) = 6x^2 + 5y^2 - xy$; $2x + y = 20$

9. $f(x, y) = x^2 + 4y^2 + 6$; $2x - 8y = 20$

10. $f(x, y) = -2x^2 + 5y^2 + 7$; $3x - 2y = 7$

11. A park district has 600 yards of fencing. Use the method of Lagrange multipliers to find the dimensions of the rectangular field of maximum area that can be enclosed by this amount of fencing.

12. A computer company produces two different computer models. If x units of the first model and y units of the second model are produced each month, the total cost is $C(x, y) = 8x^2 + 16y^2$. If it is possible to produce only 100 units of both models due to assembly line capacity, how many of each model should be produced to minimize cost? Use the method of Lagrange multipliers.

13. To fill an order for 500 units of its product, a company wants to distribute the production between its two assembly plants. The total cost function is

$$C = f(x, y) = 0.1x^2 + 7x + 15y + 1000$$

where x and y are the number of units produced at plants 1 and 2, respectively. How should the output be distributed in order to minimize cost?

14. A department store wishes to partition off a rectangular area along a wall for storage (see the figure). If only 20 feet of material are available to construct the partition, how can this be done to maximize the available storage area?

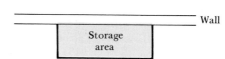

8-6 CONCLUSION: DETERMINING OPTIMAL TELEPHONE PRICING

In the Introduction to this chapter we developed a model for determining the optimal prices for rotary and push-button telephones in order to maximize total revenue, based on price–demand data. The model we developed was

Maximize
$$r(p_1, p_2) = -0.6p_1^2 - 0.4p_2^2 + 0.33p_1p_2 + 19.5p_1 + 30.1p_2$$

Since this is a function of two variables, p_1 and p_2, we can use the methods of multivariable calculus developed in this chapter to maximize this function.

To find critical points, we set the first partial derivatives equal to 0:

$$\frac{\partial r}{\partial p_1} = -1.2p_1 + 0.33p_2 + 19.5 = 0$$

$$\frac{\partial r}{\partial p_2} = -0.8p_2 + 0.33p_1 + 30.1 = 0$$

Solving the first equation for p_1, we have

$$p_1 = \frac{0.33p_2 + 19.5}{1.2}$$
$$= 0.275p_2 + 16.25$$

Substituting this expression into the second equation, we find

$$-0.8p_2 + 0.33(0.275p_2 + 16.25) + 30.1 = 0$$
$$-0.8p_2 + 0.09075p_2 + 5.3625 + 30.1 = 0$$
$$0.70925p_2 = 35.4625$$
$$p_2 = 50$$

Therefore,

$$p_1 = 0.275(50) + 16.25 = 30$$

The value of the revenue is thus $r(30, 50) = 1045$.

To determine whether the point $(30, 50, 1045)$ is indeed a relative maximum point, we employ the second derivative test. The second partial derivatives are

$$\frac{\partial^2 r}{\partial p_1^2} = -1.2$$

$$\frac{\partial^2 r}{\partial p_2^2} = -0.8$$

$$\frac{\partial^2 r}{\partial p_1 \partial p_2} = 0.33$$

Thus,

$$K = (-1.2)(-0.8) - (0.33)^2 = 0.8511$$

By the second derivative test, we see that the point $(30, 50, 1045)$ is a relative maximum point. Therefore, The Phone Store should price its rotary and push-button phones at \$30 and \$50, respectively.

8-7 CHAPTER REVIEW

1. A multivariable function is a rule that associates with each set of values of two or more independent variables a unique value of the dependent variable.
2. Graphs of two-variable functions are surfaces in the three-dimensional coordinate system.
3. The derivative of a multivariable function with respect to one independent variable, treating all other independent variables as constants, is called a partial derivative.
4. A partial derivative is the rate of change of the dependent variable with respect to a single independent variable when all other independent variables are held constant.
5. Geometrically, the value of the partial derivative $\partial z / \partial x$ is the slope of the tangent line to the surface $z = f(x, y)$ that is parallel to the xz-plane; the value of $\partial z / \partial y$ is the slope of the tangent line to the surface that is parallel to the yz-plane.

6. There are three second partial derivatives of the two-variable function $z = f(x, y)$: $\partial^2 f / \partial x^2$, $\partial^2 f / \partial y^2$, and the mixed partial derivative $\partial^2 f / \partial x \partial y$ (or $\partial^2 f / \partial y \partial x$).

7. If a function $z = f(x, y)$ has a relative maximum or minimum at $(a, b, f(a, b))$, then

$$\frac{\partial f}{\partial x}(a, b) = \frac{\partial f}{\partial y}(a, b) = 0$$

8. The second derivative test for multivariable func-

tions is used to determine whether a critical point is a relative minimum, relative maximum, or saddle point. If $K = 0$ in the test, then no conclusion can be reached.

9. The method of Lagrange multipliers is a procedure for optimizing $f(x, y)$ subject to a constraint $g(x, y) = 0$. If $g(x, y) = h(x, y) - b = 0$, the optimal value of the Lagrange multiplier represents the negative rate of change of the value of the function $f(x, y)$ with respect to an increase in b.

REVIEW PROBLEMS (Section 8-7)

Let $z = f(x, y) = 3x^2 - 2y^2 - 4xy + 15$. Evaluate the following:

1. $f(4, 0)$

2. $f(0, -3)$

3. $f(2, 3)$

4. $f(-2, -4)$

5. $f(c^2, c^3)$

6. $f(x + h, y)$

7. A closed wooden box is to be constructed with a volume of 3 cubic feet. The cost of wood is 20 cents per square foot. Express the total cost of the box as a function of the base dimensions x and y.

In Problems 8 – 11, plot each point on a three-dimensional coordinate system.

8. $(3, 1, 0)$

9. $(2, 2, 4)$

10. $(0, -5, -2)$

11. $(1, 3, 5)$

In Problems 12 – 14, sketch each function in three dimensions.

12. $z = 5 - 3x + y$

13. $z = x^2 + y - 6$

14. $z = x^2 + y^2 - 16$

15. A boat company has found that the price z (in dollars) for a boat depends on the number of boats demanded, x, and the price of inboard motors purchased from an external supplier, y. The function relating z, x, and y is

$$z = f(x, y) = -\tfrac{1}{2}x + 4y + 15{,}000$$

Find the price of a boat when $x = 50$ and $y = 450$.

In Problems 16–23, compute the partial derivatives $\partial f / \partial x$ and $\partial f / \partial y$ for each function.

16. $f(x, y) = x^3 - 2x^2 y + 3y$

17. $f(x, y) = 3x^2 + x^2 y^2 - 5y^4$

18. $f(x, y) = x^3 y^4 + 3x^2 y^3 + 5x + 2y^2 - 3xy - 6$

19. $f(x, y) = \sqrt{x - y^2}$

20. $f(x, y) = \dfrac{x^3 + 2x}{y - 4}$

21. $f(x, y) = \ln xy + e^{xy}$

22. $f(x, y) = \ln(2x^2 + 3y^2)$

23. $f(x, y) = (x^3 + 2y^2)^4$

24. Assume that a furniture store's profit is dependent on the number of salespeople, x, and the amount of inventory in stock for immediate delivery, y (in thousands of dollars). If the profit is described by the function

$$P = f(x, y) = 1200 - (10 - x)^2 - (30 - y)^2$$

find $\partial P/\partial x$ and $\partial P/\partial y$. Explain the meaning of these partial derivatives.

25. Find the slopes of the tangent lines parallel to the xz-plane and yz-plane for the function

$$z = f(x, y) = 2x^2 + 3xy + 2y^2$$

at the point $(2, 2, 28)$.

In Problems 26 – 33, find all second-order partial derivatives for each function.

26. $f(x, y) = 3x^2 + 4xy - 2y^2$

27. $f(x, y) = 2x^4 + 2x^3y + 4x^2y^2 - 5xy^3 - 3y^4$

28. $f(x, y) = e^{3xy} + 5x^2y^2$

29. $f(x, y) = 2x \ln y + e^{xy}$

30. $f(x, y) = (x^2 + 2y)^2 - (x + y)^3$

31. $f(x, y) = \ln(x^2 + y) + 8$

32. $f(x, y) = (x^2 + xy + 3)(2x^2 - xy - 4)$

33. $f(x, y) = \dfrac{(x + y)^3}{2xy}$

In Problems 34 – 36, find all critical points of each function.

34. $z = x^2 + 4y^2 - 6x + 16y$

35. $z = x^2 + 2y^2 - 6x - 3y - 5$

36. $z = x^2 + 2y^2 + 2xy - 4x + 3y - 6$

In Problems 37 – 42, determine any relative maximum, relative minimum, or saddle points and verify by the second derivative test.

37. $f(x, y) = x^2 + 3y^2 + 4x - 9y + 5$

38. $f(x, y) = -2x^2 + 8x - 3y^2 + 24y - 10$

39. $f(x, y) = 2x^2 + 2y^2 + 2xy - 18x + 4$

40. $f(x, y) = 2x^3 - 6xy + 2y^3$

41. $f(x, y) = 1.5x^2 - 2xy - 2x + 1.5y^2 - 7y$

42. $f(x, y) = 200x - 0.1x^2 + 150y - 0.25y^2$

43. The labor cost to repair an automobile is

$$C(x, y) = x^2 + y^3 - 2x - 3y - 2xy + 50$$

where x is the number of hours required by an apprentice mechanic and y is the number of hours required by a senior mechanic. Find the values of x and y that will minimize cost.

In Problems 44 – 47, use the method of Lagrange multipliers to find the critical points of the given function subject to the constraint.

44. $f(x, y) = 3x^2 + 2y^2$; $2x + 4y = -5$

45. $f(x, y) = x^2 + xy + 2y^2$; $x - 3y = 16$

46. $f(x, y) = 2x^2 + 3y^2 + xy$; $x + y = 4$

47. $f(x, y) = x^2 + 2x + 2y^2 + 3y + 2xy$; $x^2 - y = 1$

48. The total profit (in thousands of dollars) from the sales of two products is estimated to be

$$f(x, y) = 60x + 30y - 2x^2 - 3y^2$$

where x and y are the number of units (in thousands) produced and sold of products 1 and 2, respectively. If each 1000 units of product 1 requires 2000 pounds of a scarce alloy, and each 1000 units of product 2 requires 3000 pounds, find the maximum profit that can be achieved if only 30,000 pounds of the alloy can be obtained. Interpret the meaning of the value of the Lagrange multiplier.

INTEGRATION OF MULTIVARIABLE FUNCTIONS

Recall that for a nonnegative function of one variable, $y = f(x)$, the definite integral

$$\int_a^b f(x)\, dx$$

gives the area between the curve and the x-axis. The limits of integration define the *interval* over which the independent variable x varies. We integrate $f(x)$ with respect to x in order to find an antiderivative, and then apply the fundamental theorem of calculus to compute the area. In this supplement, we extend this idea to functions of two variables.

8S-1 DOUBLE INTEGRALS

A function of two variables, $z = f(x, y)$, represents a surface in three-dimensional space. The corresponding notion of a definite integral for nonnegative $f(x, y)$ gives the *volume* between the surface and the xy-plane. Since there are two independent variables, x and y, we need to integrate with respect to both of them. In this case, the limits of integration define a **region** A in the xy-plane over which the two independent variables vary. This gives the **double integral:**

$$\iint_A f(x, y)\, dx\, dy$$

For $f(x, y) \geq 0$, the value of this integral represents the volume under the surface $z = f(x, y)$ for values of x and y that are restricted to the region A in the xy-plane. Figure 8S-1 illustrates the analogy between single and double integrals.

In the simple case in which the region A is a rectangle, the double integral can be computed by a two-step process. We first consider either x or y as constant

Figure 8S-1
Geometric analogy of single and double integrals

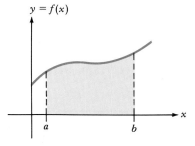

a. Area $= \int_a^b f(x)\, dx$

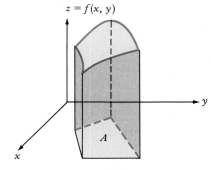

b. Volume $= \iint_A f(x, y)\, dx\, dy$

and integrate with respect to the other variable. After this integral is evaluated, we integrate with respect to the variable that was first held constant.

Example 1   Calculate the double integral:

$$\int_3^5 \left[\int_1^2 3x^2y \, dx \right] dy$$

Solution   We first consider the inner integral:

$$\int_1^2 3x^2y \, dx$$

Treating y as a constant, we obtain

$$\int_1^2 3x^2y \, dx = y \int_1^2 3x^2 \, dx = yx^3|_1^2$$
$$= 8y - y$$
$$= 7y$$

Next, we substitute this expression in the original integral and integrate with respect to y:

$$\int_3^5 7y \, dy = \frac{7y^2}{2} \Big|_3^5$$
$$= \tfrac{175}{2} - \tfrac{63}{2} = 56$$

The value of this double integral represents the volume under the surface $z = 3x^2y$ over the rectangle defined by the limits of integration, namely $1 \le x \le 2$ and $3 \le y \le 5$.   ∎

The double integral above is a special case of an iterated integral. This topic is treated more generally in the next section.

8S-2 ITERATED INTEGRALS

The limits of integration need not be constants, but may be functions of one of the variables. In this section, we consider a special type of integral called the *iterated integral*.

An **iterated integral** has the form

$$\int_a^b \left[\int_{g(x)}^{h(x)} f(x, y) \, dy \right] dx \qquad \text{or} \qquad \int_c^d \left[\int_{g(y)}^{h(y)} f(x, y) \, dx \right] dy$$

Notice that the limits on the inner integral are functions of x or y. When we evaluate

$$\int_{g(x)}^{h(x)} f(x, y)\, dy$$

x is held constant and we obtain a function of x alone, say $k(x)$. We then integrate $k(x)$ with respect to x over the interval (a, b), that is,

$$\int_a^b k(x)\, dx = \int_a^b \left[\int_{g(x)}^{h(x)} f(x, y)\, dy \right] dx$$

Example 2   Evaluate the integral:

$$\int_0^1 \left[\int_x^{x^2} (x - 2y)\, dy \right] dx$$

Solution   We treat x as a constant and evaluate the inner integral:

$$\int_x^{x^2} (x - 2y)\, dy = (xy - y^2)|_x^{x^2}$$

Note that since we have integrated with respect to y, we substitute the limits of integration for y, not for x, in this step. This yields

$$(xy - y^2)|_x^{x^2} = [x(x^2) - (x^2)^2] - [x(x) - x^2]$$
$$= x^3 - x^4 - x^2 + x^2$$
$$= x^3 - x^4$$

Finally, we evaluate

$$\int_0^1 (x^3 - x^4)\, dx = \left(\frac{x^4}{4} - \frac{x^5}{5} \right) \Big|_0^1$$
$$= \tfrac{1}{4} - \tfrac{1}{5} = \tfrac{1}{20} \qquad \blacksquare$$

The iterated integral has a useful application for computing certain volumes under surfaces. Suppose that a region A in the xy-plane is bounded by the functions $h(x)$ and $g(x)$ and the interval $a \le x \le b$, as shown in Figure 8S-2.

Figure 8S-2
Area in the xy-plane

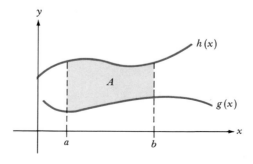

Then we can state the following result:

Let A be the region in the xy-plane bounded by the graphs of the functions $y = h(x)$, $y = g(x)$, and the lines $x = a$, $x = b$. Then if $f(x, y) \geq 0$ for (x, y) in A, the volume under the surface $z = f(x, y)$ over the region A can be computed by means of the iterated integral. That is,

$$\iint\limits_{A} f(x, y)\, dy\, dx = \int_{a}^{b} \left[\int_{g(x)}^{h(x)} f(x, y)\, dy \right] dx$$

Example 3 Compute the area under the surface $z = 6 - 2x - 3y$ and above the first quadrant of the xy-plane.

Solution This surface is shown in Figure 8S-3a. The area A in the xy-plane is shown in Figure 8S-3b. From Figure 8S-3a, we see that the line between the points $(3, 0, 0)$ and $(0, 2, 0)$ is $2x + 3y = 6$. We need to write this as a function of x:

$$3y = 6 - 2x$$
$$y = 2 - \frac{2x}{3}$$

Therefore, A is defined by the functions

$$g(x) = 0$$
$$h(x) = 2 - \frac{2x}{3}$$

and the lines $x = 0$ and $x = 3$. The volume under the surface is given by

$$\int_{0}^{3} \left[\int_{0}^{2-(2x/3)} (6 - 2x - 3y)\, dy \right] dx = \int_{0}^{3} [(6y - 2xy - \tfrac{3}{2}y^2)|_{0}^{2-(2x/3)}]\, dx$$

$$= \int_{0}^{3} \left[6\left(2 - \frac{2x}{3}\right) - 2x\left(2 - \frac{2x}{3}\right) - \frac{3}{2}\left(2 - \frac{2x}{3}\right)^2 - 0 \right] dx$$

$$= \int_{0}^{3} [12 - 4x - 4x + \tfrac{4}{3}x^2 - \tfrac{3}{2}(4 - \tfrac{8}{3}x + \tfrac{4}{9}x^2)]\, dx$$

$$= \int_{0}^{3} (6 - 4x + \tfrac{2}{3}x^2)\, dx$$

$$= (6x - 2x^2 + \tfrac{2}{9}x^3)|_{0}^{3}$$
$$= 18 - 18 + \tfrac{2}{9}(27) - 0 = 6$$

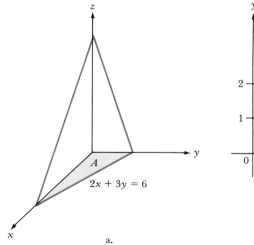

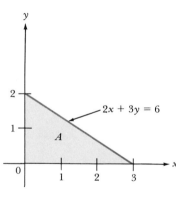

Figure 8S-3

a.

b.

■

PROBLEMS (Supplement to Chapter 8)

In Problems 1–10, evaluate each double integral.

1. $\int_3^5 \left[\int_1^2 4x^3y \, dx \right] dy$

2. $\int_5^6 \left[\int_0^3 3x^2y^2 \, dx \right] dy$

3. $\int_2^3 \left[\int_0^1 (x^2 + 3y^3) \, dx \right] dy$

4. $\int_3^5 \left[\int_1^2 (x^3 - 2y^3) \, dx \right] dy$

5. $\int_3^5 \left[\int_0^2 (x^2y - 4y) \, dx \right] dy$

6. $\int_3^4 \left[\int_0^2 (xy + 2y^2) \, dy \right] dx$

7. $\int_2^4 \left[\int_0^1 (xy - y^2) \, dx \right] dy$

8. $\int_1^4 \left[\int_2^5 (x^2 - y^2 + xy - 3) \, dx \right] dy$

9. $\int_0^2 \left[\int_{-3}^2 (x^3 + 2x^2y - y^3 + xy) \, dy \right] dx$

10. $\int_0^2 \left[\int_{-1}^1 (x^3 + x^2y + 2y^3 - xy) \, dy \right] dx$

In Problems 11–18, evaluate each iterated integral.

11. $\int_1^3 \left[\int_0^y x^2y^2 \, dx \right] dy$

12. $\int_0^1 \left[\int_{x^3}^{x^2} (x^2 - xy) \, dy \right] dx$

13. $\int_1^4 \left[\int_{\sqrt{x}}^{x^2} (x^2 + 2xy - 3y^2) \, dy \right] dx$

14. $\int_0^4 \left[\int_{1+y}^{\sqrt{y}} (x^2y + xy^2) \, dx \right] dy$

15. $\int_1^2 \left[\int_0^y 2x^2y^2 \, dx \right] dy$

16. $\int_0^2 \left[\int_{x^2}^{x^3} (x^2 + 2xy) \, dy \right] dx$

17. $\int_1^3 \left[\int_{\sqrt{x}}^{x^2} (x^2 - xy + 2y^2) \, dy \right] dx$

18. $\int_0^2 \left[\int_{1+y}^{\sqrt{y}} (x^2y - 2xy) \, dx \right] dy$

TRIGONOMETRIC FUNCTIONS

INTRODUCTION

Today it is generally recognized that long-term national security is only as strong as the long-term national economy. In turn, the national economy depends upon the extent to which a country can implement efficient high-technology methods and hardware in its industrial plants. Currently, *robotics* is one of the most rapidly developing fields of industrial technology. Industrial robots have been put to use in many industries, performing such jobs as spot welding, spray painting, and material handling. They will certainly play a central role in the manufacturing plant of the future.

Consider a robotic arm equipped with a laser scanner that rotates on a base fixed to the factory floor 6 feet from an assembly line. Suppose units are manufactured and pass down the assembly line at a rate of 12 feet per minute. The function of the robot is to track the leading edge of a unit from the point at which it is directly opposite the robot to a point 60 feet down the line (see Figure 9-1, page 432). In order to do this, the robotic arm must be programmed to rotate at the appropriate rate to keep track of the unit. What is this rate? That is, what is the rate at which the angle θ changes with time?

Referring to Figure 9-1, it can be shown that the position of the unit, x, is related to the angle θ by

$$x = 6 \tan \theta \qquad 0 \leq x \leq 60$$

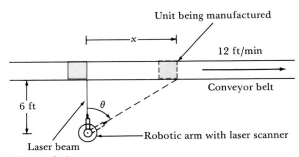

Figure 9-1
Control of a robotic laser scanner
(overhead view)

where x and θ are both increasing functions of time. The function $f(\theta) = \tan \theta$ is an example of a **trigonometric function** (the tangent function). The trigonometric functions are specifically designed to deal with questions of angles and rotation. If we differentiate both sides of this equation with respect to t (time), then we obtain (by use of the chain rule)

$$\frac{dx}{dt} = 6\left[\frac{d}{d\theta}\tan\theta\right]\frac{d\theta}{dt}$$

Since we know $dx/dt = 12$, and we wish to find $d\theta/dt$, it is necessary to know the derivative of the tangent function. In this chapter we present a systematic development of the trigonometric functions, including the calculus of the trigonometric functions.

9-1 ANGULAR MEASURE

The origins of trigonometry can be traced to ancient times, when astronomers thought the planets moved in circles and it was necessary to have some standard method of measuring an arc of a circle. The Babylonians, who pioneered trigonometry, used a sexagesimal (base 60) number system and divided the full arc of the circle into 360 ($= 6 \times 60$) standard units called **degrees.** We therefore say that the full circle contains 360 degrees, denoted 360°, the half circle 180°, and so on (see Figure 9-2).

Figure 9-2
Degree measure

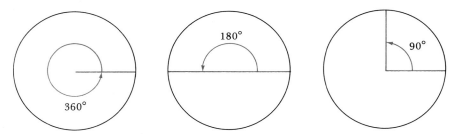

If we consider a circle of radius 1, we can also measure off arcs along the circle simply by marking off a length along the circumference of the circle. This turns out to be a much more convenient way of measuring arcs from the point of view of calculus. The unit of measure for a circular arc in this system is called a **radian.** One radian corresponds to an arc 1 unit long (that is, 1 radius) along the arc of a circle with radius 1. Since the length of the circumference of a circle of radius 1 is 2π, we see that the full arc of a circle has measure 2π radians. This simple observation gives us the basic relationship between radian measure and degree measure:

2π radians $= 360°$

Since the full arc of a circle has measure 2π radians, three-fourths of the circular arc has measure

$$\frac{3}{4} \times 2\pi = \frac{3\pi}{2} \text{ radians}$$

One-half of the circular arc has measure

$$\frac{1}{2} \times 2\pi = \pi \text{ radians}$$

One-quarter of the circular arc has measure

$$\frac{1}{4} \times 2\pi = \frac{\pi}{2} \text{ radians}$$

And so on (see Figure 9-3).

Figure 9-3
Radian measure

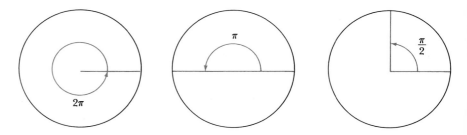

The relationship 2π radians $= 360°$ allows us to convert from radians to degrees and vice versa. If we divide both sides of this equation by 2π, we obtain

$$1 \text{ radian} = \frac{2\pi}{2\pi} \text{ radians} = \frac{360°}{2\pi} = \frac{180°}{\pi}$$

Therefore, if a is any number, then

$$a \text{ radians} = a \times \frac{180}{\pi} \text{ degrees}$$

Multiplying both sides of this equation by $\frac{\pi}{180}$ results in

$$a \text{ degrees} = a \times \frac{\pi}{180} \text{ radians}$$

Example 1 Convert $30°$, $45°$, and $315°$ to radians. Also convert $\pi/12$, $3\pi/4$, and $5\pi/9$ radians to degrees.

Solution   We use the conversion formulas above:

$$30° = 30 \times \frac{\pi}{180} \text{ radians} = \frac{\pi}{6} \text{ radian}$$

$$45° = 45 \times \frac{\pi}{180} \text{ radians} = \frac{\pi}{4} \text{ radian}$$

$$315° = 315 \times \frac{\pi}{180} \text{ radians} = \frac{7\pi}{4} \text{ radians}$$

$$\frac{\pi}{12} \text{ radian} = \frac{\pi}{12} \times \frac{180}{\pi} \text{ degrees} = 15°$$

$$\frac{3\pi}{4} \text{ radians} = \frac{3\pi}{4} \times \frac{180}{\pi} \text{ degrees} = 135°$$

$$\frac{5\pi}{9} \text{ radians} = \frac{5\pi}{9} \times \frac{180}{\pi} \text{ degrees} = 100°$$

As we mentioned earlier, we will almost always use radian measure instead of degree measure. If you use a calculator to verify examples or work problems, make sure that it is in radian mode rather than degree mode when you are using radian measure.

We will usually speak of angles rather than arcs. By an **angle** we mean the region swept out by sweeping a radial segment through a given arc. The original position of the segment is called the **initial side** of the angle, and the final position is called the **terminal side** (see Figure 9-4). We allow angles with negative as well as positive measure. Angles with positive measure are swept out from the initial side to the terminal side in the counterclockwise sense, while angles with negative measure are swept out from the initial side to the terminal side in the clockwise sense. Also, we will allow angles of any magnitude. Angles with magnitude greater than 2π radians will simply correspond to the terminal side making more than one revolution (see Figure 9-5).

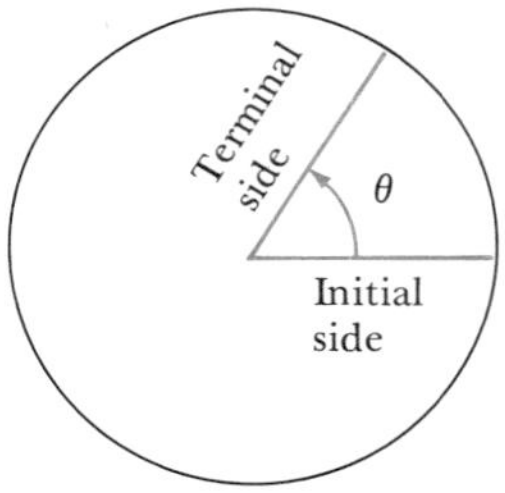

Figure 9-4
The angle θ

Figure 9-5
Positive and negative angles

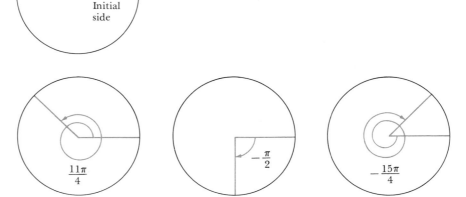

Example 2 Suppose an automobile rolls forward from left to right so that its wheel makes six and three-quarters revolutions. What angle is swept out by the valve stem on the tire?

Solution Since each revolution of the wheel is equal to 2π radians of angular measure, the magnitude of the angle is

$$6\tfrac{3}{4} \times 2\pi = \frac{27}{4} \times 2\pi = \frac{27\pi}{2} \text{ radians}$$

However, when the car moves forward the tire rotates in a clockwise sense; therefore, the angle has negative measure and the angle swept out by the valve stem is $-27\pi/2$ radians. ∎

PROBLEMS (Section 9-1)

In Problems 1–4, convert the given degree measure to radian measure.

1. a. 16°   b. 120°   c. 450°

2. a. 75°   b. 20°   c. 250°

3. a. −30°   b. 160°   c. −260°

4. a. −180°   b. −15°   c. −315°

In Problems 5–8, convert the given radian measure to degree measure.

5. a. $\dfrac{2\pi}{9}$ b. $\dfrac{5\pi}{12}$ c. $\dfrac{7\pi}{18}$

6. a. $\dfrac{9\pi}{4}$ b. $\dfrac{7\pi}{2}$ c. $\dfrac{17\pi}{6}$

7. a. $-\dfrac{11\pi}{6}$ b. $-\dfrac{7\pi}{4}$ c. $-\dfrac{\pi}{3}$

8. a. $-\dfrac{5\pi}{6}$ b. $-\dfrac{5\pi}{4}$ c. $-\dfrac{14\pi}{12}$

In Problems 9–14, give the radian measure of the indicated angle.

9.

10.

11.

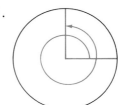

12.

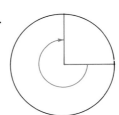

13.

14.

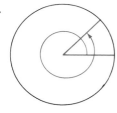

In Problems 15–18, construct an angle with the given radian measure.

15. a. $\dfrac{\pi}{4}$ b. $\dfrac{5\pi}{4}$ c. 7π

16. a. $\dfrac{11\pi}{6}$ b. $\dfrac{3\pi}{4}$ c. $\dfrac{2\pi}{3}$

17. a. $-\dfrac{5\pi}{4}$ b. -2π c. $-\dfrac{7\pi}{2}$

18. a. $-\dfrac{2\pi}{3}$ b. $-\dfrac{7\pi}{4}$ c. $-\dfrac{13\pi}{4}$

19. A bicycle wheel turns through 8 revolutions. Through how many radians does any given spoke in the wheel turn?

20. A gear wheel in a clock turns through 15° each hour. Through how many radians does the wheel turn in a week?

9-2 THE SINE AND COSINE FUNCTIONS

Consider the circle in the xy-plane that is centered at the origin and has radius 1. Given a number t, we construct an angle of radian measure t using the segment from $(0, 0)$ to $(1, 0)$ as the initial side. In this way we arrive at a definite point on the circle. We define **cos** t (the cosine of t) to be the x-coordinate of this point and **sin** t (the sine of t) to be the y-coordinate of this point (see Figure 9-6).

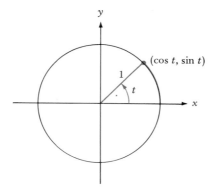

Figure 9-6
Definition of sine and cosine

Points on this circle satisfy the equation

$$x^2 + y^2 = 1$$

and since cos t is the x-coordinate of a point on the circle and sin t is the y-coordinate of a point on the circle, we obtain the fundamental *trigonometric identity*

$$\cos^2 t + \sin^2 t = 1$$

[Note that traditional notation in trigonometry places exponents on the functional symbol, that is, $\cos^2 t$ means $(\cos t)^2$, $\sin^5 t$ means $(\sin t)^5$, etc.]

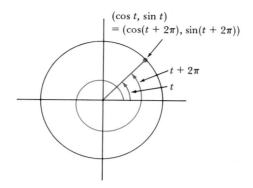

Figure 9-7
Sine and cosine are periodic

A number of other important features of the sine and cosine can be recognized directly from the definition. For example, if t is increased by 2π, we arrive at exactly the same point on the circle (see Figure 9-7); therefore,

$$\cos(t + 2\pi) = \cos t \qquad \sin(t + 2\pi) = \sin t$$

Because of this we say that the sine and cosine functions are **periodic,** with period 2π. Of course, it also follows that

$$\cos(t + 4\pi) = \cos t \qquad \sin(t + 4\pi) = \sin t$$

and indeed,

$$\cos(t + 2\pi k) = \cos t \qquad \sin(t + 2\pi k) = \sin t$$

for any integer k. From Figure 9-8 we also see that the x-coordinate of the point reached by sweeping out $-t$ radians is the same as the x-coordinate of the point reached by sweeping out t radians, that is,

$$\cos(-t) = \cos t$$

The y-coordinate of the point reached by sweeping out $-t$ radians is the negative of the y-coordinate of the point reached by sweeping out t radians:

$$\sin(-t) = -\sin t$$

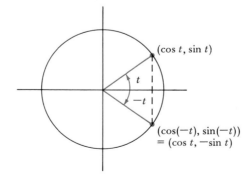

Figure 9-8
Sine and cosine symmetries

Example 3   Compute the sine and cosine of $\pi/4$ radian.

Solution   From symmetry (construct the angle $\pi/4$), it is clear that $\cos(\pi/4) = \sin(\pi/4)$. Suppose we call the common value a:

$$\cos \frac{\pi}{4} = \sin \frac{\pi}{4} = a$$

From the identity

$$\cos^2 t + \sin^2 t = 1$$

we obtain $1 = a^2 + a^2 = 2a^2$, that is, $a^2 = \frac{1}{2}$, and since a must be positive, $a = 1/\sqrt{2} = \sqrt{2}/2$. Therefore,

$$\cos \frac{\pi}{4} = \sin \frac{\pi}{4} = \frac{\sqrt{2}}{2}$$ ∎

   By using some fairly simple geometrical reasoning and a little algebra it is possible to compute the values of sine and cosine for other angles, just as we did for $\pi/4$ in Example 3. We will not do this, but we present in Table 9-1 the values of sine and cosine for various angles (expressed in radian measure). A more detailed table of sines and cosines is contained in Appendix C in the back of the book. Such tables are rapidly becoming obsolete because of the ready availability of inexpensive calculators with trigonometric keys.

Table 9-1

t	0	$\dfrac{\pi}{6}$	$\dfrac{\pi}{4}$	$\dfrac{\pi}{3}$	$\dfrac{\pi}{2}$	$\dfrac{2\pi}{3}$	$\dfrac{3\pi}{4}$	$\dfrac{5\pi}{6}$	π
$\sin t$	0	$\dfrac{1}{2}$	$\dfrac{\sqrt{2}}{2}$	$\dfrac{\sqrt{3}}{2}$	1	$\dfrac{\sqrt{3}}{2}$	$\dfrac{\sqrt{2}}{2}$	$\dfrac{1}{2}$	0
$\cos t$	1	$\dfrac{\sqrt{3}}{2}$	$\dfrac{\sqrt{2}}{2}$	$\dfrac{1}{2}$	0	$-\dfrac{1}{2}$	$-\dfrac{\sqrt{2}}{2}$	$-\dfrac{\sqrt{3}}{2}$	-1

Example 4   Find $\sin(7\pi/6)$, $\cos(7\pi/6)$, $\sin(5\pi/3)$, and $\cos(5\pi/3)$, by exploiting the symmetry in Figure 9-9 and using Table 9-1.

Figure 9-9

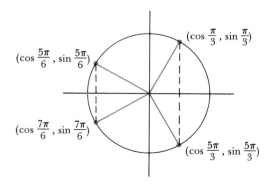

Solution From Figure 9-9 and Table 9-1, we see that

$$\cos \frac{7\pi}{6} = \cos \frac{5\pi}{6} = -\frac{\sqrt{3}}{2}$$

$$\sin \frac{7\pi}{6} = -\sin \frac{5\pi}{6} = -\frac{1}{2}$$

$$\cos \frac{5\pi}{3} = \cos \frac{\pi}{3} = \frac{1}{2}$$

$$\sin \frac{5\pi}{3} = -\sin \frac{\pi}{3} = -\frac{\sqrt{3}}{2}$$

By using symmetry as in Example 4, we can extend our table of values of sine and cosine to points in the interval $[\pi, 2\pi]$, as shown in Table 9-2. With these two tables we can plot fairly accurate graphs of the sine and cosine functions (see Figure 9-10) on the interval $[0, 2\pi]$. Since both sine and cosine are periodic, with period 2π, these cycles of the sine and cosine graphs will repeat indefinitely in the positive and negative t directions. In Figure 9-11 the periodic wavelike forms of the graphs of sine and cosine are apparent.

Table 9-2

t	$\dfrac{7\pi}{6}$	$\dfrac{5\pi}{4}$	$\dfrac{4\pi}{3}$	$\dfrac{3\pi}{2}$	$\dfrac{5\pi}{3}$	$\dfrac{7\pi}{4}$	$\dfrac{11\pi}{6}$	2π
$\sin t$	$-\dfrac{1}{2}$	$-\dfrac{\sqrt{2}}{2}$	$-\dfrac{\sqrt{3}}{2}$	-1	$-\dfrac{\sqrt{3}}{2}$	$-\dfrac{\sqrt{2}}{2}$	$-\dfrac{1}{2}$	0
$\cos t$	$-\dfrac{\sqrt{3}}{2}$	$-\dfrac{\sqrt{2}}{2}$	$-\dfrac{1}{2}$	0	$\dfrac{1}{2}$	$\dfrac{\sqrt{2}}{2}$	$\dfrac{\sqrt{3}}{2}$	1

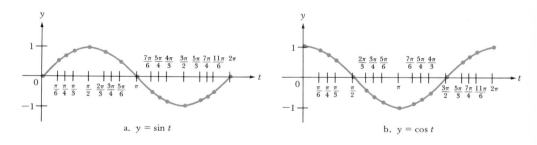

Figure 9-10
Graphs of sine and cosine

a. $y = \sin t$

b. $y = \cos t$

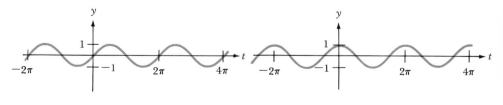

Figure 9-11
Sine and cosine are periodic

a. $y = \sin t$

b. $y = \cos t$

It seems that the graphs in Figure 9-11 are simply shifts of each other. Indeed, it appears that if the graph of the cosine function is shifted $\pi/2$ units to the right, then the graph of the sine function results. How does one find a function $g(t)$ whose graph is the graph of a function $f(t)$ shifted $\pi/2$ units to the right? It seems paradoxical, but the answer is

$$g(t) = f\left(t - \frac{\pi}{2}\right)$$

To see that this is so, consider the special case $f(t) = t$. The graph of this function is a line with slope 1 that passes through the origin. If the graph is shifted $\pi/2$ units to the right, then we obtain a line of slope 1 that has y-intercept $-\pi/2$ (see Figure 9-12), that is, we obtain the graph of the function $g(t) = t - \pi/2 = f(t - \pi/2)$. Therefore, our conjecture about the graph of the cosine function is

$$\cos\left(t - \frac{\pi}{2}\right) = \sin t$$

To see that this equation is true, consider Figure 9-13 on the facing page. In the figure, we see that the second coordinate of the point $(\cos t, \sin t)$ is the same as the first coordinate of the point $(\cos(t - \pi/2), \sin(t - \pi/2))$, which is what we wanted to show. The same figure also shows that the second coordinate of the point $(\cos(t - \pi/2), \sin(t - \pi/2))$ is the *negative* of the first coordinate of the point $(\cos t, \sin t)$, that is,

$$\sin\left(t - \frac{\pi}{2}\right) = -\cos t$$

In terms of graphs, this says that if the graph of the sine function is shifted $\pi/2$ units to the right, then the result is the graph of the function $-\cos t$, a fact which is apparent in Figure 9-11.

Figure 9-12
Shifting a graph to the right

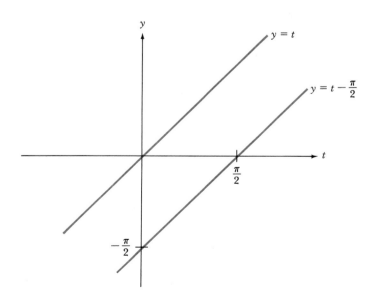

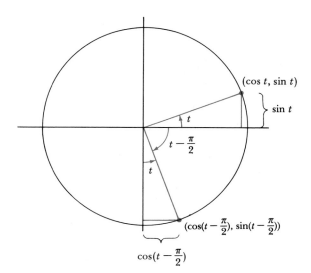

Figure 9-13

$$\cos\left(t - \frac{\pi}{2}\right) = \sin t$$

Example 5   Show that $\cos(\pi/2 - t) = \sin t$ and $\sin(\pi/2 - t) = \cos t$.

Solution   Since $\cos t = \cos(-t)$ for *any* t, we find upon replacing t by $t - \pi/2$, that

$$\cos\left(t - \frac{\pi}{2}\right) = \cos\left[-\left(t - \frac{\pi}{2}\right)\right]$$

However, $\cos(\pi/2 - t) = \cos[-(t - \pi/2)]$; therefore,

$$\cos\left(\frac{\pi}{2} - t\right) = \cos\left(t - \frac{\pi}{2}\right)$$

But, as we showed above,

$$\cos\left(t - \frac{\pi}{2}\right) = \sin t$$

Therefore,

$$\cos\left(\frac{\pi}{2} - t\right) = \sin t$$

Since $\sin(-t) = -\sin t$ for *any* t, we find upon replacing t by $t - \pi/2$, that

$$\sin\left[-\left(t - \frac{\pi}{2}\right)\right] = -\sin\left(t - \frac{\pi}{2}\right)$$

However, $\sin[-(t - \pi/2)] = \sin(\pi/2 - t)$ and, as we showed in the preceding discussion, $\sin(t - \pi/2) = -\cos t$. Therefore,

$$\sin\left(\frac{\pi}{2} - t\right) = -(-\cos t) = \cos t \qquad \blacksquare$$

There are a number of other interesting relationships, or **identities**, involving the sine and cosine functions. For example, it is true that for all numbers s and t,

$$\sin(s + t) = \sin s \cos t + \cos s \sin t$$

and

$$\cos(s + t) = \cos s \cos t - \sin s \sin t$$

A complete discussion of these identities would lead us too far astray. You can find such identities treated in detail in a good book on trigonometry or elementary mathematical analysis. For future reference we list some of the more important identities involving sine and cosine that we have discussed in this section in the box.

Identities for Sine and Cosine

$$\cos^2 t + \sin^2 t = 1$$

$$\cos(-t) = \cos t \qquad\qquad \sin(-t) = -\sin t$$

$$\cos(t + 2\pi k) = \cos t \qquad\qquad \sin(t + 2\pi k) = \sin t \qquad k \text{ any integer}$$

$$\cos\left(t - \frac{\pi}{2}\right) = \sin t \qquad\qquad \sin\left(\frac{\pi}{2} - t\right) = \cos t$$

$$\sin(s + t) = \sin s \cos t + \cos s \sin t$$

$$\cos(s + t) = \cos s \cos t - \sin s \sin t$$

Example 6   Writing $5\pi/12$ as $\pi/6 + \pi/4$, use identities and table values to compute $\cos(5\pi/12)$ and $\sin(5\pi/12)$.

Solution   From the appropriate identity and table values, we find

$$\cos\frac{5\pi}{12} = \cos\left(\frac{\pi}{6} + \frac{\pi}{4}\right) = \cos\frac{\pi}{6}\cos\frac{\pi}{4} - \sin\frac{\pi}{6}\sin\frac{\pi}{4}$$

$$= \frac{\sqrt{3}}{2}\cdot\frac{\sqrt{2}}{2} - \frac{1}{2}\cdot\frac{\sqrt{2}}{2}$$

$$= \frac{\sqrt{2}}{2}\cdot\frac{\sqrt{3}-1}{2} = \frac{\sqrt{2}(\sqrt{3}-1)}{4}$$

Similarly,

$$\sin\frac{5\pi}{12} = \sin\left(\frac{\pi}{6} + \frac{\pi}{4}\right) = \sin\frac{\pi}{6}\cos\frac{\pi}{4} + \cos\frac{\pi}{6}\sin\frac{\pi}{4}$$

$$= \frac{1}{2}\cdot\frac{\sqrt{2}}{2} + \frac{\sqrt{3}}{2}\cdot\frac{\sqrt{2}}{2}$$

$$= \frac{\sqrt{2}}{2}\cdot\frac{1+\sqrt{3}}{2} = \frac{\sqrt{2}(1+\sqrt{3})}{4} \qquad\blacksquare$$

PROBLEMS (Section 9-2)

1. Suppose $0 \leq t \leq \pi/2$ and $\cos t = \frac{1}{5}$. What is $\sin t$? [*Hint:* Use the identity $\cos^2 t + \sin^2 t = 1$.]

2. Suppose $0 \leq t \leq \pi/2$ and $\sin t = \frac{2}{7}$. What is $\cos t$?

3. Suppose $\pi/2 \leq t \leq \pi$ and $\cos t = -\frac{1}{6}$. What is $\sin t$?

4. Suppose $\pi/2 \leq t \leq \pi$ and $\sin t = \frac{4}{5}$. What is $\cos t$?

5. Suppose $-\pi \leq t \leq -\pi/2$ and $\cos t = -\frac{4}{5}$. What is $\sin t$?

6. Suppose $-\pi \leq t \leq -\pi/2$ and $\sin t = -\frac{1}{5}$. What is $\cos t$?

7. Find all values of t with $2\pi \leq t \leq 4\pi$ and $\sin t = \frac{1}{2}$.

8. Find all values of t with $2\pi \leq t \leq 4\pi$ and $\cos t = \sqrt{3}/2$.

9. Find all values of t with $-2\pi \leq t \leq 0$ and $\sin t = \sqrt{2}/2$.

10. Find all values of t with $-2\pi \leq t \leq 0$ and $\cos t = -\sqrt{2}/2$.

11. Find all values of t with $-4\pi \leq t \leq 4\pi$ and $\cos t = -\frac{1}{2}$.

12. Find all values of t with $-4\pi \leq t \leq 4\pi$ and $\sin t = -\sqrt{3}/2$.

In Problems 13–18, sketch the graph of the indicated function over the given interval.

13. $f(t) = \cos t, \quad -\pi \leq t \leq 2\pi$

14. $f(t) = \sin t, \quad -\pi \leq t \leq 2\pi$

15. $f(t) = \sin t, \quad -2\pi \leq t \leq 2\pi$

16. $f(t) = \cos t, \quad -2\pi \leq t \leq 2\pi$

17. $f(t) = 2 \sin t, \quad -\pi/2 \leq t \leq 5\pi/2$

18. $f(t) = -\cos t, \quad -\pi/2 \leq t \leq 5\pi/2$

19. Find a value of t with $-\pi/2 \leq t \leq \pi/2$ and $\cos t = \sin t$.

20. Find a value of t with $-\pi/2 \leq t \leq \pi/2$ and $\cos t = -\sin t$.

21. Sketch the graph of the function $f(t) = \cos(t - \pi)$.

22. Sketch the graph of the function $f(t) = \sin(t - 3\pi/2)$.

23. Sketch the graph of the function $f(t) = \sin(t + \pi)$.

24. Sketch the graph of the function $f(t) = \cos(t + \pi/2)$.

25. Show that $\sin 2t = 2 \sin t \cos t$. [*Hint:* Use the appropriate identity.]

26. Show that $\cos 2t = \cos^2 t - \sin^2 t$. [*Hint:* Use the appropriate identity.]

27. The identity $\cos^2 t + \sin^2 t = 1$ implies $\cos^2 t = 1 - \sin^2 t$. Substitute this expression in the identity of Problem 26 and show that

$$\sin^2 t = \frac{1 - \cos 2t}{2}$$

28. Substitute $\sin^2 t = 1 - \cos^2 t$ in the identity in Problem 26 and show that

$$\cos^2 t = \frac{1 + \cos 2t}{2}$$

29. Show that $\cos(s - t) = \cos s \cos t + \sin s \sin t$. [*Hint:* Use the appropriate identities.]

30. a. Draw the graphs of $f(t) = \cos t$ and $g(t) = \cos(t - \pi)$ on the same set of axes. Notice that the graphs are exactly out of phase, that is, f is highest when g is lowest and vice versa.

    b. It has long been known that the temperature at a time t during the yearly cycle of the seasons at a point x meters below the surface of the earth is approximately

$$ce^{-0.71x} \cos(t - 0.71x)$$

(where c is a constant) if t is given in proper units. Suppose we wish to locate a wine cellar at a certain depth x below the surface so that the cellar is coolest in summer and warmest in winter. That is, we would like the graph of the function above, which represents the temperature at depth x, to be exactly out of phase with the graph of

$$c \cos t$$

which represents the surface temperature (that is, the temperature at $x = 0$). How deep should we dig the cellar?

c. The temperature of the surface varies from $-c$ to c, depending on time. In the cellar, the temperature will vary from $-ce^{-0.71x}$ to $ce^{-0.71x}$. Therefore, the maximum temperature variation in the cellar is $e^{-0.71x}$ times what it is on the surface. For the value of x computed in part b, express $e^{-0.71x}$ as a percentage.

9-3 DERIVATIVES AND INTEGRALS OF SINE AND COSINE

Before developing formulas for the derivatives of sine and cosine functions, let us try to guess what these formulas might be. Consider first the sine function:

$$y = \sin t \qquad 0 \le t \le 2\pi$$

In Figure 9-14 we have sketched this function and indicated on the graph the apparent values of the slope of the tangent line to the graph at various points. From this figure, it seems that the slope of the tangent line to the graph of $y = \sin t$ has the values given in Table 9-3.

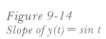

Figure 9-14
Slope of $y(t) = \sin t$

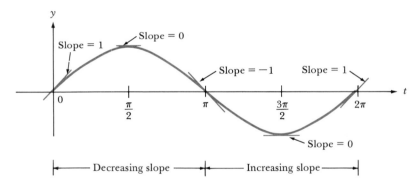

Table 9-3

t	0	$\dfrac{\pi}{2}$	π	$\dfrac{3\pi}{2}$	2π
Slope of $y = \sin t$	1	0	-1	0	1

Moreover, the slopes decrease as we move from left to right in the interval $0 \le t \le \pi$ and increase as we move from left to right in the interval $\pi \le t \le 2\pi$. What does this say about the derivative of the sine function? We know that if

$$y(t) = \sin t$$

then $y'(t) =$ slope of tangent line at t. Therefore, we have the information about the function $y'(t)$ given in Table 9-4.

Table 9-4

t	0	$\dfrac{\pi}{2}$	π	$\dfrac{3\pi}{2}$	2π
$y'(t)$	1	0	-1	0	1

$|\leftarrow y'\ decreasing \rightarrow|\leftarrow y'\ increasing \rightarrow|$

In Figure 9-15 we have sketched a probable graph for the function $y'(t)$ based on the information in Table 9-4. Does this graph remind you of something? Of course! It looks very much like the graph of the cosine function (see Figure 9-10). This leads us to *guess* that

$$\frac{d}{dt}\sin t = \cos t$$

Figure 9-15
Graph of y'(t)

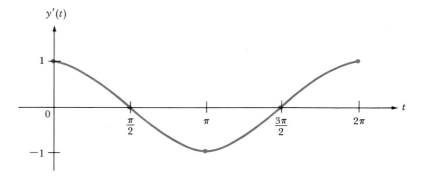

In order to verify this, recall from Chapter 2 that

$$\frac{d}{dt}\sin t = \lim_{h\to 0}\frac{\sin(t+h)-\sin t}{h}$$

To find this limit we will use a trigonometric identity for the sine:

$$\sin(t+h)=\sin t \cos h + \cos t \sin h$$

Substituting this expression in the quotient above, we have

$$\frac{\sin(t+h)-\sin t}{h}=\frac{\sin t \cos h - \sin t + \cos t \sin h}{h}$$

$$=\left(\frac{\cos h - 1}{h}\right)\sin t + \left(\frac{\sin h}{h}\right)\cos t$$

From this we see that

$$\frac{d}{dt}\sin t = \lim_{h\to 0}\frac{\sin(t+h)-\sin t}{h}$$

$$=\left(\lim_{h\to 0}\frac{\cos h - 1}{h}\right)\sin t + \left(\lim_{h\to 0}\frac{\sin h}{h}\right)\cos t$$

Therefore, to compute the derivative of sin t we must compute the two limits

$$\lim_{h\to 0}\frac{\cos h - 1}{h} \quad \text{and} \quad \lim_{h\to 0}\frac{\sin h}{h}$$

Rather than give a rigorous development of these limits, we will be content to observe the numerical results given in Table 9-5, which were obtained with a calculator.

Table 9-5

h	1	0.1	0.01	0.001	0.0001
$\dfrac{\sin h}{h}$	0.8414710	0.998334	0.99998	1	1
$\dfrac{\cos h - 1}{h}$	-0.4596977	-0.049958	-0.005	-0.0005	0

The results indicate that

$$\lim_{h\to 0}\frac{\sin h}{h} = 1 \quad \text{and} \quad \lim_{h\to 0}\frac{\cos h - 1}{h} = 0$$

Indeed, when $h = 0.0001$, the calculator is unable to distinguish the quantities from their limits, due to limitations in its accuracy. Of course, to find a limit as $h \to 0$ we must consider both positive and negative values of h. But notice that, since $\sin(-h) = -\sin h$ (see the trigonometric identities in Section 9-2),

$$\frac{\sin(-h)}{-h} = -\frac{\sin h}{-h} = \frac{\sin h}{h}$$

Therefore, calculating with the corresponding negative values of h will give the same numerical results for $\sin(-h)/(-h)$ as for $(\sin h)/h$. Also, since $\cos(-h) = \cos h$,

$$\frac{\cos(-h) - 1}{-h} = \frac{\cos h - 1}{-h} = -\frac{\cos h - 1}{h}$$

Therefore, calculating with the corresponding negative values of h will give the negatives of the numerical values we computed for $(\cos h - 1)/h$. In any case, we see that

$$\lim_{h\to 0}\frac{\sin h}{h} = 1 \quad \text{and} \quad \lim_{h\to 0}\frac{\cos h - 1}{h} = 0$$

Returning to our computation of the derivative of the sine function, we find

$$\frac{d}{dt}\sin t = \left(\lim_{h\to 0}\frac{\cos h - 1}{h}\right)\sin t + \left(\lim_{l\to 0}\frac{\sin h}{h}\right)\cos t$$
$$= 0 \cdot \sin t + 1 \cdot \cos t = \cos t$$

So our guess that the derivative of the sine function is the cosine function was a good one! We have therefore established the following formula for the derivative of the sine function:

$$\frac{d}{dt}\sin t = \cos t$$

We can combine this formula for the derivative of the sine function with the chain rule to obtain derivatives of more complicated functions involving the sine. Recall that the chain rule says that if $y = f(g(t))$, then $y' = f'(g(t))g'(t)$. In particular, if $f(u) = \sin u$, we have the differentiation rule

$$\frac{d}{dt}\sin(g(t)) = \cos(g(t)) \cdot g'(t)$$

Example 7 Differentiate $\sin(2t^2 + t + 1)$.

Solution This function has the form $\sin(g(t))$, where $g(t) = 2t^2 + t + 1$, and hence, $g'(t) = 4t + 1$. Applying the chain rule, we have

$$\frac{d}{dt}\sin(2t^2 + t + 1) = \cos(2t^2 + t + 1) \cdot (4t + 1) \qquad \blacksquare$$

Example 8
Modeling Cyclical Phenomena The periodic nature of the sine function makes it appropriate for modeling cyclical phenomena. Suppose, for example, that the sales volume (in units) of a commodity varies over a year as given by the formula

$$S(t) = 10{,}000 + 500\sin[0.0172(t + 15)] \qquad 1 \le t \le 365$$

where t represents days into the year. At what rate is the sales volume changing 90 days into the year?

Solution We require the number $S'(90)$. By the chain rule, we have

$$S'(t) = 500\cos[0.0172(t + 15)] \cdot 0.0172$$
$$= 8.6\cos[0.0172(t + 15)]$$

Therefore,

$$S'(90) = 8.6\cos[0.0172(90 + 15)]$$
$$= 8.6\cos(1.806) \approx -2.004$$

That is, sales are decreasing at a rate of about 2 units per day at this time. $\blacksquare$

The chain rule can also be used to differentiate composite functions in which the sine is the inner function. For example, if $y = f(\sin t)$, then

$$\frac{dy}{dt} = f'(\sin t) \cdot \cos t$$

Example 9 Differentiate:

    a. $e^{\sin t}$ b. $\sin^3 t$

Solution   a. We let $e^{\sin t} = f(\sin t)$, where $f(u) = e^u$. Therefore, $f'(u) = e^u$, $f'(\sin t) = e^{\sin t}$, and hence,

$$\frac{d}{dt} e^{\sin t} = e^{\sin t} \cdot \cos t$$

b. We let $\sin^3 t = f(\sin t)$, where $f(u) = u^3$ (remember our notational convention for powers of sine!). Therefore, $f'(u) = 3u^2$, $f'(\sin t) = 3 \sin^2 t$, and hence,

$$\frac{d}{dt} \sin^3 t = 3 \sin^2 t \cdot \cos t \qquad \blacksquare$$

Functions that are products or quotients involving sine functions can be differentiated by the product and quotient rules (see Chapter 4).

Example 10   Differentiate:

a. $(t^2 + 1) \sin 2t$ b. $\dfrac{t \sin t}{t + 1}$

Solution   a. The given function is the product of the function $t^2 + 1$ and the function $\sin 2t$. Therefore, by the product rule,

$$\frac{d}{dt}[(t^2 + 1) \sin 2t] = (t^2 + 1) \cdot \frac{d}{dt} \sin 2t + \frac{d}{dt}(t^2 + 1) \cdot \sin 2t$$

$$= (t^2 + 1)(\cos 2t)(2) + 2t \sin 2t$$

$$= (2t^2 + 2) \cos 2t + 2t \sin 2t$$

b. We have the quotient of the function $t \sin t$ and the function $t + 1$. Therefore, by the quotient rule,

$$\frac{d}{dt}\frac{t \sin t}{t + 1} = \frac{(t + 1)\dfrac{d}{dt}(t \sin t) - (t \sin t)\dfrac{d}{dt}(t + 1)}{(t + 1)^2}$$

$$= \frac{(t + 1)(\sin t + t \cos t) - t \sin t}{(t + 1)^2}$$

$$= \frac{t \sin t + \sin t + t(t + 1) \cos t - t \sin t}{(t + 1)^2}$$

$$= \frac{\sin t + t(t + 1) \cos t}{(t + 1)^2} \qquad \blacksquare$$

We seem to have been neglecting the cosine function. What is the derivative of $\cos t$? Fortunately, we do not need to go through a long development to find out. One way to discover quickly the answer to our question is to use the trigonometric identity

$$\cos^2 t + \sin^2 t = 1$$

and implicit differentiation (see Chapter 4). Differentiating this identity we obtain

$$2 \cos t \cdot \frac{d}{dt} \cos t + 2 \sin t \cdot \frac{d}{dt} \sin t = 0$$

that is,

$$\cos t \cdot \frac{d}{dt} \cos t + \sin t \cdot \cos t = 0$$

or

$$\frac{d}{dt} \cos t = -\sin t$$

Another way to get this result is to use the identity $\cos t = \sin(\pi/2 - t)$. By using this and other identities, we find

$$\frac{d}{dt} \cos t = \frac{d}{dt} \sin\left(\frac{\pi}{2} - t\right)$$

$$= \cos\left(\frac{\pi}{2} - t\right) \cdot (-1)$$

$$= -\cos\left[-\left(t - \frac{\pi}{2}\right)\right]$$

$$= -\cos\left(t - \frac{\pi}{2}\right) = -\sin t$$

An application of the chain rule gives

$$\frac{d}{dt} \cos(g(t)) = -\sin(g(t)) \cdot g'(t)$$

Example 11   Differentiate $\cos(e^t + t)$.

Solution   Here, $g(t) = e^t + t$, $g'(t) = e^t + 1$, and hence,

$$\frac{d}{dt} \cos(e^t + t) = -\sin(e^t + t) \cdot (e^t + 1)$$

$$= -(e^t + 1) \sin(e^t + t)$$ ∎

Also, a function of the form $y = f(\cos t)$ can be differentiated by the chain rule:

$$y' = f'(\cos t)(-\sin t)$$

Example 12   Differentiate:

a. $\ln(2 + \cos t)$ b. $\cos^3 2t$

Solution   a. The given function may be written as $f(\cos t)$, where $f(u) = \ln(2 + u)$. Therefore, $f'(u) = 1/(2 + u)$ and

$$\frac{d}{dt} \ln(2 + \cos t) = \frac{1}{2 + \cos t} (-\sin t)$$

$$= \frac{-\sin t}{2 + \cos t}$$

b. We may write $\cos^3 2t = f(\cos 2t)$, where $f(u) = u^3$ and $f'(u) = 3u^2$. Therefore,

$$\frac{d}{dt} \cos^3 2t = 3 \cos^2 2t \cdot \frac{d}{dt} \cos 2t$$

$$= 3 \cos^2 2t \cdot (-\sin 2t)(2)$$

$$= -6 \cos^2 2t \sin 2t \qquad \blacksquare$$

Example 13 **Differentiate:**

a. $e^t \cos(2t + 1)$ b. $\dfrac{t^2 + t}{\cos t}$

Solution   a. We use the product rule:

$$\frac{d}{dt}[e^t \cos(2t + 1)] = e^t \frac{d}{dt} \cos(2t + 1) + \left(\frac{d}{dt} e^t\right) \cos(2t + 1)$$

$$= e^t[-\sin(2t + 1)](2) + e^t \cos(2t + 1)$$

$$= e^t[\cos(2t + 1) - 2 \sin(2t + 1)]$$

b. The quotient rule can be used to differentiate:

$$\frac{d}{dt} \frac{t^2 + t}{\cos t} = \frac{(2t + 1) \cos t - (t^2 + t)(-\sin t)}{\cos^2 t}$$

$$= \frac{(2t + 1) \cos t + (t^2 + t) \sin t}{\cos^2 t} \qquad \blacksquare$$

The differentiation formulas

$$\frac{d}{dt} \sin t = \cos t \qquad \text{and} \qquad \frac{d}{dt} \cos t = -\sin t$$

give immediately the following integration formulas:

$$\int \cos t \, dt = \sin t + c$$

and

$$\int \sin t \, dt = -\cos t + c$$

Example 14   Find the area under the curve $y = \sin t$ from $t = 0$ to $t = \pi$.

Solution   Since $\sin t \geq 0$ for $0 \leq t \leq \pi$, this area is

$$\int_0^{\pi} \sin t \, dt = -\cos t \big|_0^{\pi}$$

$$= -\cos \pi + \cos 0$$
$$= -(-1) + 1 = 2$$

Finally, we point out that the differentiation rules

$$\frac{d}{dt} \sin(g(t)) = \cos(g(t)) \cdot g'(t) \qquad \text{and} \qquad \frac{d}{dt} \cos(g(t)) = -\sin(g(t)) \cdot g'(t)$$

immediately give the corresponding integration rules:

$$\int \cos(g(t)) \cdot g'(t) \, dt = \sin(g(t)) + c$$

and

$$\int \sin(g(t)) \cdot g'(t) \, dt = -\cos(g(t)) + c$$

These rules are illustrated in the next example.

Example 15   Compute the integrals:

a. $\displaystyle\int \cos(2t + 1) \, dt$ b. $\displaystyle\int_0^{\pi/10} 2 \sin 5t \, dt$

Solution   a. We use $g(t) = 2t + 1$, $g'(t) = 2$. Then

$$\int \cos(2t + 1) \, dt = \tfrac{1}{2} \int 2 \cos(2t + 1) \, dt = \tfrac{1}{2} \sin(2t + 1) + c$$

b. We set $g(t) = 5t$, $g'(t) = 5$. Then

$$\int_0^{\pi/10} 2 \sin 5t \, dt = \frac{2}{5} \int_0^{\pi/10} 5 \sin 5t \, dt$$

$$= -\frac{2}{5} \cos(5t) \bigg|_0^{\pi/10}$$

$$= -\frac{2}{5} \cos \frac{\pi}{2} + \frac{2}{5} \cos 0$$

$$= \frac{2}{5}$$

We summarize the important formulas of this section in the box on page 452.

The Calculus of Sine and Cosine

$$\frac{d}{dt}\sin t = \cos t \qquad\qquad \frac{d}{dt}\cos t = -\sin t$$

$$\frac{d}{dt}\sin(g(t)) = \cos(g(t))\cdot g'(t) \qquad\qquad \frac{d}{dt}f(\sin t) = f'(\sin t)\cos t$$

$$\frac{d}{dt}\cos(g(t)) = -\sin(g(t))\cdot g'(t) \qquad\qquad \frac{d}{dt}f(\cos t) = f'(\cos t)(-\sin t)$$

$$\int \cos t\, dt = \sin t + c \qquad\qquad \int \sin t\, dt = -\cos t + c$$

$$\int \cos(g(t))\cdot g'(t)\, dt = \sin(g(t)) + c \qquad\qquad \int \sin(g(t))\cdot g'(t)\, dt = -\cos(g(t)) + c$$

PROBLEMS (Section 9-3)

In Problems 1–36, compute the derivative of the function.

1. $\sin 6t$

2. $\cos 7t$

3. $2\sin 3t$

4. $4\sin(-t)$

5. $7\cos 5t$

6. $t + \sin 2t$

7. $3t^2 + \cos \pi t$

8. $e^t - \sin(-7t)$

9. $5\sin(2t+1)$

10. $-3\cos(1-2t)$

11. $\sin(t^3 + 2t)$

12. $2\cos(t^2 - 1)$

13. $4\cos(1-t)$

14. $-\sin(t+1)$

15. $3\cos e^{2t}$

16. $-5\sin e^{2t+1}$

17. $e^{2\sin 2t}$

18. $e^{2[\ln(\cos t)]}$

19. $\sin[\ln(t+1)]$

20. $-2\cos(\ln t^2)$

21. $\sin t^2$

22. $(t + \sin t)^7$

23. $(t^2 - 2\cos t)^5$

24. $e^t \cos 2t$

25. $e^{2t}\sin t$

26. $\cos^3 t^4$

27. $\sin^4 t^3$

28. $\sin^2 t \cos t$

29. $\cos 5t \sin 3t$

30. $(\ln t)(\sin t)$

31. $\ln(\sin t)$

32. $\dfrac{\cos t}{\ln t}$

33. $\dfrac{t + \sin t}{t - \cos t}$

34. $\dfrac{1}{\sin t}$

35. $\dfrac{1}{\cos t}$

36. $\dfrac{\sin t}{\cos t}$

In Problems 37–44, compute the indicated integral.

37. $\int \cos 7t \, dt$

38. $\int \sin \frac{t}{7} \, dt$

39. $\int \sin(1 + 4t) \, dt$

40. $\int \cos(1 - 4t) \, dt$

41. $\int_0^{\pi/6} \sin 2t \, dt$

42. $\int_{-3\pi}^{-\pi} \cos 6t \, dt$

43. $\int_{1/3}^{(\pi+2)/6} \sin(3t - 1) \, dt$

44. $\int_{(-\pi-1)/2}^{(\pi-2)/4} 2 \cos(2t + 1) \, dt$

45. Solve for y: $y' = 2 \cos t$, $y(\pi/2) = 3$.

46. Solve for y: $y' = \frac{1}{2} \sin 2t$, $y(\pi) = 1$.

47. Show that $y = \cos 2t$ and $y = \sin 2t$ are both solutions of the differential equation $y'' + 4y = 0$.

48. Show that $y = e^t \sin t$ and $y = e^t \cos t$ are both solutions of the differential equation $y'' - 2y' + 2y = 0$.

49. A simplified ballistics model predicts that the range of a projectile is given by

$$R(\theta) = \frac{v^2}{g} \sin 2\theta$$

where v and g are constants (the muzzle velocity and acceleration due to gravity, respectively) and θ is the angle of elevation of the cannon (see the figure). At what angle θ is the range a maximum?

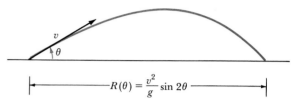

Range of a projectile

50. A study of the 30 year averages of mean monthly temperatures in Bismarck, North Dakota, shows that the mean temperature is approximately

$$f(t) = 41.4 + 30.3 \sin\left(\frac{\pi}{6} t + 31.5\right)$$

degrees Fahrenheit t months after April 1 (that is, April 1 corresponds to $t = 0$, September 1 corresponds to $t = 5$, etc.).*

a. Use a calculator to find the mean temperature for each month.
b. Use the values computed in part a to sketch the graph of f.
c. At what rate is the mean temperature changing on October 1?

9-4 THE OTHER TRIGONOMETRIC FUNCTIONS

THE TANGENT

There are four other trigonometric functions that are defined in terms of the sine and cosine. The first such function we will consider is called the **tangent** and is defined by

$$\tan t = \frac{\sin t}{\cos t}$$

* Adapted from John W. McCloskey, "A Model for Atmospheric Temperature," *The UMAP Journal*, vol. 2, pp. 5–12 (1981).

Notice that the tangent function is not defined for those t for which $\cos t = 0$, that is, $\tan t$ is defined only for $t \neq \pm\pi/2, \pm 3\pi/2, \pm 5\pi/2, \ldots$.

Since $\sin t$ and $\cos t$ are both positive in the interval $0 \leq t < \pi/2$, the tangent function is also positive in this interval. Also, since $\sin t$ approaches 1 as t approaches $\pi/2$ and $\cos t$ approaches 0 as t approaches $\pi/2$, the value $\tan t$ goes to ∞ as t approaches $\pi/2$ from the left. Therefore, the tangent function $y = \tan t$ has a vertical asymptote at $t = \pi/2$. In the interval $-\pi/2 < t \leq 0$, the graph of $y = \tan t$ can be obtained by symmetry since

$$\tan(-t) = \frac{\sin(-t)}{\cos(-t)} = -\frac{\sin t}{\cos t} = -\tan t$$

When the graphs of sin and cos are shifted π units to the left, the graphs of $-\sin$ and $-\cos$ result:

$$\sin(t + \pi) = -\sin t \qquad \cos(t + \pi) = -\cos t$$

From this we find that

$$\tan(t + \pi) = \frac{\sin(t + \pi)}{\cos(t + \pi)} = \frac{-\sin t}{-\cos t} = \frac{\sin t}{\cos t} = \tan t$$

and hence the tangent function is periodic with period π. Therefore, the entire graph of the tangent function is obtained by shifting the basic cycle in the interval $-\pi/2 < t < \pi/2$ by multiples of π units. The graph of the tangent function is shown in Figure 9-16.

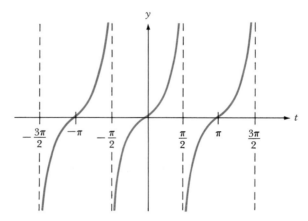

Figure 9-16
$y = tan\ t$

THE SECANT

A function that is closely related to the tangent function is the **secant** function:

$$\sec t = \frac{1}{\cos t}$$

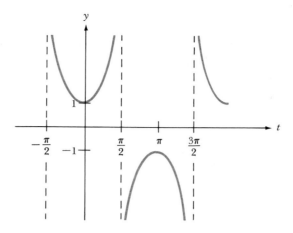

Figure 9-17
y = sec t

The graph of the secant function is shown in Figure 9-17. The relationship between the tangent and secant comes from the basic trigonometric identity

$$1 = \cos^2 t + \sin^2 t$$

If we divide this identity through by $\cos^2 t$, we obtain

$$\frac{1}{\cos^2 t} = 1 + \frac{\sin^2 t}{\cos^2 t}$$

or

$$\sec^2 t = 1 + \tan^2 t$$

DERIVATIVES

Since the tangent is, by definition, the sine divided by the cosine, we can find the derivative of the tangent function by using the quotient rule:

$$\frac{d}{dt} \tan t = \frac{d}{dt} \frac{\sin t}{\cos t}$$

$$= \frac{\cos t \dfrac{d}{dt} \sin t - \sin t \dfrac{d}{dt} \cos t}{\cos^2 t}$$

$$= \frac{(\cos t)(\cos t) - (\sin t)(-\sin t)}{\cos^2 t}$$

$$= \frac{\cos^2 t + \sin^2 t}{\cos^2 t}$$

$$= \frac{1}{\cos^2 t} = \sec^2 t$$

That is,

$$\frac{d}{dt} \tan t = \sec^2 t$$

In a similar way, using the reciprocal rule, we can find the derivative of the secant function:

$$\frac{d}{dt} \sec t = \frac{d}{dt} \frac{1}{\cos t}$$

$$= \frac{-\dfrac{d}{dt} \cos t}{\cos^2 t}$$

$$= \frac{\sin t}{\cos^2 t}$$

$$= \frac{1}{\cos t} \cdot \frac{\sin t}{\cos t} = \sec t \tan t$$

That is,

$$\frac{d}{dt} \sec t = \sec t \tan t$$

Combining these differentiation rules with the chain rule gives

$$\frac{d}{dt} \tan(g(t)) = \sec^2(g(t)) \cdot g'(t)$$

$$\frac{d}{dt} \sec(g(t)) = \sec(g(t)) \tan(g(t)) \cdot g'(t)$$

Example 16 **Differentiate:**

    a. $\tan(t^2 + 1)$ b. $\sec e^{2t}$

Solution

a. $\dfrac{d}{dt} \tan(t^2 + 1) = \sec^2(t^2 + 1) \cdot \dfrac{d}{dt}(t^2 + 1)$

$$= \sec^2(t^2 + 1) \cdot 2t$$
$$= 2t \sec^2(t^2 + 1)$$

b. $\dfrac{d}{dt} \sec e^{2t} = \sec e^{2t} \tan e^{2t} \cdot \dfrac{d}{dt} e^{2t}$

$$= \sec e^{2t} \tan e^{2t} \cdot 2e^{2t}$$
$$= 2e^{2t} \sec e^{2t} \tan e^{2t} \qquad\blacksquare$$

The chain rule also applies to compositions of functions of the type $f(\tan t)$ or $f(\sec t)$. In these cases, the chain rule gives

$$\frac{d}{dt} f(\tan t) = f'(\tan t)\, \sec^2 t$$

$$\frac{d}{dt} f(\sec t) = f'(\sec t)\, \sec t \tan t$$

Example 17   **Differentiate:**

a. $\ln(\tan t)$ b. $e^{2\sec t}$

Solution   a. We have $\ln(\tan t) = f(\tan t)$, where $f(u) = \ln u$. Therefore, $f'(u) = 1/u$ and

$$\frac{d}{dt} \ln(\tan t) = \frac{1}{\tan t} \cdot \frac{d}{dt} \tan t = \frac{\sec^2 t}{\tan t}$$

b. Here, $e^{2\sec t} = f(\sec t)$, where $f(u) = e^{2u}$. Therefore, $f'(u) = 2e^{2u}$ and

$$\frac{d}{dt} e^{2\sec t} = 2e^{2\sec t} \cdot \frac{d}{dt} \sec t = 2e^{2\sec t} \sec t \tan t \qquad\blacksquare$$

Example 18   **Differentiate:**

a. $e^{3t} \tan 2t$ b. $\dfrac{\sec^2 t}{1 + \tan t}$

Solution   a. We use the product rule:

$$\frac{d}{dt} e^{3t} \tan 2t = e^{3t} \cdot \frac{d}{dt} \tan 2t + \tan 2t \cdot \frac{d}{dt} e^{3t}$$
$$= e^{3t}(\sec^2 2t)(2) + 3e^{3t} \tan 2t$$
$$= e^{3t}(2 \sec^2 2t + 3 \tan 2t)$$

b. The quotient rule is used to differentiate:

$$\frac{d}{dt} \frac{\sec^2 t}{1 + \tan t} = \frac{(1 + \tan t) \cdot \dfrac{d}{dt} \sec^2 t - \sec^2 t \cdot \dfrac{d}{dt}(1 + \tan t)}{(1 + \tan t)^2}$$

The factor $(d/dt)\sec^2 t$ must be computed by use of the chain rule:

$$\frac{d}{dt}\sec^2 t = 2\sec t \cdot \frac{d}{dt}\sec t = 2\sec^2 t \tan t$$

Therefore,

$$\frac{d}{dt}\frac{\sec^2 t}{1 + \tan t} = \frac{(1 + \tan t)\, 2\sec^2 t \tan t - (\sec^2 t)(\sec^2 t)}{(1 + \tan t)^2}$$

$$= \frac{2\sec^2 t \tan t + 2\sec^2 t \tan^2 t - \sec^4 t}{(1 + \tan t)^2}$$

$$= \frac{2\sec^2 t \tan t + 2\sec^2 t(\sec^2 t - 1) - \sec^4 t}{(1 + \tan t)^2}$$

$$= \frac{2\sec^2 t \tan t + 2\sec^4 t - 2\sec^2 t - \sec^4 t}{(1 + \tan t)^2}$$

$$= \frac{(2\sec^2 t)(\tan t - 1) + \sec^4 t}{(1 + \tan t)^2} \qquad \blacksquare$$

Example 19   Find the absolute minimum and maximum values of the function $f(t) = \sqrt{6}\,\sin t - \sqrt{2}\,\cos t$ in the interval $0 \le t \le 2\pi$.

Solution   The first step in finding the extreme values of a function is to compute its derivative:

$$f'(t) = \sqrt{6}\,\cos t + \sqrt{2}\,\sin t$$

The critical points of f then satisfy

$$\sqrt{6}\,\cos t + \sqrt{2}\,\sin t = 0$$

$$\sqrt{6}\,\cos t = -\sqrt{2}\,\sin t$$

$$\frac{-\sqrt{6}}{\sqrt{2}} = \frac{\sin t}{\cos t}$$

or

$$-\sqrt{3} = \tan t$$

From Tables 9-1 and 9-2 (in Section 9-2) we find that $\tan t$ takes the value $-\sqrt{3}$ at $t = 2\pi/3$ and $t = 5\pi/3$. Now we can test these values to find out if f has a maximum or minimum at them. To do this we will use the second derivative test. Now,

$$f''(t) = -\sqrt{6}\,\sin t + \sqrt{2}\,\cos t$$

and from Tables 9-1 and 9-2 we find that

$$f''\left(\frac{2\pi}{3}\right) = -\sqrt{6} \cdot \frac{\sqrt{3}}{2} - \sqrt{2} \cdot \frac{1}{2} < 0$$

$$f''\left(\frac{5\pi}{3}\right) = \sqrt{6} \cdot \frac{\sqrt{3}}{2} + \sqrt{2} \cdot \frac{1}{2} > 0$$

Therefore, f has a relative maximum value of

$$f\left(\frac{2\pi}{3}\right) = \sqrt{6} \cdot \frac{\sqrt{3}}{2} + \frac{\sqrt{2}}{2}$$

$$= \frac{\sqrt{18} + \sqrt{2}}{2}$$

$$= \frac{3\sqrt{2} + \sqrt{2}}{2} = 2\sqrt{2}$$

at $t = 2\pi/3$ and a relative minimum value of

$$f\left(\frac{5\pi}{3}\right) = \sqrt{6}\left(-\frac{\sqrt{3}}{2}\right) - \frac{\sqrt{2}}{2}$$

$$= -2\sqrt{2}$$

at $t = 5\pi/3$.

Finally, note that at the endpoints of the interval $0 \le t \le 2\pi$, we have

$$f(0) = f(2\pi) = -\sqrt{2}$$

Therefore, the absolute maximum of f on the interval is $2\sqrt{2}$ and the absolute minimum of f is $-2\sqrt{2}$. ∎

THE COTANGENT AND COSECANT

The **cotangent** function is defined by

$$\cot t = \frac{\cos t}{\sin t}$$

Notice that the cotangent function is not defined when $\sin t = 0$, that is, when $t = 0, \pm\pi, \pm 2\pi, \pm 3\pi, \ldots$. Also,

$$\cot t = \frac{1}{\tan t}$$

The graph of the cotangent function is shown in Figure 9-18 on page 460.

The **cosecant** function is closely related to the cotangent function. It is defined by

$$\csc t = \frac{1}{\sin t}$$

The cosecant function has the same domain as the cotangent function. In Problem 3 at the end of this section, you are asked to show that $\cot t$ and $\csc t$ are related by the identity

$$\cot^2 t + 1 = \csc^2 t$$

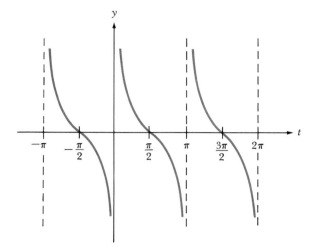

Figure 9-18
$y = \cot t$

The graph of the cosecant function is shown in Figure 9-19.

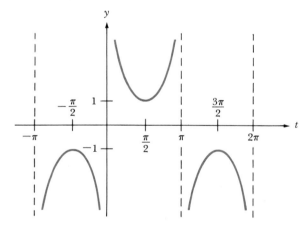

Figure 9-19
$y = \csc t$

We can compute the derivatives of the cotangent and cosecant functions by using the quotient rule. For example,

$$\frac{d}{dt} \cot t = \frac{d}{dt} \frac{\cos t}{\sin t}$$

$$= \frac{\sin t \cdot \dfrac{d}{dt} \cos t - \cos t \cdot \dfrac{d}{dt} \sin t}{\sin^2 t}$$

$$= \frac{-\sin^2 t - \cos^2 t}{\sin^2 t}$$

$$= -\frac{(\sin^2 t + \cos^2 t)}{\sin^2 t}$$

$$= \frac{-1}{\sin^2 t}$$

$$= -\csc^2 t$$

The result

$$\frac{d}{dt} \cot t = -\csc^2 t$$

immediately gives a corresponding integration rule:

$$\int \csc^2 t \, dt = -\cot t + c$$

In Problem 45 at the end of this section, we ask you to derive the rule

$$\frac{d}{dt} \csc t = -\csc t \cot t$$

Combining these differentiation rules with the chain rule, product rule, and quotient rule will enable us to differentiate complicated functions involving cot and csc. We illustrate this with the next example.

Example 20 Differentiate:

a. $e^{\cot t} \sin^2 t$ b. $\dfrac{\cot(t^2 + 1)}{1 - \csc t}$

Solution a. We use the product rule and chain rule:

$$\frac{d}{dt}(e^{\cot t} \sin^2 t) = e^{\cot t} \cdot \frac{d}{dt} \sin^2 t + \sin^2 t \cdot \frac{d}{dt} e^{\cot t}$$

$$= e^{\cot t} \cdot 2 \sin t \cos t + \sin^2 t \cdot e^{\cot t}(-\csc^2 t)$$

$$= e^{\cot t} \left(2 \sin t \cos t - \frac{1}{\sin^2 t} \cdot \sin^2 t \right)$$

$$= e^{\cot t}(2 \sin t \cos t - 1)$$

b. This function is differentiated by using the quotient rule and chain rule:

$$\frac{d}{dt} \frac{\cot(t^2 + 1)}{1 - \csc t} = \frac{(1 - \csc t) \cdot \dfrac{d}{dt} \cot(t^2 + 1) - \cot(t^2 + 1) \cdot \dfrac{d}{dt}(1 - \csc t)}{(1 - \csc t)^2}$$

$$= \frac{-\csc^2(t^2 + 1) \cdot 2t(1 - \csc t) - \cot(t^2 + 1) \cdot \csc t \cot t}{(1 - \csc t)^2} \quad \blacksquare$$

The main results of this section are summarized in the box on page 462. Note that each derivative formula has a corresponding integral formula.

The Other Four
Trigonometric Functions

Definitions

$$\tan t = \frac{\sin t}{\cos t} \qquad \sec t = \frac{1}{\cos t}$$

$$\cot t = \frac{\cos t}{\sin t} \qquad \csc t = \frac{1}{\sin t}$$

Identities

$$1 + \tan^2 t = \sec^2 t \qquad 1 + \cot^2 t = \csc^2 t$$

Derivatives

$$\frac{d}{dt} \tan t = \sec^2 t \qquad \frac{d}{dt} \sec t = \sec t \tan t$$

$$\frac{d}{dt} \cot t = -\csc^2 t \qquad \frac{d}{dt} \csc t = -\csc t \cot t$$

Integrals

$$\int \sec^2 t \, dt = \tan t + c \qquad \int \sec t \tan t \, dt = \sec t + c$$
$$\int \csc^2 t \, dt = -\cot t + c \qquad \int \csc t \cot t \, dt = -\csc t + c$$

PROBLEMS (Section 9-4)

1. Find the values of $\tan t$, $\sec t$, $\cot t$, and $\csc t$ when $t = \pi/4$, $2\pi/3$, $5\pi/6$, $5\pi/4$, and $11\pi/6$.

2. Find the values of $\tan t$, $\sec t$, $\cot t$, and $\csc t$ when $t = -\pi/4$, $\pi/6$, $-5\pi/6$, $7\pi/6$, and $7\pi/4$.

3. Prove the identity $1 + \cot^2 t = \csc^2 t$. [*Hint:* Divide the identity $\sin^2 t + \cos^2 t = 1$ by $\sin^2 t$.]

4. Prove the identity $\sec(t - \pi/2) = \csc t$. [*Hint:* Write the secant in terms of the cosine and use an appropriate identity for the cosine.]

In Problems 5–40, compute the derivative of the given function.

5. $\tan 5t$

6. $\sec 3t$

7. $\cot(-7t)$

8. $\csc(-4t + 1)$

9. $\sec(e^{6t} + 1)$

10. $\csc(2e^{3t})$

11. $2 \tan(4t^2 + 6t + 1)$

12. $3 \cot(-3t^2 + 1)$

13. $5 \tan\left(\dfrac{1}{t + 1}\right)$

14. $\ln(\tan 2t)$

15. $\ln(2 \tan t)$

16. $e^{3 \sec t^2}$

17. $4e^{-\cot t^2}$

18. $te^{\csc 2t}$

19. $t \csc\left(\dfrac{2}{t}\right)$

20. $t^2 \sec 4t$

21. $t \sec e^t$

22. $t^2 \cot e^{-t}$

23. $e^t \cot 2t$

24. $e^{2t} \csc 3t$

25. $\ln(\sec t + \tan t)$

26. $\ln(\csc 2t)$

27. $\sec[\ln(t^2 + 1)]$

28. $\tan\left(\ln \dfrac{1}{t}\right)$

29. $e^{-2t} \sec t$

30. $e^{-t} \csc 3t$

31. $\sec^3(2t^2)$

32. $\dfrac{1}{1 + \tan^2 2t}$

33. $\dfrac{1}{1 + \cot^2 2t}$

34. $\dfrac{t}{\sec t}$

35. $\dfrac{t + 1}{t \tan t}$

36. $\dfrac{\csc t}{\tan t}$

37. $\dfrac{\sec t}{\cot t}$

38. $\dfrac{e^t}{t \sec t}$

39. $\dfrac{e^t}{t \csc t}$

40. $\dfrac{\ln t}{t \csc t}$

In Problems 41–44, compute the given integral.

41. $\displaystyle\int \sec^2 2t \, dt$

42. $\displaystyle\int \sec 3t \tan 3t \, dt$

43. $\displaystyle\int_{\pi/2}^{\pi/3} \csc^2\left(\dfrac{t}{2}\right) dt$

44. $\displaystyle\int_0^{\pi/9} \sec^2 3t \, dt$

45. Show that
$$\frac{d}{dt} \csc t = -\csc t \cot t$$

[*Hint:* Write $\csc t = 1/(\sin t)$ and use the quotient rule.]

9-5 TRIANGLES AND TRIGONOMETRY

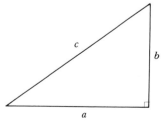

Figure 9-20
Pythagorean theorem, $c^2 = a^2 + b^2$

The values of the trigonometric functions of angles between 0 and $\pi/2$ radians can be expressed in terms of ratios of the lengths of the sides of a right triangle. The fundamental fact about right triangles is the Pythagorean theorem, which has been discussed earlier in this book. Recall that the theorem asserts that the square of the hypotenuse (the side opposite the right angle) is equal to the sum of the squares of the other two sides (see Figure 9-20). Given an angle θ between 0 and $\pi/2$ radians, we associate it with a right triangle as shown in Figure 9-21 on page 464. The side opposite the angle θ is then called the **opposite side** and the base of the triangle is called the **adjacent side.** It then follows that the sine and cosine of the angle θ are given by

$$\sin \theta = \frac{b}{c} = \frac{\text{Opposite}}{\text{Hypotenuse}}$$

$$\cos \theta = \frac{a}{c} = \frac{\text{Adjacent}}{\text{Hypotenuse}}$$

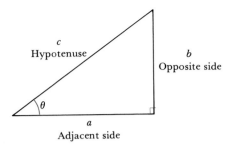

Figure 9-21

To see that this is so, simply scale the sides by a factor $1/c$, as in Figure 9-22. The hypotenuse will then be 1 unit long and will serve as the radius of a circle. Our definition of sine and cosine given in Section 9-2 then shows that $\sin\theta = b/c$ and $\cos\theta = a/c$.

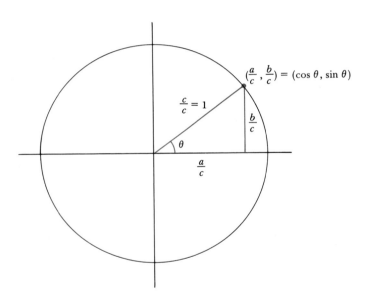

Figure 9-22

Example 21   Suppose that the adjacent side of an angle of $\pi/3$ radians in a right triangle is 4 units long. How long is the hypotenuse? How long is the opposite side?

Solution   Since $\theta = \pi/3$, we find from Table 9-1 (in Section 9-2) that $\cos\theta = \tfrac{1}{2}$. But $\cos\theta = a/c$ and therefore,

$$\frac{1}{2} = \frac{a}{c}$$

$$c = 2a$$

We know that $a = 4$, so

$$c = 2 \cdot 4 = 8$$

That is, the hypotenuse is 8 units long. The length of the opposite side can be obtained from the Pythagorean theorem:

$$c^2 = a^2 + b^2$$
$$8^2 = 4^2 + b^2$$
$$b^2 = 64 - 16$$
$$= 48$$
$$b = \sqrt{48} = 4\sqrt{3}$$

∎

Since the other four trigonometric functions are defined in terms of the sine and cosine, we can define these functions in terms of sides of right triangles also.

$$\tan \theta = \frac{\sin \theta}{\cos \theta} = \frac{b/c}{a/c} = \frac{b}{a} = \frac{\text{Opposite}}{\text{Adjacent}}$$

$$\cot \theta = \frac{1}{\tan \theta} = \frac{\text{Adjacent}}{\text{Opposite}}$$

$$\sec \theta = \frac{1}{\cos \theta} = \frac{1}{a/c} = \frac{c}{a} = \frac{\text{Hypotenuse}}{\text{Adjacent}}$$

$$\csc \theta = \frac{1}{\sin \theta} = \frac{1}{b/c} = \frac{c}{b} = \frac{\text{Hypotenuse}}{\text{Opposite}}$$

Example 22
Surveying

A surveyor stands 100 feet from the base of a building and measures the angle of elevation to the top of the building to be $30° = \pi/6$ radian (see Figure 9-23). If her sextant is 6 feet above the level ground, how tall is the building?

Figure 9-23

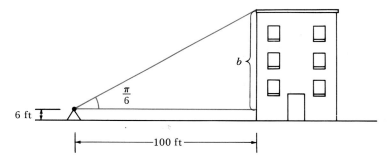

6 ft

100 ft

Solution

From Figure 9-23 we see that we know the adjacent side of the indicated triangle. Since

$$\tan \frac{\pi}{6} = \frac{\text{Opposite}}{\text{Adjacent}}$$

$$= \frac{b}{100}$$

we find that

$$b = 100 \tan \frac{\pi}{6}$$

$$= 100 \cdot \frac{1}{\sqrt{3}} \approx 57.73$$

Since the sextant is set 6 feet above the horizontal, the height of the building is

$$b + 6 \approx 63.73 \text{ feet}$$ ▪

Example 23   Suppose a 24 foot ladder leans against a vertical wall and the bottom of the ladder is pulled away from the wall at a rate of 1 foot per second. How fast is the angle that the ladder makes with the wall changing when the bottom of the ladder is 12 feet from the base of the wall?

Solution   Referring to Figure 9-24 we see that the angle θ that the ladder makes with the wall is related to the distance b by

$$\sin \theta = \frac{\text{Opposite}}{\text{Hypotenuse}}$$

$$= \frac{b}{24}$$

Figure 9-24

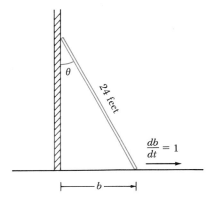

We know that $db/dt = 1$ and we would like to know the value of $d\theta/dt$ when $b = 12$. If we differentiate both sides of the equation above with respect to t, we obtain (using the chain rule)

$$\cos \theta \, \frac{d\theta}{dt} = \frac{1}{24} \frac{db}{dt} = \frac{1}{24}$$

When $b = 12$, we have

$$\sin \theta = \frac{12}{24} = \frac{1}{2}$$

and

$$\cos \theta = \sqrt{1 - \sin^2 \theta}$$

$$= \sqrt{1 - \tfrac{1}{4}} = \frac{\sqrt{3}}{2}$$

Therefore, when $b = 12$,

$$\frac{\sqrt{3}}{2} \frac{d\theta}{dt} = \frac{1}{24}$$

$$\frac{d\theta}{dt} = \frac{2}{\sqrt{3}} \cdot \frac{1}{24} \approx 0.048 \text{ radian per second}$$ ∎

Example 24   An unidentified flying object (UFO) passes over a radar station flying horizontally at an altitude of 30,000 feet. When the UFO is 50,000 feet from the station, the angle of inclination from the radar station to the UFO is decreasing at a rate of 0.03 radian per second. How fast is the UFO flying at this point?

Solution   For convenience we will measure distances in units of 10,000 feet. The situation is illustrated in Figure 9-25. From this figure, we see that

$$\tan \theta = \frac{3}{a}$$

Since θ and a are implicit functions of time, t, we may use implicit differentiation (see Chapter 4) to obtain

$$\sec^2 \theta \frac{d\theta}{dt} = -\left(\frac{3}{a^2}\right) \frac{da}{dt}$$

When $c = 5$, we use the Pythagorean theorem to obtain

$$c^2 = a^2 + 3^2$$
$$25 = a^2 + 9$$
$$4 = a$$

At this point in time,

$$\sec \theta = \frac{c}{a} = \frac{5}{4}$$

and hence,

$$\left(\frac{5}{4}\right)^2 \frac{d\theta}{dt} = -\left(\frac{3}{a^2}\right) \frac{da}{dt}$$

$$= -\frac{3}{16} \frac{da}{dt}$$

$$\frac{da}{dt} = -\left(\frac{16}{3}\right)\left(\frac{25}{16}\right) \frac{d\theta}{dt} = -\frac{25}{3} \frac{d\theta}{dt}$$

Figure 9-25
$$\frac{d\theta}{dt} = -0.03 \text{ when } c = 5$$
$$\frac{da}{dt} = ?$$

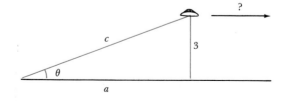

But we are given $d\theta/dt = -0.03$; therefore,

$$\frac{da}{dt} = \left(-\frac{25}{3}\right)(-0.03) = 0.25$$

That is, the speed is $0.25 \times 10{,}000 = 2500$ feet per second, or about 1700 miles per hour. ∎

PROBLEMS (Section 9-5)

For each triangle in Problems 1–8, find the values of all six trigonometric functions of the angle θ.

1.
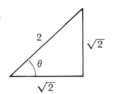

2.
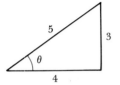

3.
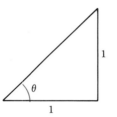

4.
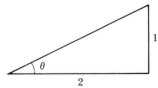

5.

6.
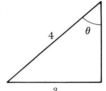

7.
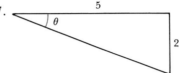

8.
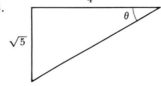

Find all sides of the triangles shown in Problems 9–16.

9.
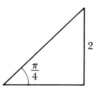

10.
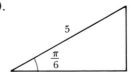

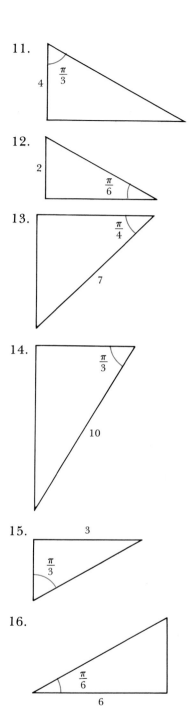

11.

12.

13.

14.

15.

16.

17. A guy wire is attached to an antenna at a point 50 feet above the level ground and makes an angle of 60° where it meets the ground (see the figure at the top of the next column). How long is the wire?

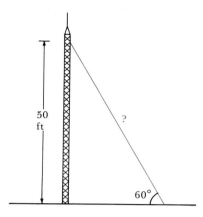

18. Suppose in Problem 17 that only 50 feet of wire is available and it is still required that the wire makes a 60° angle with the ground. How high up the antenna will the wire be attached?

19. A common method of measuring distances in inaccessible terrain is by triangulation. Suppose that we wish to measure the width of the river shown in the figure. First we sight a landmark directly across on the opposite bank and then we measure off a fixed distance parallel to the bank, say 60 feet. Using a transit we then sight the object on the opposite bank and measure the angle the line of sight makes with the line between the two points on the near shore. Suppose this angle is 70°. What is the width of the river?

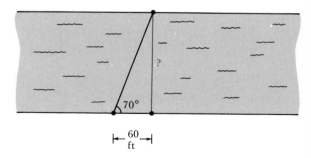

20. A loading ramp makes an angle of 15° with the horizontal ground and rests on the edge of a loading dock which is 5 feet high (see the figure). How long is the ramp?

21. A 10 foot high mine shaft is dug at an angle of 12° to the horizontal. How long is the opening at the surface (see the figure)?

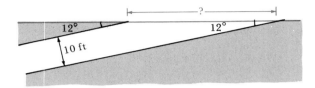

22. Suppose a reel of kite string is fastened to the ground, and the string is connected to a kite that is 100 feet above the ground and is being carried horizontally by the wind at a rate of 600 feet per minute (see the figure). How fast is the angle between the ground and the line connecting the reel and the kite changing when the kite is 100 feet from the reel, as measured along the ground?

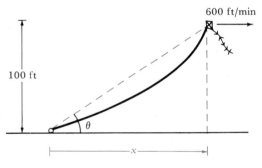

$$\frac{d\theta}{dt} = ? \text{ when } x = 100$$

23. An airplane is flying horizontally at a rate of 880 feet per second at an altitude of 20,000 feet. At what rate is the plane's angle of declination with an airport changing when the plane is 100,000 feet from the airport? (See the figure.)

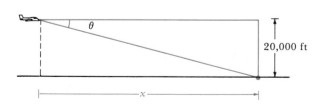

$$\frac{d\theta}{dt} = ? \text{ when } x = 100,000 \text{ feet}$$

24. A searchlight located at the center of a square prison yard that is 100 yards on a side rotates at a rate of 3 revolutions per minute, scanning the inside of the walls (see the figure). How fast is the light beam traveling along the wall when it reaches a corner of the yard?

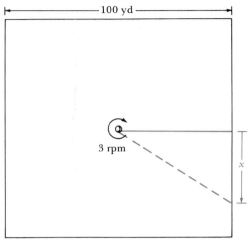

$$\frac{dx}{dt} = ? \text{ when } x = 50$$

25. A well-known hotel chain builds hotels with spectacular vertical atria and exterior glass elevators.

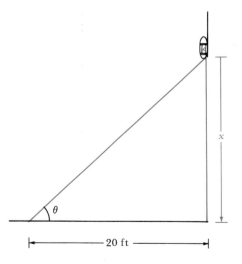

$$\frac{d\theta}{dt} = -1 \text{ degree/second when } x = 20 \text{ feet}$$

$$\frac{dx}{dt} = ?$$

Suppose the elevator is descending and you are standing in the atrium 20 feet from the base of the elevator shaft (see the figure). When the ele-vator is 20 feet above the floor, the angle of inclination is decreasing at a rate of 1° per second. How fast is the elevator descending?

9-6 CONCLUSION: CONTROLLING A ROBOTIC LASER SCANNER

We introduced this chapter with a problem in robotics. You may recall that the problem was to program a rotating robotic arm in such a way that its variable rate of rotation will allow it to track an object moving at a rate of 12 feet per minute along an assembly line 6 feet away. Referring to Figure 9-1, we see that the distance x that the unit travels past the robot forms the side opposite to θ in a right triangle. The side adjacent to θ has a length of 6 feet; therefore,

$$\tan \theta = \frac{x}{6}$$
$$6 \tan \theta = x$$

We know that $dx/dt = 12$ feet per minute, and we wish to determine $d\theta/dt$. Differentiating with respect to t (using the chain rule) gives

$$6 \sec^2 \theta \frac{d\theta}{dt} = \frac{dx}{dt}$$

However,

$$\sec^2 \theta = \tan^2 \theta + 1$$
$$= \left(\frac{x}{6}\right)^2 + 1$$

$$= \frac{x^2}{36} + 1$$

Since $dx/dt = 12$, $x = 12t$, and therefore,

$$\sec^2 \theta = \frac{(12t)^2}{36} + 1$$
$$= 4t^2 + 1$$

Substituting this expression and $dx/dt = 12$ in the equation from above, we obtain

$$6(4t^2 + 1) \frac{d\theta}{dt} = 12$$
$$\frac{d\theta}{dt} = \frac{2}{4t^2 + 1}$$

Therefore, the rate of rotation of the arm will vary from a maximum of 2 radians per minute when the object is directly opposite the robot ($x = 0$ when $t = 0$) to a minimum of $\frac{2}{101} \approx 0.02$ radian per minute when the unit is 60 feet down the line ($x = 60$ when $t = 5$) and the tracking is completed. The formula we obtained for $d\theta/dt$ can be used in a computer program to control the movement of the industrial robot automatically.

9-7 CHAPTER REVIEW

1. Arcs of circles, or angles, can be measured in degrees or radians. A full circle consists of 360° or 2π radians. To convert from degrees to radians, the formula a degrees $= a \times \pi/180$ radians is used, and to convert from radians to degrees, the formula a radians $= a \times 180/\pi$ degrees is used.

2. The sine and cosine are the fundamental trigonometric functions. The point $(\cos t, \sin t)$ is reached by starting at the point $(1, 0)$ on the unit circle centered on the origin in the xy-plane and marking off an arc of t radians in the counterclockwise sense. The functions $\cos t$ and $\sin t$ are periodic

with period 2π and satisfy the fundamental identity $\cos^2 t + \sin^2 t = 1$.

3. The derivatives of the sine and cosine functions are given by

$$\frac{d}{dt} \sin t = \cos t$$

$$\frac{d}{dt} \cos t = -\sin t$$

The corresponding integral formulas are

$$\int \cos t \, dt = \sin t + c$$

$$\int \sin t \, dt = -\cos t + c$$

4. Four other trigonometric functions can be defined in terms of the sine and cosine:

$$\tan t = \frac{\sin t}{\cos t}$$

$$\cot t = \frac{\cos t}{\sin t}$$

$$\sec t = \frac{1}{\cos t}$$

$$\csc t = \frac{1}{\sin t}$$

The tangent and secant are related by the identity

$$\tan^2 t + 1 = \sec^2 t$$

while the cotangent and cosecant are related by

$$\cot^2 t + 1 = \csc^2 t$$

The derivatives of these functions are given by

$$\frac{d}{dt} \tan t = \sec^2 t$$

$$\frac{d}{dt} \cot t = -\csc^2 t$$

$$\frac{d}{dt} \sec t = \sec t \tan t$$

$$\frac{d}{dt} \csc t = -\csc t \cot t$$

The corresponding integral formulas are

$$\int \sec^2 t \, dt = \tan t + c$$

$$\int \csc^2 t \, dt = -\cot t + c$$

$$\int \sec t \tan t \, dt = \sec t + c$$

$$\int \csc t \cot t \, dt = -\csc t + c$$

5. The trigonometric functions can also be defined in terms of the ratios of sides of a right triangle. In terms of sides of right triangles, we have

$$\sin t = \frac{\text{Opposite}}{\text{Hypotenuse}}$$

$$\cos t = \frac{\text{Adjacent}}{\text{Hypotenuse}}$$

$$\tan t = \frac{\text{Opposite}}{\text{Adjacent}}$$

$$\cot t = \frac{\text{Adjacent}}{\text{Opposite}}$$

$$\sec t = \frac{\text{Hypotenuse}}{\text{Adjacent}}$$

$$\csc t = \frac{\text{Hypotenuse}}{\text{Opposite}}$$

These relationships can be used to relate the rate at which a side changes to the rate at which an angle changes.

REVIEW PROBLEMS (Section 9-7)

1. Convert the given radian measure to degree measure:
   a. $2\pi/7$ b. $8\pi/3$ c. $-5\pi/6$ d. $-7\pi/6$

2. Convert the given degree measure to radian measure:
   a. $32°$ b. $150°$ c. $-27°$ d. $-16°$

3. Give the radian measure of the indicated angle:

a.
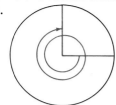

b.

c.

4. Find all values of t satisfying $-\pi \le t \le 5\pi$ and $\sin t = \sqrt{3}/2$.

5. Differentiate the following functions:
a. $\cos t^2$
b. $2 \sin 6t$
c. $\sin t \cos t - t$
d. $\dfrac{\sin 2t}{\cos t}$

6. Compute the following integrals:
a. $\displaystyle\int \cos 5t \, dt$
b. $\displaystyle\int_0^\pi \cos\left(\frac{t}{2}\right) dt$
c. $\displaystyle\int t \sin t^2 \, dt$
d. $\displaystyle\int_0^1 \sin \pi(t + 2) \, dt$

7. Differentiate:
a. $(2t + \cos t)^6$
b. $\cos e^{t^2}$
c. $e^{\cos t^2}$
d. $\ln(\sin^2 t + 1)$

8. Differentiate:
a. $\tan 2t$
b. $3t \tan 6t$
c. $\sec 3t$
d. $2e^t \sec 4t$

9. Differentiate:
a. $\cot 2t$
b. $3t \cot 6t$
c. $\csc 3t$
d. $2e^t \csc 4t$

10. Differentiate:
a. $e^{\csc 2t}$
b. $\dfrac{t \csc t}{t^2 + 1}$
c. $\dfrac{\ln(\csc 2t)}{\sin t}$
d. $\cot(e^{2t} + 1)$

11. Suppose a truck is equipped with mud shields that hang vertically 36 inches behind the rear axle and extend to within 6 inches of the road (see the figure). Using the formula

$$R(\theta) = \frac{v^2}{g} \sin 2\theta$$

(see Problem 49, Section 9-3) and assuming that gravel is scattered by the rear tire at a velocity of 10 feet per second, how far (behind the rear axle) will the gravel reach? (Use $g = 32$ ft/sec^2.)

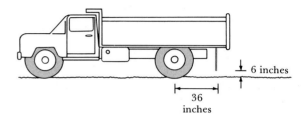

12. From a lighthouse 200 feet tall the angle of declination to a boat is 5° (see the figure). How far from the lighthouse is the boat?

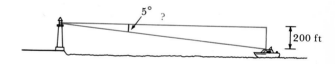

13. The angle of elevation from the base of a ski lift to the summit is 30°. If a skier rides 1200 feet from the base to the summit, how high did she rise?

14. Suppose a ship steams away at a 30° angle to a straight shore at a rate of 10 knots (nautical miles per hour). At what rate is the distance between the ship and the shore increasing?

15. A winch with a radius of 1 foot turns at 10 revolutions per minute, reeling up a cable that is attached to a boat. If the winch is on a dock and the axle of the winch is 11 feet above water level, how fast is the boat moving when 100 feet of cable stretch from it to the boat? (Ignore the change in radius due to the piling up of the cable on the winch.)

SEQUENCES, SERIES, AND APPROXIMATIONS

INTRODUCTION

Double-paned windows make an important contribution to home energy conservation. Such windows are composed of pairs of sealed glass panes with an air space (usually $\frac{1}{2}$ inch thick) in between. When a ray of sunlight falls upon a glass pane, part of it is reflected, part is absorbed by the glass, and part passes through the glass. For example, the outer pane might reflect 20% of the incident light, absorb 5% of it, and allow the remaining 75% to pass through. This diminished light beam then crosses the air space between the panes and strikes the inner pane. A portion, a_1, of the light will pass through the inner pane to the interior and a portion will be reflected back toward the outer pane. The outer pane will in turn reflect part of this beam back toward the inside and another portion, a_2, of the original beam will pass to the interior (see Figure 10-1). These reflections continue ad infinitum and give rise to an infinite list of numbers $a_1, a_2, a_3, \ldots$, representing the fractions of the original beam that pass through the inner pane on each successive reflection. Such a list of numbers is called a *sequence* in mathematics.

How much of the incident light enters the interior? The numbers $a_1, a_2, a_3, \ldots$ represent the fractions of the original beam that pass through the inner pane on each reflection. Therefore, the total fraction of the incident light that passes through is

$$a_1 + a_2 + a_3 + \cdots$$

The sum of an infinite sequence of numbers such as this is called an *infinite series.*

Sequences and infinite series are important from

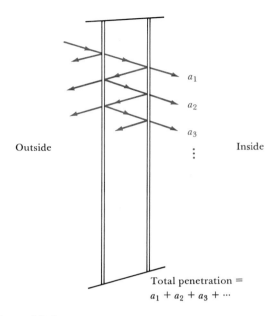

Outside

Inside

a_1

a_2

a_3

$\vdots$

Total penetration =
$a_1 + a_2 + a_3 + \cdots$

Figure 10-1
Double-pane window glazing

both a theoretical and a practical standpoint. In this chapter we will study the basics of sequences and infinite series, and we will see that they can be important tools in solving difficult problems. Many of the problems and illustrations in this chapter will involve computations that are too difficult or tedious to perform by hand — you will find a hand-held calculator indispensable.

10-1 SEQUENCES

When mathematicians speak of a **sequence** they mean a list of numbers arranged in some definite sequential order. For example, the sequence

$$2, 4, 8, 16, \ldots, 2^n, \ldots$$

is the list of all positive integer powers of 2, listed in increasing order. It is customary to use subscripts to distinguish the terms in a sequence and a notation of the form $\{a_n\}$ to indicate a general sequence

$$a_1, a_2, a_3, \ldots$$

Using this notation, the sequence above would be written $\{a_n\}$, where the **general term** of the sequence is $a_n = 2^n$, $n = 1, 2, 3, \ldots$.

Example 1   List the first five terms of each sequence $\{a_n\}$:

a. Suppose $a_n = (-1)^n n$. Then

$$a_1 = (-1)^1 \cdot 1 = -1, \qquad a_2 = (-1)^2 \cdot 2 = 2, \qquad a_3 = (-1)^3 \cdot 3 = -3,$$
$$a_4 = (-1)^4 \cdot 4 = 4, \qquad a_5 = (-1)^5 \cdot 5 = -5$$

b. Suppose $a_n = \dfrac{n(n+1)}{2}$. Then

$$a_1 = \frac{1 \cdot (1+1)}{2} = 1, \quad a_2 = \frac{2 \cdot (2+1)}{2} = 3, \quad a_3 = \frac{3 \cdot (3+1)}{2} = 6,$$

$$a_4 = \frac{4 \cdot (4+1)}{2} = 10, \quad a_5 = \frac{5 \cdot (5+1)}{2} = 15$$

c. Suppose $a_n = 1 + (0.1)^n$. Then

$$a_1 = 1 + (0.1)^1 = 1.1, \qquad a_2 = 1 + (0.1)^2 = 1.01,$$

$$a_3 = 1 + (0.1)^3 = 1.001, \qquad a_4 = 1 + (0.1)^4 = 1.0001,$$

$$a_5 = 1 + (0.1)^5 = 1.00001$$

d. Suppose $a_n = 3 + 2^{-n}$. Then

$$a_1 = 3 + 2^{-1} = 3.5, \qquad a_2 = 3 + 2^{-2} = 3.25,$$

$$a_3 = 3 + 2^{-3} = 3.125, \qquad a_4 = 3 + 2^{-4} = 3.0625,$$

$$a_5 = 3 + 2^{-5} = 3.03125$$

e. Suppose $a_n = \dfrac{2n^2}{n^2 + 1}$. Then

$$a_1 = \frac{2 \cdot 1^2}{1^2 + 1} = \frac{2}{2} = 1, \quad a_2 = \frac{2 \cdot 2^2}{2^2 + 1} = \frac{8}{5}, \quad a_3 = \frac{2 \cdot 3^2}{3^2 + 1} = \frac{18}{10},$$

$$a_4 = \frac{2 \cdot 4^2}{4^2 + 1} = \frac{32}{17}, \quad a_5 = \frac{2 \cdot 5^2}{5^2 + 1} = \frac{50}{26}$$

      It should be apparent that the terms in some of the sequences in Example 1 approach some definite numerical value as n becomes very large. We say that such sequences are **convergent**. More precisely, we say that the sequence $\{a_n\}$ **converges** to a number L, and write

$$\lim_{n \to \infty} a_n = L$$

if the terms a_n are arbitrarily close to L when n is sufficiently large. In this case we call L the **limit** of the sequence.

      For example, the sequence in Example 1c evidently converges to 1. The reason for this is that the difference between a_n and 1, that is,

$$a_n - 1 = 1 + (0.1)^n - 1 = (0.1)^n$$

is arbitrarily small if n is large enough. We therefore say that

$$\lim_{n \to \infty} [1 + (0.1)^n] = 1$$

The sequence in Example 1d converges to 3 since

$$a_n - 3 = 3 + 2^{-n} - 3 = 2^{-n}$$

can be made as small as we wish by taking n large enough. Therefore,

$$\lim_{n \to \infty} (3 + 2^{-n}) = 3$$

The five terms we computed for the sequence in Example 1e seem to indicate that it converges to 2. Indeed,

$$a_n - 2 = \frac{2n^2}{n^2 + 1} - 2$$

$$= \frac{2n^2}{n^2 + 1} - \frac{2n^2 + 2}{n^2 + 1} = -\frac{2}{n^2 + 1}$$

By taking n large enough, $-2/(n^2 + 1)$ can be made as close to 0 as we desire. Therefore,

$$\lim_{n \to \infty} \frac{2n^2}{n^2 + 1} = 2 \qquad \frac{2}{1 + \frac{1}{n^2}} = 2$$

Limits of sequences obey rules similar to those obeyed by the limits for functions discussed in Chapter 2. Namely, if $\{a_n\}$ and $\{b_n\}$ are convergent sequences, then

$$\lim_{n \to \infty}(a_n \pm b_n) = \lim_{n \to \infty} a_n \pm \lim_{n \to \infty} b_n \qquad \lim_{n \to \infty} ca_n = c \lim_{n \to \infty} a_n$$

$$\lim_{n \to \infty} a_n b_n = (\lim_{n \to \infty} a_n)(\lim_{n \to \infty} b_n) \qquad \lim_{n \to \infty} \frac{a_n}{b_n} = \frac{\lim_{n \to \infty} a_n}{\lim_{n \to \infty} b_n} \qquad \text{if } \lim_{n \to \infty} b_n \neq 0$$

Not all sequences are convergent. For example, the terms of the sequence in Example 1b become arbitrarily large as n increases and therefore do not approach any definite numerical value. Similarly, the terms with odd subscripts in Example 1a are negative and grow arbitrarily large in absolute value, while the terms with even subscripts are positive and grow arbitrarily large. Therefore, this sequence does not converge. We say that a sequence **diverges** if it does not converge. The sequences in Examples 1a and 1b are said to be **divergent.**

Example 2   The sequence

$$a_n = 5 + \frac{1}{n}$$

converges to 5. How large must n be in order that a_n will differ from 5 by less than 0.01? By less than 0.0001?

Solution   Note that $a_n > 5$ and

$$a_n - 5 = 5 + \frac{1}{n} - 5 = \frac{1}{n}$$

Therefore, a_n differs from 5 by less than 0.01 when

$$\frac{1}{n} < 0.01 = \frac{1}{100}$$

that is, $n > 100$. The term a_n will differ from 5 by less than 0.0001 when

$$\frac{1}{n} < 0.0001 = \frac{1}{10,000}$$

that is, $n > 10,000$.

Example 3   Show that the sequence

$$a_n = \frac{n + (-1)^n}{4n}$$

converges. How large must n be in order that a_n will differ from its limiting value by less than 0.001?

Solution   On dividing the numerator and denominator by n, we find

$$a_n = \frac{n + (-1)^n}{4n} = \frac{1 + (-1)^n/n}{4}$$

Since $(-1)^n/n$ approaches 0 as n grows arbitrarily large, we have

$$\lim_{n \to \infty} \frac{1 + (-1)^n/n}{4} = \frac{1}{4}$$

The term a_n is within 0.001 of $\frac{1}{4}$ if

$$\frac{1}{4} - 0.001 < a_n < \frac{1}{4} + 0.001$$

$$-0.001 < a_n - \frac{1}{4} < 0.001$$

But,

$$a_n - \frac{1}{4} = \frac{1 + (-1)^n/n}{4} - \frac{1}{4}$$

$$= \frac{1}{4} + \frac{(-1)^n}{4n} - \frac{1}{4}$$

$$= \frac{(-1)^n}{4n}$$

Therefore, a_n is within 0.001 of $\frac{1}{4}$ when

$$-0.001 < \frac{(-1)^n}{4n} < 0.001$$

or

$$\left| \frac{(-1)^n}{4n} \right| = \frac{1}{4n} < 0.001$$

that is,

$$n > \frac{1}{4(0.001)}$$
$$= 250$$

Thus, in order for a_n to differ from $\frac{1}{4}$ by less than 0.001, it is necessary that n be larger than 250.

A particular sequence that is important in probability theory is the sequence $a_n = n!$ of **factorials** of the positive integers. The symbol $n!$ is read "n factorial" and is defined by

$$n! = n(n-1) \cdot \cdots \cdot 2 \cdot 1$$

that is, $n!$ is the product of n and all smaller positive integers. For example,

$$1! = 1$$
$$3! = 3 \cdot 2 \cdot 1 = 6$$
$$6! = 6 \cdot 5 \cdot 4 \cdot 3 \cdot 2 \cdot 1 = 720$$

As a matter of notational convention we take $0! = 1$.

Example 4   Find the limit of the sequence

$$a_n = \frac{n!}{(n+1)!}$$

Solution   Since $(n+1)! = (n+1)n(n-1) \cdot \cdots \cdot 2 \cdot 1$ and $n! = n(n-1) \cdot \cdots \cdot 2 \cdot 1$, on canceling common factors, we find that

$$a_n = \frac{n!}{(n+1)!} = \frac{n(n-1) \cdot \cdots \cdot 2 \cdot 1}{(n+1)n(n-1) \cdot \cdots \cdot 2 \cdot 1} = \frac{1}{n+1}$$

This clearly converges to 0.   ∎

Another important type of sequence is a **geometric sequence,** that is, a sequence of the form

$$a_n = ar^n$$

where a and r are fixed numbers. For example, if $r = -\frac{1}{2}$ and $a = 1$, we have

$$a_1 = -\tfrac{1}{2}, \qquad a_2 = \tfrac{1}{4}, \qquad a_3 = -\tfrac{1}{8}, \qquad a_4 = \tfrac{1}{16}, \qquad a_5 = -\tfrac{1}{32}, \ldots$$

If $-1 < r < 1$, then the geometric sequence converges to 0. To see this, suppose that $a = 1$ and $0 < r < 1$, and we wish to choose n large enough so that $0 < a_n < 0.0001$. We can accomplish this by guaranteeing that

$$r^n < 0.0001$$

Taking common logarithms of both sides, we see that this is equivalent to

$$\log r^n < \log 0.0001$$
$$n \log r < -4$$
$$n > \frac{-4}{\log r}$$

(Remember that $\log r < 0$ since $r < 1$.) For instance, if $r = \frac{1}{2}$, then $\log r = \log \frac{1}{2} \approx -0.30298$ and hence, $n > -4/(-0.30298) \approx 13$. The case when $-1 < r < 0$ can be handled simply by noting that $r^n = (-1)^n s^n$ where $0 < s < 1$. In any case, we have

$$\lim_{n \to \infty} r^n = 0 \qquad \text{if} \qquad -1 < r < 1$$

Example 5 Suppose $500 is invested in an account that pays interest of 18% per annum compounded monthly. Find a sequence $\{a_n\}$ where a_n represents the balance in the account after n months.

Solution An interest rate of 18% per year is equivalent to 1.5% per month. Therefore, after 1 month the balance is

$$a_1 = 500 + 500(0.015) = 500(1.015)$$

After 2 months the balance is

$$a_2 = 500(1.015) + 500(1.015)(0.015)$$
$$= 500(1.015)(1 + 0.015) = 500(1.015)^2$$

Similarly, after 3 months the balance is

$$a_3 = 500(1.015)^2 + 500(1.015)^2(0.015) = 500(1.015)^3$$

In general, the balance after n months is

$$a_n = 500(1.015)^n$$

The sequence of monthly balances is a geometric sequence with $a = 500$ and $r = 1.015$. ∎

Example 6 Find the limit of the sequence $a_n = 2^n/n!$.

Solution The values of a_n computed in the table below seem to indicate that $\lim_{n\to\infty} a_n = 0$:

n	4	5	6	7	8	9	10
$2^n/n!$	0.6667	0.2667	0.0889	0.0254	0.0064	0.0014	0.0003

To see that the limit actually is 0, notice that for $n > 3$

$$\frac{2^n}{n!} = \frac{2}{1} \cdot \frac{2}{2} \cdot \frac{2}{3} \cdot \ldots \cdot \frac{2}{n-1} \cdot \frac{2}{n}$$
$$< 2 \cdot \frac{2}{n} \quad Since \; \frac{2}{3} < 1, \frac{2}{4} < 1, \ldots, \frac{2}{n-1} < 1$$

Therefore,

$$0 < \frac{2^n}{n!} < \frac{4}{n}$$

and $\lim_{n\to\infty}(4/n) = 0$. From this it follows that

$$\lim_{n\to\infty} \frac{2^n}{n!} = 0$$ ∎

Example 7 Suppose that $f(x) = e^{2x}$. Find a formula for the general term of the sequence $a_n = f^{(n)}(0)/n!$ [recall that $f^{(n)}(x)$ represents the nth derivative of $f(x)$].

Solution Since $f'(x) = 2e^{2x}$, we have $a_1 = 2e^{2(0)} = \frac{2}{1}$. Also, $f''(x) = 2 \cdot 2e^{2x} = 2^2e^{2x}$; therefore, $f''(0) = 2^2$ and $a_2 = 2^2/2!$. In general, we find that $f^{(n)}(x) = 2^ne^{2x}$; therefore, $a_n = f^{(n)}(0)/n! = 2^n/n!$. In Example 6 we saw that the limit of this sequence is $\lim_{n\to\infty} a_n = 0$. ∎

PROBLEMS (Section 10-1)

In Problems 1–10, compute the first five terms of the given sequence $\{a_n\}$.

1. $a_n = \dfrac{5n-4}{3n+1}$

2. $a_n = \dfrac{n^2-1}{n+1}$

3. $a_n = 1 + (-1)^{n-1}$

4. $a_n = 98 - 2(n+1)$

5. $a_n = \dfrac{\sin(n\pi/2)}{n}$

6. $a_n = \dfrac{9n^4-1}{3n^2+1}$

7. $a_n = (-1)^n n^2 2^{-n}$

8. $a_n = \dfrac{2^n}{(n-1)!}$

9. $a_n = 5$

10. $a_n = \left(1 + \dfrac{1}{n}\right)^n$

11. Find a formula for the nth term of the sequence
5, 8, 11, 14, 17, 20, 23, ...

12. Find a formula for the nth term of the sequence
$1 + \frac{1}{3}, -1 + \frac{2}{4}, 1 + \frac{3}{5}, -1 + \frac{4}{6}, 1 + \frac{5}{7}, \ldots$.

13. Find a formula for the nth term of the sequence
$1, \frac{6}{5}, \frac{9}{7}, \frac{12}{9}, \frac{15}{11}, \frac{18}{13}, \frac{21}{15}, \ldots$.

In Problems 14–35, determine whether the given sequence converges. If the sequence converges, find its limit.

14. $a_n = (-1)^n$

15. $a_n = \dfrac{n-1}{n}$

16. $a_n = \dfrac{1}{n^3}$

17. $a_n = -2 + (0.1)^n$

18. $a_n = 3 - \dfrac{4}{n}$

19. $a_n = \dfrac{1}{n} - \dfrac{1}{n+1}$

20. $a_n = \dfrac{3^{n+1}}{4^{n+2}}$

21. $a_n = \dfrac{2n^2-1}{3n^2+2}$

22. $a_n = n^n$

23. $a_n = (-1)^n \dfrac{n-2}{n+1}$

24. $a_n = \sin\dfrac{n\pi}{2}$

25. $a_n = \dfrac{n}{n+1} - \dfrac{n+1}{n}$

26. $a_n = \dfrac{1}{n^{5/2}}$

27. $a_n = \dfrac{7n^2 - 4n + 1}{5n^2 + 2}$

28. $a_n = \dfrac{\sqrt{n^2 + 1}}{2n}$

29. $a_n = \dfrac{1}{n^3(n^2 + 1)}$

30. $a_n = \dfrac{(-1)^n}{n + 1}$

31. $a_n = 1 + (-1)^n 2$

32. $a_n = \dfrac{(n - 3)!}{(n - 1)!}$

33. $a_n = \dfrac{4n^2 - 9}{2n + 3}$

34. $a_n = \dfrac{2n}{\sqrt{2n^2 + 1}}$

35. $a_n = \dfrac{(-1)^n n!}{n}$

36. Suppose \$1000 is deposited into an account that pays 14% per year (compounded at the end of each year). Find a formula for the amount a_n in the account after n years.

37. Let a_n stand for the sum of the first n positive integers, that is,
$$a_n = 1 + 2 + \cdots + (n - 1) + n$$
a. Write a_n in the reverse order
$$a_n = n + (n - 1) + \cdots + 2 + 1$$
Add this to a_n above to conclude
$$2a_n = (n + 1) + (n + 1) + \cdots + (n + 1) + (n + 1) = n(n + 1)$$
and hence, $a_n = n(n + 1)/2$.
b. Compute $a_1, a_2, a_3, a_4,$ and a_5, and verify in each case that a_n is the sum of the first n positive integers.

38. Suppose a manufacturer currently spends \$1 million each year on advertising and plans to increase its advertising budget by 4% each year.

Find a formula for a_n, the advertising budget after n years.

39. Suppose a colony of bacteria contains 1000 individuals and reproduces in such a way as to double its numbers every 8 hours. Find a formula for the number a_n of bacteria in the colony after n days.

40. Suppose r is a fixed number, $r \neq 1$. Let
$$a_n = 1 + r + r^2 + \cdots + r^{n-1} + r^n$$
a. Multiply the above equation by r to obtain
$$ra_n = r + r^2 + r^3 + \cdots + r^n + r^{n+1}$$
Now subtract the second equation from the first, canceling common terms, to obtain
$$a_n - ra_n = 1 - r^{n+1}$$
Conclude that
$$a_n = \dfrac{1 - r^{n+1}}{1 - r}$$
b. Suppose $r = 2$. Compute a_{10}.

41. Suppose $a_1 = 1$ and $a_2 = 1$, and for $n > 1$
$$a_{n+1} = a_n + a_{n-1}$$
a. Compute the first twelve terms of this sequence.
b. Let $b_n = a_{n+1}/a_n$ and suppose $\lim_{n \to \infty} b_n = b$. Show that $b_n = 1 + (1/b_{n-1})$.
c. Conclude that $b = (1 + \sqrt{5})/2$. (This number was called the "divine proportion" by Renaissance writers.)
d. Compute $b_1, b_2, b_3, b_4, b_5,$ and b_6, and compare each with b.

The sequence in this exercise has been of interest to mathematicians since the Middle Ages and is known as the *Fibonacci sequence* (after Leonardo of Pisa alias Leonardo Fibonacci, a 12th century scholar).

42. Suppose that \$50 is deposited in an account that earns 18% per year compounded monthly.
a. Find a formula for a_n, the amount in the account after n months.
b. Compute the first five terms of this sequence.

10-2 ROOT-FINDING SEQUENCES

One of the most common mathematical problems is that of solving equations. Many of the equations that occur in practical problems are difficult or impossible to solve exactly. For example, the equation

$$e^{-x} = \sin x$$

cannot be solved by simple algebraic means. If a solution cannot be determined exactly, then the next best thing is to obtain a good *approximation* to the solution. In this section we investigate two methods of generating *sequences* that approximate solutions of equations. The first method, which is based on a fundamental property of continuous functions, is called the *bisection method*. The second method makes use of the derivative of the function and is called *Newton's method*.

THE BISECTION METHOD

The **bisection method** can be used to solve an equation of the form

$$f(x) = 0$$

where f is a continuous function. A solution of $f(x) = 0$ is called a **root** of the function $f(x)$. Suppose, for example, we wish to solve the equation

$$x^3 = 2$$

We rewrite this equation in the form $f(x) = 0$ as

$$f(x) = x^3 - 2 = 0$$

The bisection method produces two sequences of numbers that "bracket" the root. That is, at each step of the bisection method we will produce numbers a_n and b_n such that

$$a_n \leq s \leq b_n$$

where s is a root of $f(x)$. The first pair of numbers is obtained by simply choosing numbers a_1 and b_1 so that the function has a sign change between a_1 and b_1, that is, $f(a_1)f(b_1) < 0$. For example, since

$$f(1) = 1^3 - 2 = -1 < 0$$

and

$$f(2) = 2^3 - 2 = 6 > 0$$

the function $f(x) = x^3 - 2$ has a sign change between $a_1 = 1$ and $b_1 = 2$. Since the function is continuous, is negative at $x = 1$, and is positive at $x = 2$, it must take the value 0 somewhere between $a_1 = 1$ and $b_1 = 2$ (see Figure 10-2).

We now narrow the interval in which the root is trapped by bisecting it (that is, cutting it in half). The midpoint of the interval $[a_1, b_1] = [1, 2]$ is the point

$$c = \frac{a_1 + b_1}{2} = \frac{3}{2} = 1.5$$

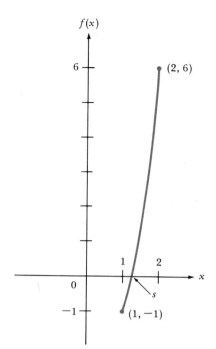

Figure 10-2
Bracketing a root of f(x)

Next, we find the sign of $f(c)$:

$$f(c) = f(1.5) = 1.5^3 - 2 = 1.375 > 0$$

Therefore, $a_1 < c < b_1$, and

$$f(a_1) < 0 \qquad f(c) > 0 \qquad f(b_1) > 0$$

The function now changes sign on the smaller interval $[a_1, c] = [1, 1.5]$. We therefore set $a_2 = a_1 = 1$ and $b_2 = c = 1.5$ and begin the process again on the smaller interval $[a_2, b_2] = [1, 1.5]$. Again calling the midpoint of this interval c, we have

$$c = \frac{1 + 1.5}{2} = 1.25$$

and

$$f(c) = f(1.25) = (1.25)^3 - 2 \approx -0.0469 < 0$$

The sign change is now on the interval $[c, b_2] = [1.25, 1.5]$. We therefore set $a_3 = c = 1.25$ and $b_3 = b_2 = 1.5$ and we have the solution trapped in the interval $[a_3, b_3] = [1.25, 1.5]$. Continuing to bisect the interval successively in this way, we obtain two sequences $\{a_n\}$ and $\{b_n\}$ such that the solution s satisfies $a_n \leq s \leq b_n$ and $\lim_{n \to \infty}(b_n - a_n) = 0$. The process is illustrated in Figure 10-3 on page 486. If we continue the process on our example for five more steps, we obtain the results in Table 10-1.

Figure 10-3
The bisection method

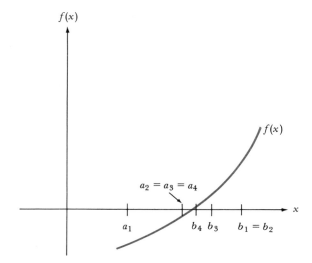

Table 10-1

n	a_n	b_n
1	1	2
2	1	1.5
3	1.25	1.5
4	1.25	1.375
5	1.25	1.3125
6	1.25	1.28125
7	1.25	1.265625
8	1.257813	1.265625

From these results we see that the solution lies between 1.257813 and 1.265625. The actual root is

$$\sqrt[3]{2} \approx 1.259921$$

To sum up, the bisection method works like this: Starting with an interval at the endpoints of which the function has different signs, successively bisect the interval, always keeping the half on which the function has a change of sign. A simplified flowchart for the bisection method is given in Figure 10-4.

Example 8
Electrostatic Repulsion

If a spherical soap bubble of radius 2 cm is charged with electric charges, then due to the mutual repulsion of these charges the bubble will expand to a new radius R where

$$q^2 = R(R^2 + 2R + 4)$$

and q is proportional to the total electric charge. Suppose $q^2 = 26.4$. Approximate the new radius R to two decimal places.

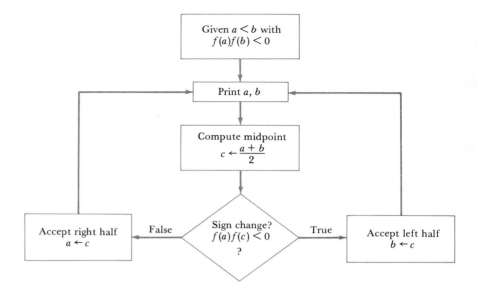

Figure 10-4
Flowchart for the bisection method

Solution   The radius R satisfies

$$R(R^2 + 2R + 4) - q^2 = 0$$

That is,

$$f(R) = R(R^2 + 2R + 4) - 26.4 = 0$$

Now,

$$f(2) = 2(4 + 4 + 4) - 26.4 = -2.4 < 0$$

and

$$f(2.2) = 2.2(13.24) - 26.4 = 2.728 > 0$$

Therefore, there is a root R between 2 and 2.2, and we take $a_1 = 2$, $b_1 = 2.2$:

$$a_1 = 2 < R < 2.2 = b_1$$

Now set

$$c = \frac{a_1 + b_1}{2} = 2.1$$

Then

$$f(c) = 2.1(2.1^2 + 2 \cdot 2.1 + 4) - 26.4 = 0.081 > 0$$

and we have a sign change between $a_1 = 2$ and $c = 2.1$. Accordingly, we take $a_2 = 2$, $b_2 = 2.1$, and compute

$$c = \frac{a_2 + b_2}{2} = \frac{4.1}{2} = 2.05$$

Now,

$$f(c) \approx -1.180 < 0$$

Hence, there is a sign change between $c = 2.05$ and $b_2 = 2.1$, and we set $a_3 = 2.05$, $b_3 = 2.1$. The midpoint of this interval is

$$c = \frac{2.05 + 2.1}{2} = 2.075$$

Checking the sign at this midpoint, we find that

$$f(c) = f(2.075) \approx -0.555 < 0$$

Therefore, there is a sign change between $c = 2.075$ and $b_3 = 2.1$, and we take $a_4 = 2.075$, $b_4 = 2.1$. Again, the midpoint is computed:

$$c = \frac{2.075 + 2.1}{2} = 2.0875$$

and the sign of the function is checked:

$$f(c) = f(2.0875) \approx -0.238 < 0$$

Therefore, there is a sign change between $c = 2.0875$ and $b_4 = 2.1$. We take $a_5 = 2.0875$, $b_5 = 2.1$ and bisecting again, we find

$$c = \frac{2.0875 + 2.1}{2} = 2.09375$$

$$f(c) = f(2.09375) \approx -0.0789 < 0$$

Hence, $a_6 = 2.09375$ and $b_6 = 2.1$. Therefore, the radius R satisfies

$$2.09 < R < 2.10$$ ■

NEWTON'S METHOD

Newton's method gives a single sequence of approximations that often converges quite rapidly to the root. The reason for the fast convergence is that Newton's method uses information not only about the function but also about its derivative. The idea is illustrated in Figure 10-5. Suppose x_1 is an approximation to the root of a function $f(x)$. Consider the tangent line to the graph of

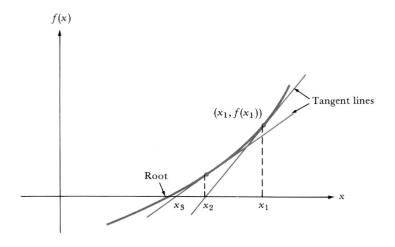

Figure 10-5
Newton's method

$f(x)$ at the point $(x_1, f(x_1))$. This line has slope $f'(x_1)$ and therefore its equation is

$$y - f(x_1) = f'(x_1)(x - x_1)$$

Since this tangent line approximates the graph of $f(x)$, the root of this line (that is, the value of x that makes $y = 0$) should approximate the root of the function $f(x)$. Let us call this root of the tangent line x_2. Then when we substitute $x = x_2$ into the equation above, we get $y = 0$:

$$0 - f(x_1) = f'(x_1)(x_2 - x_1)$$

$$x_2 = x_1 - \frac{f(x_1)}{f'(x_1)}$$

This formula allows us to generate a new approximation x_2 to the root given the old approximation x_1. Once we have x_2, we can generate a third approximation in a similar way:

$$x_3 = x_2 - \frac{f(x_2)}{f'(x_2)}$$

In general, the terms of the Newton sequence are given by

Newton's Method for Solving
$f(x) = 0$

$$x_{n+1} = x_n - \frac{f(x_n)}{f'(x_n)}$$

Example 9
Electrostatic Repulsion

Using $R_1 = 2$ we will generate the next two Newton approximations to the root of the function

$$f(R) = R(R^2 + 2R + 4) - 26.4$$

of Example 8. The derivative is

$$f'(R) = 3R^2 + 4R + 4$$

Therefore, the next two Newton approximations are

$$R_2 = R_1 - \frac{f(R_1)}{f'(R_1)}$$

$$= 2 - \frac{f(2)}{f'(2)}$$

$$= 2 - \frac{-2.4}{24} = 2.1$$

$$R_3 = R_2 - \frac{f(R_2)}{f'(R_2)}$$

$$= 2.1 - \frac{f(2.1)}{f'(2.1)}$$

$$\approx 2.0968$$ ∎

Example 10
Nerve Axons

In the nervous system, electrical impulses travel along nerve fibers which make up the *axon*. The axon is surrounded by an insulating layer called the *myelin sheath.* Both the axon and the myelin sheath may be regarded as thin circular cylinders of diameter d and D, respectively (see Figure 10-6). The velocity of the impulse along the axon is given by

$$v = -k \left(\frac{d}{D} \right)^2 \ln \frac{d}{D}$$

where k is a constant of proportionality. Suppose this velocity is measured and it is found that $v/k = 0.18$. What is the ratio, d/D, of the diameter of the axon to the diameter of the myelin sheath?

Figure 10-6
Nerve axon

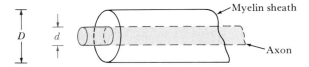

Myelin sheath

D d

Axon

Solution   Suppose we let $x = d/D$. Then the equation above can be rewritten as

$$0.18 = \frac{v}{k} = -x^2 \ln x$$

Therefore, we would like to find a root of the equation

$$f(x) = 0.18 + x^2 \ln x = 0$$

We then have

$$f'(x) = 2x \ln x + x$$

and Newton's method gives

$$x_{n+1} = x_n - \frac{f(x_n)}{f'(x_n)}$$

$$= x_n - \frac{x_n^2 \ln x_n + 0.18}{2x_n \ln x_n + x_n}$$

Since the axon is contained within the sheath, it is clear that $x = d/D < 1$. As a first approximation we will take $x_1 = 1$. Substituting this into the formula above, we obtain

$$x_2 = 1 - \frac{1^2 \ln 1 + 0.18}{2 \cdot 1 \cdot \ln 1 + 1}$$

$$= 1 - 0.18 = 0.82$$

Using Newton's method successively and carrying the computations to five decimal places, we obtain

$$x_3 = 0.72585$$
$$x_4 = 0.68293$$
$$x_5 = 0.66976$$
$$x_6 = 0.66830$$
$$x_7 = 0.66828$$

The last two approximations indicate that the ratio of the diameters is 0.668 to three decimal places.   ■

Example 11
Discount Rate of an Annuity

The present value P of a \$20,000 per year annuity that is to be received over a 15 year period beginning immediately is

$$P = \int_0^{15} e^{-rt} 20{,}000 \, dt$$

$$= \frac{-20{,}000}{r} e^{-rt} \bigg|_{t=0}^{t=15}$$

$$= \frac{20{,}000}{r} (1 - e^{-15r})$$

where r is the *discount rate*. Suppose you are told that the present value of the annuity is \$155,375. What is the discount rate?

Solution

Since $P = \$155{,}375$, we wish to solve the equation

$$155{,}375 = \frac{20{,}000}{r} (1 - e^{-15r})$$

$$\frac{155{,}375}{20{,}000} \cdot r = 1 - e^{-15r}$$

$$7.7688r = 1 - e^{-15r}$$

Now, this is equivalent to

$$f(r) = e^{-15r} + 7.7688r - 1 = 0$$

and to use Newton's method we compute $f'(r)$:

$$f'(r) = -15e^{-15r} + 7.7688$$

The formula for Newton's method is then

$$r_{n+1} = r_n - \frac{e^{-15r_n} + 7.7688r_n - 1}{-15e^{-15r_n} + 7.7688}$$

As a first approximation, we use a discount rate of 5%, that is, $r_1 = 0.05$. Substituting into the formula above and rounding to three decimal places, we obtain

$$r_2 \approx 0.129$$

Two more applications of the Newton formula yield

$$r_3 \approx 0.103 \quad \text{and} \quad r_4 \approx 0.100$$

Therefore, to the nearest percent, the discount rate is 10%.   ■

SQUARE ROOTS

Newton's method is behind a popular algorithm for computing square roots. For example, suppose we would like to compute $\sqrt{3}$. In order to do so by using Newton's method, we must interpret $\sqrt{3}$ as the solution of some equation of the form $f(x) = 0$. It does not take long to see that $f(x) = x^2 - 3$ is the appropriate function:

$$f(\sqrt{3}) = \sqrt{3}^2 - 3 = 0$$

Since $f'(x) = 2x$, the formula for Newton's method for computing $\sqrt{3}$ is

$$
\begin{aligned}
x_{n+1} &= x_n - \frac{f(x_n)}{f'(x_n)} \\
&= x_n - \frac{x_n^2 - 3}{2x_n} \\
&= x_n - \frac{x_n}{2} + \frac{3}{2x_n} \\
&= \frac{1}{2}\left(x_n + \frac{3}{x_n}\right)
\end{aligned}
$$

Example 12   Using $x_1 = 1$ we will compute the next four terms in the Newton sequence above for $\sqrt{3}$. Carrying out the computations to five decimal places, we have

$$x_2 = \frac{1}{2}\left(x_1 + \frac{3}{x_1}\right) = \frac{1}{2}\left(1 + \frac{3}{1}\right) = 2$$

$$x_3 = \frac{1}{2}\left(x_2 + \frac{3}{x_2}\right) = \frac{1}{2}\left(2 + \frac{3}{2}\right) = 1.75$$

$$x_4 = \frac{1}{2}\left(x_3 + \frac{3}{x_3}\right) = \frac{1}{2}\left(1.75 + \frac{3}{1.75}\right) \approx 1.73214$$

$$x_5 = \frac{1}{2}\left(x_4 + \frac{3}{x_4}\right) = \frac{1}{2}\left(1.73214 + \frac{3}{1.73214}\right) \approx 1.73205$$

This last approximation agrees with the actual value of $\sqrt{3}$ to five decimal places. ∎

PROBLEMS (Section 10-2)

1. Show that the equation $e^x = 4x$ has a solution between 0 and 1. Use the bisection method to approximate this solution to two decimal places.

2. Show that the equation $\ln x = -x$ has a solution between $\frac{1}{2}$ and 1. Use the bisection method to approximate this solution to two decimal places.

3. Show that the equation $5x^2 - 2 = 0$ has a solution between 0 and 1. Use the bisection method to approximate this solution to two decimal places.

4. Show that the equation $3x^3 - x = 1$ has a solution between 0 and 1. Use the bisection method to approximate this root to two decimal places.

5. Using $x_1 = 2$, compute the next four approximations to $\sqrt{5}$ obtained by Newton's method (see Example 12).

6. Using $x_1 = 3$, compute the next four approximations to $\sqrt{11}$ obtained by Newton's method (see Example 12).

7. Explain how Newton's method can be used to approximate $\sqrt[3]{6}$, that is, to find the solution of $x^3 - 6 = 0$. Compute the approximation x_5 to $\sqrt[3]{6}$, starting with $x_1 = 1$.

8. Explain how Newton's method can be used to approximate $\sqrt[5]{11}$, that is, to find the solution of $x^5 - 11 = 0$. Compute the approximation x_5 to $\sqrt[5]{11}$, starting with $x_1 = 2$.

9. Verify that the equation $x^3 - 2x - 5 = 0$ has a solution between 2 and 3. Start with $x_1 = 2$ and compute x_4 by Newton's method.

10. Verify that the equation $x^3 + x^2 - x + 1 = 0$ has a solution between -2 and -1. Start with $x_1 = -2$ and compute x_7 by Newton's method.

11. Show that the equation $\cos x = x$ has a solution between 0 and $\pi/2$. Use Newton's method to approximate this solution to four decimal places.

12. Show that the equation $x = 2 \sin x$ has a solution between $\pi/2$ and π. Use Newton's method to approximate this solution to four decimal places.

13. One of the earliest uses of Newton's method by Newton himself was to solve Kepler's equation of astronomy. This equation has the form

$$E = \theta - \varepsilon \sin \theta$$

where ε is the *eccentricity* of an elliptical orbit and $E = 2\pi t/T$, where T is the period of revolution and t is the time since perihelion passage (see the figure). For the earth's orbit, $\varepsilon = 0.017$ and $T = 365.25$ days. Use Newton's method to find the central angle θ, 30 days after perihelion passage (that is, when $t = 30$).

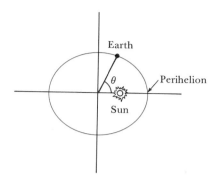

The earth's orbit

14. The equation $\theta = 1.1 \sin \theta$ occurs in experimental biology. Use Newton's method to find a positive solution of this equation accurate to three decimal places.

15. The **method of successive substitution** for solving an equation of the type $\theta = f(\theta)$ is defined by $\theta_{n+1} = f(\theta_n)$. Solve the equation in Problem 14 using the method of successive substitution and the same initial value you used in Problem 14. Compare the relative rates of convergence of Newton's method and the method of successive substitution for this problem.

16. The probability that the electron in an unexcited hydrogen atom lies between a radius of r and $r + \Delta r$ angstroms of the nucleus is given by $P(r) \Delta r$, where

$$P(r) = \frac{r^2}{0.037} e^{-r/0.265}$$

Use Newton's method to approximate to two decimal places the radius r such that $P(r) = 0.6$. [*Hint:* Note that $P(1) \approx 0.62$.]

10-3 INFINITE SERIES

An **infinite series** results when all the terms of a sequence are added:

$$a_1 + a_2 + a_3 + \cdots$$

For example, each of the following is an infinite series:

$$\tfrac{1}{10} + \tfrac{1}{100} + \tfrac{1}{1000} + \cdots$$
$$(-1) + 1 + (-1) + 1 + (-1) + \cdots$$
$$2 + 2 + 2 + 2 + 2 + \cdots$$

It is possible to associate a finite "sum" with certain infinite series. For example, the first infinite series above actually sums to a finite number. To see this, consider the sequence of partial sums of the series:

$$\tfrac{1}{10} = 0.1$$
$$\tfrac{1}{10} + \tfrac{1}{100} = 0.11$$
$$\tfrac{1}{10} + \tfrac{1}{100} + \tfrac{1}{1000} = 0.111$$
$$\tfrac{1}{10} + \tfrac{1}{100} + \tfrac{1}{1000} + \tfrac{1}{10,000} = 0.1111$$

It seems that these sums approach a decimal fraction of the form 0.1111111. . . . If you divide 1 by 9 on a calculator, you will find that this decimal fraction is $\tfrac{1}{9}$. We will show later that the sum represented by the first infinite series above actually is $\tfrac{1}{9}$ in the sense that the sequence of partial sums of the series converges to $\tfrac{1}{9}$. In this case, we say that the infinite series **converges** to $\tfrac{1}{9}$.

Consider the sequence of partial sums of the second infinite series above:

$$-1 = -1$$
$$-1 + 1 = 0$$
$$-1 + 1 - 1 = -1$$
$$-1 + 1 - 1 + 1 = 0$$
$$-1 + 1 - 1 + 1 - 1 = -1$$

In this case, it is clear that the sequence of partial sums oscillates between -1 and 0 and approaches no definite value. This infinite series therefore does not converge, and we say that it **diverges.**

The sequence of partial sums of the third infinite series above grows without bound:

$$2 = 2$$
$$2 + 2 = 4$$
$$2 + 2 + 2 = 6$$
$$2 + 2 + 2 + 2 = 8$$

Therefore, this infinite series also diverges.

SIGMA NOTATION   In dealing with series, it is convenient to have a compact notation for sums. The capital Greek letter sigma, Σ, is traditionally used to denote summation. We have already used this notation in our discussion of Riemann sums in Chapter 5.

For example, the sum

$$a_1 + a_2 + a_3 + a_4$$

may be denoted

$$\sum_{k=1}^{4} a_k$$

where k is called the **index of summation,** and the notation indicates that this index takes the values 1, 2, 3, 4 and the resulting terms are summed. That is,

$$\sum_{k=1}^{4} a_k = a_1 + a_2 + a_3 + a_4$$

For example,

$$\sum_{k=1}^{3} \frac{1}{k} = \frac{1}{1} + \frac{1}{2} + \frac{1}{3} = \frac{11}{6}$$

$$\sum_{k=2}^{4} 2^{-k} = 2^{-2} + 2^{-3} + 2^{-4}$$

$$= \frac{1}{4} + \frac{1}{8} + \frac{1}{16} = \frac{7}{16}$$

We can write a given sum in terms of sigma notation in many different ways. For example, the sum

$$a_1 + a_2 + a_3 + a_4 + a_5$$

may be written

$$\sum_{k=1}^{5} a_k \quad \text{or} \quad \sum_{i=1}^{5} a_i \quad \text{or} \quad \sum_{j=5}^{9} a_{j-4}$$

In the case of an infinite series, the index of summation takes on arbitrarily large values and we write

$$a_1 + a_2 + a_3 + \cdots = \sum_{k=1}^{\infty} a_k$$

Example 13   Write each infinite series in sigma notation:

a. $1 + \dfrac{1}{2} + \dfrac{1}{4} + \dfrac{1}{8} + \cdots = \displaystyle\sum_{k=0}^{\infty} \left(\dfrac{1}{2}\right)^k$

b. $1 + \dfrac{1}{3} + \dfrac{1}{5} + \dfrac{1}{7} + \cdots = \displaystyle\sum_{k=0}^{\infty} \dfrac{1}{2k+1}$

c. $\dfrac{1}{1} + \dfrac{2}{1} + \dfrac{4}{2} + \dfrac{8}{3 \cdot 2} + \dfrac{16}{4 \cdot 3 \cdot 2} + \cdots = \displaystyle\sum_{k=0}^{\infty} \dfrac{2^k}{k!}$

d. $-1 + 2^2 - 3^2 + 4^2 - \cdots = \displaystyle\sum_{k=1}^{\infty} (-1)^k k^2$

e. $\dfrac{1}{2} + \dfrac{2}{3} + \dfrac{3}{4} + \dfrac{4}{5} + \cdots = \displaystyle\sum_{k=1}^{\infty} \dfrac{k}{k+1}$

GEOMETRIC SERIES

The sum of the terms of a geometric sequence (see Section 10-1) is called a **geometric series.** That is, a geometric series is an infinite series of the form

$$\sum_{k=0}^{\infty} ar^k$$

where a and r are nonzero constants. Notice that for $r = 1$ the partial sums of the geometric series are

$$a, \quad a + a, \quad a + a + a, \quad a + a + a + a, \quad \ldots$$

which is clearly a divergent sequence. Therefore, the geometric series diverges for $r = 1$. Similarly, it is easy to see that the geometric series diverges when $r = -1$. In this case, the partial sums are

$$a, \quad 0, \quad a, \quad 0, \quad a, \quad \ldots$$

What happens for other values of r? Suppose that $r \neq \pm 1$ and let S_n stand for the partial sum

$$S_n = a + ar + ar^2 + \cdots + ar^{n-1} + ar^n$$

Then

$$S_n \cdot r = ar + ar^2 + ar^3 + \cdots + ar^n + ar^{n+1}$$

Subtracting the second equation from the first and canceling like terms, we obtain

$$S_n \cdot (1 - r) = a - ar^{n+1}$$

or, equivalently,

$$S_n = \frac{a}{1 - r} - \frac{ar^{n+1}}{1 - r}$$

Now we know that $\lim_{n \to \infty} r^{n+1} = 0$ if $|r| < 1$, while $\lim_{n \to \infty} r^{n+1}$ does not exist if $|r| > 1$. Therefore, the partial sums of the geometric series diverge if $|r| > 1$ or $r = \pm 1$, that is, $|r| \geq 1$, and converge to the value $a/(1 - r)$ if $|r| < 1$. We summarize these important facts in the box.

Geometric Series

The geometric series

$$\sum_{k=0}^{\infty} ar^k \quad \text{diverges} \qquad \text{if} \quad |r| \geq 1$$

while

$$\sum_{k=0}^{\infty} ar^k = \frac{a}{1 - r} \qquad \text{if} \quad |r| < 1$$

Example 14

The infinite series $\sum_{k=0}^{\infty} 5(-3)^{-k}$ is a geometric series with $a = 5$ and $r^k = (-3)^{-k} = (-\frac{1}{3})^k$, that is, $r = -\frac{1}{3}$. Since $|-\frac{1}{3}| = \frac{1}{3} < 1$, the series converges and, in fact,

$$\sum_{k=0}^{\infty} 5(-3)^{-k} = \frac{a}{1-r} = \frac{5}{1-(-\frac{1}{3})} = \frac{15}{4}$$

■

You have probably heard before that an infinite repeating decimal fraction is a rational number, that is, a quotient of integers. The reason for this is that an infinite repeating decimal is actually the sum of a geometric series. Consider, for example, the decimal fraction

0.111111 . . .

As we mentioned at the beginning of this section, this number represents the sum

$$\frac{1}{10} + \frac{1}{100} + \frac{1}{1000} + \cdots = \frac{1}{10} + \left(\frac{1}{10}\right)^2 + \left(\frac{1}{10}\right)^3 + \cdots$$

$$= \sum_{k=0}^{\infty} \left(\frac{1}{10}\right)\left(\frac{1}{10}\right)^k$$

The series given by this last expression is a geometric series with $a = \frac{1}{10}$ and $r = \frac{1}{10}$; therefore,

$$\sum_{k=0}^{\infty} \left(\frac{1}{10}\right)\left(\frac{1}{10}\right)^k = \frac{\frac{1}{10}}{1-\frac{1}{10}} = \frac{1}{9}$$

and

$$0.1111 \ldots = \sum_{k=0}^{\infty} \left(\frac{1}{10}\right)\left(\frac{1}{10}\right)^k = \frac{1}{9}$$

Example 15   Write the repeating decimal $7.23\overline{23}$ (the bar indicates that the corresponding digits are repeated indefinitely) as a rational fraction.

Solution   Consider first the repeating part:

$$0.2323\overline{23} = \frac{23}{100} + \frac{23}{10,000} + \frac{23}{1,000,000} + \cdots$$

$$= \left(\frac{23}{100}\right)\left(1 + \frac{1}{100} + \frac{1}{10,000} + \cdots\right)$$

$$= \left(\frac{23}{100}\right) \sum_{k=0}^{\infty} \left(\frac{1}{100}\right)^k$$

The series is a geometric series with $a = \frac{23}{100}$ and $r = \frac{1}{100}$; therefore,

$$\frac{23}{100} \sum_{k=0}^{\infty} \left(\frac{1}{100}\right)^k = \frac{\frac{23}{100}}{1-\frac{1}{100}} = \frac{23}{99}$$

We then find that

$$7.2323\overline{23} = 7 + \frac{23}{99} = \frac{716}{99}$$

■

Example 16
Value of a Perpetuity

In banking, a **perpetuity** is an account that makes a periodic sequence of payments that continues forever. The **value** of a perpetuity is the present value of all future payments. Suppose the interest rate is 9% compounded monthly, that is, $\frac{9}{12}\% = 0.75\%$ per month, and the monthly payment is $1000. What is the value of the perpetuity?

Solution

Let P be the present value of $1000 which is to be paid 1 month in the future. Then, under the terms of the contract,

$$P + P \cdot (0.0075) = 1000$$

that is,

$$P = 1000 \cdot (1.0075)^{-1}$$

Similarly, the present value of $1000 paid 2 months in the future is $1000 \cdot (1.0075)^{-2}$. In general, the present value of $1000 paid k months in the future is $1000 \cdot (1.0075)^{-k}$. The value of the perpetuity (assuming the first payment is made immediately) is then

$$1000 + 1000 \cdot (1.0075)^{-1} + 1000 \cdot (1.0075)^{-2} + \cdots$$
$$= \sum_{k=0}^{\infty} 1000 \cdot (1.0075)^{-k}$$

This is a geometric series with $a = 1000$ and $r = (1.0075)^{-1}$; therefore, the value of the perpetuity is

$$\frac{1000}{1 - (1.0075)^{-1}} = \frac{1007.5}{0.0075} \approx \$134,333 \qquad \blacksquare$$

Example 17

Suppose that a government decides to mint 10 million pennies per year and that each year 5% of the coins are taken out of circulation. This means that 95% of the 10 million coins minted in one year, or $10(0.95)$ million, survive into the next year; $10(0.95)(0.95)$ million survive into the third year; and so on. If this policy continues, how many pennies will eventually be in circulation?

Solution

At the beginning of the second year there will be 10 million new coins in circulation and $10(0.95)$ million coins that have survived the first year. At the beginning of the third year there will be 10 million new coins, $10(0.95)$ million 1 year old coins, and $10(0.95)(0.95) = 10(0.95)^2$ million 2 year old coins. In general, at the beginning of the $(n + 1)$st year there will be

$$\sum_{k=0}^{n} 10(0.95)^k \text{ million}$$

coins in circulation. In the long run the number of coins would therefore approach

$$\sum_{k=0}^{\infty} 10(0.95)^k \text{ million}$$

a geometric series with $a = 10$ and $r = 0.95$. Therefore, the number of pennies in circulation eventually would approach

$$\frac{a}{1-r} = \frac{10}{1-0.95} = 200 \text{ million}$$ ∎

Example 18
The Multiplier Effect
in Banking

Banks take deposits from customers and reinvest these deposits in the economy. A certain percentage of the total deposits is kept as a cash reserve and the remaining portion is invested. Different banks keep different fractions of their total deposits as cash reserves, but suppose for simplicity that all banks keep a 17% reserve and spend the remaining 83% on loans and investments. This means that $1 invested in a given bank will give a second set of banks $0.83 \times 1 = 0.83$ dollar in new deposits. These banks will in turn keep 17% of this 0.83 dollar and reinvest $(0.83)(0.83)$ dollar in other banks. Continuing in this way we find that the total amount of new deposits generated by the original dollar is

$$1 + 0.83 + (0.83)^2 + (0.83)^3 + \cdots = \sum_{k=0}^{\infty} (0.83)^k$$

This is a geometric series with $a = 1$ and $r = 0.83$; therefore, the total effect of the $1 deposit is

$$\frac{1}{1-0.83} \approx 5.88 \text{ dollars}$$

of new deposits. ∎

PROBLEMS (Section 10-3)

In Problems 1–10, write the series in sigma notation.

1. $a_0 + a_1 + a_2 + a_3 + \cdots$

2. $a_2 + a_3 + a_4 + a_5 + \cdots$

3. $\frac{1}{2} + \frac{1}{3} + \frac{1}{4} + \cdots$

4. $-\frac{1}{3} + \frac{1}{5} - \frac{1}{7} + \frac{1}{9} - \cdots$

5. $-1 + \frac{1}{5} - \frac{1}{25} + \frac{1}{125} - \cdots$

6. $2 + \frac{2}{3} + \frac{2}{9} + \frac{2}{27} + \cdots$

7. $2 - \frac{4}{3} + \frac{8}{9} - \frac{16}{27} + \cdots$

8. $\frac{1}{2} + \frac{4}{3} + \frac{9}{4} + \frac{16}{5} + \cdots$

9. $3 + \frac{9}{2} + \frac{27}{4} + \frac{81}{8} + \frac{243}{16} + \cdots$

10. $7 - 1.4 + 0.28 - 0.056 + 0.0112 - \cdots$

In Problems 11–16, convert the infinite decimal expansion into a rational fraction.

11. $0.999\overline{9} \ldots$

12. $0.1212\overline{12} \ldots$

13. $0.123123\overline{123} \ldots$

14. $0.191191\overline{191} \ldots$

15. $4.2121\overline{21} \ldots$

16. $12.322\overline{2} \ldots$

In Problems 17–26, determine whether the infinite series converges. If it converges, find its sum.

17. $\displaystyle\sum_{k=0}^{\infty} (-1)^k$

18. $\displaystyle\sum_{k=0}^{\infty}\left(\frac{1}{2}\right)^k$

19. $\displaystyle\sum_{k=0}^{\infty}3\left(\frac{3}{4}\right)^k$

20. $\displaystyle\sum_{k=0}^{\infty}4\left(-\frac{2}{3}\right)^k$

21. $\displaystyle\sum_{k=0}^{\infty}5\left(\frac{4}{3}\right)^{2k}$

22. $\displaystyle\sum_{k=0}^{\infty}2\left(\frac{2}{3}\right)^{3k}$

23. $\displaystyle\sum_{k=0}^{\infty}2(-4)^{-k}$

24. $\displaystyle\sum_{k=0}^{\infty}4\left(-\frac{1}{2}\right)^{-k}$

25. $\displaystyle\sum_{k=1}^{\infty}\left(\frac{1}{5}\right)^k$

26. $\displaystyle\sum_{k=0}^{\infty}\frac{1}{5^{k+1}}$

27. What is the value of a $500 per month perpetuity at 12% per annum interest compounded monthly?

28. What is the value of a $100 per month perpetuity at 18% per annum interest compounded monthly?

29. Suppose a ball is dropped from a height of 20 feet. At each bounce it rebounds to 80% of its previous height. What is the total distance traveled by the ball?

30. Suppose that a human body can eliminate 95% of a certain drug each day. If 50 milligrams of the drug is administered to a patient, how much is left in the bloodstream after 1 day? 2 days? 3 days? n days?

31. Suppose that the patient in Problem 30 is administered 50 milligrams of the drug each day. To what level will the drug eventually rise in the patient's bloodstream?

32. What would be the total effect on deposits of a $10,000 deposit if all banks keep a 20% cash reserve? (See Example 18.)

10-4 TAYLOR SERIES

Many functions $f(x)$ can be represented by an infinite series of the form

$$f(x) = a_0 + a_1x + a_2x^2 + a_3x^3 + \cdots$$

$$= \sum_{n=0}^{\infty} a_n x^n$$

The numbers $a_0, a_1, a_2, \ldots$ are called the **coefficients** of the representation. It is quite easy to find a formula for the general term in the sequence $\{a_n\}$. First note that upon substituting $x = 0$ we obtain

$$f(0) = a_0 + a_1 \cdot 0 + a_2 \cdot 0^2 + \cdots = a_0$$

Therefore, $a_0 = f(0)$. Differentiating the representation yields

$$f'(x) = a_1 + 2a_2x + 3a_3x^2 + \cdots$$

Substituting $x = 0$ results in

$$f'(0) = a_1 + 2a_2 \cdot 0 + 3a_3 \cdot 0^2 + \cdots = a_1$$

That is, $a_1 = f'(0)$. Differentiating again we obtain

$$f''(x) = 2a_2 + 6a_3 x + \cdots$$

and

$$f''(0) = 2a_2 + 6a_3 \cdot 0 + \cdots = 2a_2$$

That is, $a_2 = f''(0)/2$. Continuing in this way, we obtain

$$f^{(n)}(0) = n! a_n$$

$$a_n = \frac{f^{(n)}(0)}{n!}$$

We are therefore led to define the **Taylor series** of the function $f(x)$ at $x = 0$.

Taylor Series

> The **Taylor series** at $x = 0$ of a function $f(x)$ is the series
>
> $$f(0) + f'(0)x + \frac{f''(0)}{2!} x^2 + \frac{f^{(3)}(0)}{3!} x^3 + \cdots$$

It can be shown (by methods that are beyond the scope of this text) that many functions $f(x)$ are *represented* by their Taylor series in the sense that

$$f(x) = f(0) + f'(0)x + \frac{f''(0)}{2!} x^2 + \frac{f^{(3)}(0)}{3!} x^3 + \cdots$$

for each x in the domain of the function. We will deal only with functions for which this is true.

Example 19   Find the Taylor series of the function $f(x) = e^x$ at $x = 0$.

Solution   Since $f'(x) = e^x, f''(x) = e^x, f'''(x) = e^x, \ldots$, we find that $f^{(n)}(0) = e^0 = 1$ and

$$a_n = \frac{f^{(n)}(0)}{n!} = \frac{1}{n!}$$

Therefore, the Taylor series for e^x at $x = 0$ is

$$e^x = 1 + x + \frac{x^2}{2!} + \frac{x^3}{3!} + \cdots$$

$$= \sum_{k=0}^{\infty} \frac{x^k}{k!}$$

Example 20   Find the Taylor series for the functions $\sin x$ and $\cos x$ at $x = 0$.

Solution   First suppose $f(x) = \sin x$. Then

$$f'(x) = \cos x, \quad f''(x) = -\sin x, \quad f'''(x) = -\cos x, \quad f^{(4)}(x) = \sin x, \ldots$$
$$f'(0) = 1, \qquad f''(0) = 0, \qquad f'''(0) = -1, \qquad f^{(4)}(0) = 0, \ldots$$

Since $f(0) = 0$, we have the series representation

$$\sin x = 0 + 1 \cdot x + 0 \cdot x^2 - \frac{1}{3!} \cdot x^3 + 0 \cdot x^4 + \frac{1}{5!} \cdot x^5 - \cdots$$

$$= x - \frac{x^3}{3!} + \frac{x^5}{5!} - \cdots$$

Suppose now that $f(x) = \cos x$. Then

$$f'(x) = -\sin x, \quad f''(x) = -\cos x, \quad f'''(x) = \sin x, \quad f^{(4)}(x) = \cos x, \ldots$$
$$f'(0) = 0, \qquad f''(0) = -1, \qquad f'''(0) = 0, \qquad f^{(4)}(0) = 1, \ldots$$

Since $\cos 0 = 1$, we have

$$\cos x = 1 + 0 \cdot x - \frac{1}{2!} \cdot x^2 + 0 \cdot x^3 + \frac{1}{4!} \cdot x^4 + \cdots$$

$$= 1 - \frac{x^2}{2!} + \frac{x^4}{4!} - \cdots$$ ∎

One of the most important uses of Taylor series is in obtaining numerical approximations to values of functions. Suppose, for instance, that we wish to compute $e^{-0.5}$. If this is punched into a calculator that calculates to five decimal places, there is a slight delay and then the calculator returns

$$e^{-0.5} \approx 0.60653$$

How does the calculator do it? The calculator might be programmed to approximate e^x by its Taylor series. It goes something like this. As we saw in Example 19, the Taylor series for e^x is

$$e^x = 1 + x + \frac{x^2}{2!} + \frac{x^3}{3!} + \frac{x^4}{4!} + \cdots$$

If we substitute $x = -0.5$ and calculate the partial sums of this series (to seven decimal places), we obtain

$$1$$

$$1 - 0.5 = 0.5$$

$$1 - 0.5 + \frac{(-0.5)^2}{2} = 0.625$$

$$1 - 0.5 + \frac{(-0.5)^2}{2} + \frac{(-0.5)^3}{3!} = 0.6041667$$

$$1 - 0.5 + \frac{(-0.5)^2}{2} + \frac{(-0.5)^3}{3!} + \frac{(-0.5)^4}{4!} = 0.6067709$$

$$1 - 0.5 + \frac{(-0.5)^2}{2} + \frac{(-0.5)^3}{3!} + \frac{(-0.5)^4}{4!} + \frac{(-0.5)^5}{5!} = 0.6065105$$

$$1 - 0.5 + \frac{(-0.5)^2}{2} + \frac{(-0.5)^3}{3!} + \frac{(-0.5)^4}{4!} + \frac{(-0.5)^5}{5!} + \frac{(-0.5)^6}{6!} = 0.6065310$$

Therefore, the value 0.60653 is the result of summing the first seven terms of the series and rounding to five decimal places.

Example 21   Approximate $\cos(\frac{1}{2})$ by summing the first three nonzero terms of its Taylor series.

Solution   From Example 20, we find that taking the first three nonzero terms yields the approximation

$$\cos x \approx 1 - \frac{x^2}{2} + \frac{x^4}{24}$$

Therefore,

$$\cos\left(\frac{1}{2}\right) \approx 1 - \frac{(\frac{1}{2})^2}{2} + \frac{(\frac{1}{2})^4}{24}$$

$$= 1 - \frac{1}{8} + \frac{1}{384} = \frac{337}{384} \approx 0.8776 \quad \blacksquare$$

Taylor series also have important theoretical implications. For example, consider the Taylor series for $\sin x$ found in Example 20:

$$\sin x = x - \frac{1}{3!} \cdot x^3 + \frac{1}{5!} \cdot x^5 - \cdots$$

If we differentiate, we find that

$$\frac{d}{dx} \sin x = 1 - \frac{1}{3!} \cdot 3x^2 + \frac{1}{5!} \cdot 5x^4 - \cdots$$

$$= 1 - \frac{x^2}{2!} + \frac{x^4}{4!} - \cdots$$

which is the Taylor series for $\cos x$. In this way we obtain an independent verification of the differentiation formula

$$\frac{d}{dx} \sin x = \cos x$$

Taylor series also can be integrated termwise to give useful approximations to difficult definite integrals.

Example 22   Definite integrals of the form $\int_a^b e^{-x^2} \, dx$ are important in probability theory. Approximate $\int_0^1 e^{-x^2} \, dx$ by using four terms in the Taylor series approximation:

$$e^x \approx 1 + x + \frac{x^2}{2} + \frac{x^3}{6}$$

Solution   Replacing x by $-x^2$, we obtain

$$e^{-x^2} \approx 1 - x^2 + \frac{x^4}{2} - \frac{x^6}{6}$$

and therefore,

$$\int_0^1 e^{-x^2}\, dx \approx \int_0^1 \left(1 - x^2 + \frac{x^4}{2} - \frac{x^6}{6}\right) dx$$

$$= \left(x - \frac{x^3}{3} + \frac{x^5}{10} - \frac{x^7}{42}\right)\Big|_0^1$$

$$= 1 - \frac{1}{3} + \frac{1}{10} - \frac{1}{42} \approx 0.7429$$

Up till now we have based all of our Taylor series at the point $x = 0$. There is a more general form of the Taylor series, called a *Taylor series centered at* $x = c$:

$$f(x) = f(c) + \frac{f'(c)}{1!}(x - c)^1 + \frac{f''(c)}{2!}(x - c)^2 + \frac{f'''(c)}{3!}(x - c)^3 + \cdots$$

We have to use this form if f or one of its derivatives is not defined at $x = 0$, for example, if $f(x) = \ln x$.

Example 23   Calculate the first four nonzero terms of the Taylor series of $f(x) = \ln x$ at $x = 1$ and use these to approximate $\ln 2$.

Solution   First, notice that

$$f'(x) = \frac{1}{x}, \quad f''(x) = -\frac{1}{x^2}, \quad f'''(x) = \frac{2}{x^3}, \quad f^{(4)}(x) = -\frac{6}{x^4}$$

$$f'(1) = 1, \quad f''(1) = -1, \quad f'''(1) = 2, \quad f^{(4)}(1) = -6$$

Therefore,

$$\ln x = f(x) \approx f(1) + \frac{f'(1)}{1!}(x - 1) + \frac{f''(1)}{2!}(x - 1)^2 + \frac{f'''(1)}{3!}(x - 1)^3$$

$$+ \frac{f^{(4)}(1)}{4!}(x - 1)^4$$

$$= 0 + 1(x - 1) - \frac{1}{2}(x - 1)^2 + \frac{2}{3!}(x - 1)^3 - \frac{6}{4!}(x - 1)^4$$

$$= (x - 1) - \frac{1}{2}(x - 1)^2 + \frac{1}{3}(x - 1)^3 - \frac{1}{4}(x - 1)^4$$

Setting $x = 2$, we obtain the approximation

$$\ln 2 \approx 1 - \tfrac{1}{2}(1)^2 + \tfrac{1}{3}(1)^3 - \tfrac{1}{4}(1)^4$$

$$= 1 - \tfrac{1}{2} + \tfrac{1}{3} - \tfrac{1}{4} \approx 0.58$$

PROBLEMS (Section 10-4)

In Problems 1–8, find the Taylor series of the given function at x = 0.

1. $(x + 1)^3$

2. $(1 - x)^3$

3. $\ln(1 + x)$

4. $\ln(1 + 2x)$

5. e^{-x}

6. e^{2x}

7. xe^x

8. x^4

9. Find the Taylor series for $f(x) = 1/(1 - x)$ at $x = 0$. Compare with the geometric series (see page 496).

10. Use Taylor series to show that

$$\frac{d}{dx} \cos x = -\sin x$$

11. Use Taylor series to show that

$$\frac{d}{dx} e^x = e^x$$

12. Find the Taylor series for $\sin 2x$ at $x = 0$ by replacing x by $2x$ in the Taylor series for $\sin x$ (see Example 20).

13. Find the Taylor series for $x \cos 2x$ at $x = 0$ by replacing x by $2x$ in the Taylor series for $\cos x$ (see Example 20) and multiplying by x.

14. Find the Taylor series for $e^{x/2}$ at $x = 0$ by replacing x by $x/2$ in the Taylor series for e^x (see Example 19).

15. Find the Taylor series for $x^2 e^{-x}$ at $x = 0$ by replacing x by $-x$ in the Taylor series for e^x (see Example 19) and multiplying by x^2.

16. Suppose $f(x) = x^7 e^{2x}$. What is $f^{(4)}(0)$? [*Hint:* First find the Taylor series (see Problems 14 and 15) and then relate $f^{(4)}(0)$ to the coefficient of x^4.]

17. Suppose $f(x) = x^5 \cos 2x$. What is $f^{(4)}(0)$?

18. Find the Taylor series for $f(x) = \sqrt{x}$ at $x = 1$. Use the first five terms to approximate $\sqrt{2}$.

19. Find the Taylor series for $f(x) = 1/x$ at $x = 1$.

20. It is known that

$$\int_0^1 \frac{1}{1 + x^2} \, dx = \frac{\pi}{4}$$

Find the first six terms of the Taylor series for $1/(1 + x^2)$ at $x = 0$ by replacing x by $-x^2$ in the Taylor series for $1/(1 - x)$ (see Problem 9). Integrate these terms from 0 to 1 and sum the results to obtain an approximation to $\pi/4$.

10-5 CONCLUSION: DETERMINING LIGHT PENETRATION

We now solve the problem of the double-paned window posed in the Introduction to this chapter. Recall that the glass panes in a double-glazed window (see Figure 10-1) reflect 20% of the incident light and absorb 5% of it. We would like to know what fraction of the incident light on the outside pane can penetrate into the interior.

When a light beam passes through the outside pane, only 0.75 of it penetrates into the air space. This diminished beam strikes the inside pane and 20% of it, that is, $(0.2)(0.75)$, is reflected back to the outside pane, while 75% of it, that is, $(0.75)(0.75)$, penetrates into the interior. Referring to Figure 10-1, we have

$$a_1 = (0.75)(0.75) = 0.5625$$

The reflected beam [with intensity $(0.2)(0.75)$] is then reflected back off the outside pane, giving a beam of intensity $(0.2)(0.2)(0.75)$ which again strikes the inside pane. The inside pane allows 75% of this beam to pass,

$$a_2 = (0.2)(0.2)(0.75)(0.75)$$
$$= (0.04)(0.5625)$$

while 20% of the beam, that is, $(0.2)(0.2)(0.2)(0.75)$, is reflected back to the outside pane. The outside pane again reflects 20%, that is, $(0.2)(0.2)(0.2)(0.2)(0.75)$, back, and 75% of this beam passes through the inner pane, that is,

$$a_3 = (0.2)(0.2)(0.2)(0.2)(0.75)(0.75)$$
$$= (0.04)^2(0.5625)$$

In general, the total fraction of the incident beam that passes into the interior is given by the geometric series

$$a_1 + a_2 + a_3 + \cdots = 0.5625 + (0.5625)(0.04) + (0.5625)(0.04)^2 + \cdots$$
$$= \sum_{k=0}^{\infty} (0.5625)(0.04)^k$$

In this geometric series, $a = 0.5625$ and $r = 0.04$. Therefore, the sum of the series is

$$\frac{0.5625}{1 - 0.04} \approx 0.57$$

That is, about 57% of the incident light penetrates into the interior.

10-6 CHAPTER REVIEW

1. A sequence is an infinite list of numbers of the form $a_1, a_2, a_3, \ldots$. We say that the sequence $\{a_n\}$ converges to the number L and write $\lim_{n\to\infty} a_n = L$ if the numbers a_n can be made as close to L as we please by taking n large enough. A sequence that does not converge is said to diverge.

2. A pair of sequences $\{a_n\}$ and $\{b_n\}$ that bracket a root of the equation $f(x) = 0$ can be obtained by the bisection method. The idea is to begin with a pair of numbers a_1, b_1 for which $f(a_1)f(b_1) < 0$ and successively bisect the interval $[a_1, b_1]$, keeping at each stage the half of the interval on which $f(x)$ changes sign.

3. Newton's method is another technique for approximating a root of $f(x) = 0$. In this method an initial approximation x_1 is chosen, and then further approximations are generated by the formula

$$x_{n+1} = x_n - \frac{f(x_n)}{f'(x_n)}$$

4. An infinite series is a sum of the terms of a sequence:

$$a_1 + a_2 + a_3 + \cdots$$

An infinite series is said to converge to a number L if its sequence of partial sums converges to L. An important infinite series is the geometric series $\sum_{k=0}^{\infty} ar^k$. This series converges when $|r| < 1$, and we have

$$\sum_{k=0}^{\infty} ar^k = \frac{a}{1 - r} \qquad |r| < 1$$

5. The Taylor series of a function $f(x)$ at $x = c$ is

$$f(x) = f(c) + \frac{f'(c)}{1!}(x - c) + \frac{f''(c)}{2!}(x - c)^2 + \cdots$$

For suitable functions, this series can be used to approximate numerical values of the function as well as values of its derivatives or definite integrals.

REVIEW PROBLEMS (Section 10-6)

In Problems 1–6, determine whether the sequence converges. If it does, give its limit.

1. $\dfrac{(-1)^k}{k}$

2. $100(0.3)^k$

3. k^k

4. $\dfrac{k-1}{k}$

5. $(-2)^k$

6. $\sin\dfrac{k\pi}{2}$

In Problems 7–10, find the sum of the given geometric series.

7. $\displaystyle\sum_{k=0}^{\infty} 22(0.4)^k$

8. $\displaystyle\sum_{k=0}^{\infty} (-0.2)^k$

9. $\displaystyle\sum_{k=2}^{\infty} 2\left(\dfrac{1}{2}\right)^k$

10. $\displaystyle\sum_{k=-1}^{\infty} 3\left(\dfrac{1}{3}\right)^k$

11. Write the following repeating decimals as rational fractions:
   a. $1.1212\overline{12}$ b. $2.14343\overline{43}$

12. What is the value of a \$200 per month perpetuity at an interest rate of 12% per annum compounded monthly? (See Example 16 in Section 10-3.)

13. If banks keep 18% of deposits as a cash reserve, what is the total effect of a \$100,000 deposit? (See Example 18 in Section 10-3.)

14. Using $x_1 = 2$, compute the next five terms in the Newton sequence for approximating $\sqrt[3]{9}$.

15. Use the bisection method to approximate the positive solution of the equation $2 + x = e^x$ to two decimal places of accuracy.

16. Write the Taylor series for $f(x) = e^{-3x}$ at $x = 0$.

17. Use five terms in the Taylor series of Problem 16 to approximate e^{-3}.

18. Find the Taylor series for $f(t) = 1/t$ at $t = 1$. Integrate from 1 to x to obtain a Taylor series for $\ln x$. Approximate $\ln 2$ by using five terms of this series.

11

PROBABILITY AND CALCULUS

OUTLINE

INTRODUCTION

Suppose that a deadly infectious disease breaks out in a city. The public health service decides to test every member of the population by taking a small blood sample in order to identify and treat those who have contracted the disease. Obviously, testing individual samples drawn from each member of the population of a large city (perhaps several million) would be an enormous and costly task. An alternative procedure would be to pool the individual blood samples into groups of a certain fixed size and test the pooled samples. If the test on a pooled sample is negative, then all the individuals in the group are cleared, while if the test is positive, each member of the group must be individually tested. Since we would expect whole groups to be cleared by a single test, it seems that this pooling procedure is more efficient than individual

screening. How efficient is it? More precisely, what group size will maximize the relative efficiency of the pooling technique in the sense of minimizing its cost?

The efficiency of the pooling technique depends on the likelihood that one or more infected individuals will be members of a given test group. This likelihood is expressed as a number, between 0 and 1, called a *probability*. Values close to 0 indicate that there is little chance that a test group will contain an infected individual, while values close to 1 indicate a high likelihood that the group contains an infected individual. The larger the probability, the more individual tests can be expected to be required and the smaller the relative efficiency of the pooling procedure will be.

Suppose that the total size of the population is N.

Then N tests must be performed if each individual is tested separately. However, if blood samples are pooled in groups of size n, then at least N/n tests must be performed even if every test is negative. But if p is the probability that a given group contains an infected individual, then p is the fraction of these groups for which it is necessary to retest each of the individuals in the group. This increases the number of required tests by

$$(\text{Number per group}) \times \left(\begin{array}{c}\text{Number of groups} \\ \text{requiring retest}\end{array}\right)$$

$$= n\left(\frac{N}{n}\right) p = Np$$

Therefore, the *expected* number of tests using the pooling technique is

$$\frac{N}{n} + Np$$

The relative cost of the pooling technique (in terms of expected number of tests) compared to individual tests is then

$$C = \frac{N/n + Np}{N} = \frac{1}{n} + p$$

We would like to choose n to minimize this relative cost.

In some of the examples of Chapter 5 we have already alluded to the notions of likelihood and probability density functions. Here, we will make the concepts of probability and expected value more precise and study several models of discrete and continuous random situations.

11-1 SAMPLE SPACES, EVENTS, AND PROBABILITY

The origins of probability theory can be traced back to 16th and 17th century Italy and France, when some of the greatest scholars of the time (including Galileo, Fermat, and Pascal) began in a systematic way to quantify games of chance such as dice and card games. Today, its importance extends far beyond mere games. For example, in planning a difficult operation, a surgeon must make an assessment of the relative probabilities of success of various alternative surgical procedures. A marketing director for a large firm must plan the marketing campaign for his company in such a way as to maximize the expected revenue. A military commander must decide between various tactics so as to minimize his expected losses. Each of these problems is much more important than a dice game, but, like a dice game, each can be analyzed by using probability theory.

Probability can most easily be developed in terms of **experiments.** We use the word *experiment* in a special sense to mean the observation of some occurrence. For example, we may speak of the experiment of flipping a coin, meaning that the coin is tossed and it is observed whether a head or tail comes up.

> The **sample space** for an experiment is the set of all possible outcomes of the experiment.

If we let H stand for heads and T stand for tails, then the sample space for the coin toss experiment is {H, T}.

Example 1   Consider the experiment of tossing a standard die and observing the number on the top face. The sample space for this experiment is {1, 2, 3, 4, 5, 6}. ∎

Example 2   Suppose a coin is flipped twice in succession. The sample space for this experiment is {HH, HT, TH, TT}, where HH represents the outcome that the coin comes up heads on each flip, HT means the coin came up heads on the first flip and tails on the second flip, etc. ∎

Example 3   A fast-food restaurant has a maximum capacity of 100 people. Suppose the manager observes the number of patrons in the restaurant at a given time. The sample space for the manager's experiment is {0, 1, 2, 3, . . . , 99, 100}. ∎

Example 4   A factory is connected to a power plant by an overhead power line 10 miles long. The power line breaks at a certain point and a repair crew leaves the power plant to find the point at which the line broke. The sample space for this experiment is $\{x \mid 0 \leq x \leq 10\}$, where x represents the number of miles from the power plant to the point at which the break occurred. ∎

> By an **event**, we mean some subset of the sample space.

An event is said to occur if the outcome of the experiment is a member of this subset. In Example 1 the subset {1, 4} represents the event that a 1 or a 4 comes up when the die is tossed. The subset {20, 21, 22} of the sample space in Example 3 represents the event that the manager observed 20, 21, or 22 people in the restaurant. The subset $\{x \mid 2 \leq x \leq 5\}$ of the sample space in Example 4 represents the event that the break in the power line occurred between 2 and 5 miles from the power plant.

Example 5   Suppose two coins are flipped. Give the sample space for this experiment and identify the event that at least one of the coins came up tails.

Solution   We can, for example, let HT stand for the first coin coming up heads and the second coin coming up tails. Then the sample space is {HH, HT, TH, TT} and the event that at least one of the coins came up tails is represented by the subset {HT, TH, TT}. ∎

Example 6   An automobile dealership has an inventory of 65 cars. Consider the experiment of observing the number of cars sold in a given day. Give the sample space for this experiment and represent the event that fewer than 10 cars were sold.

Solution   The sample space may be expressed as {0, 1, 2, 3, . . . , 63, 64, 65}. The event that fewer than 10 cars were sold in a given day is represented by the subset {0, 1, 2, 3, 4, 5, 6, 7, 8, 9} of the sample space. ∎

With each event we associate a number called the **probability** of the event. A probability will always be a number between 0 and 1 which represents the likelihood that the event will occur when the experiment is performed. The greater the probability of an event, the more likely it is to occur, and the smaller the probability of an event, the less likely it is to occur.

Probabilities can be assigned to events in either an intuitive or an empirical way. For instance, in the experiment of flipping a perfectly balanced and symmetrical coin we intuitively feel that heads and tails are equally likely to occur. That is, in a long sequence of tosses we expect heads to occur half the time and tails to occur half the time. We therefore say that the probability of heads is $\frac{1}{2}$ and the probability of tails is $\frac{1}{2}$. The symbol P is used to denote probability, and we write

$$P(\{H\}) = \tfrac{1}{2} \qquad P(\{T\}) = \tfrac{1}{2}$$

Example 7   A coin is flipped and a die is tossed. What is the probability that the coin comes up heads and the die comes up 3?

Solution   The individual points in the sample space may be represented by pairs such as (T, 6), which indicates that the coin comes up tails and the die comes up 6. The sample space is then

$$\{(H, 1), (T, 1), (H, 2), (T, 2), (H, 3), (T, 3), (H, 4), (T, 4), (H, 5), (T, 5), (H, 6), (T, 6)\}$$

Each of the points in this sample space is equally likely to occur, and since there are 12 points in the sample space,

$$P(\{(H, 3)\}) = \tfrac{1}{12}$$ ∎

Example 8   In Example 4, what is the probability that the break in the power line occurred within 3 miles of the power plant?

Solution   The event in question may be represented as $\{x | 0 \le x \le 3\}$. Assuming that the break is equally likely to occur at any point along the 10 mile long line, we have

$$P(\{x | 0 \le x \le 3\}) = \tfrac{3}{10} = 0.3$$ ∎

Often, it is not possible to assign probabilities to events in an intuitive way. In such cases, we can make an **empirical** assignment of probability by repeating the experiment a large number of times and noting how many times the event in question occurred. The relative frequency of occurrence of the event, that is, the number of occurrences of the event divided by the total number of trials, is then taken as its probability. Suppose, for example, that electric toasters are tested at the factory before they are packaged. If 212 toasters out of 10,000 are found to be defective, then we say that the probability of a given toaster being defective is $212/10,000 = 0.0212$.

Example 9   In a survey of 20,000 college freshmen, 12,126 indicated they were registered Democrats, 5021 indicated they were registered Republicans, and 2853 indicated no party affiliation or an affiliation with a party other than the Democrats or Republicans. If a college freshman is chosen at random from this group, what is the probability that this student is a member of one of the two major parties?

Solution   Since $12,126 + 5021 = 17,147$ of the 20,000 students surveyed indicated they belong to a major party, we can say that the probability that a randomly chosen member of the group belongs to one of the major parties is $17,147/20,000 \approx 0.8574$.   ∎

Example 9 illustrates a fundamental property of probability. Namely, if A and B are *mutually exclusive* events, that is, they contain no common outcomes, then the **addition law** for probability holds.

Addition Law for Probability

> If A and B are mutually exclusive, then
>
> $$P(A \text{ or } B) = P(A) + P(B)$$

To illustrate the addition law, let A represent the event that the student is a Democrat in Example 9. Then $P(A) = 12{,}126/20{,}000$. If B represents the event that the student is a Republican, then $P(B) = 5021/20{,}000$. As we saw in the example,

$$P(A \text{ or } B) = \frac{12{,}126}{20{,}000} + \frac{5021}{20{,}000} = \frac{17{,}147}{20{,}000}$$

The sample space of an experiment must account for every possible outcome of the experiment. When the experiment is performed, some outcome always occurs, so we always assign a probability of 1 to the sample space S itself: $P(S) = 1$. If A is an event in the sample space, then so is the nonoccurrence of A, called *not A*. Since A and not A are mutually exclusive and either A or not A always occurs, we have

$$1 = P(S) = P(A \text{ or not } A) = P(A) + P(\text{not } A)$$

From this we immediately obtain the **law of complementary probabilities:**

Law of Complementary Probabilities

> $$P(\text{not } A) = 1 - P(A)$$

We also note that if $S = \{a_1, a_2, \ldots, a_m\}$, then since the events $\{a_1\}$, $\{a_2\}$, $\ldots$, $\{a_m\}$ are mutually exclusive,

$$1 = P(\{a_1\}) + P(\{a_2\}) + \cdots + P(\{a_m\})$$

Example 10   A die is tossed. What is the probability that a 3 does not come up?

Solution   Let $\{3\}$ stand for the event that the 3 comes up on the toss of the die. Since ea of the six faces of the die is equally likely to come up, $P(\{3\}) = \frac{1}{6}$. Therefor

$$P(\text{not } \{3\}) = 1 - P(\{3\}) = 1 - \tfrac{1}{6} = \tfrac{5}{6}$$

Of course, we can see this in a different way. The event not $\{3\}$ is the sa the event $\{1, 2, 4, 5, 6\}$, and by the addition law:

$$P(\{1, 2, 4, 5, 6\}) = P(\{1\}) + P(\{2\}) + P(\{4\}) + P(\{5\}) + P(\{6\})$$
$$= \tfrac{1}{6} + \tfrac{1}{6} + \tfrac{1}{6} + \tfrac{1}{6} + \tfrac{1}{6} = \tfrac{5}{6}$$

We summarize the basic laws of probability in the box on page 51

Laws of Probability

Suppose S is a sample space and A is an event of S. Then

$$P(S) = 1$$
$$0 \le P(A) \le 1$$
$$P(\text{not } A) = 1 - P(A)$$

If A and B are mutually exclusive events, then

$$P(A \text{ or } B) = P(A) + P(B)$$

PROBLEMS (Section 11-1)

1. A number is chosen at random from the first ten positive integers. Set up a sample space for this experiment.

2. Set up a sample space for the experiment of flipping three coins.

3. A die is in the form of a regular tetrahedron (see the figure) with one through four dots on each of its four faces. Set up a sample space for the experiment of tossing this die.

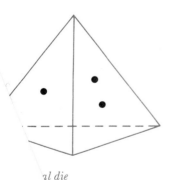

ch
e,
al die

ne as   'e are assigned at random to four
               bs. Set up a sample space for this

               'at random from a square that is
               e. Set up a sample space for this

4.

6. Set up a sample space for the experiment of turning on a light bulb and measuring the time it takes to burn out.

7. Represent the event that an even number is chosen in the experiment of Problem 1.

8. Represent the event that two heads turn up in the experiment of Problem 2.

9. Suppose two dice are tossed.
   a. Set up a sample space for this experiment.
   b. Represent the event that the sum of the dots on the two faces that turn up is 5.

10. A bowl contains four each of red, white, yellow, and green jelly beans. Four beans are chosen at random from the bowl.
    a. Set up a sample space for this experiment.
    b. Identify the event that all but one of the chosen beans is red.

11. A bin contains a large number of nuts, bolts, and washers, all of the same size. Three objects are taken from the bin at random.
    a. Describe the sample space for this experiment.
    b. Represent the event that one each of a nut, bolt, and washer is chosen.

12. Find the probability of the event in Problem 8.

13. Find the probability of the event in Problem 9b.

14. Find the probability of the event in Problem 10b.

15. What is the probability that on the toss of two dice the sum of the numbers on the top faces is not 5?

16. A point is chosen at random from a square that is 1 unit on a side. What is the probability that the point lies in the inscribed circle of diameter 1?

17. A card is chosen at random from a well-shuffled poker deck. What is the probability that it is:
    a. An ace?   b. Not a face card?
    c. Not a club?

18. Suppose that {a, b, c, d, e} is the sample space of some experiment and the probabilities of the outcomes are given in the table. Let $A = \{a, d\}$ and $B = \{b, c\}$.

Outcome	a	b	c	d	e
Probability	0.20	0.10	0.40	0.06	0.24

    a. What is $P(A)$?   b. What is $P(\text{not } A)$?
    c. What is $P(B)$?   d. What is $P(A \text{ or } B)$?

19. During a promotional campaign, a cigarette company gave away sample packs of cigarettes in a hotel lobby. Of 2000 people offered cigarettes, 326 declined the offer, 283 chose menthol cigarettes, and all those remaining chose regular cigarettes.
    a. What is the probability that a person who was approached did not choose menthol cigarettes?
    b. What is the probability that a person who was approached accepted cigarettes?

20. The table represents the observed waiting time of a patron arriving at a fast-food restaurant.

Wait (in Minutes)	Number of Patrons
Less than 1	568
Between 1 and 2	232
Between 2 and 3	164
Between 3 and 4	103
More than 4	62

    a. What is the probability that a patron must wait between 1 and 2 minutes?

    b. What is the probability that a patron must wait less than 2 minutes?
    c. What is the probability that a patron must wait more than 3 minutes?

21. The table summarizes data collected from a population relating eye color and blood factor.

Blood	Eye Color		
	Blue	Brown	Green
Rh+	2514	3641	1642
Rh−	2213	2954	1856

    a. What is the probability that a person in this population has Rh+ blood?
    b. What is the probability that a brown-eyed person in this population has Rh− blood?
    c. What is the probability that a person in this population who has Rh+ blood also has green eyes?

22. A survey of a certain population resulted in the data on income and number of automobiles in a family given in the table.

Income (Dollars)	Number of Cars				
	0	1	2	3	More than 3
Less than 10,000	1242	675	192	12	0
10,000–20,000	64	893	741	214	22
20,000–30,000	40	1067	2115	679	54
30,000–40,000	26	261	912	543	112
40,000–50,000	32	111	416	629	154
Over 50,000	6	49	371	492	201

    a. What is the probability that a family in this population owns more than 3 cars?
    b. What is the probability that a family in this population which earns between $10,000 and $20,000 per year owns fewer than 3 cars?
    c. What is the probability that a family in this population which owns 3 cars earns more than $10,000?

11-2 DISCRETE RANDOM VARIABLES

PROB

In this section we consider experiments in which an outcome can be specified as a member of a discrete set (usually of whole numbers). If the outcome of such an experiment is denoted by the letter X, then we call X a **discrete random variable.** For instance, consider the experiment of tossing two dice and observing the sum of the numbers that come up on the top faces. If we call this sum X, then X is a random variable that can take on the values 2, 3, 4, 5, 6, 7, 8, 9, 10, 11, and 12. Of course, X does not take on each of these values with the same probability. For example, the event $X = 2$ (called "snake eyes" in the game of craps) can occur in only one way, namely by a 1 turning up on each die. Since the first die can fall in 6 ways and for each result on the first die there are 6 possible results on the second die, there are $6 \times 6 = 36$ possible outcomes on tossing two dice. Assuming that the dice are not "loaded," each of these outcomes is equally likely, and hence,

$$P(X = 2) = \tfrac{1}{36}$$

The event $X = 3$, however, can occur in two ways: a 2 on the first die and a 1 on the second or vice versa. Therefore,

$$P(X = 3) = \tfrac{2}{36} = \tfrac{1}{18}$$

In the same way we can compute $P(X = k)$ for each possible value of k. The results are given in Table 11-1.

Probabilities also may be represented as areas. The probability $P(X = k)$ may be represented as the area of a rectangle of width 1 centered on the value k in a **probability density histogram** (see Figure 11-1).

Table 11-1

k	$P(X = k)$
2	$\tfrac{1}{36}$
3	$\tfrac{2}{36}$
4	$\tfrac{3}{36}$
5	$\tfrac{4}{36}$
6	$\tfrac{5}{36}$
7	$\tfrac{6}{36}$
8	$\tfrac{5}{36}$
9	$\tfrac{4}{36}$
10	$\tfrac{3}{36}$
11	$\tfrac{2}{36}$
12	$\tfrac{1}{36}$

Figure 11-1
Probability density histogram for toss of two dice ($p_k = P(X = k)$)

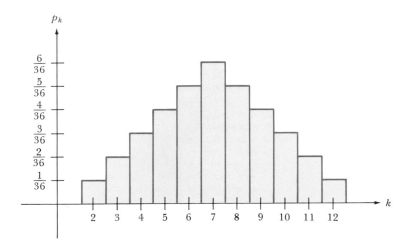

Example 11   A coin is flipped three times. Let the random variable X be the total number of heads on the three flips. Find $p_k = P(X = k)$ for each possible value k of X and draw the probability density histogram for X.

Solution   The possible values for X are 0, 1, 2, and 3. The total number of possible outcomes in the experiment is $2 \times 2 \times 2 = 8$. The event $X = 0$ can occur in only one way, namely the coin comes up tails each time; therefore, $p_0 = P(X = 0) = \frac{1}{8}$. The event $X = 1$ occurs when the coin comes up heads on exactly one flip. This can occur in 3 ways: HTT, THT, TTH. Therefore, $p_1 = P(X = 1) = \frac{3}{8}$. There are 3 ways for the event $X = 2$ to occur, and the event $X = 3$ can occur in only one way. Therefore, $P(X = 2) = \frac{3}{8}$ and $P(X = 3) = \frac{1}{8}$. The probability density histogram for X is drawn in Figure 11-2.

Figure 11-2

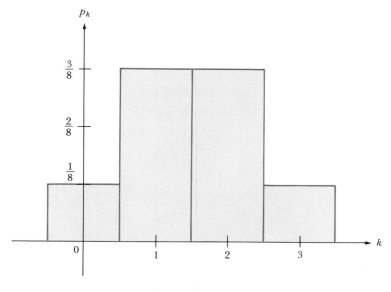

Example 12   Table 11-2 represents the observed numbers of live births in a litter collected from a population of 1000 dogs. Suppose a dog is chosen at random from this

Table 11-2

Live Births	Number of Dogs
0	10
1	25
2	40
3	80
4	200
5	180
6	245
7	115
8	90
9	10
10	5

population, and let X stand for the number of live births in that dog's litter. Draw a probability density histogram for the random variable X.

Solution   First, we convert the raw data into a probability density by computing the empirical probability of each event $X = k$. We do this by dividing by the total population size of 1000 to obtain the probabilities given in Table 11-3.

Table 11-3

k	$P(X = k)$
0	0.01
1	0.025
2	0.040
3	0.080
4	0.20
5	0.18
6	0.245
7	0.115
8	0.09
9	0.01
10	0.005

The probability density histogram for X is shown in Figure 11-3.

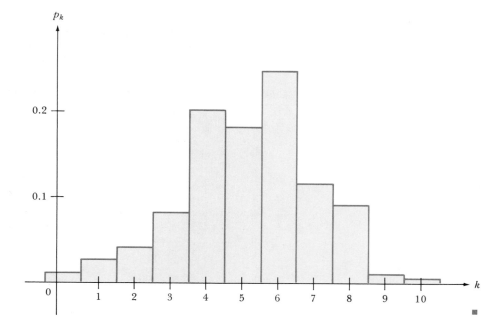

Figure 11-3
Number of live births in a litter

Probability density histograms provide a useful way of envisioning an event of the form $a \leq X \leq b$. For instance, in Example 12 the event $1 \leq X \leq 3$ means that there are between 1 and 3 live births in the litter. The probability of this event is

$$P(1 \leq X \leq 3) = P(X = 1) + P(X = 2) + P(X = 3)$$
$$= p_1 + p_2 + p_3$$

In terms of areas, this is the sum of the areas of those rectangles over the values 1, 2, 3 (see Figure 11-3). In general, the area of that portion of the histogram over the values between k_1 and k_2 represents the probability $P(k_1 \leq X \leq k_2)$. This idea will become particularly important when we consider continuous random variables.

EXPECTED VALUE

How many live births should we expect in a litter in Example 12? A glance at Figure 11-3 should convince you that a litter should contain between 4 and 6 births. But how do we give precise meaning to the *expected value* of the random variable X that measures the size of the litter? The idea is to take into account all possible outcomes and to weight each outcome by the probability that it occurs. For example, based on our data, a litter of 3 occurs with probability 0.08 (or 8% of the time). Therefore, in a long sequence of trials we should expect "on the average" 8% of the litters to contain 3 puppies. Similarly, 5 puppies will be born 18% of the time, and so on. So, "on the average" we should expect

$$0(0.01) + 1(0.025) + 2(0.04) + 3(0.08) + 4(0.20) + 5(0.18) + 6(0.245)$$
$$+ 7(0.115) + 8(0.09) + 9(0.01) + 10(0.005) = 5.180$$

puppies in a litter. Note that this example shows that the expected value of a random variable is not necessarily a possible value. Obviously, we cannot expect 5.180 puppies in a litter, but 5 puppies would be a "good bet."

In general, we define the **expected value** of a discrete random variable X as

$$E(X) = \sum_k k p_k$$

where $p_k = P(X = k)$ and the sum is extended over all possible values k of X.

Example 13   Two dice are tossed. If the sum of the values that come up is even, you win an amount in dollars equal to this sum; otherwise, you lose an amount equal to this sum. What is the expected value of your winnings?

Solution   Let the random variable X be the winnings. For example, if snake eyes comes up, $X = 2$; while if the sum of the faces on the dice is 3, then you lose $3 and hence, $X = -3$ (that is, your winnings are negative). The expected value of X can be computed using the probabilities that were given in Table 11-1:

$$E(X) = 2(\tfrac{1}{36}) - 3(\tfrac{2}{36}) + 4(\tfrac{3}{36}) - 5(\tfrac{4}{36}) + 6(\tfrac{5}{36}) - 7(\tfrac{6}{36})$$
$$+ 8(\tfrac{5}{36}) - 9(\tfrac{4}{36}) + 10(\tfrac{3}{36}) - 11(\tfrac{2}{36}) + 12(\tfrac{1}{36})$$

$$= \frac{126 - 126}{36} = 0$$

The expected winnings are 0. Of course, this does not mean that there are no winnings each time the game is played. But it does mean that if the game is played a large number of times, then losses and wins should be balanced and you should break even. A game of this type in which the expected winnings are 0 is called a **fair game.**

Example 14
Comparing Armor Plating Effectiveness

The army must decide on a type of armored plating to be used in its tanks. Two types of plating are considered, type A and type B. One hundred tanks are outfitted with type A plating and 100 tanks are outfitted with type B plating. All 200 tanks go through an identical combat simulation test and the number of penetrations of the armor is observed. The results are given in Tables 11-4 and 11-5. In each table, the left-hand column gives the number of antitank shells that penetrated the armor and the right-hand column gives the number of tanks that sustained the given number of penetrations. Which type of armor should the army use?

Table 11-4
Type A

Penetrations	Tanks
0	60
1	17
2	10
3	10
4	3
5	0

Table 11-5
Type B

Penetrations	Tanks
0	65
1	15
2	15
3	1
4	2
5	2

Solution

Let X_A stand for the number of penetrations of a tank equipped with type A armor. From Table 11-4 we see that

$$E(X_A) = 0(0.60) + 1(0.17) + 2(0.10) + 3(0.10) + 4(0.03) + 5(0)$$
$$= 0.79$$

That is, if type A armor is used, then the expected number of penetrations is 0.79.

Similarly, if X_B represents the number of penetrations of type B armor, then

$$E(X_B) = 0(0.65) + 1(0.15) + 2(0.15) + 3(0.01) + 4(0.02) + 5(0.02)$$
$$= 0.66$$

Therefore, if the criterion for choice is to minimize the expected number of penetrations, then type B armor should be used.

VARIANCE

The *variance* of a random variable gives a measure of the spread of a probability density histogram about the expected value. Suppose X is a random variable with expected value $m = E(X)$. The **variance** of X is defined to be the expected value of the square of the deviation of X from m:

$$\mathrm{Var}(X) = E((X - m)^2) = \sum_k (k - m)^2 p_k$$

where $p_k = P(X = k)$. The next example gives a graphic illustration of relative sizes of variance.

Example 15   Consider two random variables X and Y, each of which can take on the values 1, 2, 3, 4, 5. Suppose the probabilities of the various values of X and Y are given in Table 11-6.

Table 11-6

Value of Random Variable	1	2	3	4	5
Probability X Takes This Value	0.3	0.1	0.2	0.1	0.3
Probability Y Takes This Value	0.1	0.2	0.4	0.2	0.1

The expected values are then

$$E(X) = 1(0.3) + 2(0.1) + 3(0.2) + 4(0.1) + 5(0.3)$$
$$= 3$$
$$E(Y) = 1(0.1) + 2(0.2) + 3(0.4) + 4(0.2) + 5(0.1)$$
$$= 3$$

The variances are

$$\mathrm{Var}(X) = \sum_{k=1}^{5} (k - 3)^2 P(X = k)$$
$$= (1 - 3)^2(0.3) + (2 - 3)^2(0.1)$$
$$\qquad + (3 - 3)^2(0.2) + (4 - 3)^2(0.1) + (5 - 3)^2(0.3)$$
$$= 2.6$$
$$\mathrm{Var}(Y) = \sum_{k=1}^{5} (k - 3)^2 P(Y = k)$$
$$= (1 - 3)^2(0.1) + (2 - 3)^2(0.2)$$
$$\qquad + (3 - 3)^2(0.4) + (4 - 3)^2(0.2) + (5 - 3)^2(0.1)$$
$$= 1.2$$

The probability density histograms of X and Y appear in Figure 11-4 on page 522. These histograms show clearly that the spread of the histogram (weighted with probability) about the expected value is proportional to the variance.

Figure 11-4

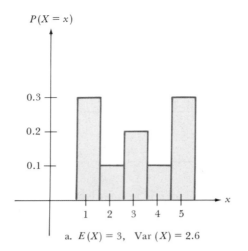

a. $E(X) = 3$, Var $(X) = 2.6$

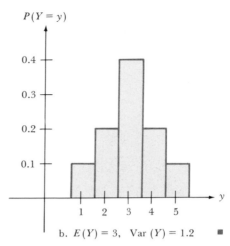

b. $E(Y) = 3$, Var $(Y) = 1.2$ ■

PROBLEMS (Section 11-2)

1. Suppose a random variable X takes on the values $-2, -1, 0, 1, 2$ with probabilities given in the following table:

Value	-2	-1	0	1	2
Probability	0.2	0.1	0.4	0.1	0.2

   a. Draw the probability density histogram for X.
   b. Shade the region representing the probability $P(X \geq 0)$.
   c. Compute $E(X)$.

2. Suppose a random variable Y takes on values 2, 4, 5, 7 with probabilities 0.2, 0.3, 0.3, 0.2, respectively.
   a. Draw a probability density histogram for Y.
   b. Shade the region that represents the probability $P(Y \leq 6)$.
   c. Compute $E(Y)$.

3. Three coins are flipped. If an odd number of heads comes up, you win \$1. If all tails comes up, you win \$2. Otherwise, you lose \$1.50. What are your expected winnings?

4. A single die is tossed. If the value on the top face comes up even, you lose an amount equal to that value; otherwise, you win an amount equal to double the value on the top face. What is the expected value of your winnings?

5. A municipal motor pool contains 156 cars. The approximate fuel consumption of each car is checked, and the results are given in the table. A car is chosen at random from the motor pool. What is the expected fuel consumption rate for the car?

Consumption Rate (Miles / Gallon)	Number of Cars with this Consumption Rate
17	10
18	12
19	30
20	41
21	26
22	22
23	15

6. The output of a manufacturing plant is measured each day of the year, and the results are given in the table at the top of the next page. What is the expected output on a randomly given day in the year?

Output (Units)	Number of Days with This Output
125	64
127	70
130	80
132	80
133	6
140	65

7. Fifty children in an elementary school are tested in a long jump competition, with the results given in the table. What is the expected length of a jump for a child chosen at random from this group?

Jump (in Feet)	Number of Children with This Jump
6	4
7	8
8	11
9	18
10	6
11	3

8. The table gives the observed repair times for 100 computer terminals of the same type. What is the expected repair time for a terminal of this type?

Repair Time (Hours)	Number of Terminals
0.5	5
1.0	17
1.5	26
2.0	41
2.5	7
3	4

9. The tensile strengths of 25 steel wires were measured, giving the following results (in pounds): 120, 119, 120, 121, 118, 119, 120, 122, 124, 120, 122, 117, 121, 118, 122, 119, 122, 117, 118, 119, 118, 120, 118, 117, 121. What is the expected tensile strength of a wire in this group?

10. John Graunt (1620–1674), a London haberdasher, is considered by many to be the father of applied statistics. By studying the Bills of Mortality (death lists), he arrived at what he called a life table for 64 people who had reached the age of 6 years. The table gives the number of people reaching the given age. Based on Graunt's figures, what is the life expectancy of a person who passes his or her sixth birthday?

Age	6	16	26	36	46	56	66	76	86
Number	64	40	25	16	10	6	3	1	0

11. Suppose a trusted TV meteorologist predicts a 65% chance of rain on Saturday. You feel that it will not rain, but a friend believes that it will. If you wager $5 that it will not rain, how much must your friend wager so that the bet is fair?

12. You bet $8 against a friend's $3 in a fair bet. What is the probability that you will win?

13. A chicken farmer anticipates revenues of $60,000 from his crop. However, bacteria are found in the chicken coops, which, with a probability of 0.3, will destroy 30% of the stock, reducing his revenue to $42,000. For $6000 he can treat the coops, eliminate the bacteria, and save virtually all of his stock. Should he do it?

14. Compute the variance of the random variable that indicates the sum of the top faces on the toss of two dice.

15. Consider two random variables X and Y that take on values with the probabilities given in the table.

Value	−2	−1	0	1	2
Probability (X)	0.4	0.2	0	0.2	0.2
Probability (Y)	0.1	0.2	0.2	0.2	0.3

a. Draw probability density histograms for X and Y.
b. Compute $\text{Var}(X)$ and $\text{Var}(Y)$ and relate these quantities to the histograms.

11-3 SEQUENCES, SERIES, AND PROBABILITY

There are many experiments in which the possible outcomes form a sequence. The sample space for such an experiment is then a discrete, but infinite, set. A simple example of such an experiment is that of successively tossing a coin and observing the number of tails occurring before the first head appears. The possible outcomes of this experiment are 0, 1, 2, 3, . . . , corresponding to the first head appearing on the first, second, third, fourth, . . . , toss, respectively. Notice that there is no upper bound on the sequence of outcomes, which corresponds to the fact that an arbitrarily long run of tails is possible before the first head appears.

Another experiment with an infinite sequence of possible outcomes would be the observation of the number of calls made to a police station in a given month. The possible outcomes of this experiment are 0, 1, 2, 3, ... Or consider the number of meteors that strike a satellite in a given year. This number can be 0, 1, 2, 3, ...

If we let X denote a random variable that can take on an infinite sequence of values, then with each such value we may associate a probability — the probability that X assumes the given value:

$$P(X = k) = p_k$$

In this way an infinite sequence $\{p_k\}$ of probabilities is associated with the random variable X. Since each of the numbers p_k is a probability, we must have $0 \leq p_k \leq 1$ for each k. Also,

$$\sum_k p_k = 1$$

where the sum is extended over all possible values k of the random variable X. (Do you see why this follows from the law of complementary probabilities?)

The expected value of a random variable X that takes infinitely many values is defined in exactly the same way as for a random variable that takes only finitely many values:

$$E(X) = \sum_k k p_k$$

But now it is important to realize that, since there are infinitely many possible values k for the random variable X, the sum in the definition above is actually an *infinite sum*, that is, an infinite series.

In Chapter 10 we saw that the geometric series

$$\frac{1}{1-x} = 1 + x + x^2 + x^3 + \cdots \qquad |x| < 1$$

is a very important infinite series. In order to compute expectations in this section we will find a certain allied series very useful. Namely, if we differentiate each side of the above formula with respect to x, we obtain

$$\frac{d}{dx}\frac{1}{1-x} = \frac{d}{dx}(1 + x + x^2 + x^3 + \cdots)$$

or

$$\frac{1}{(1-x)^2} = 1 + 2x + 3x^2 + \cdots \qquad |x| < 1$$

We can also obtain this formula by expanding the function $f(x) = 1/(1-x)^2$ in a Taylor series.

Example 16   Suppose a random variable X takes on the values $0, 1, 2, 3, \ldots$, with probabilities $p_k = P(X = k) = \frac{2}{5}(0.6)^k, k = 0, 1, 2, 3, \ldots$. The expected value of X is then

$$E(X) = \sum_{k=0}^{\infty} kp_k = \sum_{k=0}^{\infty} k\tfrac{2}{5}(0.6)^k$$

This expectation can be calculated by setting $x = 0.6$ in the formula above:

$$\sum_{k=0}^{\infty} k\tfrac{2}{5}(0.6)^k = 0 + \tfrac{2}{5}(0.6)[1 + 2(0.6) + 3(0.6)^2 + \cdots]$$

$$= \left(\frac{1.2}{5}\right)\left[\frac{1}{(1-0.6)^2}\right]$$

$$= \left(\frac{1.2}{5}\right)\left(\frac{1}{0.16}\right) = 1.5$$

Suppose the expected value of the random variable X is m, that is, $m = E(X)$. As before, we define the variance of X by

$$\mathrm{Var}(X) = \sum_k (k - m)^2 p_k$$

where $p_k = P(X = k)$. Again, the sum is in general an infinite series, since X may take on infinitely many values. How do we calculate the value of such an infinite series? A little algebra gives

$$\sum_k (k - m)^2 p_k = \sum_k (k^2 - 2mk + m^2)p_k$$

$$= \sum_k [k(k-1) + (1 - 2m)k + m^2]p_k$$

$$= \sum_k [k(k-1)p_k] + (1 - 2m)\sum_k kp_k + m^2\sum_k p_k$$

However, $m = E(X) = \sum_k kp_k$ and $\sum_k p_k = 1$; therefore,

$$\begin{aligned}
\text{Var}(X) &= \sum_k (k - m)^2 p_k \\
&= \sum_k [k(k-1)p_k] + (1 - 2m)m + m^2 \\
&= \sum_k [k(k-1)p_k] + m - m^2
\end{aligned}$$

So the calculation of Var(X) is reduced to finding the sum of the infinite series

$$\sum_k k(k-1)p_k$$

In many cases this can be done by differentiating both sides of the formula

$$\frac{1}{(1-x)^2} = 1 + 2x + 3x^2 + 4x^3 + \cdots \qquad |x| < 1$$

with respect to x. In fact,

$$\frac{d}{dx}\frac{1}{(1-x)^2} = 2 + 3 \cdot 2x + 4 \cdot 3x^2 + \cdots$$

$$= \sum_{k=0}^{\infty} k(k-1)x^{k-2}$$

But

$$\frac{d}{dx}\frac{1}{(1-x)^2} = \frac{2}{(1-x)^3}$$

and therefore,

$$\frac{2}{(1-x)^3} = \sum_{k=0}^{\infty} k(k-1)x^{k-2} \qquad |x| < 1$$

Example 17   We found above that

$$\text{Var}(X) = \sum_k [k(k-1)p_k] + m - m^2$$

where $m = E(X)$. Now we will use this formula to compute the variance of the random variable in Example 16. In that example, $p_k = \frac{2}{5}(0.6)^k$ and we found that $m = 1.5$. Therefore,

$$\text{Var}(X) = \sum_{k=0}^{\infty} [k(k-1)\tfrac{2}{5}(0.6)^k] + 1.5 - (1.5)^2$$

$$= \tfrac{2}{5}(0.6)^2 \sum_{k=0}^{\infty} [k(k-1)(0.6)^{k-2}] - 0.75$$

But, substituting $x = 0.6$ in the formula

$$\sum_{k=0}^{\infty} k(k-1)x^{k-2} = \frac{2}{(1-x)^3}$$

we obtain

$$\sum_{k=0}^{\infty} k(k-1)(0.6)^{k-2} = \frac{2}{(0.4)^3}$$

Therefore,

$$\text{Var}(X) = \frac{2}{5}(0.6)^2 \cdot \frac{2}{(0.4)^3} - 0.75$$
$$= 3.75$$

POISSON VARIABLES

A random variable X that takes the values $0, 1, 2, 3, \ldots$, is said to be a **Poisson random variable** or simply a **Poisson variable** if

$$p_k = P(X = k) = \frac{\lambda^k e^{-\lambda}}{k!}$$

where λ is a positive parameter. Such random variables occur frequently in practice. Some examples include:

1. The number of calls arriving at a telephone switchboard during a given time interval
2. The number of alpha particles emitted by a radioactive isotope during a given time interval
3. The number of automobiles crossing an intersection during the rush hour
4. The number of suicides in a given month
5. The number of bacteria in a given volume of pond water
6. The number of newspapers sold on a given street corner during a day

In each of these examples, the number in question is unbounded and can take any of the values $0, 1, 2, 3, \ldots$.

Note that $0 \leq p_k \leq 1$ and, by the Taylor series expansion for e^{λ} (see Example 19, Chapter 10), we have

$$\sum_{k=0}^{\infty} p_k = e^{-\lambda} \sum_{k=0}^{\infty} \frac{\lambda^k}{k!}$$
$$= e^{-\lambda} e^{\lambda} = e^0 = 1$$

We can calculate the expected value of a Poisson variable by a similar technique:

$$E(X) = \sum_{k=0}^{\infty} k p_k = \sum_{k=1}^{\infty} k \cdot \frac{\lambda^k}{k!} \cdot e^{-\lambda}$$

$$= e^{-\lambda} \lambda \sum_{k=1}^{\infty} \frac{\lambda^{k-1}}{(k-1)!}$$

$$= e^{-\lambda} \lambda \left(1 + \lambda + \frac{\lambda^2}{2!} + \frac{\lambda^3}{3!} + \cdots \right)$$

$$= e^{-\lambda} \lambda e^{\lambda} = \lambda$$

In a similar way (see Problem 17 at the end of this section), we can show that $Var(X) = \lambda$. These results are summarized as follows:

If X is a Poisson variable, then

$$P(X = k) = \frac{\lambda^k e^{-\lambda}}{k!} \qquad k = 0, 1, 2, \ldots$$

$$E(X) = \lambda \qquad Var(X) = \lambda$$

Example 18   Suppose that during the lunch hour customers arrive at a restaurant at an average rate of 6 per minute. Assuming that the number of arrivals is a Poisson variable, what is the probability that 3 customers will arrive during a 1 minute period?

Solution   Let X be the random variable that represents the number of arrivals during a 1 minute period. Then the expected number of arrivals is

$$E(X) = \lambda = 6$$

The probability of 3 arrivals is then

$$P(X = 3) = \frac{6^3}{3!} \cdot e^{-6} = 36e^{-6} \approx 0.089$$ ∎

Example 19   Suppose that the number of sales at an automobile dealership during a given hour is a Poisson variable with $\lambda = 2$. What is the probability that more than two cars are sold during the hour?

Solution   If X represents the number of cars sold during the hour, then

$$P(X = k) = \frac{2^k}{k!} \cdot e^{-2}$$

By the law of complementary probabilities, the probability of more than two sales is

$$P(X > 2) = 1 - P(X \le 2)$$
$$= 1 - [P(X = 0) + P(X = 1) + P(X = 2)]$$
$$= 1 - \left(e^{-2} + \frac{2}{1!} \cdot e^{-2} + \frac{2^2}{2!} \cdot e^{-2} \right)$$
$$= 1 - 5e^{-2} \approx 0.323$$ ∎

PROBLEMS (Section 11-3)

1. Suppose a coin is tossed repeatedly and let X denote the number of tails that appear before the first head. What is $P(X = 0)$? $P(X = 2)$? $P(X = 5)$?

2. Let X be the random variable in Problem 1. What is $P(X > 3)$?

3. What is the probability that X is even, where X is the random variable in Problem 1?

4. What is the probability that X is odd, where X is the random variable in Problem 1?

5. Suppose Y is a random variable that takes on the values 0, 1, 2, 3, . . . , with probabilities $P(Y = k) = (0.7)(0.3)^k$. What is $P(Y \leq 3)$? $P(Y \geq 4)$?

6. What is $P(3 \leq Y < 7)$, where Y is the random variable in Problem 5?

7. Suppose X is a random variable taking values 0, 1, 2, 3, . . . , with probabilities $P(X = k) = a(0.4)^{k+1}$. What is a?

8. Suppose Y is a random variable taking values 0, 1, 2, 3, . . . , with probabilities $P(Y = k) = \frac{1}{4} \cdot b^k$. What is b?

9. Suppose that the daily number of telephone calls to a certain police station is a Poisson variable with $\lambda = 12$. What is the probability that exactly one call is received in a day? What is the probability that at most 3 calls are received in a day?

10. Let X denote the number of errors made in typing a page of text. Suppose that X is a Poisson variable with $\lambda = 3$. What is the probability that at most one error is made, that is, $P(X \leq 1)$?

11. A book of 1000 pages contains 800 misprints. What is the probability that a given page contains at least one misprint? [*Hint:* Let X be the number of misprints on a given page; assume X is a Poisson variable.]

12. Suppose that the monthly number of claims filed in an insurance company is a Poisson variable with $\lambda = 50$. What is the probability that at most 5 claims are filed in a given month?

13. In an experiment, the number of alpha particles emitted from a radioactive isotope and recorded on a Geiger counter during a fixed period of time is a Poisson variable with $\lambda = 3.5$. What is the probability that the counter will detect at most one alpha particle during the given time period?

14. Suppose computer chips are packaged 50 to a case by a manufacturer, and the probability that an individual chip is defective is 0.005. What is the probability that a given case contains more than one defective chip? [*Hint:* Assume the number of defective chips in a case is a Poisson variable.]

15. Suppose the probability of hitting a target with an individual shot is 0.01. What is the probability of getting at least 3 hits in 500 shots? [*Hint:* Assume the number of hits in 500 shots is a Poisson variable.]

16. Suppose a technician on a college campus is responsible for the maintenance of 25 computer terminals. The probability that a given terminal will break down during a day is 0.01. What is the probability that in a given day at least three machines will require his attention?

17. Show that if X is a Poisson variable with parameter λ then $\text{Var}(X) = \lambda$ [*Hint:* $\text{Var}(X) = \lambda - \lambda^2 + e^{-\lambda} \sum_{k=0}^{\infty} k(k-1)\lambda^k/k!$, cancel appropriate terms, and use the Taylor expansion for e^{λ}.]

11-4 CONTINUOUS RANDOM VARIABLES

A random variable that assumes a continuous range of values is called a **continuous random variable.** For example, earlier we considered the situation of a random break in a 10 mile long power line. The random variable that gives the distance from the power plant to the point at which the break occurred is a continuous random variable that takes values in the interval [0, 10]. If an automobile power train is warranted for 50,000 miles and there is a power train failure during the warranty period, then the mileage at which the failure occurred can be considered a continuous random variable taking values in the interval [0, 50,000].

PROBABILITY DENSITY FUNCTION

Suppose X is a continuous random variable taking values in an interval $[a, b]$. We can associate with X a nonnegative function $f(x)$ such that for $a \le c \le d \le b$,

$$P(c \le X \le d) = \int_c^d f(x)\, dx$$

The function $f(x)$ is called a **probability density function** for X, and its graph is the continuous analog of the probability density histogram of a discrete random variable (see Figure 11-5).

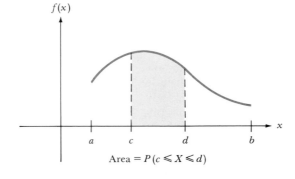

Figure 11-5
Probability density function of X

Area $= P(c \le X \le d)$

Example 20   Find a probability density function for the random variable that gives the distance from the power plant to a random break in a 10 mile long power line.

Solution   Assuming that the break is equally likely to occur at any point along the line, we assign a probability to a break in a given segment proportional to the length of the segment. For example, $P(3 \le X \le 5) = (5 - 3)/10 = 0.2$. In general, for $0 \le c \le d \le 10$,

$$P(c \le X \le d) = \frac{d - c}{10}$$

We must find a nonnegative function $f(x)$ such that

$$\int_c^d f(x)\, dx = \frac{d-c}{10}$$

Of course, such a function is $f(x) = \frac{1}{10}$. ∎

Example 21 A mortar shell is randomly fired into a circle of radius 1 mile in a desert test range. Let X be the continuous random variable that gives the distance from the center at which the shell strikes. Find a probability density function for X.

Solution If $c \leq X \leq d$, then the shell strikes in the shaded region shown in Figure 11-6. We take the probability of this event to be the ratio of the area of this region to the area of the unit circle:

$$P(c \leq X \leq d) = \frac{\pi d^2 - \pi c^2}{\pi} = d^2 - c^2$$

Therefore, we must find a function $f(x)$ such that

$$d^2 - c^2 = \int_c^d f(x)\, dx$$

for each c and d with $0 \leq c \leq d \leq 1$. In this case, we see that $f(x) = 2x$ does the trick.

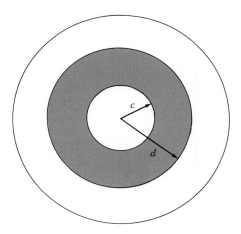

Figure 11-6

If X is a continuous random variable that takes values in the interval $[a, b]$ and $f(x)$ is a probability density function for X, then since $\{a \leq X \leq b\}$ is the sure event and its probability is 1, we have

$$1 = P(a \leq X \leq b) = \int_a^b f(x)\, dx$$

For instance, in Example 20,

$$\int_0^{10} f(x)\, dx = \int_0^{10} \frac{1}{10}\, dx = \frac{x}{10}\Big|_0^{10} = \frac{10}{10} - 0 = 1$$

In Example 21,

$$\int_0^1 f(x)\, dx = \int_0^1 2x\, dx = x^2 \Big|_0^1 = 1$$

The properties of a probability density function may be summarized as follows:

Properties of Probability Density Functions

If $f(x)$ is the probability density function of a random variable X taking values in $[a, b]$, then

$$f(x) \geq 0 \quad \text{for} \quad a \leq x \leq b$$

$$\int_a^b f(x)\, dx = 1 \qquad P(c \leq X \leq d) = \int_c^d f(x)\, dx$$

Example 22   Suppose $f(x) = ke^{-2x}$ is the probability density function of a random variable X taking values in the interval $[1, 3]$. What is k?

Solution   Since $f(x)$ is a probability density function, we must have

$$1 = \int_1^3 f(x)\, dx = \int_1^3 ke^{-2x}\, dx$$

$$= \frac{-k}{2} \cdot e^{-2x} \Big|_1^3$$

$$= \frac{k}{2}(-e^{-6} + e^{-2})$$

Therefore,

$$k = \frac{2}{-e^{-6} + e^{-2}} \approx 15.054$$

CUMULATIVE DISTRIBUTION FUNCTION

If X is a random variable taking values in $[a, b]$ and $f(x)$ is the probability density function for X, then the **cumulative distribution function** for X is the function

$$F(x) = \int_a^x f(t)\, dt$$

Notice that this function simply gives the probability that $X \leq x$:

$$P(a \leq X \leq x) = \int_a^x f(t)\, dt = F(x)$$

See Figure 11-7.

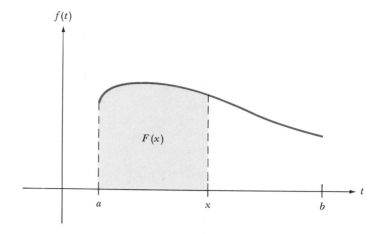

Figure 11-7
Cumulative distribution function

Also, by using the additive property of the integral (see Chapter 5), we see that probabilities of events of the form $c \leq X \leq d$ can be computed by taking a difference of the cumulative distribution function:

$$P(c \leq X \leq d) = \int_c^d f(t)\, dt$$

$$= \int_a^d f(t)\, dt - \int_a^c f(t)\, dt$$

$$= F(d) - F(c)$$

See Figure 11-8.

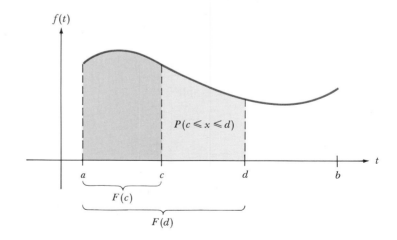

Figure 11-8
$P(c \leq X \leq d) = F(d) - F(c)$

Example 23   Suppose X takes values in $[0, 1]$ and has probability density function $f(x) = 4x^3$. Find the cumulative distribution function of X. What is $P(X \leq 0.2)$? $P(0.1 \leq X \leq 0.2)$?

Solution   First, notice that $f(x)$ actually is a probability density function since $f(x) \geq 0$ and

$$\int_0^1 f(x)\, dx = \int_0^1 4x^3\, dx = x^4 \Big|_0^1 = 1$$

The cumulative distribution function for X is

$$F(x) = \int_0^x f(t)\, dt = 4\int_0^x t^3\, dt = x^4$$

We then have

$$P(X \leq 0.2) = F(0.2) = (0.2)^4 = 0.0016$$

and

$$P(0.1 \leq X \leq 0.2) = F(0.2) - F(0.1) = 0.0016 - 0.0001 = 0.0015 \qquad \blacksquare$$

Notice that

$$F(a) = \int_a^a f(t)\, dt = 0$$

and therefore,

$$F(x) - F(a) = \int_a^x f(t)\, dt$$

That is, $F(x)$ is an antiderivative of $f(x)$ (see the fundamental theorem of calculus, Chapter 5). This gives us another relationship between the probability density function and the cumulative distribution function:

$$F'(x) = f(x)$$

Example 24   Suppose $F(x) = 4x^3 - 3x^4$ is the cumulative distribution function of a random variable taking values in $[0, 1]$. Then the probability density function of this random variable is

$$f(x) = F'(x) = 12x^2 - 12x^3 = 12x^2(1 - x) \qquad \blacksquare$$

EXPECTED VALUE AND VARIANCE

Suppose X is a continuous random variable taking values in the interval $[a, b]$. How should we define the expected value of X? Our definition should be consistent with the definition of the expected value of a discrete random variable. One way to arrange this is to approximate X in some way by a discrete random variable. Suppose $f(x)$ is the probability density function of X and we divide the interval $[a, b]$ into n intervals of length $\Delta x = (b - a)/n$. If we call the right endpoint of the first such interval x_1, the right endpoint of the second such interval x_2, and so on, then we can think of a discrete random variable X_n that takes on the values $x_1, x_2, \ldots$. Suppose that

$$p_k = P(X_n = x_k) \approx f(x_k)\, \Delta x$$

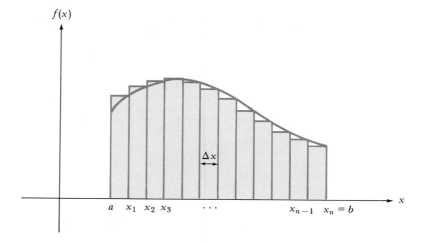

Figure 11-9
*Histogram approximation to a
density function*

Then the probability density histogram for X_n approximates the probability density plot for X, as illustrated in Figure 11-9. The expected value of X_n is then

$$E(X_n) = \sum_{k=1}^{n} x_k p_k$$

$$\approx \sum_{k=1}^{n} x_k f(x_k)\,\Delta x$$

and it seems that as $n \to \infty$ the random variable X_n approximates X more and more closely. Therefore, it is reasonable to define

$$E(X) = \lim_{n \to \infty} \sum_{k=1}^{n} x_k f(x_k)\,\Delta x$$

But this is just the limit of Riemann sums for the integral of the function $xf(x)$ (see Chapter 5); therefore,

$$E(X) = \int_a^b xf(x)\,dx$$

In a similar way, we define the variance in terms of an integral:

$$\text{Var}(X) = \int_a^b (x - m)^2 f(x)\,dx \qquad \text{where } m = E(X)$$

Example 25 Find the expected value and variance of the random variable in Example 21.

Solution   Recall that the random variable took values in $[0, 1]$ and had probability density function $f(x) = 2x$. Therefore,

$$E(X) = \int_0^1 x \cdot 2x \, dx = \tfrac{2}{3} \cdot x^3 \Big|_0^1 = \tfrac{2}{3}$$

and

$$Var(X) = \int_0^1 (x - \tfrac{2}{3})^2 \cdot 2x \, dx$$

$$= 2 \int_0^1 (x^3 - \tfrac{4}{3} \cdot x^2 + \tfrac{4}{9} \cdot x) \, dx$$

$$= 2(\tfrac{1}{4} \cdot x^4 - \tfrac{4}{9} \cdot x^3 + \tfrac{4}{18} \cdot x^2) \Big|_0^1$$

$$= 2(\tfrac{1}{4} - \tfrac{4}{9} + \tfrac{4}{18}) = \tfrac{1}{18} \qquad \blacksquare$$

An alternate way of computing the variance is sometimes more convenient. From the formula above, we have

$$Var(X) = \int_a^b (x - m)^2 f(x) \, dx$$

$$= \int_a^b x^2 f(x) \, dx - 2m \int_a^b x f(x) \, dx + m^2 \int_a^b f(x) \, dx$$

But

$$\int_a^b x f(x) \, dx = m \qquad \text{and} \qquad \int_a^b f(x) \, dx = 1$$

Therefore,

$$Var(X) = \int_a^b x^2 f(x) \, dx - 2m \cdot m + m^2 \cdot 1$$

$$= \int_a^b x^2 f(x) \, dx - m^2$$

That is,

$$\boxed{\; Var(X) = \int_a^b x^2 f(x) \, dx - m^2 \qquad \text{where } m = E(X) \;}$$

Example 26   The random variable X in Example 23 had probability density function $f(x) = 4x^3$, $0 \le x \le 1$. The expected value is therefore

$$m = E(X) = \int_0^1 x f(x) \, dx$$

$$= \int_0^1 x \cdot 4x^3 \, dx = \tfrac{4}{5} \cdot x^5 \Big|_0^1 = \tfrac{4}{5}$$

The variance is then

$$\text{Var}(X) = \int_0^1 x^2 \cdot 4x^3 \, dx - (\tfrac{4}{5})^2$$

$$= \tfrac{2}{3} \cdot x^6 \Big|_0^1 - (\tfrac{4}{5})^2$$

$$= \tfrac{2}{3} - \tfrac{16}{25} = \tfrac{2}{75}$$

■

PROBLEMS (Section 11-4)

1. A point is chosen at random in the unit square $0 \le x \le 1$, $0 \le y \le 1$, and the distance from this point to the origin is measured. What is the range of values taken on by the random variable that gives this distance?

2. A certain automobile can travel at most 360 miles on a tank of fuel. Suppose that the fuel tank has a capacity of 12 gallons and an experiment is performed in which the tank is filled and the car is run until it stops. What is the range of values for the random variable that gives the average mileage of the car in the experiment?

In Problems 3–6, show that the given function is a probability density function.

3. $f(x) = \tfrac{1}{2} \cdot x$, $\quad 0 \le x \le 2$

4. $f(x) = \dfrac{e^{3-x}}{e^3 - 1}$, $\quad 0 \le x \le 3$

5. $f(x) = \sin x$, $\quad \pi/2 \le x \le \pi$

6. $f(x) = \dfrac{1}{2x}$, $\quad \dfrac{1}{2} \le x \le \dfrac{e^2}{2}$

In Problems 7–10, find the value of the constant a that makes the given function a probability density function.

7. $f(x) = ax^3$, $\quad 1 \le x \le 3$

8. $f(x) = \dfrac{a}{1+x}$, $\quad 0 \le x \le e-1$

9. $f(x) = ax(1-x)^2$, $\quad 0 \le x \le 1$

10. $f(x) = \dfrac{a}{\sqrt{x}}$, $\quad 1 \le x \le 9$

In Problems 11–14, find the cumulative distribution function of the random variable whose probability density function is given.

11. $f(x) = 6x(1-x)$, $\quad 0 \le x \le 1$

12. $f(x) = \sin x$, $\quad 0 \le x \le \pi/2$

13. $f(x) = e^x$, $\quad \ln 2 \le x \le \ln 3$

14. $f(x) = 12x^2(1-x)$, $\quad 0 \le x \le 1$

In Problems 15–18, find the probability density function of the random variable whose cumulative distribution function is given.

15. $F(x) = \tfrac{1}{6} \cdot x - \tfrac{1}{3}$, $\quad 2 \le x \le 8$

16. $F(x) = 3x^2 - 2x^3$, $\quad 0 \le x \le 1$

17. $F(x) = e^{2(x-1)} - e^{-2}$, $\quad 0 \le x \le 1 + \tfrac{1}{2}\ln(1 + e^{-2})$

18. $F(x) = \sin x$, $\quad 0 \le x \le \pi/2$

19. Suppose a continuous random variable X has probability density function $f(x) = x/2$ on the interval $0 \le x \le 2$. Sketch the graph of $f(x)$ and indicate the area under the graph that represents:

   a. $P(X \le 1)$ b. $P(1 \le X \le 1.5)$

20. The probability density function of a continuous random variable X is $f(x) = 6x(1-x)$, $0 \le x \le 1$. Sketch the graph of $f(x)$ and indicate the area under the graph that represents:

   a. $P(X \ge 0.5)$ b. $P(0.25 \le X \le 0.5)$

21. Find $P(X \le 2)$, where X is a continuous random variable whose probability density function is given in Problem 6.

22. Find $P(X \geq 3\pi/4)$, where X is the continuous random variable whose probability density function is given in Problem 5.

23. Find $P(X \leq 0.5)$, where X is the continuous random variable whose probability density function is given in Problem 11.

24. Find $P(X \geq 1)$, where X is the continuous random variable whose probability density function is given in Problem 13.

25. Find $P(X \leq 5)$, where X is the continuous random variable whose cumulative distribution function is given in Problem 15.

26. Find $P(\frac{1}{2} \leq X \leq \frac{2}{3})$, where X is the continuous random variable whose cumulative distribution function is given in Problem 16.

27. A continuous random variable X has cumulative distribution function $F(x) = (x - 1)/8$, $1 \leq x \leq 9$. Find:

    a. $P(X \leq 2)$ b. $P(X > 4)$ c. $P(2 \leq X \leq 5)$

28. A continuous random variable X has cumulative distribution function $F(x) = (x^4 - 1)/15$, $1 \leq x \leq 2$. Find:

    a. $P(X \leq 1.5)$ b. $P(X > 1.5)$
    c. $P(1.2 \leq X \leq 1.5)$

29. A continuous random variable X has probability density function $f(x) = \frac{1}{6}$, $1 \leq x \leq 7$. Find k such that $P(X \leq k) = 0.3$.

30. A continuous random variable X has probability density function $f(x) = x/48$, $2 \leq x \leq 10$. Find k such that $P(X \geq k) = 0.4$.

31. A continuous random variable X has cumulative distribution function $F(x) = (x - 2)/6$, $2 \leq x \leq 8$. Find k such that $P(3 \leq X \leq k) = 0.6$.

32. A continuous random variable X has cumulative distribution function $F(x) = \sin x$, $0 \leq x \leq \pi/2$. Find k such that $P(X \leq k) = 1/\sqrt{2}$.

33. Find $E(X)$ and $\mathrm{Var}(X)$, where X is a continuous random variable whose probability density function is given in Problem 3.

34. Find $E(X)$ and $\mathrm{Var}(X)$, where X is a continuous random variable whose probability density function is given in Problem 11.

35. Find $E(X)$ and $\mathrm{Var}(X)$, where X is a continuous random variable whose probability density function is given in Problem 13.

11-5 CONCLUSION: OPTIMIZING BLOOD SAMPLE TESTING

We opened this chapter with a discussion of a pooling technique for blood tests in large populations. Suppose the probability that a randomly selected individual from the population has the disease is $\frac{1}{16}$. Then the probability that a randomly selected individual is free of the disease is $\frac{15}{16}$. If the total population is N, then N tests are required if each individual is tested separately, while at least N/n tests are required if the blood samples are pooled into groups of n. It can be shown that the probability that such a group of n individuals is disease-free is $(\frac{15}{16})^n$. Therefore, the probability that a group contains at least one diseased individual is, by the law of complementary probability, $1 - (\frac{15}{16})^n$. That is, the probability that each member of a group must be individually retested is $p = 1 - (\frac{15}{16})^n$. We would like to know how many addi-

tional individual tests we should expect to perform on a group. Let Y represent this random variable. Then Y takes value 0 (no additional tests required) with probability $(\frac{15}{16})^n$, and Y takes value n (that is, the test on the pooled sample is positive and each individual in the group of n must be retested) with probability $1 - (\frac{15}{16})^n$. Therefore,

$$E(Y) = 0 \cdot (\tfrac{15}{16})^n + n[1 - (\tfrac{15}{16})^n]$$
$$= n[1 - (\tfrac{15}{16})^n]$$

Since there are N/n groups, we should expect to perform

$$\left(\frac{N}{n}\right) n \left[1 - \left(\frac{15}{16}\right)^n\right] = N\left[1 - \left(\frac{15}{16}\right)^n\right]$$

additional tests. That is, altogether, we should expect to perform

$$\frac{N}{n} + N\left[1 - \left(\frac{15}{16}\right)^n\right]$$

tests using the pooling procedure.

Assuming a constant cost per test, we may take as a measure of the cost of this technique (relative to a program of individually testing all N members of the population) the ratio

$$C(n) = \frac{(N/n) + N[1 - (\frac{15}{16})^n]}{N} = \frac{1}{n} + 1 - \left(\frac{15}{16}\right)^n$$

Note that this relative cost is independent of the popu-

lation size N. Our task is thus to choose n to minimize $C(n)$.

It turns out that trying to minimize $C(n)$ by the techniques of calculus leads to an equation that is very difficult to solve. (Try it!) However, the function $C(n)$ is easily tabulated:

n	2	3	4	5	6
$C(n)$	0.6211	0.5094	0.4775	0.4758	0.4877

From this we see that the minimum of C is around $n = 5$ and that pooling the samples into groups of 5 will result in a relative cost savings of about 48% over an individual testing program.

11-6 CHAPTER REVIEW

1. The set of all possible outcomes of an experiment is called a sample space. Subsets of the sample space are called events. Probabilities can be assigned to events in either an intuitive or an empirical way.
2. The probability of an event A is a number $P(A)$ satisfying $0 \le P(A) \le 1$, $P(\text{not } A) = 1 - P(A)$, and $P(A \text{ or } B) = P(A) + P(B)$ if A and B are mutually exclusive events.
3. A variable X that takes on discrete numerical values associated with the outcomes of an experiment is called a discrete random variable. The expected value of X is the number

   $$E(X) = \sum_k kP(X = k)$$

   where the sum is extended over all possible values k of the random variable X. The variance of X is given by

   $$\text{Var}(X) = \sum_k (k - m)^2 P(X = k)$$

   where $m = E(X)$.
4. A random variable X taking on the values $0, 1, 2, 3, \ldots$ is called a Poisson variable if

   $$P(X = k) = \frac{\lambda^k e^{-\lambda}}{k!} \qquad k = 0, 1, 2, 3, \ldots$$

   where $\lambda > 0$. If X is a Poisson variable, then $E(X) = \lambda$ and $\text{Var}(X) = \lambda$.
5. A random variable that takes on a continuous range of values is called a continuous random variable. If X is a continuous random variable taking values in the interval $a \le x \le b$, then the probability density function of X is the nonnegative function $f(x)$ satisfying

   $$P(c \le X \le d) = \int_c^d f(x)\, dx$$

   where $a \le c \le d \le b$.
6. The cumulative distribution function of X is defined by

   $$F(x) = \int_a^x f(t)\, dt$$

   In particular, $F'(x) = f(x)$ and $P(c \le X \le d) = F(d) - F(c)$.
7. The expected value and variance of a continuous random variable X are given by

   $$E(X) = \int_a^b xf(x)\, dx$$

   $$\text{Var}(X) = \int_a^b (x - m)^2 f(x)\, dx$$

   respectively, where $m = E(X)$.

REVIEW PROBLEMS (Section 11-6)

1. A shopper in a supermarket selects a one-dozen carton of eggs and observes the number of eggs that are cracked or broken. What is the sample space of this experiment?

2. A number is selected at random from the first 100 positive integers. What is the sample space for this experiment? Describe the event that the number selected is a multiple of 17.

3. A point is selected at random from the square

$$\{(x, y)|0 \le x \le 1, 0 \le y \le 1\}$$

   What is the probability that $y \le x^2$?

4. A police department reports the following breakdowns for calls it received during a given year: homicides 5, assaults 102, domestic violence 222, burglary 260, car theft 196, robbery 124, other 312. What is the probability that a given call during this year concerned an assault? What is the probability that a given call concerned a nonviolent crime (that is, not assault, homicide, or domestic violence)?

5. The table represents the number of failures in a sample of 500 radial tires during the indicated span of mileage.

Mileage	Number of Failures
Less than 10,000	12
Less than 20,000	56
Less than 30,000	122
Less than 40,000	256
Less than 50,000	382
Less than 60,000	500

Suppose a tire is chosen at random and let X be the number of miles (in units of 10,000 miles) until failure. Sketch the probability density histogram for X. Find $P(2 \le X \le 4)$ and indicate the area of the histogram that corresponds to this probability.

6. Find the life expectancy of a tire in Problem 5.

7. Suppose a random variable X takes on the values indicated with the given probabilities. Find $E(X)$ and $\text{Var}(X)$.

Value	−2	−1	0	1	2	3	4
Probability	0.1	0.2	0.1	0.05	0.3	0.2	0.05

8. Suppose X is a random variable with density function $f(x) = \frac{1}{5}$ on the interval $-2 \le x \le 3$. Find $E(X)$ and $\text{Var}(X)$.

9. Suppose that $f(x) = 6x^\alpha(1 - x)$, $0 \le x \le 1$ ($\alpha > 0$), is a probability density function (it is called a *beta* density). Find the expected value of a random variable having this density function.

10. What is the cumulative distribution function of the density function in Problem 9?

11. Suppose X is a continuous random variable taking values in the interval $2 \le x \le 5$ and having cumulative distribution function $F(x) = (x^3 - 8)/117$.
    a. Find $P(3 \le X \le 4)$.
    b. Find the probability density function of X.

12. Suppose that a firm's yearly advertising costs (in thousands of dollars) is a random variable X with cumulative distribution function $F(x) = (x - 1)^2/16$, $1 \le x \le 5$.
    a. What can the firm expect to pay during the year for advertising?
    b. What is the probability that the firm will pay more than $4000 a year for advertising?

13. The annual family income (in thousands of dollars) in a certain community is a random variable X with probability density function $f(x) = x/1178$, $12 \le x \le 50$. An education at a local private college costs $6000 per year and a family can afford to spend at most 20% of its income on a college education for a child. What percentage of families in this community can afford to send a child to the local college?

14. Suppose that the weekly study time (in hours) required to pass a certain course is a random variable X with cumulative distribution function $F(x) = 1 - (20 - x)^2/400$, $0 \le x \le 20$. How many hours per week should the "average" student expect to study in order to pass?

15. Suppose that in a certain city the monthly number X of suicides is a Poisson variable with $\lambda = 2$. What is the probability of more than 3 suicides in a given month?

16. A biologist observes the number X of diatoms (in units of a hundred) in a given volume of pond water. She knows that X is a Poisson variable and that "on the average" such a volume contains 220 diatoms. What is the probability that the volume contains less than 200 diatoms?

THE UNIFORM, EXPONENTIAL, AND NORMAL DENSITY FUNCTIONS

Many practical problems involving continuous random variables can be modeled in terms of uniform, exponential, or normal density functions. It is common practice in probability theory to use the term *density* in place of *probability density function;* we will do this throughout this supplement. Uniform densities are involved when a given interval of values of the random variable is assigned a probability proportional to its length, as in the power line break problem in Chapter 11. Waiting times and times to failure are often modeled with exponential densities. Situations in which an effect is caused by a large number of independent random factors are modeled by normal densities. In this supplement we define and illustrate these three important densities.

11S-1 UNIFORM DENSITY

A continuous random variable X taking values in the interval $[a, b]$ is said to have a **uniform** probability density (or to be **uniformly distributed**) if its probability density function is given as follows:

Uniform Density

$$f(x) = \frac{1}{b-a} \qquad a \leq x \leq b$$

Uniformly distributed random variables model situations in which probabilities are uniformly distributed over the interval $[a, b]$ in the sense that the probability of any event of the type $c \leq X \leq d$ is equal to the relative size of the interval $[c, d]$:

$$P(c \leq X \leq d) = \int_c^d \frac{1}{b-a} \, dx = \frac{d-c}{b-a}$$

The problem of the power line break (Example 4 of Chapter 11) gives a good illustration of a uniformly distributed random variable. Another example follows.

Example 1
Tensile Strength

Suppose the tensile strength of a steel wire is a uniformly distributed random variable taking values between 30 and 45 pounds per square millimeter. Industry standards require a tensile strength of at least 34 pounds per square millimeter. What is the probability that the standard is met?

Solution   Let X represent the tensile strength of a given wire. We would like to know $P(34 \leq X \leq 45)$. Since X is uniformly distributed on $[30, 45]$, its probability density function is $f(x) = 1/(45 - 30) = \frac{1}{15}$. Therefore,

$$P(34 \leq X \leq 45) = \int_{34}^{45} \frac{1}{15}\, dx = \frac{45 - 34}{15} = \frac{11}{15} \approx 0.73$$

■

11S-2 EXPONENTIAL DENSITY

Situations involving failure times are often effectively modeled by use of the exponential density. Consider, for example, the experiment of observing the time it takes for a certain machine, say a personal computer, to fail. We can consider this to be a continuous random variable X that takes on nonnegative values. Since we do not wish to arbitrarily set a maximum possible running time, X may take on arbitrarily large values, that is, it takes values in the interval $[0, \infty)$. Suppose we denote the probability that the machine will continue to operate over a time interval of length t by $p(t)$. By the law of complementary probabilities, the probability of failure during a time interval of length t is then $1 - p(t)$. It is reasonable to assume that the longer the length of an interval, the greater the probability of a failure in that interval. In particular, suppose that the probability of a failure during a small time interval of length Δt is proportional to Δt:

$$1 - p(\Delta t) = \lambda \Delta t$$

where λ is a positive constant of proportionality. A basic assumption is that the probability that the machine does not fail during a time interval of length $t + \Delta t$ is equal to the product of the probability that it does not fail in an interval of time t and the probability that it does not fail in an interval of time Δt, that is,

$$p(t + \Delta t) = p(t)p(\Delta t)$$

The justification of this assumption requires the notion of independence, which we will not go into. Using this relationship and the preceding equation, we obtain

$$p(t + \Delta t) = p(t)p(\Delta t)$$
$$= p(t)(1 - \lambda \Delta t)$$
$$\frac{p(t + \Delta t) - p(t)}{\Delta t} = -\lambda p(t)$$

Taking limits as $\Delta t \to 0$, we find

$$p'(t) = \lim_{\Delta t \to 0} \frac{p(t + \Delta t) - p(t)}{\Delta t} = -\lambda p(t)$$

That is, the function $p(t)$ satisfies the differential equation

$$p'(t) = -\lambda p(t)$$

This is a linear first-order differential equation that may be solved by the method of Chapter 7 to yield

$$p(t) = ce^{-\lambda t} \qquad t \geq 0$$

Notice that t is allowed to take on arbitrarily large values. The constant c is determined by requiring that p is a probability density function:

$$\int_0^\infty p(t)\, dt = 1$$

where the definite integral is interpreted as

$$\int_0^\infty p(t)\, dt = \lim_{b \to \infty} \int_0^b p(t)\, dt$$

(see the supplement to Chapter 5). We then find that

$$
\begin{aligned}
1 &= \lim_{b \to \infty} \int_0^b ce^{-\lambda t}\, dt \\
&= c \lim_{b \to \infty} \left(-\frac{1}{\lambda}\right) e^{-\lambda t} \Big|_0^b \\
&= \frac{c}{\lambda} \lim_{b \to \infty} (-e^{-\lambda b} + 1) \\
&= \frac{c}{\lambda} \\
c &= \lambda
\end{aligned}
$$

Therefore,

$$p(t) = \lambda e^{-\lambda t} \qquad t \geq 0$$

This leads us to make the following definition: We say that a random variable X taking values in the interval $[0, \infty)$ is **exponentially distributed** if its probability density function has the following form:

Exponential Density

$$f(x) = \lambda e^{-\lambda x} \qquad x \geq 0$$

where λ is a positive constant.

The expected value of an exponentially distributed random variable X is given by

$$
\begin{aligned}
E(X) &= \int_0^\infty xp(x)\, dx \\
&= \lim_{b \to \infty} \int_0^b \lambda x e^{-\lambda x}\, dx
\end{aligned}
$$

This integral may be computed by integration by parts (see the supplement to Chapter 5). Setting $u = x$ and $dv = \lambda e^{-\lambda x}\, dx$, we have

$$\int_0^b \lambda x e^{-\lambda x}\, dx = x(-e^{-\lambda x})\Big|_0^b - \int_0^b (-e^{-\lambda x})\, dx$$

$$= -be^{-\lambda b} - \frac{1}{\lambda}\cdot e^{-\lambda x}\Big|_0^b$$

$$= -be^{-\lambda b} - \frac{1}{\lambda}\cdot e^{-\lambda b} + \frac{1}{\lambda}$$

However,

$$\lim_{b\to\infty} be^{-\lambda b} = 0 \qquad \text{and} \qquad \lim_{b\to\infty} e^{-\lambda b} = 0$$

Therefore,

$$E(X) = \frac{1}{\lambda}$$

To put it another way, the parameter λ is the reciprocal of the expected value of an exponentially distributed random variable.

Example 2 Suppose that the time (in days) that a new automobile is in use before it requires service is an exponentially distributed random variable with $\lambda = 0.02$. What is the probability that the car will require service during its second week in use?

Solution Let X be the random variable in question. Then we require $P(7 \le X \le 14)$. The probability density function for X is $f(x) = 0.02e^{-0.02x}$. Therefore,

$$P(7 \le X \le 14) = \int_7^{14} 0.02e^{-0.02x}\, dx = -e^{-0.02x}\Big|_7^{14}$$

$$= e^{-0.14} - e^{-0.28} \approx 0.114 \qquad\blacksquare$$

Example 3 Connectors in oil pipelines sometimes fail due to improper welding. Suppose that tests over many years indicate that the lifetime of a connector is an exponentially distributed random variable and that the expected life is 10.2 years. What is the probability that a connector will exceed its life expectancy?

Solution Let X be the time (in years) to failure of a connector. Then $E(X) = 10.2 = 1/\lambda$. Therefore,

$$P(X \ge 10.2) = \int_{10.2}^{\infty} \lambda e^{-\lambda x}\, dx = -e^{-\lambda x}\Big|_{10.2}^{\infty}$$

$$= e^{-\lambda(10.2)} = e^{-1} \approx 0.368 \qquad\blacksquare$$

11S-3 NORMAL DENSITY

A continuous random variable X taking on values in the interval $-\infty < x < \infty$ is said to have a **standard normal distribution** if its density function is given by:

Standard Normal Density

$$f(x) = \frac{1}{\sqrt{2\pi}}\, e^{-x^2/2} \qquad -\infty < x < \infty$$

The graph of this density function is the *bell-shaped curve* shown in Figure 11S-1. The expected value of X is 0, since

$$E(X) = \int_{-\infty}^{\infty} xf(x)\, dx$$

$$= \frac{1}{\sqrt{2\pi}} \int_{-\infty}^{\infty} xe^{-x^2/2}\, dx$$

$$= \frac{1}{\sqrt{2\pi}} \left(-e^{-x^2/2}\, \Big|_{-\infty}^{\infty} \right)$$

$$= \frac{1}{\sqrt{2\pi}}\, (0 - 0) = 0$$

Figure 11S-1
The standard normal density
$$f(x) = \frac{1}{\sqrt{2\pi}}\, e^{-x^2/2}$$

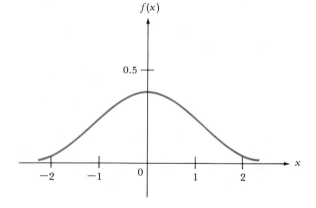

By using techniques that are beyond the scope of this book, it can be shown that $\text{Var}(X) = 1$.

A general type of normal random variable that has a probability density is given as follows:

A random variable X taking values in the interval $-\infty < x < \infty$ is said to be **normally distributed** with **mean** μ and **standard deviation** σ if its probability density function is

$$f(x) = \frac{1}{\sigma\sqrt{2\pi}}\, e^{-(x-\mu)^2/2\sigma^2} \qquad -\infty < x < \infty$$

The **mean** of a random variable is the same as its expected value; that is, for the random variable above,

$$E(X) = \mu$$

The **standard deviation** of a random variable is defined to be the square root of its variance. Thus, for the normal random variable above,

$$\sqrt{\text{Var}(X)} = \sigma$$
$$\text{Var}(X) = \sigma^2$$

(Proving this requires techniques beyond the scope of this book.) Some normal densities with mean μ and various standard deviations σ are sketched in Figure 11S-2. Notice that each graph is symmetric about the mean μ and has a "peakedness" that is in inverse proportion to the standard deviation σ.

Figure 11S-2
Normal probability densities

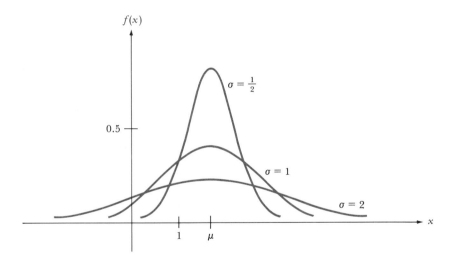

Many careful empirical investigations have established that a normal distribution is appropriate for modeling random variables that may be considered to result from a large number of independent random factors. For example, in manufacturing precision machine parts the deviation from actual specifications may result from temperature differences, random surges in electrical supplies, work schedules, psychological factors of the machine operator, variability in material quality, and many other independent factors. In cases such as this, a normal distribution has been found to be effective in modeling the deviation.

The cumulative distribution function of a random variable with a standard normal density will be denoted by $\Phi(x)$:

$$\Phi(x) = \frac{1}{\sqrt{2\pi}} \int_{-\infty}^{x} e^{-t^2/2} \, dt$$

This integral cannot be computed analytically; however, values of the function $\Phi(x)$ are tabulated in Appendix D. Notice that since the total area under the graph of the standard normal density is 1, the area under the graph to the right of the point x is equal to $1 - \Phi(x)$ (see Figure 11S-3). From the symmetry of the graph, we then see that $\Phi(-x)$, the area under the graph to the left of $-x$, is equal to $1 - \Phi(x)$:

$$\Phi(-x) = 1 - \Phi(x)$$

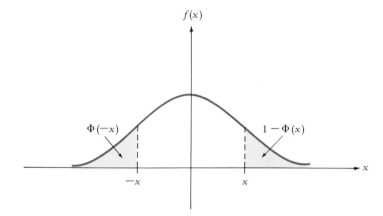

Figure 11S-3
$\Phi(-x) = 1 - \Phi(x)$

Example 4   Suppose X is a random variable with a standard normal density. Find:

a. $P(X \leq 0.1)$ b. $P(X \leq -0.3)$ c. $P(-0.1 \leq X \leq 0.4)$

Solution   a. From Appendix D we find that

$$P(X \leq 0.1) = \Phi(0.1) = 0.5398$$

b. Using the relationship between $\Phi(-x)$ and $\Phi(x)$, we obtain

$$P(X \leq -0.3) = \Phi(-0.3) = 1 - \Phi(0.3) = 1 - 0.6179 = 0.3821$$

c. Finally,

$$\begin{aligned} P(-0.1 \leq X \leq 0.4) &= \Phi(0.4) - \Phi(-0.1) \\ &= \Phi(0.4) - [1 - \Phi(0.1)] \\ &= \Phi(0.4) + \Phi(0.1) - 1 \\ &= 0.6554 + 0.5398 - 1 = 0.1952 \end{aligned}$$ ∎

Suppose X is a random variable that has a normal density with mean μ and standard deviation σ and we would like to compute the probability $P(a \leq X \leq b)$, that is,

$$\frac{1}{\sigma\sqrt{2\pi}} \int_a^b e^{-(x-\mu)^2/2\sigma^2}\, dx$$

Suppose we make the substitution

$$t = g(x) = \frac{x - \mu}{\sigma} \qquad g'(x) = \frac{1}{\sigma}$$

in the integral. Then

$$P(a \le X \le b) = \frac{1}{\sigma\sqrt{2\pi}} \int_a^b e^{-(x-\mu)^2/2\sigma^2} \, dx$$

$$= \frac{1}{\sqrt{2\pi}} \int_{(a-\mu)/\sigma}^{(b-\mu)/\sigma} e^{-t^2/2} \, dt$$

But this last integral is equal to

$$P\left(\frac{a-\mu}{\sigma} \le Z \le \frac{b-\mu}{\sigma}\right)$$

where Z is the standard normal random variable. Therefore:

If X is a normal random variable with mean μ and standard deviation σ, then

$$P(a \le X \le b) = \Phi\left(\frac{b-\mu}{\sigma}\right) - \Phi\left(\frac{a-\mu}{\sigma}\right)$$

In this way, problems involving general normal random variables can be converted to problems involving standard normal random variables.

Example 5
Quality Control   Suppose that the diameter of a cylindrical machine part is a normal random variable with mean 4 centimeters and standard deviation 0.07 centimeter. An acceptable part must have a diameter of between 3.86 and 4.14 centimeter. If 1000 such parts are manufactured, about how many are acceptable?

Solution   Let X represent the diameter of a given part. This part is acceptable if and only if $3.86 \le X \le 4.14$. The probability that a part is acceptable is then

$$P(3.86 \le X \le 4.14) = \Phi\left(\frac{4.14-4}{0.07}\right) - \Phi\left(\frac{3.86-4}{0.07}\right)$$

$$= \Phi\left(\frac{14}{7}\right) - \Phi\left(-\frac{14}{7}\right)$$

$$= \Phi(2) - [1 - \Phi(2)]$$

$$= 2\Phi(2) - 1$$

$$= 2(0.9772) - 1$$

$$= 0.8544$$

Therefore, of 1000 such parts, 1000×0.8544, or about 854 will be acceptable. ∎

PROBLEMS (Supplement to Chapter 11)

1. Suppose X is a uniformly distributed random variable taking values in $[a, b]$. Compute $E(X)$ and $\mathrm{Var}(X)$.

2. Calculate the cumulative distribution function for a uniformly distributed random variable X on the interval $[a, b]$.

3. Compute $P(|X - (a + b)/2| \leq (b - a)/4)$, where X is the random variable in Problem 2.

4. Suppose the response time of an emergency paramedic squad is uniformly distributed between 5 and 22 minutes. In 1000 calls to the squad, about how many will obtain a response in less than average time?

5. The grades on a certain test are uniformly distributed from 60 to 100. If 200 students took the test, about how many received grades higher than 95?

6. Suppose X is exponentially distributed with parameter λ. What is $\mathrm{Var}(X)$?

7. What is the cumulative distribution function of the random variable in Problem 6?

8. Suppose that the time required to serve a customer at a fast-food restaurant is an exponentially distributed random variable with expected value 1.5 minutes. What is the probability that a customer will require more than 3 minutes to be serviced?

9. Suppose that the time between arrivals of consecutive telephone calls at an office is an exponentially distributed random variable with expected value 30 seconds. What is the probability that between 2 and 3 minutes elapse between consecutive telephone calls?

10. Suppose the lifetime of a gear in a turbine engine is exponentially distributed and the probability that the gear will last at most 1000 hours is 0.5. What is the life expectancy of the gear?

11. Suppose that extensive tests performed on a certain type of solar cell have determined that the lifetime of the cell is exponentially distributed and that, in testing a large number of cells, the average lifetime of a cell is 800 hours. What is the probability that the lifetime of such a cell exceeds 1000 hours?

12. Suppose that tests on an automobile shock absorber indicate that the lifetime of the shock absorber is an exponentially distributed random variable and that the life expectancy of a given shock absorber is 30,000 miles. Suppose 10,000 such shock absorbers are installed on new cars. About how many shock absorbers can be expected to be operating satisfactorily after 40,000 miles?

13. Referring to Problem 12, suppose that reports from the field indicate that 2500 of the shock absorbers have failed before 10,000 miles. Should the manufacturer issue a recall? (That is, is this event unexpected?)

14. Suppose X is an exponentially distributed random variable. Let A be the number such that $P(X \leq A) = 0.5$. Show that $A \leq E(X)$. Interpret this result.

In Problems 15–18, the given function is the probability density function of a normally distributed random variable. In each case, find the mean and standard deviation by inspection.

15. $f(x) = \dfrac{1}{2\sqrt{8\pi}}\, e^{-x^2/32}$

16. $f(x) = \dfrac{1}{\sqrt{2\pi}}\, e^{-(x + 2)^2/2}$

17. $f(x) = \dfrac{1}{3\sqrt{2\pi}}\, e^{-(x - 1)^2/18}$

18. $f(x) = \dfrac{1}{\sqrt{4\pi}}\, e^{-(x^2 + 2x + 1)/4}$

19. Suppose X is a standard normal random variable. Find:
    a. $P(X \geq 2.2)$ b. $P(-1 \leq X \leq 3.2)$
    c. $P(X \geq -1.5)$

20. Suppose X is a normal random variable with mean -2 and standard deviation 0.3. Find:
    a. $P(X \leq -1.58)$ b. $P(1.46 \leq X \leq 3.21)$
    c. $P(X \geq 0.27)$

21. The sales price of houses in a particular city is a normally distributed random variable with mean $65,000 and standard deviation $6000. You can spend at most $75,000 on a house and answer an ad to inspect a house with the agreement that the owner will tell you the price only in person. What is the probability that your trip will be wasted?

22. Suppose an automobile is advertised as getting 38 miles per gallon (mpg) in city driving. Actually, the mileage is a normally distributed random variable with mean 38 mpg and standard deviation 4 mpg. What percentage of the cars attain less than 35 mpg?

23. Suppose that the net weight of a bag of a certain brand of potato chips is a normally distributed random variable with mean 8 ounces and standard deviation 0.25 ounce. What is the probability that a given bag of chips contains more than 8.5 ounces?

24. Suppose that the life span of a certain automobile spark plug is a normally distributed random variable with mean 10,000 miles and standard devia-

tion 1000 miles. What is the probability that such a spark plug will last between 8000 and 12,000 miles?

25. The scores on a certain standardized test are normally distributed with mean 500 and standard deviation 80. What is the minimum score necessary to rank in the 90th percentile (that is, the top 10%)?

26. Suppose that X is a normally distributed random variable with mean μ and standard deviation σ. Show that the probability that X is within 2 standard deviations of the mean is about 0.95, that is, $P(-2\sigma \leq X - \mu \leq 2\sigma) \approx 0.95$.

27. The number of raisins in a box of breakfast cereal is a normal random variable with mean 220 and standard deviation 50. Suppose a scoop holds 100 raisins. What is the probability that a box of the cereal contains less than 2 scoops of raisins?

28. A coffee machine dispenses coffee in 6.5 ounce cups. The actual amount of coffee dispensed is a normal random variable with mean 6 and standard deviation 0.4. What percentage of the cups overflow?

29. Suppose X is a normal random variable with mean 0 and $P(-3 \leq X \leq 3) = 0.6826$. Estimate the standard deviation of X.

12

COMPUTERS AND CALCULUS

INTRODUCTION

There is no question that computers have changed the way we live. Besides the advances in technology that we see around us every day, computers have enabled business managers, economists, and social and physical scientists to analyze and solve many complicated or previously intractable problems. Even in pure mathematics, computers have assisted mathematicians in proving certain theorems that have eluded traditional proof techniques. In this chapter, we present a number of computer programs that are designed either to demonstrate concepts of calculus or to solve many of the problems encountered in this text. Each section is independent and can be used in conjunction with one of the previous chapters or supplements. Table 12-1 on page 554 gives a list of the programs presented in this chapter and the chapters to which they apply.

There are many reasons to use a computer in learning and applying calculus: manual calculations are often tedious or difficult to perform; there is a need for repetitive solution; many problems cannot be solved with analytical methods. For example, computers can quickly and accurately evaluate functions. Thus, they can be used to help you understand fundamental concepts of limits and their role in differentiation and integration. The programs LIMITS, DERIV, INTEGRAT, and TRAPEZD are primarily demonstration programs to assist you in understanding many of the important concepts introduced in this text. The programs TABLE and GRAPH will allow you to examine the behavior of functions in order to interpret their meaning.

Many applied calculus problems require solving an equation. It may be difficult or even impossible to do this analytically. Hence, numerical methods become necessary. Programs NEWTON, GOLDEN, and MULTIVAR are computer implementations of numerical techniques that can solve many difficult problems. Programs of this type are used extensively in practice.

All programs were written in Microsoft BASIC for the IBM Personal Computer* and can easily be adapted to other microcomputer systems. Listings of the programs are given in the appendix to this chapter. We assume that the user is familiar with the "DEF FN" statement in BASIC. This is used in most programs to enter the function being considered. In all programs requiring a function as input, the user is prompted as to whether the appropriate line contains the function. If the answer is negative, the program is terminated. The user must then edit the program file to include the function and rerun the program. All other required information is prompted by the programs.

Table 12-1
Computer programs for calculus techniques and applications

Program Name	Purpose	Chapter Reference
TABLE	To construct a table of function values	1
GRAPH*	To graph a function	1
LIMITS	To approximate limits of functions	2, supplement 2
DERIV	To approximate the derivative at a point	2
NEWTON	To approximate the roots of functions	3
GOLDEN	To approximate maximum/minimum points	3
INTEGRAT	To evaluate integrals by rectangular approximation	5
TRAPEZD	To evaluate integrals by trapezoidal approximation	5
DIFFEQ	To approximate solutions to differential equations	7
MULTIVAR	To approximate maximum/minimum points of multivariable functions	8
PROB	To compute expected value and variance of a discrete random variable	11

* This program requires color/graphics capability.

12-1 GRAPHS OF FUNCTIONS

The most common method used to graph a function $y = f(x)$ is to select a number of values for x and evaluate $f(x)$ at each one. Such a table of values can be plotted on a coordinate system and a smooth curve can be drawn. In this

---

* An Apple version is also available (there are some differences in the BASIC language between them).

section, we present a simple program for computing such a table, as well as a program that actually draws the curve if graphics capability is available.

PROGRAM TABLE

This program evaluates $f(x)$ for any number of points between specified minimum and maximum values. The user enters the minimum and maximum values of x and the number of points to be evaluated, n (see shaded areas in the program run below). The program then evaluates $f(x)$ at n points spaced over equal intervals between (and including) the minimum and maximum values.

To illustrate the use of this program, consider the function

$$f(x) = x^4 - x^3 + 3x$$

For the interval $-3 \le x \le 3$ and $n = 7$, the program yields the following results:

```
THIS PROGRAM COMPUTES A TABLE OF FUNCTION VALUES OVER AN INTERVAL

LINE 70 MUST CONTAIN YOUR FUNCTION AS 'DEF FNF(X) ='
DOES LINE 70 CONTAIN YOUR FUNCTION? (Y/N)   Y

ENTER MINIMUM, MAXIMUM VALUES FOR X   -3,3
ENTER THE NUMBER OF POINTS TO BE EVALUATED   7

     X                F(X)
    ----------------------------
    -3                99
    -2                18
    -1                -1
     0                 0
     1                 3
     2                14
     3                63
```

PROGRAM GRAPH

The graphics capabilities of personal computers allow us to graph functions easily. Program GRAPH must be run with advanced BASIC (BASICA) on machines with color/graphics capability. The user is prompted to enter minimum and maximum values for x and y in order to define the scale on the coordinate system. It is useful to run TABLE first in order to determine the range of y over the specified interval for x. For the function

$$f(x) = x^4 - x^3 + 3x$$

we see from the output of TABLE above that y varies from -1 to 99. Therefore, for plotting purposes, the range of y was selected as $-10 \le y \le 90$. The run of the program is shown at the top of the next page. Figure 12-1 shows the graph that is drawn on the monitor.

FUNCTION PLOTTER

THIS PROGRAM REQUIRES ADVANCED BASIC
AND COLOR/GRAPHICS

LINE 100 MUST CONTAIN YOUR FUNCTION AS 'DEF FNF(X) ='
DOES LINE 100 CONTAIN YOUR FUNCTION? (Y/N)   y

ENTER MINIMUM, MAXIMUM VALUES FOR X   -3,3
ENTER MINIMUM, MAXIMUM VALUES FOR Y   -10,90

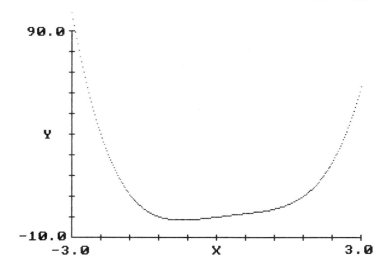

Figure 12-1
Graph of $f(x) = x^4 - x^3 + 3x$

PROBLEMS *(Section 12-1)*

For Problems 1–10, use program TABLE to plot points and draw the graph of each function.

1. $y = -3x + 5$

2. $y = x + 4$

3. $y = -x^2 - 2x + 5$

4. $y = \sqrt{3x + 4}$

5. $y = \dfrac{1}{x} + 3$ [*Hint:* Select intervals in such a way to avoid division by 0.]

6. $y = \dfrac{2}{x + 1}$

7. $y = \sqrt{4 - x^2}$

8. $y = x^3 + 3x^2 + 2$

9. $y = \frac{2}{3}x^3 - 6x^2 + 3$

10. $y = 2x^3 - 3x^2 + 4\sqrt{x} + 1$

In Problems 11–20, use program GRAPH to plot the function over the specified interval.

11. $f(x) = 4\sqrt{x}, \quad 0 \le x \le 9$

12. $f(x) = x^3 + 3x^2 - 9x + 5, \quad -3 \le x \le 3$

13. $y = (x + 1)^2, \quad -2 \le x \le 2$

14. $y = \sqrt{1 - x^2}, \quad -1 \le x \le 1$

15. $f(x) = x^3 - 6x^2, \quad -1 \le x \le 3$

16. $y = \dfrac{x}{x^2 + 4}, \quad -2 \le x \le 4$

17. $f(x) = \dfrac{x^4}{4} - 5x^3 + \dfrac{63x^2}{2} - 81x, \quad -1 \le x \le 12$

18. $y = \dfrac{1.467x}{x + (x^2/20) + 12}, \quad 0 \le x \le 60$

19. $y = \dfrac{1}{5\sqrt{x}}, \quad 1 \le x \le 25$

20. $f(x) = \frac{1}{3}x^2 + 50, \quad -10 \le x \le 10$

21. *Programming Problem:* Combine the two programs TABLE and GRAPH so that you only need to input the interval for x and the program will automatically scale the y-axis to the appropriate range.

12-2 LIMITS AND THE DERIVATIVE

In this section, we present programs for approximating limits of functions, and in particular, estimating the derivative of a function at a specific point in its domain or showing that the derivative does not exist.

PROGRAM LIMITS

LIMITS is a program that approximates

$$\lim_{x \to a} f(x)$$

The value of a must be finite; that is, the program will not approximate a limit at plus or minus infinity. Also, it will approximate a one-sided limit only. The user must enter a starting value for x, the value of a, and the number of points in the sequence. The distance between the starting value of x and a is successively halved. For example, to approximate

$$\lim_{x \to 2} \frac{x - 2}{x^2 - 4}$$

starting at $x = 7$ and using 15 points, we would have the following:

```
LIMIT EVALUATION OF F(X) AS X APPROACHES A
A CANNOT BE PLUS OR MINUS INFINITY

LINE 140 MUST CONTAIN YOUR FUNCTION AS '140 DEF FNF(X)='

DOES LINE 140 CONTAIN YOUR FUNCTION (Y/N)? y

ENTER INITIAL VALUE OF X  7
ENTER THE VALUE OF 'A'  2
ENTER THE NUMBER OF POINTS IN THE LIMIT SEQUENCE 15
```

```
X                    F(X)
 7                    .1111111
 4.5                  .1538462
 3.25                 .1904762
 2.625                .2162162
 2.3125               .2318841
 2.15625              .2406015
 2.078125             .2452107
 2.039063             .2475822
 2.019531             .2487852
 2.009766             .2493912
 2.004883             .2496952
 2.002441             .2498536
 2.001221             .2499268
 2.00061              .2499512
 2.000305             .25
 2.000153             .25
```

The approximations approach 0.25. Analytically, we see that

$$\lim_{x \to 2} \frac{x-2}{x^2-4} = \lim_{x \to 2} \frac{x-2}{(x-2)(x+2)}$$

$$= \lim_{x \to 2} \frac{1}{x+2} = \frac{1}{4}$$

As a second example, we evaluate

$$\lim_{x \to 9} \frac{\sqrt{x^3} - 9\sqrt{x}}{x - 9}$$

using 10 points starting at $x = 0$. The results are shown below:

```
LIMIT EVALUATION OF F(X) AS X APPROACHES A
A CANNOT BE PLUS OR MINUS INFINITY

LINE 140 MUST CONTAIN YOUR FUNCTION AS '140 DEF FNF(X)='

DOES LINE 140 CONTAIN YOUR FUNCTION (Y/N)? y

ENTER INITIAL VALUE OF X  0
ENTER THE VALUE OF 'A'  9
ENTER THE NUMBER OF POINTS IN THE LIMIT SEQUENCE 10

X                    F(X)
 0                    0
 4.5                  2.12132
 6.75                 2.598077
 7.875                2.806244
 8.4375               2.904738
 8.71875              2.952752
 8.859375             2.976481
```

```
8.929688          2.988281
8.964844          2.994141
8.982422          2.99707
8.991211          2.998481
```

PROGRAM DERIV   The program DERIV approximates the derivative of a function $f(x)$ at a point $x = a$. That is, it approximates $f'(a)$, the slope of the tangent line to the graph of $f(x)$ at $x = a$, if it exists. The method is based on evaluating the difference quotient

$$\frac{f(a + h) - f(a)}{h} = \text{Slope of secant line}$$

at some starting value of h. This value is successively halved, and the difference quotient is recomputed until h is less than 0.0001. The limits of the difference quotient are evaluated from both the left and the right so that existence of the derivative at $x = a$ can be confirmed.

The user is prompted to enter the value of a and an initial value of h. The initial value of h can be either positive or negative.

The function $f(x)$ must be entered in the program in line 80. For example, if $f(x) = 6x^3 - 3x^2 + 4x$ and $a = 3$, running the program will give the following results:

```
EVALUATION OF THE DERIVATIVE OF F(X) AT X = A

LINE 80 MUST CONTAIN YOUR FUNCTION AS 'DEF FNF(X) = '

DOES LINE 80 CONTAIN YOUR FUNCTION (Y/N)? y

ENTER THE VALUE OF 'A'  3
ENTER STARTING VALUE OF H  2

LIMITING VALUE OF DIFFERENCE QUOTIENT AS H APPROACHES ZERO FROM THE RIGHT

H                   (F(A+H) - F(A))/H
 2                      274
 1                      205
 .5                     175
 .25                    161.125
 .125                   154.4688
 .0625                  151.2109
 .03125                 149.5996
 .015625                148.7988
 .0078125               148.3984
 3.90625E-03            148.1992
 1.953125E-03           148.1016
 9.765625E-04           148.0469
 4.882813E-04           148.0313
 2.441406E-04           148
 1.220703E-04           148
```

```
LIMITING VALUE OF DIFFERENCE QUOTIENT AS H APPROACHES ZERO FROM THE LEFT

H                      (F(A+H) - F(A))/H
-2                         70
-1                         103
-.5                        124
-.25                       135.625
-.125                      141.7188
-.0625                     144.8359
-.03125                    146.4121
-.015625                   147.2051
-.0078125                  147.6016
-3.90625E-03               147.8008
-1.953125E-03              147.8984
-9.765625E-04              147.9531
-4.882813E-04              147.9688
-2.441406E-04              148
-1.220703E-04              148
```

We see that the value of the difference quotient approaches the value 148 from both the left and the right. To verify this result, we notice that

$$f'(x) = 18x^2 - 6x + 4$$
$$f'(3) = 18(9) - 6(3) + 4$$
$$= 148$$

As a second example, consider the function $f(x) = x^{2/3}$. To avoid an "illegal function call" in BASIC, line 80 must be coded as

```
DEF FNF(X) = (X^2)^(1/3)
```

At $a = 0$, we have the following:

```
EVALUATION OF THE DERIVATIVE OF F(X) AT X = A

LINE 80 MUST CONTAIN YOUR FUNCTION AS 'DEF FNF(X) = '

DOES LINE 80 CONTAIN YOUR FUNCTION (Y/N)? y

ENTER THE VALUE OF 'A'   0
ENTER STARTING VALUE OF H   1

LIMITING VALUE OF DIFFERENCE QUOTIENT AS H APPROACHES ZERO FROM THE RIGHT

H                      (F(A+H) - F(A))/H
 1                         1
 .5                        1.259921
 .25                       1.587401
 .125                      2
 .0625                     2.519843
 .03125                    3.174802
```

.015625	4
.0078125	5.039685
3.90625E-03	6.349606
1.953125E-03	8
9.765625E-04	10.07937
4.882813E-04	12.69921
2.441406E-04	16
1.220703E-04	20.15873

LIMITING VALUE OF DIFFERENCE QUOTIENT AS H APPROACHES ZERO FROM THE LEFT

H	(F(A+H) - F(A))/H
-1	-1
-.5	-1.259921
-.25	-1.587401
-.125	-2
-.0625	-2.519843
-.03125	-3.174802
-.015625	-4
-.0078125	-5.039685
-3.90625E-03	-6.349606
-1.953125E-03	-8
-9.765625E-04	-10.07937
-4.882813E-04	-12.69921
-2.441406E-04	-16
-1.220703E-04	-20.15873

We easily see that the limits from either side are not approaching a common value, indicating that the derivative does not exist. To see this, note that

$$\lim_{h \to 0^+} \frac{f(0 + h) - f(0)}{h} = \infty$$

$$\lim_{h \to 0^-} \frac{f(0 + h) - f(0)}{h} = -\infty$$

PROBLEMS (Section 12-2)

In Problems 1 – 10, use program LIMITS to find the limits as x approaches a for each function, or determine that the limit does not exist.

1. $y = 6 + \dfrac{4}{x}$ for $a = 4$

2. $y = 3x^3 + 4$ for $a = -2$

3. $y = 5x^3 - 2x^2 + 4x + 1$ for $a = 1$

4. $y = \dfrac{x^2 + x}{x}$ for $a = 0$

5. $y = \dfrac{1}{x^2}$ for $a = 0$

6. $y = \dfrac{x^2 - 4}{x - 2}$ for $a = 2$

7. $y = \dfrac{x^2 - 25}{x + 5}$ for $a = -5$

8. $y = \dfrac{2x^2 - x - 3}{x + 1}$ for $a = -1$

9. $y = \dfrac{3x + 2x^2}{4x - x^3}$ for $a = 0$

10. $y = \sqrt{5x - 5}$ for $a = 6$

In Problems 11–20, estimate the derivative or determine that it does not exist using program DERIV. Check your results analytically to verify the accuracy of the program.

11. $f(x) = \dfrac{2}{x}$ at $x = 3$

12. $f(x) = 3x^2 - 1$ at $x = 1$

13. $f(x) = 2x^2 + 13$ at $x = 3$

14. $f(x) = x^2 + 2x - 1$ at $x = 1$

15. $f(x) = \dfrac{3}{x^2}$ at $x = 0$

16. $f(x) = \dfrac{x + 2}{x - 1}$ at $x = 0$

17. $f(x) = 2x^4 + 4x^3 + x^2 - 2x$ at $x = -1$

18. $f(x) = \sqrt{x} + \tfrac{5}{2}x^2$ at $x = 4$

19. $f(x) = 3\sqrt{x} - \dfrac{1}{x - 1}$ at $x = 1$

20. $f(x) = x^3 - 2x^2 + 3x + 1$ at $x = 2$

21. *Programming Problem:* Modify program DERIV so that the derivative is automatically computed for a range of x values. Use this program to develop a table of derivative values (similar to program TABLE) so that the derivative function can be plotted.

12-3 OPTIMIZATION OF SINGLE-VARIABLE FUNCTIONS

In Chapter 3 we considered finding the maximum or minimum of a function. The procedure we used was to take the derivative and set it equal to 0 in order to find critical points. In this section, we present a program for finding roots of functions. This can be used to solve $f'(x) = 0$ to find critical points. In addition, we present a numerical method that does not depend on taking derivatives, but will find the absolute minimum or maximum of certain functions over a closed interval.

**PROGRAM
NEWTON**

Newton's method is a procedure that approximates a root of a function. The idea behind Newton's method is illustrated in Figure 12-2. Suppose that x_1 is an approximation to the root of $f(x)$. Consider the tangent line at $(x_1, f(x_1))$. From Figure 12-2 we can see that x_2 (where the tangent line crosses the x-axis) is a better approximation to the root of $f(x)$. Since the slope of the tangent line is $f'(x_1)$ and the tangent line passes through $(x_1, f(x_1))$ and $(x_2, 0)$, we have

$$f'(x_1) = \frac{0 - f(x_1)}{x_2 - x_1}$$

$$x_2 = x_1 - \frac{f(x_1)}{f'(x_1)}$$

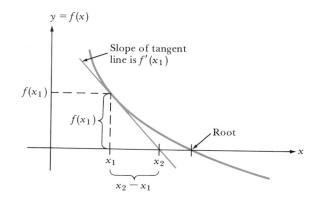

Figure 12-2
Illustration of Newton's method

Thus, x_2 is determined by knowing x_1 and its derivative. Once we have x_2, we can get a better approximation by finding

$$x_3 = x_2 - \frac{f(x_2)}{f'(x_2)}$$

and so on. This method usually converges very rapidly to the root.

The program NEWTON requires input of both $f(x)$ and $f'(x)$, an initial point x_1, and the maximum difference between successive approximations allowed before termination. The program stops when the difference is within this range, or until 100 points have been generated. In the latter case, the method probably did not converge. This often happens for certain functions if the initial approximation is very far from the root. One other caution: If the value of the derivative gets very close to 0, an "overflow" error message may result because of the division by the derivative in the Newton formula.

For example, consider finding a root of the function

$$f(x) = x^3 + \frac{x^2}{2} - 2x + 4$$

The first derivative is

$$f'(x) = 3x^2 + x - 2$$

Starting at $x = -2$, we obtain the following results:

```
FINDING ROOTS OF EQUATIONS USING NEWTON'S METHOD

LINE 90 MUST CONTAIN YOUR FUNCTION AS `DEF FNF(X) = `
LINE 100 MUST CONTAIN THE DERIVATIVE OF YOUR FUNCTION AS
     `DEF FNDER(X) =`
DO LINES 90 AND 100 CONTAIN YOUR FUNCTION AND ITS DERIVATIVE? (Y/N) Y

ENTER STARTING VALUE FOR X? -2
ENTER MAXIMUM ERROR ALLOWED? .001

ESTIMATED ROOT
X = -2.216505
```

However, if we let $x_1 = 2$ to begin with, the method does not converge:

```
FINDING ROOTS OF EQUATIONS USING NEWTON'S METHOD

LINE 90 MUST CONTAIN YOUR FUNCTION AS 'DEF FNF(X) = '
LINE 100 MUST CONTAIN THE DERIVATIVE OF YOUR FUNCTION AS
    'DEF FNDER(X) ='
DO LINES 90 AND 100 CONTAIN YOUR FUNCTION AND ITS DERIVATIVE? (Y/N) y

ENTER STARTING VALUE FOR X? 2
ENTER MAXIMUM ERROR ALLOWED? .001
METHOD DID NOT CONVERGE
CURRENT SOLUTION 1.14669    RESIDUAL .8313065
```

This can be explained by the graph of the function in Figure 12-3. From the tangent lines drawn, we can trace the first four points generated by Newton's method. These are

$$x_1 = 2.000$$
$$x_2 = 1.167$$
$$x_3 = -0.004$$
$$x_4 = 1.962$$

The value of x_4 is very close to x_1. We would expect that x_5 will thus be very close to x_2, and so on. In fact, the method keeps cycling around these four points and does not converge toward the root. Therefore, you must be careful in choosing good initial approximations to the roots.

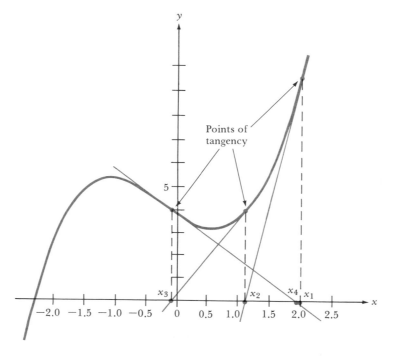

Figure 12-3
Nonconvergence of Newton's method

Newton's method is often used to approximate roots of numbers. For example, to find $\sqrt{3}$, we note that this is the solution to the equation

$$x^2 = 3 \qquad \text{or} \qquad f(x) = x^2 - 3 = 0$$

Using Newton's method beginning at $x_1 = 5$, we get the following:

```
FINDING ROOTS OF EQUATIONS USING NEWTON'S METHOD

LINE 90 MUST CONTAIN YOUR FUNCTION AS 'DEF FNF(X) = '
LINE 100 MUST CONTAIN THE DERIVATIVE OF YOUR FUNCTION AS
    'DEF FNDER(X) ='
DO LINES 90 AND 100 CONTAIN YOUR FUNCTION AND ITS DERIVATIVE? (Y/N) y

ENTER STARTING VALUE FOR X? 5
ENTER MAXIMUM ERROR ALLOWED? .00001

ESTIMATED ROOT
X =   1.732051
```

PROGRAM GOLDEN

An efficient method for finding an absolute maximum or minimum point of a function over an interval is called the **golden section method.** This method does not rely on taking derivatives and setting them equal to 0, which is convenient in many situations where dealing with derivatives may be difficult or result in an equation that is very complicated. The idea behind the golden section method is illustrated in Figure 12-4. Suppose that we wish to *maximize* a function over the interval $[a, b]$ and assume that there is at most one relative maximum that occurs at $x = x^*$ on this interval. Further assume that the function is increasing on $[a, x^*]$ and decreasing on $[x^*, b]$. Under these assumptions, the function must assume one of the shapes given in Figure 12-4. These assumptions are necessary

Figure 12-4
Illustration of the golden section method

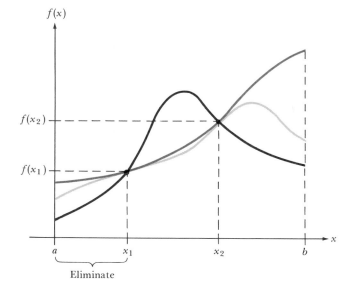
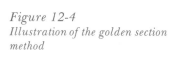

in order for the method to identify the absolute maximum. We select two points in the interval $[a, b]$, say x_1 and x_2, with $x_1 < x_2$. If $f(x_1) < f(x_2)$, then we can state that the maximum point *cannot* be less than x_1. That is, the absolute maximum must be between x_1 and x_2, between x_2 and b, or at b. Therefore, we can eliminate the interval from a to x_1. By similar reasoning, if $f(x_1) > f(x_2)$, then we can eliminate the interval from x_2 to b. We then eliminate that portion of the interval and repeat the procedure on the smaller, remaining interval. We continue this process until the interval is reduced to a predetermined size. The point at which the maximum occurs will lie somewhere in the final interval.

Of course, the golden section method can also be used to minimize a function. To do so, we simply note that the minimum of a function $f(x)$ occurs at the same point as the maximum of the function $g(x) = -f(x)$. Therefore, by maximizing $g(x) = -f(x)$ we minimize $f(x)$. Program GOLDEN can be used to find either the maximum or the minimum of a function over a closed interval.

For example, consider the function $f(x) = x^2 + 2x$, which we want to minimize over the interval $[-3, 5]$. The program yields the following results:

```
ESTIMATION OF MINIMUM OR MAXIMUM VALUE OF F(X) OVER THE INTERVAL [A,B]

LINE 80 MUST CONTAIN YOUR FUNCTION AS 'DEF FNF(X) = '

DOES LINE 80 CONTAIN YOUR FUNCTION? (Y/N) y

TYPE 1 IF THIS IS A MINIMIZATION PROBLEM
TYPE 2 IF THIS IS A MAXIMIZATION PROBLEM
? 1
ENTER THE VALUE OF A   -3
ENTER THE VALUE OF B   5
ENTER THE SIZE OF THE FINAL INTERVAL  .1
ESTIMATE OF OPTIMAL POINT

X =            -1.013756
F(X) =         -.9998108
```

As a second illustration, consider finding both the absolute maximum and the absolute minimum for the function

$$f(x) = \frac{x^4}{4} - \frac{x^3}{3} - x^2 + 5$$

over the interval $-1 \le x \le 1$. (This is Example 10 in Chapter 3.) The runs of the program are shown below, and a graph of the function is shown in Figure 12-5.

```
ESTIMATION OF MINIMUM OR MAXIMUM VALUE OF F(X) OVER THE INTERVAL [A,B]

LINE 80 MUST CONTAIN YOUR FUNCTION AS 'DEF FNF(X) = '

DOES LINE 80 CONTAIN YOUR FUNCTION? (Y/N) y

TYPE 1 IF THIS IS A MINIMIZATION PROBLEM
```

```
TYPE 2 IF THIS IS A MAXIMIZATION PROBLEM
? 1
ENTER THE VALUE OF A   -1
ENTER THE VALUE OF B   1
ENTER THE SIZE OF THE FINAL INTERVAL  .001
ESTIMATE OF OPTIMAL POINT

X =            .9995466
F(X) =         3.917574


ESTIMATION OF MINIMUM OR MAXIMUM VALUE OF F(X) OVER THE INTERVAL [A,B]

LINE 80 MUST CONTAIN YOUR FUNCTION AS 'DEF FNF(X) =

DOES LINE 80 CONTAIN YOUR FUNCTION? (Y/N)  y

TYPE 1 IF THIS IS A MINIMIZATION PROBLEM
TYPE 2 IF THIS IS A MAXIMIZATION PROBLEM
? 2
ENTER THE VALUE OF A   -1
ENTER THE VALUE OF B   1
ENTER THE SIZE OF THE FINAL INTERVAL  .001
ESTIMATE OF OPTIMAL POINT

X =            -3.233719E-04
F(X) =         5
```

Figure 12-5
Graph of
$$y = f(x) = \frac{x^4}{4} - \frac{x^3}{3} - x^2 + 5$$

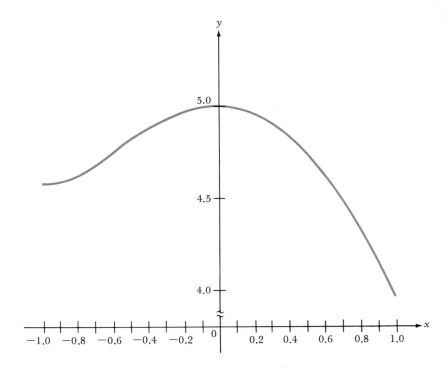

If there is some doubt as to whether a function satisfies the assumption of having at most one relative optimum, you should divide the given interval into smaller subintervals and run the program for each subinterval. The best solution found over all subintervals will probably be the absolute optimum.

PROBLEMS (Section 12-3)

In Problems 1–5, use program NEWTON to find roots of each equation starting at the indicated points. Terminate when successive approximations differ by less than 0.001.

1. $y = x^3 - 2x - 5 = 0$; $x_1 = 2$

2. $y = x^3 + x^2 - x + 1 = 0$; $x_1 = -2$

3. $y = 4x^3 + 1 = 0$; $x_1 = -1$

4. $y = 3x^2 - 18x + 24 = 0$; $x_1 = 1$ and $x_1 = 5$

5. $y = x^2 + 6x = 0$; $x_1 = 1$, $x_1 = -5$, and $x_1 = -10$

In Problems 6–10, use program NEWTON to compute each root. Terminate when successive approximations differ by less than 0.0001.

6. $\sqrt{2}$; $x_1 = 1$

7. $\sqrt{5}$; $x_1 = 2$

8. $\sqrt{11}$; $x_1 = 3$

9. $\sqrt[3]{6}$; $x_1 = 1$

10. $\sqrt[5]{11}$; $x_1 = 2$

In Problems 11–15, use program GOLDEN to find both the absolute minimum and the absolute maximum over the given interval.

11. $f(x) = (x + 1)^2$, $-2 \le x \le 2$

12. $f(x) = \sqrt{1 - x^2}$, $-1 \le x \le 1$

13. $f(x) = x^3 - 6x^2$, $-1 \le x \le 3$

14. $f(x) = 2x^3 + 9x^2 - 24x + 3$, $-2 \le x \le 2$

15. $f(x) = \dfrac{x}{x^2 - 1}$, $2 \le x \le 6$

16. *Programming Problem:* Combine the two programs DERIV and NEWTON so that the derivative of the function is computed at each iteration and $f'(x)$ need not be entered explicitly into the program.

12-4 NUMERICAL INTEGRATION

In this section, we present two programs for numerical evaluation of integrals. The first program approximates the integral using rectangular approximation, as discussed in Chapter 5; the second, based on the trapezoidal rule, provides a more accurate approximation.

PROGRAM INTEGRAT

In Chapter 5, we discussed how the area under a curve between $x = a$ and $x = b$ can be approximated by dividing the interval $[a, b]$ into n subintervals $[x_0, x_1]$, $[x_1, x_2]$, . . . , $[x_{n-1}, x_n]$, where $x_0 = a$ and $x_n = b$, and computing the sum

$$\sum_{i=1}^{n} (x_i - x_{i-1})f(x_i)$$

Recall that this sum is the area of a set of rectangles having width $(x_i - x_{i-1})$ and height $f(x_i)$. The limit of this sum as n approaches infinity is the integral

$$\int_a^b f(x) \, dx$$

The values for a, b, and the number of rectangles to be used must be specified by the user.

To illustrate, consider the definite integral

$$\int_1^5 (x^3 + 3x^2 - 4x) \, dx$$

The runs below show the results using 10 and 100 rectangles:

```
EVALUATION OF THE DEFINITE INTEGRAL OF F(X) FROM X=A TO X=B

LINE 80 MUST CONTAIN YOUR FUNCTION AS 'DEF FNF(X) = '

DOES LINE 80 CONTAIN YOUR FUNCTION? (Y/N)? y

ENTER THE VALUE FOR A    1
ENTER THE VALUE FOR B    5
ENTER THE NUMBER OF RECTANGLES    10

SIZE OF SUBINTERVAL IS          .4

THE VALUE OF THE INTEGRAL IS               269.28

EVALUATION OF THE DEFINITE INTEGRAL OF F(X) FROM X=A TO X=B

LINE 80 MUST CONTAIN YOUR FUNCTION AS 'DEF FNF(X) = '

DOES LINE 80 CONTAIN YOUR FUNCTION? (Y/N)? y

ENTER THE VALUE FOR A    1
ENTER THE VALUE FOR B    5
ENTER THE NUMBER OF RECTANGLES    100

SIZE OF SUBINTERVAL IS          .04

THE VALUE OF THE INTEGRAL IS               235.6128
```

The true value of the integral is

$$\int_1^5 (x^3 + 3x^2 - 4x) \, dx = \left(\frac{x^4}{4} + x^3 - 2x^2 \right)\Bigg|_1^5$$
$$= 232$$

We see that even with 100 rectangles, the error is fairly substantial. However, a simple modification of this program can dramatically improve the accuracy of this general procedure. This is discussed next.

**PROGRAM
TRAPEZD**

A more accurate evaluation of the definite integral is obtained by the *trapezoidal rule*. This is illustrated in Figure 12-6. In the trapezoidal rule, instead of using rectangles to approximate the area under the curve, we compute the total area of approximating trapezoids. From Figure 12-6, you can easily see that this results in a better approximation to the area. Recall from geometry that a trapezoid is a four-sided figure with two parallel sides.

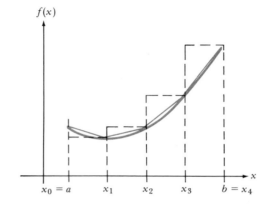

Figure 12-6
*Trapezoidal versus rectangular
approximation of integrals*

To compute the area between x_{i-1} and x_i, refer to Figure 12-7. The area of the lower rectangle is $f(x_i)(x_i - x_{i-1})$. The area of the upper triangle is $\frac{1}{2}[f(x_{i-1}) - f(x_i)][x_i - x_{i-1}]$. Adding these terms, we find that the area of the trapezoid is given by

$$\frac{f(x_{i-1}) + f(x_i)}{2}(x_i - x_{i-1})$$

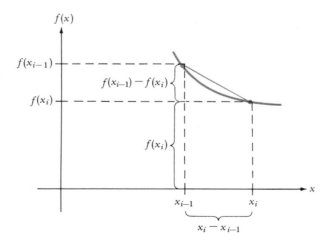

Figure 12-7
*Computing the area of a trapezoid
approximation for the integral*

Therefore, an approximation of the integral is obtained by summing these areas over all trapezoids, that is,

$$\int_a^b f(x)\, dx \approx \sum_{i=1}^n \frac{f(x_{i-1}) + f(x_i)}{2}(x_i - x_{i-1})$$

Let us find the approximate value of the integral

$$\int_1^5 (x^3 + 3x^2 - 4x)\, dx$$

using program TRAPEZD for both 10 and 100 subintervals. The results are given below:

```
EVALUATION OF THE DEFINITE INTEGRAL USING THE TRAPEZOIDAL RULE

LINE 80 MUST CONTAIN YOUR FUNCTION AS 'DEF FNF(X) = '

DOES LINE 80 CONTAIN YOUR FUNCTION? (Y/N)? y

ENTER THE VALUE FOR A   1
ENTER THE VALUE FOR B   5
ENTER THE NUMBER OF TRAPEZOIDS   10

SIZE OF SUBINTERVAL IS          .4

THE VALUE OF THE INTEGRAL IS                    233.28


EVALUATION OF THE DEFINITE INTEGRAL USING THE TRAPEZOIDAL RULE

LINE 80 MUST CONTAIN YOUR FUNCTION AS 'DEF FNF(X) = '

DOES LINE 80 CONTAIN YOUR FUNCTION? (Y/N)? y

ENTER THE VALUE FOR A   1
ENTER THE VALUE FOR B   5
ENTER THE NUMBER OF TRAPEZOIDS   100

SIZE OF SUBINTERVAL IS          .04

THE VALUE OF THE INTEGRAL IS                    232.0128
```

Comparing these results to those obtained using program INTEGRAT, we note a significant improvement.

Numerical integration is useful and necessary since many functions are difficult or impossible to integrate analytically. An important function that is used in probability and statistics is

$$f(x) = \frac{1}{\sqrt{2\pi}}\, e^{-x^2/2}$$

where $e \approx 2.71828$ is the base of the natural logarithm. The integral of this function cannot be expressed in terms of elementary functions. Thus, a numerical method must be used to evaluate it.

Using TRAPEZD, we evaluate the integral

$$\int_0^1 \frac{1}{\sqrt{2\pi}} e^{-x^2/2} \, dx$$

in the program run below:

```
EVALUATION OF THE DEFINITE INTEGRAL USING THE TRAPEZOIDAL RULE

LINE 80 MUST CONTAIN YOUR FUNCTION AS 'DEF FNF(X) = '

DOES LINE 80 CONTAIN YOUR FUNCTION? (Y/N)? y

ENTER THE VALUE FOR A   0
ENTER THE VALUE FOR B   1
ENTER THE NUMBER OF TRAPEZOIDS   50

SIZE OF SUBINTERVAL IS          .02

THE VALUE OF THE INTEGRAL IS              .3413368
```

PROBLEMS (Section 12-4)

In Problems 1–10, approximate each definite integral using both INTEGRAT and TRAPEZD. Compare the results for 10 and 100 subintervals with the exact integral.

1. $\int_0^3 x^2 \, dx$

2. $\int_4^{10} \frac{2}{x} \, dx$

3. $\int_1^8 \frac{5}{\sqrt[3]{x}} \, dx$

4. $\int_{-1}^3 (3x^2 - 2x) \, dx$

5. $\int_{-1}^2 (x^2 - 2x + 5) \, dx$

6. $\int_2^5 (x + 3)(x + 2) \, dx$

7. $\int_0^4 (2x^2 + \sqrt{x} - 3) \, dx$

8. $\int_1^4 (5x + 4) \, dx$

9. $\int_0^2 x(x^2 - 4)^3 \, dx$

10. $\int_2^3 \frac{2x}{x^2 + 5} \, dx$

In Problems 11–15, find the area between the curves using TRAPEZD and 1000 subintervals.

11. $y = x^2$ and $y = 2x + 8$

12. $y = 2x - x^2$ and $y = x - 2$

13. $y = x^2 - 4$ and $y = -x^2 + 4$

14. $y = x^3$ and $y = 4x$

15. $y = 5 - x^2$ and $y = x^2 - 3$

12-5 SOLVING DIFFERENTIAL EQUATIONS

Many differential equations that model real-life phenomena cannot be solved using analytical methods, and numerical approximations to the differential equation must therefore be used. In this section, we present a program based on the *modified Euler method* for solving differential equations.

PROGRAM DIFFEQ   Consider the differential equation

$$\frac{dy}{dx} = f(x, y)$$

where f is a function of two variables, with the initial condition $y(x_0) = y_0$. That is, we seek a function $y(x)$ whose derivative is $f(x, y)$. A simple method for approximating the solution of such a differential equation is called **Euler's method** (see the supplement to Chapter 7). For a given value of h (the *step size*), the method generates values y_1, y_2, y_3, . . . , which approximate the values $y(x_0 + h)$, $y(x_0 + 2h)$, $y(x_0 + 3h)$, . . . of the true solution $y(x)$. To understand this, consider Figure 12-8. Since the slope of the tangent line at (x_0, y_0) is $f(x_0, y_0)$, an approximation to the curve at $x_1 = x_0 + h$ is

$$y_1 = y_0 + hf(x_0, y_0)$$

That is, the point (x_1, y_1) is an approximation to the point $(x_1, y(x_1))$ on the graph of the function $y(x)$. Once we have this point, we may let $x_2 = x_1 + h$ and approximate $y(x_2)$ by

$$y_2 = y_1 + hf(x_1, y_1)$$

and so on.

Figure 12-8
Illustration of Euler's method

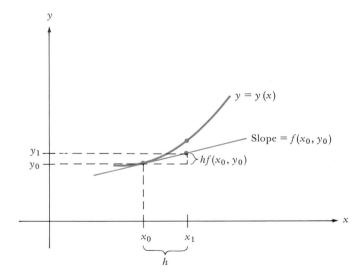

A better approximation can be found by modifying this procedure slightly. We first approximate the slope at (x_0, y_0) and also at (x_1, y_1) using the equations in Euler's method. However, these slopes are then *averaged* and used to determine a better approximation for $y(x_1)$. At (x_0, y_0), we approximate the slope by $f(x_0, y_0)$. Since $x_1 = x_0 + h$ and $y_1 = y_0 + hf(x_0, y_0)$, we approximate the slope at (x_1, y_1) by $f(x_0 + h, y_0 + hf(x_0, y_0))$. We compute the next value of y_1 using the average of these slopes, that is,

$$y_1 = y_0 + \tfrac{1}{2}h[f(x_0, y_0) + f(x_0 + h, y_0 + hf(x_0, y_0))]$$

We repeat this procedure to compute y_2, and so on.

This method is called the **modified Euler method** and usually gives superior results to the original Euler method. Program DIFFEQ is based on this method. The user must input the differential equation, the initial condition, the step size h, and the maximum value of x. To illustrate, consider the differential equation

$$\frac{dy}{dx} = 3y \qquad y(0) = 2$$

The program yields the following results:

```
THIS PROGRAM SOLVES ELEMENTARY DIFFERENTIAL EQUATIONS

LINE 80 MUST CONTAIN THE DIFFERENTIAL EQUATION DY/DX = F(X,Y)
    AS `DEF FNDEQ(X,Y) = '
DOES LINE 80 CONTAIN YOUR FUNCTION?  (Y/N)   y

ENTER INITIAL VALUES FOR X,Y   0,2
ENTER THE STEP SIZE   .1
ENTER THE MAXIMUM VALUE OF X   2

X                 Y
.1                2.69
.2                3.61805
.3                4.866278
.4                6.545144
.5                8.803219
.6                11.84033
.7000001          15.92524
.8000001          21.41945
.9000001          28.80916
1                 38.74832
1.1               52.1165
1.2               70.09668
1.3               94.28004
1.4               126.8067
1.5               170.555
1.6               229.3964
1.7               308.5382
1.8               414.9838
1.9               558.1532
2                 750.7161
```

We know that the solution to this equation is

$$y = 2e^{3x}$$

Figure 12-9 shows a plot of this function and the numerical approximation generated by the program. For values of x close to the initial condition, the approximation is quite good.

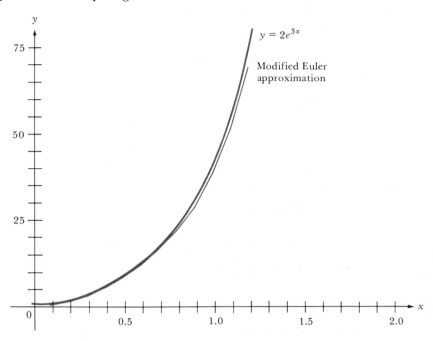

Figure 12-9
Numerical approximation to
$\dfrac{dy}{dx} = 3y, \; y(0) = 2$

As a second example, consider the equation

$$\frac{dy}{dx} = -xy^2 \qquad y(1) = \frac{2}{3}$$

The program gives the following results:

```
THIS PROGRAM SOLVES ELEMENTARY DIFFERENTIAL EQUATIONS

LINE 80 MUST CONTAIN THE DIFFERENTIAL EQUATION DY/DX = F(X,Y)
    AS 'DEF FNDEQ(X,Y) = '
DOES LINE 80 CONTAIN YOUR FUNCTION? (Y/N)   Y

ENTER INITIAL VALUES FOR X,Y   1,.6667
ENTER THE STEP SIZE   .05
ENTER THE MAXIMUM VALUE OF X   2

    X                  Y
    1.05               .6446849
    1.1                .623106
    1.15               .6020183
    1.2                .5814655
```

1.25	.5614811
1.3	.5420897
1.35	.5233081
1.4	.5051459
1.45	.4876075
1.5	.4706919
1.55	.4543941
1.599999	.4387057
1.649999	.4236156
1.699999	.4091104
1.749999	.3951751
1.799999	.3817932
1.849999	.3689474
1.899999	.3566197
1.949999	.3447918
1.999999	.3334452
2.049999	.3225614

In this case, we used a smaller step size and a smaller interval. The exact solution to this differential equation is

$$y = \frac{2}{t^2 + 2}$$

If you compare this function to the numerical results, you will find that the approximation is accurate to 3 decimal places in this range of x.

PROBLEMS (Section 12-5)

In Problems 1–5, approximate the solution to the differential equation over the stated range for the two given values of h and compare the results.

1. $\dfrac{dy}{dx} = 10y$, $y(1) = 0.01$; $1 \le x \le 10$; $h = 1$,

   $h = 0.5$

2. $\dfrac{dy}{dx} = xy^2$, $y(1) = -1$; $1 \le x \le 2$; $h = 0.5$,

   $h = 0.1$

3. $\dfrac{dy}{dx} = (x - 1)y$, $y(0) = 1$; $0 \le x \le 1$; $h = 0.2$,

   $h = 0.1$

4. $\dfrac{dy}{dx} = e^x + y$, $y(0) = 1$; $0 \le x \le 2$; $h = 0.5$,

   $h = 0.25$

5. $\dfrac{dy}{dx} = y(1 - y)$, $y(0) = 0.25$; $0 \le x \le 3$;

   $h = 0.5$, $h = 0.25$

In Problems 6–10, the exact solution is given. Find the numerical approximation and use program TABLE to compute the exact values. Draw graphs similar to Figure 12-9 to compare the results.

6. $\dfrac{dy}{dx} = \dfrac{2y}{x}$, $y(1) = 1$; $1 \le x \le 2$; $h = 0.1$;

   $y = x^2$ is the exact solution

7. $\dfrac{dy}{dx} = \dfrac{3(y+1)}{x} - 4$, $y(1) = 2$; $1 \le x \le 2$;

   $h = 0.2$; $y = x^3 + 2x - 1$ is the exact solution

8. $\dfrac{dy}{dx} = \dfrac{(1+x)y}{x}$, $y(1) = e$; $1 \le x \le 3$; $h = 0.25$;

   $y = xe^x$ is the exact solution

9. $\dfrac{dy}{dx} = \dfrac{x+y}{x}$, $y(1) = 0$; $1 \le x \le 2$; $h = 0.1$;

   $y = x \ln x$ is the exact solution

10. $\dfrac{dy}{dx} = 0.5y$, $y(0) = 1$; $0 \le x \le 1$; $h = 0.2$;

    $y = e^{0.5x}$ is the exact solution

11. *Programming Problem:* Modify DIFFEQ to include the capability of inputting the exact solution and printing out the error between the exact and approximate solution for each value of x.

12-6 *OPTIMIZATION OF MULTIVARIABLE FUNCTIONS*

In this section, we present a program for minimizing functions of more than one variable.

PROGRAM MULTIVAR

This program finds a relative minimum for an unconstrained function of two or more variables. The user must enter the function in line 170 using variable names X(1), X(2), X(3), and so on, which correspond to the independent variables in the function. For example, the function

$$f(x, y) = (x - 2)^2 + (y - 1)^2$$

must be coded as*

```
170 DEF FNF#(X) = (X(1) - 2)^2 + (X(2) - 1)^2
```

The program is designed to minimize only. In order to maximize a function, simply multiply the function you wish to minimize by -1. For example, to maximize $3x_1^2 - 4x_1x_2$, we would enter

```
170 DEF FNF#(X) = -3*X(1)^2 + 4*X(1)*X(2)
```

Note that the optimal value of the function that is actually minimized is the negative of the value of the original function we wish to maximize.

Any initial point can be chosen; however, the program will run faster if a point close to the optimum is selected initially. The user has the option of printing either initial and final solutions only, or the results of every iteration. If the latter is chosen, the path to the optimum can be traced.

The method used is a multidimensional search procedure that does not rely on differentiation. Essentially, all but one variable are held constant and the program minimizes the function over that one variable. This process is repeated

---

* In FNF#, the "#" indicates double-precision arithmetic. This is necessary for algorithms of this type to run effectively.

for each variable consecutively. When this has been done for all variables, one iteration is completed. If this new point is not significantly different from the previous point, then the program stops. Otherwise, we repeat the process of minimizing over each variable individually and move to a new point.

To illustrate the use of this program, consider minimizing $f(x_1, x_2) = (x_1 - 2)^2 + (x_2 - 1)^2$ starting from the point (0, 0). The results are given below:

```
THIS PROGRAM FINDS THE MINIMUM OF AN UNCONSTRAINED FUNCTION
OF SEVERAL VARIABLES

LINE 170 MUST CONTAIN YOUR FUNCTION AS 'DEF FNF#(X) ='

DOES LINE 170 CONTAIN YOUR FUNCTION? (Y/N)   y

NUMBER OF VARIABLES? 2
INITIAL POINT, VARIABLE # 1
? 0
INITIAL POINT, VARIABLE # 2
? 0

INPUT PRINT PARAMETER
   0 - INITIAL AND FINAL SOLUTIONS ONLY
   1 - DETAILED SUMMARY
? 1


      ******INITIAL SOLUTION******

VARIABLE        INITIAL VALUE
   1                0
   2                0
FUNCTION VALUE =    5


***** ITERATION 1  *****

VARIABLE         CURRENT VALUE
   1             1.979890257120132
   2             .992992103099823

FUNCTION VALUE =   4.535123589448631D-04

***** ITERATION 2  *****

VARIABLE         CURRENT VALUE
   1             2.001890255138278
   2             1.002992102876306

FUNCTION VALUE =   1.252574384125182D-05
```

```
***** ITERATION 3  *****

VARIABLE         CURRENT VALUE
   1             1.999690255150199
   2             1.000352103146724

FUNCTION VALUE =   2.199184905293805D-07

***** ITERATION 4  *****

VARIABLE         CURRENT VALUE
   1             1.999954255159537
   2             .9999881031399127

FUNCTION VALUE =   2.234125551225929D-09

        ******OPTIMAL SOLUTION******

FUNCTION VALUE =   2.234125551225929D-09

VARIABLE         OPTIMAL VALUE
   1             1.999954255159537
   2             .9999881031399127
```

PROBLEMS (Section 12-6)

In Problems 1–5, use program MULTIVAR to locate any relative optima, starting from the given initial points.

1. Minimize $f(x, y) = x^2 + y^2 - 3x$;   (2, 1)

2. Minimize $f(x, y) = x^2 + 3y^2 - 4x - 3y$;   (1, 1)

3. Maximize $f(x, y) = 2x - 4y - x^2 - 6y^2 - 4xy$; (0, 0)

4. Minimize $f(x, y) = x^2 + \frac{1}{4}y^4 + 6xy$; (0, 0), (10, -5), (-10, 5)

5. Minimize $f(x, y) = xy + \frac{4}{x} + \frac{2}{y}$;   (1, 1)

12-7 PROBABILITY COMPUTATIONS

In Chapter 11 and its supplement we introduced basic concepts of probability. In this section we present a program for computing the expected value and variance of a discrete random variable and discuss how integration programs presented earlier can be used to compute probabilities associated with continuous random variables.

PROGRAM PROB   Program PROB computes the expected value and variance for a discrete random variable using the techniques described in Section 11-2. It also produces a
table of probabilities for a set of observations of values of a random variable.
The user is prompted first to enter the number of values of the random variable.
Next, the value of the random variable and the number of observations of this
value are input to the program. The program then prints a table giving the input
data and the probability that the random variable takes on each value. Finally,
the expected value and variance are computed and printed.

To illustrate the program, consider Example 12 in Section 11-2. The observed number of live births in a litter for a population of 1000 dogs gives the
following results:

```
THIS PROGRAM COMPUTES PROBABILITIES AND THE EXPECTED VALUE
 AND VARIANCE OF A SET OF OBSERVATIONS OF A RANDOM VARIABLE

ENTER THE NUMBER OF VALUES OF THE RANDOM VARIABLE 11

ENTER THE VALUE OF EACH RANDOM VARIABLE, NUMBER OF OBSERVATIONS
? 0,10
? 1,25
? 2,40
? 3,80
? 4,200
? 5,180
? 6,245
? 7,115
? 8,90
? 9,10
? 10,5
```

VALUE OF RANDOM VARIABLE	NO. OF OBSERVATIONS	PROBABILITY
0	10	.01
1	25	.025
2	40	.04
3	80	.08
4	200	.2
5	180	.18
6	245	.245
7	115	.115
8	90	9.000001E-02
9	10	.01
10	5	.005

```
EXPECTED VALUE =  5.180001

VARIANCE =  3.2976
```

The computations in Chapter 11 verify these results.

PROBABILITIES AND CONTINUOUS RANDOM VARIABLES

In Chapter 11 we showed that the probability that the random variable X, with density function $f(x)$, takes on a value between c and d, that is, $P(c \leq X \leq d)$, is given by the integral

$$\int_c^d f(t)\, dt$$

We know that this integral can be evaluated easily provided that an antiderivative of $f(t)$, $F(t)$, is known. That is,

$$\int_c^d f(t)\, dt = F(d) - F(c)$$

We may perform this computation numerically and obtain an estimate of the probability using one of the integration programs discussed earlier in this chapter. For instance, in Example 23 in Section 11-4, we computed the probability $P(0.1 \leq X \leq 0.2)$ when the probability density function was given by

$$f(x) = 4x^3 \qquad 0 \leq x \leq 1$$

Using the program INTEGRAT, we estimate this probability to be as follows:

```
EVALUATION OF THE DEFINITE INTEGRAL OF F(X) FROM X=A TO X=B

LINE 80 MUST CONTAIN YOUR FUNCTION AS 'DEF FNF(X) = '

DOES LINE 80 CONTAIN YOUR FUNCTION? (Y/N)? y

ENTER THE VALUE FOR A   .1
ENTER THE VALUE FOR B   .2
ENTER THE NUMBER OF RECTANGLES   500

SIZE OF SUBINTERVAL IS         .0002

THE VALUE OF THE INTEGRAL IS               1.502802E-03
```

The exact value was shown to be 0.0015.

For many functions, however, a closed form solution to the antiderivative may not exist. This is indeed the case for the normal probability density function discussed in the supplement to Chapter 11. Therefore, we must use numerical techniques of integration in order to perform probability calculations. For the normal density function

$$f(x) = \frac{1}{\sqrt{2\pi}}\, e^{-x^2/2}$$

we estimate the probability $P(-0.1 \leq X \leq 0.4)$ using the program TRAPEZD as shown below:

```
EVALUATION OF THE DEFINITE INTEGRAL USING THE TRAPEZOIDAL RULE

LINE 80 MUST CONTAIN YOUR FUNCTION AS 'DEF FNF(X) = '
```

```
DOES LINE 80 CONTAIN YOUR FUNCTION? (Y/N)? y

ENTER THE VALUE FOR A   -.1
ENTER THE VALUE FOR B   .4
ENTER THE NUMBER OF TRAPEZOIDS   1000

SIZE OF SUBINTERVAL IS          .0005

THE VALUE OF THE INTEGRAL IS              .1952496
```

This is an accurate result to four decimal places. Numerical integration can be employed in this fashion to compute probabilities for a variety of other continuous density functions.

PROBLEMS (Section 12-7)

Use program PROB to compute probabilities, expected values, and variances for the data in the following problems in Section 11-2.

1. Problem 5

2. Problem 6

3. Problem 7

4. Problem 8

5. Problem 9

6. Problem 10

7. Consider the probability density function

$$f(x) = 6x(1 - x) \qquad 0 \leq x \leq 1$$

Compute $P(0.25 \leq X \leq 0.5)$ and $P(X > 0.5)$ using program INTEGRAT with 1000 subintervals.

8. A continuous random variable has density function

$$f(x) = \frac{x}{48} \qquad 2 \leq x \leq 10$$

Show numerically that this is indeed a probability density function. What is $P(X \leq 4)$?

9. The *exponential density* function is given by

$$f(x) = \lambda e^{-\lambda x} \qquad x \geq 0$$

Suppose that $\lambda = 1$. Using numerical integration, find $P(X \leq 1)$ and $P(3 \leq X \leq 8)$.

10. For the probability density function

$$f(x) = \frac{x - 1}{8} \qquad 1 \leq x \leq 5$$

compute the expected value using numerical integration. What is the probability that X is greater than 4?

For Problems 11–15, assume a normal density function with parameters μ and σ, and compute the specified probabilities.

11. $\mu = 0, \sigma = 1; \quad P(-0.3 \leq X \leq 0.3)$

12. $\mu = 4, \quad \sigma = 2; \quad P(0 \leq X \leq 5)$

13. $\mu = 0, \quad \sigma = 3; \quad P(-3 \leq X \leq 0)$

14. $\mu = 0, \quad \sigma = 1; \quad P(X \geq 0)$ (How do you choose the upper limit of integration?)

15. $\mu = -5, \quad \sigma = 0.2; \quad P(-9 \leq X \leq 4)$

12-8 CALCULUS OF TRIGONOMETRIC FUNCTIONS

We have not ignored trigonometric functions in this chapter. The BASIC programming language (indeed, most other high-level languages) have built-in routines to evaluate trigonometric functions. The most common are

COS(X)   cosine of angle x, x in radians
SIN(X)   sine of angle x, x in radians
TAN(X)   tangent of angle x, x in radians

These functions can be used in any of the programs presented in this chapter. Below we give a few examples.

Example 1   Graph the function $y(x) = \sin x$, $0 \le x \le 2\pi$.

Solution   Using the program GRAPH, we obtain the following printout and graph:

```
FUNCTION PLOTTER

THIS PROGRAM REQUIRES ADVANCED BASIC
AND COLOR/GRAPHICS

LINE 100 MUST CONTAIN YOUR FUNCTION AS 'DEF FNF(X) ='
DOES LINE 100 CONTAIN YOUR FUNCTION? (Y/N)   Y

ENTER MINIMUM, MAXIMUM VALUES FOR X   0,6.28318
ENTER MINIMUM, MAXIMUM VALUES FOR Y   -1.5,1.5
```

Figure 12-10
Graph of sin x

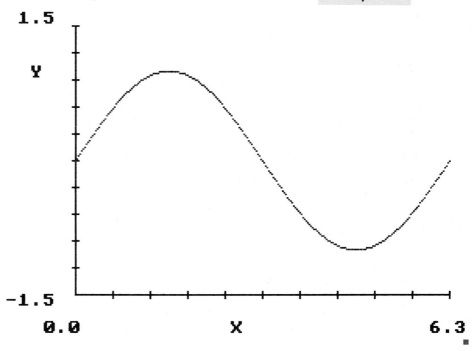

Example 2   Find the derivative of sin x at $x = 3\pi/2$.

Solution   With program DERIV, we find:

```
EVALUATION OF THE DERIVATIVE OF F(X) AT X = A

LINE 80 MUST CONTAIN YOUR FUNCTION AS 'DEF FNF(X) = '

DOES LINE 80 CONTAIN YOUR FUNCTION (Y/N)? y

ENTER THE VALUE OF `A'   4.712385
ENTER STARTING VALUE OF H   1

LIMITING VALUE OF DIFFERENCE QUOTIENT AS H APPROACHES ZERO FROM THE RIGHT

H                    (F(A+H) - F(A))/H
 1                        .4596945
 .5                       .2448312
 .25                      .1243465
 .125                     6.241465E-02
 .0625                    .0312357
 .03125                   1.561737E-02
 .015625                  7.804871E-03
 .0078125                 3.890991E-03
 3.90625E-03              1.953125E-03
 1.953125E-03             9.765625E-04
 9.765625E-04             3.66211E-04
 4.882813E-04             1.220703E-04
 2.441406E-04             0
 1.220703E-04             0


LIMITING VALUE OF DIFFERENCE QUOTIENT AS H APPROACHES ZERO FROM THE LEFT

H                    (F(A+H) - F(A))/H
-1                       -.4597009
-.5                      -.2448385
-.25                     -.1243539
-.125                    -.0624218
-.0625                   -3.124237E-02
-.03125                  -1.562691E-02
-.015625                 -.0078125
-.0078125                -3.898621E-03
-3.90625E-03             -1.953125E-03
-1.953125E-03            -9.460449E-04
-9.765625E-04            -4.882813E-04
-4.882813E-04            -2.441406E-04
-2.441406E-04             0
-1.220703E-04             0
```

The value of the derivative appears to be 0. Is this correct? ∎

Example 3   Compute the integral $\int_0^{\pi/2} \cos x\, dx$.

Solution   By using the program TRAPEZD, we obtain:

```
EVALUATION OF THE DEFINITE INTEGRAL USING THE TRAPEZOIDAL RULE

LINE 80 MUST CONTAIN YOUR FUNCTION AS 'DEF FNF(X) = '

DOES LINE 80 CONTAIN YOUR FUNCTION? (Y/N)? y

ENTER THE VALUE FOR A   0
ENTER THE VALUE FOR B   1.570795
ENTER THE NUMBER OF TRAPEZOIDS   1000

SIZE OF SUBINTERVAL IS            1.570795E-03

THE VALUE OF THE INTEGRAL IS                .9999992
```

PROBLEMS (Section 12-8)

In Problems 1–5, use programs TABLE and GRAPH to determine graphs of the given functions.

1. $f(t) = \cos t, \quad -\pi \le t \le 2\pi$

2. $f(t) = \sin t, \quad -\pi \le t \le 2\pi$

3. $f(t) = 2 \sin t, \quad -\pi/2 \le t \le 5\pi/2$

4. $f(t) = -\cos t, \quad -\pi/2 \le t \le 5\pi/2$

5. $f(t) = \cos(t - \pi)$ (Select an appropriate interval for t.)

6. Use program NEWTON to find any roots of the function
$$f(t) = \sin t \qquad 0 \le t \le 2\pi$$
Use starting values of $t = 1$ and $t = 3$.

7. Use program GOLDEN to find a minimum point of the function
$$f(t) = \cos t \qquad 0 \le t \le 2\pi$$

8. Estimate the derivative of $(t^2 + 1) \sin 2t$ at $t = \pi$. Compare the numerical answer with that obtained using the result of Example 10 in Section 9-3.

9. Approximate the value of the integral
$$\int_0^\pi \cos \frac{t}{2}\, dt$$

10. Approximate the value of the integral
$$\int_0^1 \sin \pi(t + 2)\, dt$$

PROGRAM TABLE

```
10  CLS
20  PRINT "THIS PROGRAM COMPUTES A TABLE OF FUNCTION VALUES OVER AN INTERVAL"
30  PRINT
40  PRINT "LINE 70 MUST CONTAIN YOUR FUNCTION AS 'DEF FNF(X) ='"
50  INPUT "DOES LINE 70 CONTAIN YOUR FUNCTION? (Y/N)  ",ANS$
60  IF ANS$ = "N" OR ANS$ ="n" THEN END
65  REM***FUNCTION MUST BE DEFINED IN LINE 70
70  DEF FNF(X) = X^4-X^3+3*X
80  PRINT
90  INPUT "ENTER MINIMUM, MAXIMUM VALUES FOR X  ",XMIN,XMAX
100 INPUT "ENTER THE NUMBER OF POINTS TO BE EVALUATED  ",N
110 INC = (XMAX-XMIN)/(N-1)            'COMPUTE INTERVAL BETWEEN POINTS
120 PRINT
130 PRINT "   X","F(X)"
140 PRINT "--------------------------------"
150 FOR I = 0 TO N-1
160 X = XMIN + INC*I
170 PRINT X,FNF(X)                     'LOOP TO EVALUATE FUNCTION
180 NEXT I
190 END
```

PROGRAM GRAPH

```
10 CLS
20 PRINT "FUNCTION PLOTTER"
30 PRINT
40 PRINT "THIS PROGRAM REQUIRES ADVANCED BASIC"
50 PRINT "AND COLOR/GRAPHICS"
60 PRINT
70 PRINT "LINE 100 MUST CONTAIN YOUR FUNCTION AS 'DEF FNF(X)  = '"
80 INPUT "DOES LINE 100 CONTAIN YOUR FUNCTION? (Y/N)   ",ANS$
90 IF ANS$ = "N" OR ANS$ = "n" THEN END
95 REM***FUNCTION MUST BE DEFINED IN LINE 100
100 DEF FNF(X) =X^4-X^3+3*X
110 PRINT
120 INPUT "ENTER MINIMUM, MAXIMUM VALUES FOR X   ",XMIN,XMAX
130 INPUT "ENTER MINIMUM, MAXIMUM VALUES FOR Y   ", YMIN,YMAX
140 SCREEN 1,0:CLS:KEY OFF
150 DEF FNX(X) = 52+240*X/100
160 DEF FNY(Y)=16+ (100-Y)*160/100
170 LINE (FNX(0),FNY(100))-(FNX(0),FNY(0))        'DRAW AXES
180 LINE -(FNX(100),FNY(0))
190 LOCATE 2,1:PRINT USING "###.#";YMAX;          'PRINT AXIS LABELS
200 LOCATE 6,4:PRINT "Y"
210 LOCATE 25,34:PRINT USING "###.#";XMAX;
220 LOCATE 23,1: PRINT USING "###.#";YMIN;
230 LOCATE 25,3: PRINT USING "###.#";XMIN;
240 LOCATE 25,20: PRINT "X";
250 FOR Y=10 TO 100 STEP 10
260 LINE (50,FNY(Y))-(54,FNY(Y))                  'DRAW HASH MARKS
270 NEXT Y
280 FOR X=10 TO 100 STEP 10
290 LINE (FNX(X),174)-(FNX(X),178)
300 NEXT X
310 FOR I=1 TO N: X=(XX(I)-XMIN)*100/(XMAX-XMIN):Y=(YY(I)-YMIN)*100/(YMAX-YMIN)
320 LINE (FNX(X)-2,FNY(Y)-2)-STEP(4,4),1,B
330 NEXT I
340 FOR X = 52 TO 292                             'MAIN LOOP TO PLOT PIXELS FOR GRAPH
350 XC = XMIN + (X-52)*(XMAX-XMIN)/240
360 YC = FNF(XC)
370 Y = 16+160*(YMAX-YC)/(YMAX-YMIN)
380 PSET(X,Y),2
390 NEXT
400 IF INKEY$=""THEN 400
410 END
```

PROGRAM LIMITS

```
10 CLS
20 PRINT "LIMIT EVALUATION OF F(X) AS X APPROACHES A"
30 PRINT "A CANNOT BE PLUS OR MINUS INFINITY"
40 PRINT
50 PRINT    "LINE 140 MUST CONTAIN YOUR FUNCTION AS '140 DEF FNF(X)='"
60 PRINT
70 INPUT "DOES LINE 140 CONTAIN YOUR FUNCTION (Y/N)";ANS$
80 IF ANS$="N" OR ANS$="n" THEN END
90 PRINT
100 INPUT "ENTER INITIAL VALUE OF X  ", XINIT
110 INPUT "ENTER THE VALUE OF 'A' ",A
120 INPUT "ENTER THE NUMBER OF POINTS IN THE LIMIT SEQUENCE ",NP
130 PRINT
135 REM*** FUNCTION MUST BE DEFINED IN LINE 140
140 DEF FNF(X) = (X-2)/(X^2-4)
150 X=XINIT
160 PRINT "X","F(X)"
170 GOSUB 240
180 FOR J=1 TO NP     'LOOP TO INCREMENT (DECREMENT) VALUE OF X
190 IF A<XINIT THEN X = A+.5*(X-A) ELSE X=A-.5*(A-X)
200 GOSUB 240
210 NEXT J
220 END
230 REM***PRINT SUBROUTINE
240 PRINT X,FNF(X)
250 RETURN
```

PROGRAM DERIV

```
10 CLS
20 PRINT "EVALUATION OF THE DERIVATIVE OF F(X) AT X = A"
30 PRINT
40 PRINT "LINE 80 MUST CONTAIN YOUR FUNCTION AS 'DEF FNF(X) = '"
50 PRINT
60 INPUT "DOES LINE 80 CONTAIN YOUR FUNCTION (Y/N)";ANS$
70 IF ANS$ = "N" OR ANS$ = "n" THEN END
75 REM**FUNCTION MUST BE DEFINED IN LINE 140
80 DEF FNF(X) = 6*X^3 - 3*X^2 +4*X
90 PRINT
100 INPUT "ENTER THE VALUE OF 'A' ",A
110 INPUT "ENTER STARTING VALUE OF H ",HINIT
120 IF HINIT < 0 THEN HINIT = -HINIT    'FIRST EVALUATE FROM RIGHT OF A
130 H = HINIT
140 PRINT
150 PRINT "LIMITING VALUE OF DIFFERENCE QUOTIENT AS H APPROACHES ZERO FROM THE R
IGHT"
160 PRINT
170 PRINT "H"," (F(A+H) - F(A))/H"
180 WHILE ABS(H) > .0001    'STOP WHEN H IS WITHIN .0001 OF A
190 VALUE = FNF(A+H)-FNF(A)    'COMPUTE DIFFERENCE QUOTIENT
200 VALUE = VALUE/H
210 PRINT H;TAB(20);VALUE
220 H = .5*H
230 WEND
240 IF HINIT < 0 THEN END    'NOW EVALUATE FROM LEFT OF A
250 HINIT=-HINIT
260 PRINT:PRINT
270 H=HINIT
280 PRINT "LIMITING VALUE OF DIFFERENCE QUOTIENT AS H APPROACHES ZERO FROM THE L
EFT"
290 PRINT
300 GOTO 170
310 END
```

PROGRAM NEWTON

```
10 CLS
20 PRINT "FINDING ROOTS OF EQUATIONS USING NEWTON'S METHOD"
30 PRINT
40 PRINT "LINE 90 MUST CONTAIN YOUR FUNCTION AS 'DEF FNF(X) = '"
50 PRINT "LINE 100 MUST CONTAIN THE DERIVATIVE OF YOUR FUNCTION AS"
60 PRINT "   'DEF FNDER(X) = '"
70 INPUT "DO LINES 90 AND 100 CONTAIN YOUR FUNCTION AND ITS DERIVATIVE? (Y/N) ",
   ANS$
80 IF ANS$="N" OR ANS$="n" THEN END
85 REM***FUNCTION AND ITS DERIVATIVE MUST BE DEFINED IN LINES 90 AND 100
90 DEF FNF(X) = X^3+X^2/2-2*X+4
100 DEF FNDER(X) = 3*X^2+X-2
110 PRINT
120 INPUT "ENTER STARTING VALUE FOR X";X
130 INPUT "ENTER MAXIMUM ERROR ALLOWED";EPS
140 FOR I=1 TO 100          'DO 100 ITERATIONS MAXIMUM
150 A=FNF(X)
160 B=FNDER(X)
170 IF B><0 THEN 190
180 PRINT "DERIVATIVE IS ZERO--PROGRAM TERMINATED":END     'DENOMINATOR=0
190 DEV = A/B
200 X=X-DEV                 'COMPUTE NEW VALUE FOR X
210 IF ABS(DEV)<EPS THEN 260     'STOP IF WITHIN ALLOWABLE ERROR
220 NEXT
230 PRINT "METHOD DID NOT CONVERGE"
240 PRINT "CURRENT SOLUTION";X, "RESIDUAL";DEV
250 END
260 PRINT: PRINT "ESTIMATED ROOT"
270 PRINT "X = ";X
280 END
```

PROGRAM GOLDEN

```
10 CLS
20 PRINT "ESTIMATION OF MINIMUM OR MAXIMUM VALUE OF F(X) OVER THE INTERVAL [A,B]
30 PRINT
40 PRINT "LINE 80 MUST CONTAIN YOUR FUNCTION AS 'DEF FNF(X) = '"
50 PRINT
60 INPUT "DOES LINE 80 CONTAIN YOUR FUNCTION? (Y/N)   ",ANS$
70 IF ANS$ = "N" OR ANS$ = "n" THEN END
75 REM**FUNCTION MUST BE DEFINED IN LINE 80
80 DEF FNF(X) = X^4/4-X^3/3-X^2+5
90 PRINT
100 PRINT "TYPE 1 IF THIS IS A MINIMIZATION PROBLEM"
110 PRINT "TYPE 2 IF THIS IS A MAXIMIZATION PROBLEM"
120 INPUT TYPE
130 INPUT "ENTER THE VALUE OF A ",A
140 INPUT "ENTER THE VALUE OF B ",B
150 INPUT "ENTER THE SIZE OF THE FINAL INTERVAL ",L
160 ALPHA = .618
170 P1 = A +(1-ALPHA)*(B-A)              'LOCATE INITIAL POINTS
180 P2 = A + ALPHA*(B-A)
190 F1 = FNF(P1)
200 F2 = FNF(P2)
```

```
210 IF TYPE = 2 THEN F1=-F1      'CHANGE SIGN OF FUNCTION FOR MAXIMIZATION
220 IF TYPE = 2 THEN F2=-F2
230 IF B-A < L THEN GOTO 390     'WITHIN SPECIFIED FINAL INTERVAL : STOP
240 IF F1>F2 THEN GOTO 320
250 B=P2                          'ELIMINATE LEFT INTERVAL
260 P2=P1
270 P1=A +(1-ALPHA)*(B-A)
280 F2=F1
290 IF TYPE=2 THEN F2=-F2
300 F1=FNF(P1)
310 GOTO 210
320 A=P1                          'ELIMINATE RIGHT INTERVAL
330 P1=P2
340 P2=A+ALPHA*(B-A)
350 F1=F2
360 IF TYPE=2 THEN F1=-F1
370 F2=FNF(P2)
380 GOTO 210
390 X=(B+A)/2
400 PRINT "ESTIMATE OF OPTIMAL POINT"
410 PRINT
420 PRINT "X = ",X
430 PRINT "F(X) = ",FNF(X)
440 END
```

PROGRAM INTEGRAT

```
10 CLS
20 PRINT "EVALUATION OF THE DEFINITE INTEGRAL OF F(X) FROM X=A TO X=B"
30 PRINT
40 PRINT "LINE 80 MUST CONTAIN YOUR FUNCTION AS 'DEF FNF(X) = '"
50 PRINT
60 INPUT "DOES LINE 80 CONTAIN YOUR FUNCTION? (Y/N)";ANS$
70 IF ANS$ = "N" OR ANS$ = "n" THEN END
75 REM***FUNCTION MUST BE DEFINED IN LINE 80
80 DEF FNF(X) = X^3 + 3*X^2 - 4*X
90 PRINT
100 INPUT "ENTER THE VALUE FOR A  ",A
110 INPUT "ENTER THE VALUE FOR B  ",B
120 INPUT "ENTER THE NUMBER OF RECTANGLES  ",N
130 DELTA = (B-A)/N          'COMPUTE SIZE OF SUBINTERVALS
140 PRINT:PRINT "SIZE OF SUBINTERVAL IS ",DELTA
150 SUM = 0
160 FOR I = 1 TO N           'LOOP TO EVALUATE AREA OF EACH RECTANGLE
170 X = A + I*DELTA
180 SUM = SUM + DELTA*FNF(X)
190 NEXT I
200 PRINT: PRINT "THE VALUE OF THE INTEGRAL IS  ",SUM
210 END
```

PROGRAM TRAPEZD

```
10 CLS
20 PRINT "EVALUATION OF THE DEFINITE INTEGRAL USING THE TRAPEZOIDAL RULE"
30 PRINT
40 PRINT "LINE 80 MUST CONTAIN YOUR FUNCTION AS 'DEF FNF(X) = '"
50 PRINT
60 INPUT "DOES LINE 80 CONTAIN YOUR FUNCTION? (Y/N) ";ANS$
70 IF ANS$ = "N" OR ANS$ = "n" THEN END
75 REM***FUNCTION MUST BE DEFINED IN LINE 80
80 DEF FNF(X) = X^3 + 3*X^2 - 4*X
90 PRINT
100 INPUT "ENTER THE VALUE FOR A  ",A
110 INPUT "ENTER THE VALUE FOR B  ",B
120 INPUT "ENTER THE NUMBER OF TRAPEZOIDS  ",N
130 DELTA = (B-A)/N                    'COMPUTE SIZE OF SUBINTERVALS
140 PRINT:PRINT "SIZE OF SUBINTERVAL IS ",DELTA
150 SUM = 0
160 FOR I = 1 TO N                     'LOOP TO EVALUATE AREA OF EACH TRAPEZOID
170 X = A + I*DELTA
180 XO = X - DELTA
190 SUM = SUM + DELTA*(FNF(XO)+FNF(X))/2
200 NEXT I
210 PRINT: PRINT "THE VALUE OF THE INTEGRAL IS  ",SUM
220 END
```

PROGRAM DIFFEQ

```
10 CLS
20 PRINT "THIS PROGRAM SOLVES ELEMENTARY DIFFERENTIAL EQUATIONS"
30 PRINT
40 PRINT "LINE 80 MUST CONTAIN THE DIFFERENTIAL EQUATION DY/DX = F(X,Y)"
50 PRINT "   AS 'DEF FNDEQ(X,Y) = '"
60 INPUT "DOES LINE 80 CONTAIN YOUR FUNCTION? (Y/N)   ",ANS$
70 IF ANS$="N" OR ANS$ ="n" THEN END
75 REM***FUNCTION MUST BE DEFINED IN LINE 80
80 DEF FNDEQ(X,Y) = 3*Y
90 PRINT
100 INPUT "ENTER INITIAL VALUES FOR X,Y ",X,Y
110 INPUT "ENTER THE STEP SIZE ",H
120 INPUT "ENTER THE MAXIMUM VALUE OF X ",XLAST
130 PRINT
140 PRINT " X"," Y"
150 K1 = H*FNDEQ(X,Y)          'MODIFIED EULER METHOD
160 K2 = H*FNDEQ(X+H,Y+K1)
170 Y = Y+.5*(K1+K2)
180 X = X+H
190 PRINT X,Y
200 IF X - XLAST < 0 THEN 150          'STOPPING CRITERION
210 END
```

PROGRAM PROB

```
10 CLS
20 PRINT "THIS PROGRAM COMPUTES PROBABILITIES AND THE EXPECTED VALUE"
30 PRINT " AND VARIANCE OF A SET OF OBSERVATIONS OF A RANDOM VARIABLE"
40 PRINT
50 INPUT "ENTER THE NUMBER OF VALUES OF THE RANDOM VARIABLE ",N
60 PRINT:PRINT "ENTER THE VALUE OF EACH RANDOM VARIABLE, NUMBER OF OBSERVATIONS
70 TOTAL=0
80 DIM VARBLE(N),OBS(N)
90 FOR I=1 TO N
100 INPUT VARBLE(I),OBS(I)
110 TOTAL=TOTAL+OBS(I)
120 NEXT I
130 PRINT:PRINT "VALUE OF RANDOM VARIABLE","NO. OF OBSERVATIONS","PROBABILITY"
140 PRINT "----------------------------------------------------------------"
150 FOR I = 1 TO N
160 LOCATE ,5:PRINT VARBLE(I),,OBS(I),,OBS(I)/TOTAL
170 NEXT I
180 REM  COMPUTE EXPECTED VALUE
190 EV=0
200 FOR I=1 TO N
210 EV=EV+VARBLE(I)*OBS(I)/TOTAL
220 NEXT I
230 REM  COMPUTE VARIANCE
240 VARIANCE=0
250 FOR I=1 TO N
260 VARIANCE=VARIANCE+(VARBLE(I)-EV)^2*OBS(I)/TOTAL
270 NEXT I
280 PRINT:PRINT "EXPECTED VALUE = ";EV
290 PRINT:PRINT "VARIANCE = ";VARIANCE
300 END
```

PROGRAM MULTIVAR

```
10 CLS
20 PRINT "THIS PROGRAM FINDS THE MINIMUM OF AN UNCONSTRAINED FUNCTION"
30 PRINT "OF SEVERAL VARIABLES"
40 PRINT
50 PRINT "LINE 170 MUST CONTAIN YOUR FUNCTION AS `DEF FNF#(X) ='"
60 PRINT
70 INPUT "DOES LINE 170 CONTAIN YOUR FUNCTION? (Y/N)   ",ANS$
80 IF ANS$="N" OR ANS$="n" THEN END
90 DEFDBL Z,L,V,H,G,X,Y,P
100 DATA .1,.00001,.1,.0001,1.2
110 READ DELTA, DELMIN,BETA,EPSILON,ALPHA
120 PRINT: INPUT "NUMBER OF VARIABLES? ",N
130 DIM X(N),Y(N),YO(N)
140 FOR J=1 TO N
150 PRINT "INITIAL POINT, VARIABLE #";J
160 INPUT Y(J):X(J)=Y(J):NEXT
165 REM***FUNCTION MUST BE DEFINED IN LINE 170
170 DEF FNF#(X)=(X(1)-2)^2+(X(2)-1)^2
180 PRINT: PRINT "INPUT PRINT PARAMETER"
190 PRINT "  0 - INITIAL AND FINAL SOLUTIONS ONLY"
200 PRINT "  1 - DETAILED SUMMARY"
210 INPUT PPARM%
220 ZSTAR=FNF#(X)           'BEST VALUE THUS FAR
230 PRINT
240 PRINT "    ******INITIAL SOLUTION******"
250 PRINT
260 PRINT "VARIABLE","INITIAL VALUE"
270 FOR J=1 TO N
280 PRINT J,X(J):NEXT
290 PRINT "FUNCTION VALUE =  ";FNF#(X)
300 PRINT
310 ITER%=1
320 REM CYCLIC COORDINATE OPTIMIZATION
330 FOR J=1 TO N            'SAVE CURRENT POINT
340 YO(J)=Y(J):NEXT
350 FOR I=1 TO N            'FOR EACH VARIABLE, FIX OTHERS AND MINIMIZE
360 GOSUB 780
370 Y(I)=Y(I)+MU            'MU IS DISTANCE MOVED ALONG COORDINATE AXIS I
380 NEXT
390 SUM=0                   'DETERMINE IF BETTER POINT WAS FOUND
400 FOR J=1 TO N
410 SUM=SUM+ABS(YO(J)-Y(J)):NEXT
420 IF SUM >= EPSILON THEN GOTO 330
430 IF PPARM% >< 0 THEN GOSUB 640
440 BD=BETA*DELTA           'DECREASE STEPSIZE FOR FUTURE ITERATIONS
450 IF DELMIN > BD THEN GOTO 530        'STOP IF STEPSIZE < DELMIN
460 DELTA=BD
470 GOTO 330
```

```
480 FOR J=1 TO N              'RECORD NEW POINT AND FUNCTION VALUE
490 X(J)=Y(J)
500 YO(J)=Y(J):NEXT
510 ZSTAR=FNF#(X)
520 GOTO 350
530 PRINT "    ******OPTIMAL SOLUTION******"
540 PRINT
550 FOR J=1 TO N: X(J)=Y(J): NEXT
560 PRINT "FUNCTION VALUE =   ";FNF#(X)
570 PRINT
580 PRINT "VARIABLE","OPTIMAL VALUE"
590 FOR J=1 TO N
600 PRINT J,X(J):NEXT
610 END
620 REM*********************************************
630 REM ITERATION PRINT SUBROUTINE
640 PRINT
650 PRINT "***** ITERATION"; ITER%; " *****"
660 PRINT
670 ITER%=ITER%+1
680 PRINT "VARIABLE","CURRENT VALUE"
690 FOR J=1 TO N
700 X(J)=Y(J)
710 PRINT J,X(J):NEXT
720 PRINT
730 PRINT "FUNCTION VALUE =   ";FNF#(X)
740 PRINT
750 RETURN
760 REM*********************************************
770 REM***LINE SEARCH
780 MU=0
790 S=1                       'SEARCH IN POSITIVE DIRECTION FROM CURRENT POINT
800 FIRST=1
810 MU=MU+S*DELTA
820 FOR J=1 TO N
830 IF I=J THEN X(I)=Y(I)+MU ELSE X(J)=Y(J)
840 NEXT
850 Z=FNF#(X)
860 IF Z>=ZSTAR THEN GOTO 910
870 FIRST=0                   'BETTER POINT FOUND, CONTINUE IN SAME DIRECTION
880 ZSTAR=Z
890 S=ALPHA*S
900 GOTO 810
910 IF FIRST=0 THEN GOTO 960
920 MU=MU-S*DELTA
930 S=-1                      'REPEAT SEARCH IN NEGATIVE DIRECTION
940 FIRST=0
950 GOTO 810
960 MU=MU-S*DELTA
970 RETURN
```

A PRACTICAL CASE APPLICATION OF CALCULUS*

In the supplement to Chapter 2, we discussed the idea of continuous compound interest. This is used extensively in financial modeling. We showed that the value of P dollars 1 year into the future compounded at an annual interest rate r is Pe^r. We can easily extend this to any future time period, as stated below:

> The value of P dollars t years into the future is given by
>
> $$V = Pe^{rt}$$

Now, let us pose a related question. How much would V dollars t years from now be worth today at an interest rate r? The answer is found by solving the equation above for P:

$$V = Pe^{rt}$$

Multiplying both sides by e^{-rt} yields

$$P = Ve^{-rt}$$

> The **present value** of V dollars t years from now at an annual interest rate r is given by
>
> $$P = Ve^{-rt}$$

In other words, if we wished to have V dollars available in t years, we would only have to deposit P dollars today, assuming an annual interest rate r.

Now suppose that we wish to determine the present value of several future cash flows, for example, $100 at $t = 1$, $100 at $t = 2$, and $200 at $t = 5$.

The present value is equal to the sum of the present values:

$$P = 100e^{-1r} + 100e^{-2r} + 200e^{-5r}$$

If $r = 0.10$, for instance, then we would have

$$P = 100e^{-0.10} + 100e^{-0.20} + 200e^{-0.50}$$
$$\approx 90.48 + 81.87 + 121.31$$
$$= \$293.66$$

A general formula can be developed as follows. Let $V_1, V_2, \ldots, V_n$ be cash flows acquired at times $t_1 = 0$, $t_2, \ldots, t_n = T$, respectively. Then the present value is given by

$$P = \sum_{i=1}^{n} V_i e^{-rt_i}$$

If n is large and the V_i's follow some pattern, then we might be able to approximate the V_i's by a continuous function $V(t)$, as was done with the typewriting learning data in Chapter 2. The function $V(t)$ is a rate with dimensions of dollars per unit time. The amount of the cash flow over the interval $(t, t + \Delta t)$ is approximately $V(t)\Delta t$ (dollars per unit time multiplied by an increment of time equals dollars). The present value of this is

$$V(t)\Delta t e^{-rt}$$

If we add these values for each Δt over the interval $[0, T]$, assuming equal time increments Δt, we have (see Figure 12S-1, page 602)

$$P = \sum_{i=1}^{T/\Delta t} V(t_i)\Delta t e^{-rt_i}$$

where $t_i = i(\Delta t)$. This is a Riemann sum as defined in Chapter 5. As Δt gets smaller, the number of intervals

---

* Adapted from J. R. Evans, "An 'Integrating' Application of Mathematics," *International Journal of Mathematical Education in Science and Technology*, vol. 12, no. 5, pp. 557–560 (1981).

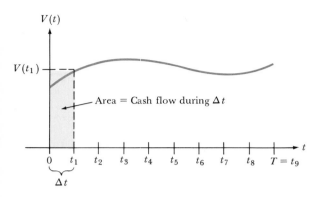

Figure 12S-1
An illustration of computation of
the present value of a continuous
cash flow

grows larger, and

$$P_c = \lim_{\Delta t \to 0} \sum_{i=1}^{T/\Delta t} V(t_i)e^{-rt_i}\Delta t$$

$$= \int_0^T V(t)e^{-rt}\,dt$$

where P_c represents the present value of a continuous cash flow and is a formula used extensively in financial modeling.

Although most cash flows are discrete, the use of a continuous function approximation simplifies many of the mathematical calculations. Financial analysts need to be concerned with errors that arise from using a continuous function to model a discrete process. In this supplement, we show how many of the ideas presented in this book can be used in studying such a problem.

12S-1 MODELING UNIFORM CASH FLOWS

We already have seen that the present value of v dollars 1 year from now at an interest rate r is

$$P = ve^{-r}$$

Suppose this amount is paid in k equal installments over the year; then the time between payments is $\Delta t = 1/k$. For example, with quarterly payments, $k = 4$ and $\Delta t = 0.25$. To compute the present value, we must compute the present value of each payment and sum them up. We should be careful, however, and realize that each payment will be $v\Delta t$ and that the interest rate applied to each payment will now be $r\Delta t$. Thus, the present value of four quarterly payments is

$$P = \sum_{i=1}^{4} v\Delta t e^{-r\Delta t_i}$$
$$= 0.25ve^{-0.25r} + 0.25ve^{-0.5r} + 0.25ve^{-0.75r} + 0.25ve^{-r}$$

Thus, for an arbitrary number of k payments, each of $v\Delta t$, the present value is

$$P_d = \sum_{i=1}^{k} v\Delta t e^{-r\Delta t_i}$$
$$= v\Delta t \sum_{i=1}^{k} e^{-r\Delta t_i}$$

where $\Delta t = 1/k$.

The question we are addressing is: For a fixed value of k, how much does P_c (the present value of the continuous approximation) differ from P_d (the discrete present value)? For a large value of k, computing P_d directly is rather tedious. We are able to express it in a simple closed form by using a little "trick" of mathematics:

THEOREM

$$P_d = v\Delta t \left(\frac{1 - e^{-r}}{e^{r\Delta t} - 1} \right)$$

We will prove this theorem here, since the proof is rather simple. Let

$$S = \sum_{i=1}^{k} e^{-r\Delta t_i}$$
$$= e^{-r\Delta t} + e^{-2r\Delta t} + \cdots + e^{-kr\Delta t}$$

Now factor $e^{-r\Delta t}$ from each term of S:

$$S = e^{-r\Delta t}(1 + e^{-r\Delta t} + \cdots + e^{-(k-1)r\Delta t})$$

Note that the expression in parentheses is almost $1 + S$, but without the term $e^{-kr\Delta t}$. Therefore, we have

$$S = e^{-r\Delta t}(1 + S + e^{-kr\Delta t})$$

Solving this equation for S yields

$$S = \frac{e^{-r\Delta t}(1 - e^{-kr\Delta t})}{1 - e^{-r\Delta t}}$$

Since $k = 1/\Delta t$, multiplying both the numerator and denominator by $e^{r\Delta t}$, we get

$$S = \frac{1 - e^{-r}}{e^{r\Delta t} - 1}$$

which proves the theorem.

If $V(t) = v$, then P_c can be integrated (with $T = 1$) as follows:

$$P_c = \int_0^1 v e^{-rt} \, dt$$
$$= v \left(\frac{e^{-rt}}{-r} \right) \Big|_0^1$$
$$= \frac{v e^{-r}}{-r} - \frac{v}{-r}$$
$$= \frac{v(1 - e^{-r})}{r}$$

Now, if all our mathematics is correct, P_d should approach P_c as $t \to 0$. This is a good opportunity to use limits. To verify our result, consider

$$\lim_{t \to 0} P_d = \lim_{t \to 0} v\Delta t \frac{1 - e^{-r}}{e^{r\Delta t} - 1}$$
$$= v(1 - e^{-r}) \lim_{t \to 0} \frac{\Delta t}{e^{r\Delta t} - 1}$$

Multiplying the expression in the limit by r/r, we have

$$\lim_{t \to 0} P_d = v(1 - e^{-r}) \lim_{t \to 0} \frac{r\Delta t}{r(e^{r\Delta t} - 1)}$$

One of the fundamental results we presented in Section 6-3 regarding exponential functions is that

$$\lim_{h \to 0} \frac{e^h - 1}{h} = 1$$

Letting $h = r\Delta t$, we see that $h \to 0$ as $\Delta t \to 0$. Therefore, this limit is simply $1/r$. Thus, P_d approaches

$$\frac{v(1 - e^{-r})}{r} = P_c$$

We can now compute the percentage error that arises from using the continuous approximation. This is given by

$$\text{Percentage error} = \frac{P_c - P_d}{P_d} 100$$

If we substitute the expressions we have derived for P_c and P_d and simplify, we obtain

$$\text{Percentage error} = \left(\frac{e^{r\Delta t} - 1}{r\Delta t} - 1 \right) 100$$

Figure 12S-2 exhibits a family of curves of percentage error as a function of Δt for various values of r. We see that if $r = 0.15$, discrete flows must be distributed

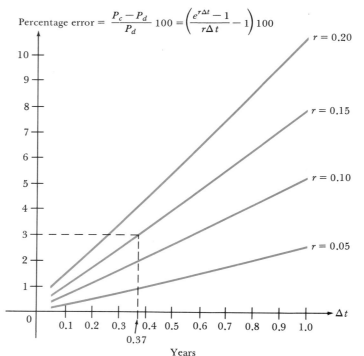

$$\text{Percentage error} = \frac{P_c - P_d}{P_d} 100 = \left(\frac{e^{r\Delta t} - 1}{r\Delta t} - 1 \right) 100$$

Figure 12S-2
Constant cash flow rate error curves

about 0.37 year (135 days) or less for the error to be no greater than 3%. Financial analysts can use these graphs to determine errors of this type.

PROBLEMS (Supplement to Chapter 12)

1. Use programs TABLE and GRAPH to draw error curves for interest rates of 0.075, 0.125, 0.175, and 0.25.

Use program TRAPEZD with 1000 subintervals to approximate P_c and compare this to the exact value for the following cash flows and interest rates:

2. $v = 50, \quad r = 0.10$

3. $v = 200, \quad r = 0.08$

4. $v = 150, \quad r = 0.15$

5. Use program LIMITS to verify that

$$\lim_{t \to 0} \frac{\Delta t}{e^{r\Delta t} - 1} = \frac{1}{r}$$

for $r = 0.05$ and for $r = 0.15$.

6. *Programming Problem:* Make all changes necessary in program TABLE so that you can input r and Δt and compute P_d using the theorem in this chapter.

Absolute maximum of a function A number $f(c)$ such that $f(c) \geq f(x)$ for all x in the domain of $f(x)$

Absolute minimum of a function A number $f(c)$ such that $f(c) \leq f(x)$ for all x in the domain of $f(x)$

Algebraic equation An equation involving algebraic expressions

Algebraic expression An expression consisting of sums, differences, products, quotients, powers, and roots of variables and constants

Antiderivative of a function $f(x)$ A function $F(x)$ such that $F'(x) = f(x)$

Antidifferentiation The process of finding a function whose derivative is known

Average rate of change of a function $f(x)$ between $x = a$ and $x = b$ The ratio of the change in $f(x)$ to the change in x, that is, $[f(b) - f(a)]/(b - a)$

Axiom An assumption that is accepted to be true

Bisection method A bracketing method for approximating a root of an equation

Concave down A property of a graph for which the first derivative is decreasing over an interval

Concave up A property of a graph for which the first derivative is increasing over an interval

Constant A quantity that has a fixed value

Continuous function A function that is continuous at every point in its domain

Critical point of a function $f(x)$ A point $(c, f(c))$ where either $f'(c) = 0$ or the function is defined but the derivative does not exist at that point

Cumulative distribution function An indefinite integral of a probability density function

Definite integral of $f(x)$ on the interval $[a, b]$ The number $\int_a^b f(x)\, dx = \lim_{\Delta x_j \to 0} \sum_{j=1}^{n} f(x_j) \Delta x_j$

Degree A measure of the extent of a circular arc

Derivative of the function $f(x)$ The function $f'(x) = \lim_{h \to 0} [f(x + h) - f(x)]/h$, provided the limit exists

Differentiable function A function that is differentiable at every point in its domain

Differential The quantity $dy = f'(x)\Delta x$

Differential equation An equation involving a function and one or more of its derivatives

Dimensions   Units of measurement

Domain of a function   The set of permissible values that the independent variable may take on

Euler's method   A simple method for obtaining numerical approximations to the solution of a differential equation

Event   A subset of a sample space

Expected value   The mean or average value of a random variable

Exponential function   A function of the form $f(x) = b^x$, where $b > 0$ and $b \neq 1$

Function   A rule that assigns to each value of an independent variable exactly one value of a dependent variable

Graph of the function $f(x)$ The set of points $(x, f(x))$ where x is any value in the domain of the function

Horizontal asymptote   A line $y = b$ having the property that the y-coordinates approach the value b as the x-coordinates become arbitrarily large in magnitude

Hypotenuse   The side opposite the right angle in a right triangle

Indefinite integral of a function $f(x)$ The general form of the antiderivative of $f(x)$, that is, $\int f(x)\, dx = F(x) + c$, where $F'(x) = f(x)$

Indeterminate form   A meaningless mathematical quantity such as $0/0$, ∞/∞, $\infty - \infty$, or $(\infty)(0)$

Infinite series   The sum of the terms of a sequence

Inflection point   A point where the curve changes from concave up to concave down or vice versa

Initial condition   An additional constraint imposed on the solution of a differential equation by specifying the value of the solution or its derivatives at a given point

Instantaneous rate of change of a function $f(x)$ at $x = a$ The value of the derivative $f'(a)$

Integration   The process of determining an indefinite integral

Iterated integral   An integral of the form $\int_a^b [\int_{g(x)}^{h(x)} f(x, y)\, dy]\, dx$ or $\int_c^d [\int_{g(y)}^{h(y)} f(x, y)\, dx]\, dy$

Linear first-order differential equation   A differential equation of the form $y' + ay = g$

Linear function   A function that can be written in the form $f(x) = mx + b$

Logarithmic function   The inverse of an exponential function, that is, $f(x) = \log_b x$ if and only if $b^{f(x)} = x$

Marginal value of the function $f(x)$ at $x = a$ The value of the derivative $f'(a)$, representing the approximate change in $f(x)$ for a 1 unit increase in the value of x at $x = a$

Mathematical model   A representation of a problem using mathematical symbols and relations

Mathematical modeling   The process of translating a problem statement into a mathematical representation

Method of Lagrange multipliers   A general procedure for optimizing a function $f(x, y)$ subject to a constraint $g(x, y) = 0$

Multivariable function   A function of more than one variable

Newton's method   A method, using derivatives, for approximating the root of an equation

Nonlinear function   A function that cannot be written as $f(x) = mx + b$

Objective function   The expression to be minimized or maximized in an optimization problem

Optimization problem   A problem in which the goal is to minimize or maximize some quantity

Order   The highest derivative of the unknown function in a differential equation

Partial derivative   The derivative of a multivariable function with respect to one independent variable, treating all other independent variables as constants

Polynomial   A function of the form $p(x) = a_n x^n + a_{n-1} x^{n-1} + \cdots + a_1 x + a_0$, where the a's are constants and n is a nonnegative integer

Probability   A function defined on events, taking values between 0 and 1, that represents the likelihood of an event

Probability density function   A function that indicates the probability that a random variable takes its value in a given interval, also simply called a density

Quadratic function   A function that has the form $f(x) = ax^2 + bx + c$

Radian   A measure of the extent of a circular arc; π radians $= 180$ degrees

Random variable   A numerical-valued function defined on a sample space

Range of a function   The set of possible values that the dependent variable may assume

Rate of change of a function   How fast the dependent variable changes as the value of the independent variable changes

Relative maximum of a function   A number $f(c)$ such that $f(c) \geq f(x)$ for all x in some open interval (a, b) in the domain of $f(x)$ containing c

Relative minimum of a function   A number $f(c)$ such that $f(c) \leq f(x)$ for all x in some open interval (a, b) in the domain of $f(x)$ containing c

Relative rate of change of $f(x)$ with respect to x   The ratio $f'(x)/f(x)$

Riemann sum of a function $f(x)$ An expression of the form $\sum_{j=1}^{n} f(x_j)\Delta x_j$

Sample space   The set of all possible outcomes of an experiment

Secant line   A line joining two points on the graph of a function

Separable differential equation   A differential equation of the following form: $g(y)\,(dy/dt) = f(t)$

Sequence   A list of numbers arranged in a definite sequential order; equivalently, a real-valued function whose domain is the set of positive integers

Tangent line to a curve at a point   A line that touches the curve at that point and approximates the curve well in a small region surrounding the point

Taylor series   An infinite series that sometimes can be used to approximate a given function

Term   An expression consisting only of products, quotients, or roots of variables and constants

Theorem   A mathematical truth that logically follows from a set of axioms; any proposition that can be proven from accepted assumptions and premises

Variable   A quantity that may assume a variety of values

Variance   A measure of the dispersion of a random variable about its expected value

Vertical asymptote   A line $x = a$ having the property that the y-coordinates of points on a graph become arbitrarily large in magnitude as the x-coordinates approach the value a

x-intercept   The x-coordinate of any point on a graph whose y-coordinate is 0

y-intercept   The y-coordinate of any point on a graph whose x-coordinate is 0

Values of e^x and e^{-x}

x	e^x	e^{-x}	x	e^x	e^{-x}	x	e^x	e^{-x}	x	e^x	e^{-x}
.00	1.00000	1.00000	.50	1.64872	.60653	1.00	2.71828	.36788	1.50	4.48169	.22313
.01	1.01005	.99005	.51	1.66529	.60050	1.01	2.74560	.36422	1.51	4.52673	.22091
.02	1.02020	.98020	.52	1.68203	.59452	1.02	2.77319	.36059	1.52	4.57223	.21871
.03	1.03045	.97045	.53	1.69893	.58860	1.03	2.80107	.35701	1.53	4.61818	.21654
.04	1.04081	.96079	.54	1.71601	.58275	1.04	2.82922	.35345	1.54	4.66459	.21438
.05	1.05127	.95123	.55	1.73325	.57695	1.05	2.85765	.34994	1.55	4.71147	.21225
.06	1.06184	.94176	.56	1.75067	.57121	1.06	2.88637	.34646	1.56	4.75882	.21014
.07	1.07251	.93239	.57	1.76827	.56553	1.07	2.91538	.34301	1.57	4.80665	.20805
.08	1.08329	.92312	.58	1.78604	.55990	1.08	2.94468	.33960	1.58	4.85496	.20598
.09	1.09417	.91393	.59	1.80399	.55433	1.09	2.97427	.33622	1.59	4.90375	.20393
.10	1.10517	.90484	.60	1.82212	.54881	1.10	3.00417	.33287	1.60	4.95303	.20190
.11	1.11628	.89583	.61	1.84043	.54335	1.11	3.03436	.32956	1.61	5.00281	.19989
.12	1.12750	.88692	.62	1.85893	.53794	1.12	3.06485	.32628	1.62	5.05309	.19790
.13	1.13883	.87810	.63	1.87761	.53259	1.13	3.09566	.32303	1.63	5.10387	.19593
.14	1.15027	.86936	.64	1.89648	.52729	1.14	3.12677	.31982	1.64	5.15517	.19398
.15	1.16183	.86071	.65	1.91554	.52205	1.15	3.15819	.31664	1.65	5.20698	.19205
.16	1.17351	.85214	.66	1.93479	.51685	1.16	3.18993	.31349	1.66	5.25931	.19014
.17	1.18530	.84366	.67	1.95424	.51171	1.17	3.22199	.31037	1.67	5.31217	.18825
.18	1.19722	.83527	.68	1.97388	.50662	1.18	3.25437	.30728	1.68	5.36556	.18637
.19	1.20925	.82696	.69	1.99372	.50158	1.19	3.28708	.30422	1.69	5.41948	.18452
.20	1.22140	.81873	.70	2.01375	.49659	1.20	3.32012	.30119	1.70	5.47395	.18268
.21	1.23368	.81058	.71	2.03399	.49164	1.21	3.35348	.29820	1.71	5.52896	.18087
.22	1.24608	.80252	.72	2.05443	.48675	1.22	3.38719	.29523	1.72	5.58453	.17907
.23	1.25860	.79453	.73	2.07508	.48191	1.23	3.42123	.29229	1.73	5.64065	.17728
.24	1.27125	.78663	.74	2.09594	.47711	1.24	3.45561	.28938	1.74	5.69734	.17552
.25	1.28403	.77880	.75	2.11700	.47237	1.25	3.49034	.28650	1.75	5.75460	.17377
.26	1.29693	.77105	.76	2.13828	.46767	1.26	3.52542	.28365	1.76	5.81244	.17204
.27	1.30996	.76338	.77	2.15977	.46301	1.27	3.56085	.28083	1.77	5.87085	.17033
.28	1.32313	.75578	.78	2.18147	.45841	1.28	3.59664	.27804	1.78	5.92986	.16864
.29	1.33643	.74826	.79	2.20340	.45384	1.29	3.63279	.27527	1.79	5.98945	.16696
.30	1.34986	.74082	.80	2.22554	.44933	1.30	3.66930	.27253	1.80	6.04965	.16530
.31	1.36343	.73345	.81	2.24791	.44486	1.31	3.70617	.26982	1.81	6.11045	.16365
.32	1.37713	.72615	.82	2.27050	.44043	1.32	3.74342	.26714	1.82	6.17186	.16203
.33	1.39097	.71892	.83	2.29332	.43605	1.33	3.78104	.26448	1.83	6.23389	.16041
.34	1.40495	.71177	.84	2.31637	.43171	1.34	3.81904	.26185	1.84	6.29654	.15882
.35	1.41907	.70469	.85	2.33965	.42741	1.35	3.85743	.25924	1.85	6.35982	.15724
.36	1.43333	.69768	.86	2.36316	.42316	1.36	3.89619	.25666	1.86	6.42374	.15567
.37	1.44773	.69073	.87	2.38691	.41895	1.37	3.93535	.25411	1.87	6.48830	.15412
.38	1.46228	.68386	.88	2.41090	.41478	1.38	3.97490	.25158	1.88	6.55350	.15259
.39	1.47698	.67706	.89	2.43513	.41066	1.39	4.01485	.24908	1.89	6.61937	.15107
.40	1.49182	.67032	.90	2.45960	.40657	1.40	4.05520	.24660	1.90	6.68589	.14957
.41	1.50682	.66365	.91	2.48432	.40252	1.41	4.09596	.24414	1.91	6.75309	.14808
.42	1.52196	.65705	.92	2.50929	.39852	1.42	4.13712	.24171	1.92	6.82096	.14661
.43	1.53726	.65051	.93	2.53451	.39455	1.43	4.17870	.23931	1.93	6.88951	.14515
.44	1.55271	.64404	.94	2.55998	.39063	1.44	4.22070	.23693	1.94	6.95875	.14370
.45	1.56831	.63763	.95	2.58571	.38674	1.45	4.26311	.23457	1.95	7.02869	.14227
.46	1.58407	.63128	.96	2.61170	.38289	1.46	4.30596	.23224	1.96	7.09933	.14086
.47	1.59999	.62500	.97	2.63794	.37908	1.47	4.34924	.22993	1.97	7.17068	.13946
.48	1.61607	.61878	.98	2.66446	.37531	1.48	4.39295	.22764	1.98	7.24274	.13807
.49	1.63232	.61263	.99	2.69123	.37158	1.49	4.43710	.22537	1.99	7.31553	.13670

Values of e^x and e^{-x} (continued)

x	e^x	e^{-x}	x	e^x	e^{-x}	x	e^x	e^{-x}	x	e^x	e^{-x}
2.00	7.38906	.13534	**2.50**	12.18249	.08208	**3.00**	20.08554	.04979	**3.50**	33.11545	.03020
2.01	7.46332	.13399	2.51	12.30493	.08127	3.01	20.28740	.04929	3.51	33.44827	.02990
2.02	7.53832	.13266	2.52	12.42860	.08046	3.02	20.49129	.04880	3.52	33.78443	.02960
2.03	7.61409	.13134	2.53	12.55351	.07966	3.03	20.69723	.04832	3.53	34.12397	.02930
2.04	7.69061	.13003	2.54	12.67967	.07887	3.04	20.90524	.04783	3.54	34.46692	.02901
2.05	7.76790	.12873	2.55	12.80710	.07808	3.05	21.11534	.04736	3.55	34.81332	.02872
2.06	7.84597	.12745	2.56	12.93582	.07730	3.06	21.32756	.04689	3.56	35.16320	.02844
2.07	7.92482	.12619	2.57	13.06582	.07654	3.07	21.54190	.04642	3.57	35.51659	.02816
2.08	8.00447	.12493	2.58	13.19714	.07577	3.08	21.75840	.04596	3.58	35.87354	.02788
2.09	8.08491	.12369	2.59	13.32977	.07502	3.09	21.97708	.04550	3.59	36.23408	.02760
2.10	8.16617	.12246	**2.60**	13.46374	.07427	**3.10**	22.19795	.04505	**3.60**	36.59823	.02732
2.11	8.24824	.12124	2.61	13.59905	.07353	3.11	22.42104	.04460	3.61	36.96605	.02705
2.12	8.33114	.12003	2.62	13.73572	.07280	3.12	22.64638	.04416	3.62	37.33757	.02678
2.13	8.41487	.11884	2.63	13.87377	.07208	3.13	22.87398	.04372	3.63	37.71282	.02652
2.14	8.49944	.11765	2.64	14.01320	.07136	3.14	23.10387	.04328	3.64	38.09184	.02625
2.15	8.58486	.11648	2.65	14.15404	.07065	3.15	23.33606	.04285	3.65	38.47467	.02599
2.16	8.67114	.11533	2.66	14.29629	.06995	3.16	23.57060	.04243	3.66	38.86134	.02573
2.17	8.75828	.11418	2.67	14.43997	.06925	3.17	23.80748	.04200	3.67	39.25191	.02548
2.18	8.84631	.11304	2.68	14.58509	.06856	3.18	24.04675	.04159	3.68	39.64639	.02522
2.19	8.93521	.11192	2.69	14.73168	.06788	3.19	24.28843	.04117	3.69	40.04485	.02497
2.20	9.02501	.11080	**2.70**	14.87973	.06721	**3.20**	24.53253	.04076	**3.70**	40.44730	.02472
2.21	9.11572	.10970	2.71	15.02928	.06654	3.21	24.77909	.04036	3.71	40.85381	.02448
2.22	9.20733	.10861	2.72	15.18032	.06587	3.22	25.02812	.03996	3.72	41.26439	.02423
2.23	9.29987	.10753	2.73	15.33289	.06522	3.23	25.27966	.03956	3.73	41.67911	.02399
2.24	9.39333	.10646	2.74	15.48698	.06457	3.24	25.53372	.03916	3.74	42.09799	.02375
2.25	9.48774	.10540	2.75	15.64263	.06393	3.25	25.79034	.03877	3.75	42.52108	.02352
2.26	9.58309	.10435	2.76	15.79984	.06329	3.26	26.04954	.03839	3.76	42.94843	.02328
2.27	9.67940	.10331	2.77	15.95863	.06266	3.27	26.31134	.03801	3.77	43.38006	.02305
2.28	9.77668	.10228	2.78	16.11902	.06204	3.28	26.57577	.03763	3.78	43.81604	.02282
2.29	9.87494	.10127	2.79	16.28102	.06142	3.29	26.84286	.03725	3.79	44.25640	.02260
2.30	9.97418	.10026	**2.80**	16.44465	.06081	**3.30**	27.11264	.03688	**3.80**	44.70118	.02237
2.31	10.07442	.09926	2.81	16.60992	.06020	3.31	27.38512	.03652	3.81	45.15044	.02215
2.32	10.17567	.09827	2.82	16.77685	.05961	3.32	27.66035	.03615	3.82	45.60421	.02193
2.33	10.27794	.09730	2.83	16.94546	.05901	3.33	27.93834	.03579	3.83	46.06254	.02171
2.34	10.38124	.09633	2.84	17.11577	.05843	3.34	28.21913	.03544	3.84	46.52547	.02149
2.35	10.48557	.09537	2.85	17.28778	.05784	3.35	28.50273	.03508	3.85	46.99306	.02128
2.36	10.59095	.09442	2.86	17.46153	.05727	3.36	28.78919	.03474	3.86	47.46535	.02107
2.37	10.69739	.09348	2.87	17.63702	.05670	3.37	29.07853	.03439	3.87	47.94238	.02086
2.38	10.80490	.09255	2.88	17.81427	.05613	3.38	29.37077	.03405	3.88	48.42421	.02065
2.39	10.91349	.09163	2.89	17.99331	.05558	3.39	29.66595	.03371	3.89	48.91089	.02045
2.40	11.02318	.09072	**2.90**	18.17414	.05502	**3.40**	29.96410	.03337	**3.90**	49.40245	.02024
2.41	11.13396	.08982	2.91	18.35680	.05448	3.41	30.26524	.03304	3.91	49.89895	.02004
2.42	11.24586	.08892	2.92	18.54129	.05393	3.42	30.56941	.03271	3.92	50.40044	.01984
2.43	11.35888	.08804	2.93	18.72763	.05340	3.43	30.87664	.03239	3.93	50.90698	.01964
2.44	11.47304	.08716	2.94	18.91585	.05287	3.44	31.18696	.03206	3.94	51.41860	.01945
2.45	11.58835	.08629	2.95	19.10595	.05234	3.45	31.50039	.03175	3.95	51.93537	.01925
2.46	11.70481	.08543	2.96	19.29797	.05182	3.46	31.81698	.03143	3.96	52.45732	.01906
2.47	11.82245	.08458	2.97	19.49192	.05130	3.47	32.13674	.03112	3.97	52.98453	.01887
2.48	11.94126	.08374	2.98	19.68782	.05079	3.48	32.45972	.03081	3.98	53.51703	.01869
2.49	12.06128	.08291	2.99	19.88568	.05029	3.49	32.78595	.03050	3.99	54.05489	.01850

Values of e^x and e^{-x} (continued)

x	e^x	e^{-x}	x	e^x	e^{-x}	x	e^x	e^{-x}	x	e^x	e^{-x}
4.00	54.59815	.01832	**4.50**	90.01713	.01111	**5.00**	148.41316	.00674	**10.00**	22026.46313	.00005
4.01	55.14687	.01813	4.51	90.92182	.01100	5.10	164.02190	.00610	10.10	24343.00708	.00004
4.02	55.70110	.01795	4.52	91.83560	.01089	5.20	181.27224	.00552	10.20	26903.18408	.00004
4.03	56.26091	.01777	4.53	92.75856	.01078	5.30	200.33680	.00499	10.30	29732.61743	.00003
4.04	56.82634	.01760	4.54	93.69080	.01067	5.40	221.40641	.00452	10.40	32859.62500	.00003
4.05	57.39745	.01742	4.55	94.63240	.01057	5.50	244.69192	.00409	10.50	36315.49854	.00003
4.06	57.97431	.01725	4.56	95.58347	.01046	5.60	270.42640	.00370	10.60	40134.83350	.00002
4.07	58.55696	.01708	4.57	96.54411	.01036	5.70	298.86740	.00335	10.70	44355.85205	.00002
4.08	59.14547	.01691	4.58	97.51439	.01025	5.80	330.29955	.00303	10.80	49020.79883	.00002
4.09	59.73989	.01674	4.59	98.49443	.01015	5.90	365.03746	.00274	10.90	54176.36230	.00002
4.10	60.34029	.01657	**4.60**	99.48431	.01005	**6.00**	403.42877	.00248	**11.00**	59874.13477	.00002
4.11	60.94671	.01641	4.61	100.48415	.00995	6.10	445.85775	.00224	11.10	66171.15430	.00002
4.12	61.55924	.01624	4.62	101.49403	.00985	6.20	492.74903	.00203	11.20	73130.43652	.00001
4.13	62.17792	.01608	4.63	102.51406	.00975	6.30	544.57188	.00184	11.30	80821.63379	.00001
4.14	62.80282	.01592	4.64	103.54435	.00966	6.40	601.84502	.00166	11.40	89321.72168	.00001
4.15	63.43400	.01576	4.65	104.58498	.00956	6.50	665.14159	.00150	11.50	98715.75879	.00001
4.16	64.07152	.01561	4.66	105.63608	.00947	6.60	735.09516	.00136	11.60	109097.78906	.00001
4.17	64.71545	.01545	4.67	106.69774	.00937	6.70	812.40582	.00123	11.70	120571.70605	.00001
4.18	65.36585	.01530	4.68	107.77007	.00928	6.80	897.84725	.00111	11.80	133252.34570	.00001
4.19	66.02279	.01515	4.69	108.85318	.00919	6.90	992.27469	.00101	11.90	147266.62109	.00001
4.20	66.68633	.01500	**4.70**	109.94717	.00910	**7.00**	1096.63309	.00091			
4.21	67.35654	.01485	4.71	111.05216	.00900	7.10	1211.96703	.00083			
4.22	68.03348	.01470	4.72	112.16825	.00892	7.20	1339.43076	.00075			
4.23	68.71723	.01455	4.73	113.29556	.00883	7.30	1480.29985	.00068			
4.24	69.40785	.01441	4.74	114.43420	.00874	7.40	1635.98439	.00061			
4.25	70.10541	.01426	4.75	115.58428	.00865	7.50	1808.04231	.00055			
4.26	70.80998	.01412	4.76	116.74592	.00857	7.60	1998.19582	.00050			
4.27	71.52163	.01398	4.77	117.91924	.00848	7.70	2208.34796	.00045			
4.28	72.24044	.01384	4.78	119.10435	.00840	7.80	2440.60187	.00041			
4.29	72.96647	.01370	4.79	120.30136	.00831	7.90	2697.28226	.00037			
4.30	73.69979	.01357	**4.80**	121.51041	.00823	**8.00**	2980.95779	.00034			
4.31	74.44049	.01343	4.81	122.73161	.00815	8.10	3294.46777	.00030			
4.32	75.18863	.01330	4.82	123.96509	.00807	8.20	3640.95004	.00027			
4.33	75.94429	.01317	4.83	125.21096	.00799	8.30	4023.87219	.00025			
4.34	76.70754	.01304	4.84	126.46935	.00791	8.40	4447.06665	.00022			
4.35	77.47846	.01291	4.85	127.74039	.00783	8.50	4914.76886	.00020			
4.36	78.25713	.01278	4.86	129.02420	.00775	8.60	5431.65906	.00018			
4.37	79.04363	.01265	4.87	130.32091	.00767	8.70	6002.91180	.00017			
4.38	79.83803	.01253	4.88	131.63066	.00760	8.80	6634.24371	.00015			
4.39	80.64042	.01240	4.89	132.95357	.00752	8.90	7331.97339	.00014			
4.40	81.45087	.01228	**4.90**	134.28978	.00745	**9.00**	8103.08295	.00012			
4.41	82.26946	.01216	4.91	135.63941	.00737	9.10	8955.29187	.00011			
4.42	83.09628	.01203	4.92	137.00261	.00730	9.20	9897.12830	.00010			
4.43	83.93141	.01191	4.93	138.37951	.00723	9.30	10938.01868	.00009			
4.44	84.77494	.01180	4.94	139.77024	.00715	9.40	12088.38049	.00008			
4.45	85.62694	.01168	4.95	141.17496	.00708	9.50	13359.72522	.00007			
4.46	86.48751	.01156	4.96	142.59379	.00701	9.60	14764.78015	.00007			
4.47	87.35672	.01145	4.97	144.02688	.00694	9.70	16317.60608	.00006			
4.48	88.23467	.01133	4.98	145.47438	.00687	9.80	18033.74414	.00006			
4.49	89.12144	.01122	4.99	146.93642	.00681	9.90	19930.36987	.00005			

Natural logarithms, ln x

x		0	1	2	3	4	5	6	7	8	9
1.0	0.0	0000	0995	1980	2956	3922	4879	5827	6766	7696	8618
1.1		9531	*0436	*1333	*2222	*3103	*3976	*4842	*5700	*6551	*7395
1.2	0.1	8232	9062	9885	*0701	*1511	*2314	*3111	*3902	*4686	*5464
1.3	0.2	6236	7003	7763	8518	9267	*0010	*0748	*1481	*2208	*2930
1.4	0.3	3647	4359	5066	5767	6464	7156	7844	8526	9204	9878
1.5	0.4	0547	1211	1871	2527	3178	3825	4469	5108	5742	6373
1.6		7000	7623	8243	8858	9470	*0078	*0682	*1282	*1879	*2473
1.7	0.5	3063	3649	4232	4812	5389	5962	6531	7098	7661	8222
1.8		8779	9333	9884	*0432	*0977	*1519	*2058	*2594	*3127	*3658
1.9	0.6	4185	4710	5233	5752	6269	6783	7294	7803	8310	8813
2.0		9315	9813	*0310	*0804	*1295	*1784	*2271	*2755	*3237	*3716
2.1	0.7	4194	4669	5142	5612	6081	6547	7011	7473	7932	8390
2.2		8846	9299	9751	*0200	*0648	*1093	*1536	*1978	*2418	*2855
2.3	0.8	3291	3725	4157	4587	5015	5442	5866	6289	6710	7129
2.4		7547	7963	8377	8789	9200	9609	*0016	*0422	*0826	*1228
2.5	0.9	1629	2028	2426	2822	3216	3609	4001	4391	4779	5166
2.6		5551	5935	6317	6698	7078	7456	7833	8208	8582	8954
2.7		9325	9695	*0063	*0430	*0796	*1160	*1523	*1885	*2245	*2604
2.8	1.0	2962	3318	3674	4028	4380	4732	5082	5431	5779	6126
2.9		6471	6815	7158	7500	7841	8181	8519	8856	9192	9527
3.0		9861	*0194	*0526	*0856	*1186	*1514	*1841	*2168	*2493	*2817
3.1	1.1	3140	3462	3783	4103	4422	4740	5057	5373	5688	6002
3.2		6315	6627	6938	7248	7557	7865	8173	8479	8784	9089
3.3		9392	9695	9996	*0297	*0597	*0896	*1194	*1491	*1788	*2083
3.4	1.2	2378	2671	2964	3256	3547	3837	4127	4415	4703	4990
3.5		5276	5562	5846	6130	6413	6695	6976	7257	7536	7815
3.6		8093	8371	8647	8923	9198	9473	9746	*0019	*0291	*0563
3.7	1.3	0833	1103	1372	1641	1909	2176	2442	2708	2972	3237
3.8		3500	3763	4025	4286	4547	4807	5067	5325	5584	5841
3.9		6098	6354	6609	6864	7118	7372	7624	7877	8128	8379
4.0		8629	8879	9128	9377	9624	9872	*0118	*0364	*0610	*0854
4.1	1.4	1099	1342	1585	1828	2070	2311	2552	2792	3031	3270
4.2		3508	3746	3984	4220	4456	4692	4927	5161	5395	5629
4.3		5862	6094	6326	6557	6787	7018	7247	7476	7705	7933
4.4		8160	8367	8614	8840	9065	9290	9515	9739	9962	*0185
4.5	1.5	0408	0630	0851	1072	1293	1513	1732	1951	2170	2388
4.6		2606	2823	3039	3256	3471	3687	3902	4116	4330	4543
4.7		4756	4969	5181	5393	5604	5814	6025	6235	6444	6653
4.8		6862	7070	7277	7485	7691	7898	8104	8309	8515	8719
4.9		8924	9127	9331	9534	9737	9939	*0141	*0342	*0543	*0744
5.0	1.6	0944	1144	1343	1542	1741	1939	2137	2334	2531	2728
5.1		2924	3120	3315	3511	3705	3900	4094	4287	4481	4673
5.2		4866	5058	5250	5441	5632	5823	6013	6203	6393	6582
5.3		6771	6959	7147	7335	7523	7710	7896	8083	8269	8455
5.4		8640	8825	9010	9194	9378	9562	9745	9928	*0111	*0293

*The first two digits are those at the beginning of the next row.

Natural logarithms, ln x (continued)

x		0	1	2	3	4	5	6	7	8	9
5.5	1.7	0475	0656	0838	1019	1199	1380	1560	1740	1919	2098
5.6		2277	2455	2633	2811	2988	3166	3342	3519	3695	3871
5.7		4047	4222	4397	4572	4746	4920	5094	5267	5440	5613
5.8		5786	5958	6130	6302	6473	6644	6815	6985	7156	7326
5.9		7495	7665	7834	8002	8171	8339	8507	8675	8842	9009
6.0	1.7	9176	9342	9509	9675	9840	*0006	*0171	*0336	*0500	*0665
6.1	1.8	0829	0993	1156	1319	1482	1645	1808	1970	2132	2294
6.2		2455	2616	2777	2938	3098	3258	3418	3578	3737	3896
6.3		4055	4214	4372	4530	4688	4845	5003	5160	5317	5473
6.4		5630	5786	5942	6097	6253	6408	6563	6718	6872	7026
6.5		7180	7334	7487	7641	7794	7947	8099	8251	8403	8555
6.6		8707	8858	9010	9160	9311	9462	9612	9762	9912	*0061
6.7	1.9	0211	0360	0509	0658	0806	0954	1102	1250	1398	1545
6.8		1692	1839	1986	2132	2279	2425	2571	2716	2862	3007
6.9		3152	3297	3442	3586	3730	3874	4018	4162	4305	4448
7.0		4591	4734	4876	5019	5161	5303	5445	5586	5727	5869
7.1		6009	6150	6291	6431	6571	6711	6851	6991	7130	7269
7.2		7408	7547	7685	7824	7962	8100	8238	8376	8513	8650
7.3		8787	8924	9061	9198	9334	9470	9606	9742	9877	*0013
7.4	2.0	0148	0283	0418	0553	0687	0821	0956	1089	1223	1357
7.5		1490	1624	1757	1890	2022	2155	2287	2419	2551	2683
7.6		2815	2946	3078	3209	3340	3471	3601	3732	3862	3992
7.7		4122	4252	4381	4511	4640	4769	4898	5027	5156	5284
7.8		5412	5540	5668	5796	5924	6051	6179	6306	6433	6560
7.9		6686	6813	6939	7065	7191	7317	7443	7568	7694	7819
8.0		7944	8069	8194	8318	8443	8567	8691	8815	8939	9063
8.1		9186	9310	9433	9556	9679	9802	9924	*0047	*0169	*0291
8.2	2.1	0413	0535	0657	0779	0900	1021	1142	1263	1384	1505
8.3		1626	1746	1866	1986	2106	2226	2346	2465	2585	2704
8.4		2823	2942	3061	3180	3298	3417	3535	3653	3771	3889
8.5		4007	4124	4242	4359	4476	4593	4710	4827	4943	5060
8.6		5176	5292	5409	5524	5640	5756	5871	5987	6102	6217
8.7		6332	6447	6562	6677	6791	6905	7020	7134	7248	7361
8.8		7475	7589	7702	7816	7929	8042	8155	8267	8380	8493
8.9		8605	8717	8830	8942	9054	9165	9277	9389	9500	9611
9.0		9722	9834	9944	*0055	*0166	*0276	*0387	*0497	*0607	*0717
9.1	2.2	0827	0937	1047	1157	1266	1375	1485	1594	1703	1812
9.2		1920	2029	2138	2246	2354	2462	2570	2678	2786	2894
9.3		3001	3109	3216	3324	3431	3538	3645	3751	3858	3965
9.4		4071	4177	4284	4390	4496	4601	4707	4813	4918	5024
9.5		5129	5234	5339	5444	5549	5654	5759	5863	5968	6072
9.6		6176	6280	6384	6488	6592	6696	6799	6903	7006	7109
9.7		7213	7316	7419	7521	7624	7727	7829	7932	8034	8136
9.8		8238	8340	8442	8544	8646	8747	8849	8950	9051	9152
9.9		9253	9354	9455	9556	9657	9757	9858	9958	*0058	*0158
10.0	2.3	0259	0358	0458	0558	0658	0757	0857	0956	1055	1154

The trigonometric functions (θ in radians)

θ	$\sin \theta$	$\cos \theta$	$\tan \theta$	$\cot \theta$	$\sec \theta$	$\csc \theta$	θ	$\sin \theta$	$\cos \theta$	$\tan \theta$	$\cot \theta$	$\sec \theta$	$\csc \theta$
.00	.0000	1.0000	.0000	. . . .	1.000	. . . .	.50	.4794	.8776	.5463	1.830	1.139	2.086
.01	.0100	1.0000	.0100	99.997	1.000	100.00	.51	.4882	.8727	.5594	1.788	1.146	2.048
.02	.0200	.9998	.0200	49.993	1.000	50.00	.52	.4969	.8678	.5726	1.747	1.152	2.013
.03	.0300	.9996	.0300	33.323	1.000	33.34	.53	.5055	.8628	.5859	1.707	1.159	1.978
.04	.0400	.9992	.0400	24.987	1.001	25.01	.54	.5141	.8577	.5994	1.668	1.166	1.945
.05	.0500	.9988	.0500	19.983	1.001	20.01	.55	.5227	.8525	.6131	1.631	1.173	1.913
.06	.0600	.9982	.0601	16.647	1.002	16.68	.56	.5312	.8473	.6269	1.595	1.180	1.883
.07	.0699	.9976	.0701	14.262	1.002	14.30	.57	.5396	.8419	.6310	1.560	1.188	1.853
.08	.0799	.9968	.0802	12.473	1.003	12.51	.58	.5480	.8365	.6552	1.526	1.196	1.825
.09	.0899	.9960	.0902	11.081	1.004	11.13	.59	.5564	.8309	.6696	1.494	1.203	1.797
.10	.0998	.9950	.1003	9.967	1.005	10.02	.60	.5646	.8253	.6841	1.462	1.212	1.771
.11	.1098	.9940	.1104	9.054	1.006	9.109	.61	.5729	.8196	.6989	1.431	1.220	1.746
.12	.1197	.9928	.1206	8.293	1.007	8.353	.62	.5810	.8139	.7139	1.401	1.229	1.721
.13	.1296	.9916	.1307	7.649	1.009	7.714	.63	.5891	.8080	.7291	1.372	1.238	1.697
.14	.1395	.9902	.1409	7.096	1.010	7.166	.64	.5972	.8021	.7445	1.343	1.247	1.674
.15	.1494	.9888	.1511	6.617	1.011	6.692	.65	.6052	.7961	.7602	1.315	1.256	1.652
.16	.1593	.9872	.1614	6.197	1.013	6.277	.66	.6131	.7900	.7761	1.288	1.266	1.631
.17	.1692	.9856	.1717	5.826	1.015	5.911	.67	.6210	.7838	.7923	1.262	1.276	1.610
.18	.1790	.9838	.1820	5.495	1.016	5.586	.68	.6288	.7776	.8087	1.237	1.286	1.590
.19	.1889	.9820	.1923	5.200	1.018	5.295	.69	.6365	.7712	.8253	1.212	1.297	1.571
.20	.1987	.9801	.2027	4.933	1.020	5.033	.70	.6442	.7648	.8423	1.187	1.307	1.552
.21	.2085	.9780	.2131	4.692	1.022	4.797	.71	.6518	.7584	.8595	1.163	1.319	1.534
.22	.2182	.9759	.2236	4.472	1.025	4.582	.72	.6594	.7518	.8771	1.140	1.330	1.517
.23	.2280	.9737	.2341	4.271	1.027	4.386	.73	.6669	.7452	.8949	1.117	1.342	1.500
.24	.2377	.9713	.2447	4.086	1.030	4.207	.74	.6743	.7358	.9131	1.095	1.354	1.483
.25	.2474	.9689	.2553	3.916	1.032	4.042	.75	.6816	.7317	.9316	1.073	1.367	1.467
.26	.2571	.9664	.2660	3.759	1.035	3.890	.76	.6889	.7248	.9505	1.052	1.380	1.452
.27	.2667	.9638	.2768	3.613	1.038	3.749	.77	.6961	.7179	.9697	1.031	1.393	1.437
.28	.2764	.9611	.2876	3.478	1.041	3.619	.78	.7033	.7109	.9893	1.011	1.407	1.422
.29	.2860	.9582	.2984	3.351	1.044	3.497	.79	.7104	.7038	1.009	.9908	1.421	1.408
.30	.2955	.9553	.3093	3.223	1.047	3.384	.80	.7174	.6967	1.030	.9712	1.435	1.394
.31	.3051	.9523	.3203	3.122	1.050	3.278	.81	.7243	.6895	1.050	.9520	1.450	1.381
.32	.3146	.9492	.3314	3.018	1.053	3.179	.82	.7311	.6822	1.072	.9331	1.466	1.368
.33	.3240	.9460	.3425	2.920	1.057	3.086	.83	.7379	.6749	1.093	.9146	1.482	1.355
.34	.3335	.9428	.3537	2.827	1.061	2.999	.84	.7446	.6675	1.116	.8964	1.498	1.343
.35	.3429	.9394	.3650	2.740	1.065	2.916	.85	.7513	.6600	1.138	.8785	1.515	1.331
.36	.3523	.9359	.3764	2.657	1.068	2.839	.86	.7578	.6524	1.162	.8609	1.533	1.320
.37	.3616	.9323	.3879	2.578	1.073	2.765	.87	.7643	.6448	1.185	.8437	1.551	1.308
.38	.3709	.9287	.3994	2.504	1.077	2.696	.88	.7707	.6372	1.210	.8267	1.569	1.297
.39	.3802	.9249	.4111	2.433	1.081	2.630	.89	.7771	.6294	1.235	.8100	1.589	1.287
.40	.3894	.9211	.4228	2.365	1.086	2.568	.90	.7833	.6216	1.260	.7936	1.609	1.277
.41	.3986	.9171	.4346	2.301	1.090	2.509	.91	.7895	.6137	1.286	.7774	1.629	1.267
.42	.4078	.9131	.4466	2.239	1.095	2.452	.92	.7956	.6058	1.313	.7615	1.651	1.257
.43	.4169	.9090	.4586	2.180	1.100	2.399	.93	.8016	.5978	1.341	.7458	1.673	1.247
.44	.4259	.9048	.4708	2.124	1.105	2.348	.94	.8076	.5898	1.369	.7303	1.696	1.238
.45	.4350	.9004	.4831	2.070	1.111	2.299	.95	.8134	.5817	1.398	.7151	1.719	1.229
.46	.4439	.8961	.4954	2.018	1.116	2.253	.96	.8192	.5735	1.428	.7001	1.744	1.221
.47	.4529	.8916	.5080	1.969	1.122	2.208	.97	.8249	.5653	1.459	.6853	1.769	1.212
.48	.4618	.8870	.5206	1.921	1.127	2.166	.98	.8305	.5570	1.491	.6707	1.795	1.204
.49	.4706	.8823	.5334	1.875	1.133	2.125	.99	.8360	.5487	1.524	.6563	1.823	1.196

The trigonometric functions (θ in radians) (continued)

θ	sin θ	cos θ	tan θ	cot θ	sec θ	csc θ	θ	sin θ	cos θ	tan θ	cot θ	sec θ	csc θ
1.00	.8415	.5403	1.557	.6421	1.851	1.118	1.30	.9636	.2675	3.602	.2776	3.738	1.038
1.01	.8468	.5319	1.592	.6281	1.880	1.181	1.31	.9662	.2579	3.747	.2669	3.878	1.035
1.02	.8521	.5234	1.628	.6142	1.911	1.174	1.32	.9687	.2482	3.903	.2562	4.029	1.032
1.03	.8573	.5148	1.665	.6005	1.942	1.166	1.33	.9711	.2385	4.072	.2456	4.193	1.030
1.04	.8624	.5062	1.704	.5870	1.975	1.160	1.34	.9735	.2288	4.256	.2350	4.372	1.027
1.05	.8674	.4976	1.743	.5736	2.010	1.153	1.35	.9757	.2190	4.455	.2245	4.566	1.025
1.06	.8724	.4889	1.784	.5604	2.046	1.146	1.36	.9779	.2092	4.673	.2140	4.779	1.023
1.07	.8772	.4801	1.827	.5473	2.083	1.140	1.37	.9799	.1994	4.913	.2035	5.014	1.021
1.08	.8820	.4713	1.871	.5344	2.122	1.134	1.38	.9819	.1896	5.177	.1931	5.273	1.018
1.09	.8866	.4625	1.917	.5216	2.162	1.128	1.39	.9837	.1798	5.471	.1828	5.561	1.017
1.10	.8912	.4536	1.965	.5090	2.205	1.122	1.40	.9854	.1700	5.798	.1725	5.883	1.015
1.11	.8957	.4447	2.014	.4964	2.249	1.116	1.41	.9871	.1601	6.165	.1622	6.246	1.013
1.12	.9001	.4357	2.066	.4840	2.295	1.111	1.42	.9887	.1502	6.581	.1519	6.657	1.011
1.13	.9044	.4267	2.120	.4718	2.344	1.106	1.43	.9901	.1403	7.055	.1417	7.126	1.010
1.14	.9086	.4176	2.176	.4596	2.395	1.101	1.44	.9915	.1304	7.602	.1315	7.667	1.009
1.15	.9128	.4085	2.234	.4475	2.448	1.096	1.45	.9927	.1205	8.238	.1214	8.299	1.007
1.16	.9168	.3993	2.296	.4356	2.504	1.091	1.46	.9939	.1106	8.989	.1113	9.044	1.006
1.17	.9208	.3902	2.360	.4237	2.563	1.086	1.47	.9949	.1006	9.887	.1011	9.938	1.005
1.18	.9246	.3809	2.427	.4120	2.625	1.082	1.48	.9959	.0907	10.983	.0910	11.029	1.004
1.19	.9284	.3717	2.498	.4003	2.691	1.077	1.49	.9967	.0807	12.350	.0810	12.390	1.003
1.20	.9320	.3624	2.572	.3888	2.760	1.073	1.50	.9975	.0707	14.101	.0709	14.137	1.003
1.21	.9356	.3530	2.650	.3773	2.833	1.069	1.51	.9982	.0608	16.428	.0609	16.458	1.002
1.22	.9391	.3436	2.733	.3659	2.910	1.065	1.52	.9987	.0508	19.670	.0508	19.965	1.001
1.23	.9425	.3342	2.820	.3546	2.992	1.061	1.53	.9992	.0408	24.498	.0408	24.519	1.001
1.24	.9458	.3248	2.912	.3434	3.079	1.057	1.54	.9995	.0308	32.461	.0308	32.476	1.000
1.25	.9490	.3153	3.010	.3323	3.171	1.054	1.55	.9998	.0208	48.078	.0208	48.089	1.000
1.26	.9521	.3058	3.113	.3212	3.270	1.050	1.56	.9999	.0108	92.620	.0108	92.626	1.000
1.27	.9551	.2963	3.224	.3102	3.375	1.047	1.57	1.0000	.0008	1255.8	.0008	1255.8	1.000
1.28	.9580	.2867	3.341	.2993	3.488	1.044	1.58	1.0000	−.0092	−108.65	−.0092	−108.65	1.000
1.29	.9608	.2771	3.467	.2884	3.609	1.041	1.59	.9998	−.0192	−52.067	−.0192	−52.08	1.000
							1.60	.9996	−.0292	−34.233	−.0292	−34.25	1.000

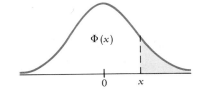

The cumulative normal distribution

x	.00	.01	.02	.03	.04	.05	.06	.07	.08	.09
.0	.5000	.5040	.5080	.5120	.5160	.5199	.5239	.5279	.5319	.5359
.1	.5398	.5438	.5478	.5517	.5557	.5596	.5636	.5675	.5714	.5753
.2	.5793	.5832	.5871	.5910	.5948	.5987	.6026	.6064	.6103	.6141
.3	.6179	.6217	.6255	.6293	.6331	.6368	.6406	.6443	.6480	.6517
.4	.6554	.6591	.6628	.6664	.6700	.6736	.6772	.6808	.6844	.6879
.5	.6915	.6950	.6985	.7019	.7054	.7088	.7123	.7157	.7190	.7224
.6	.7257	.7291	.7324	.7357	.7389	.7422	.7454	.7486	.7517	.7549
.7	.7580	.7611	.7642	.7673	.7704	.7734	.7764	.7794	.7823	.7852
.8	.7881	.7910	.7939	.7967	.7995	.8023	.8051	.8078	.8106	.8133
.9	.8159	.8186	.8212	.8238	.8264	.8289	.8315	.8340	.8365	.8389
1.0	.8413	.8438	.8461	.8485	.8508	.8531	.8554	.8577	.8599	.8621
1.1	.8643	.8665	.8686	.8708	.8729	.8749	.8770	.8790	.8810	.8830
1.2	.8849	.8869	.8888	.8907	.8925	.8944	.8962	.8980	.8997	.9015
1.3	.9032	.9049	.9066	.9082	.9099	.9115	.9131	.9147	.9162	.9177
1.4	.9192	.9207	.9222	.9236	.9251	.9265	.9279	.9292	.9306	.9319
1.5	.9332	.9345	.9357	.9370	.9382	.9394	.9406	.9418	.9429	.9441
1.6	.9452	.9463	.9474	.9484	.9495	.9505	.9515	.9525	.9535	.9545
1.7	.9554	.9564	.9573	.9582	.9591	.9599	.9608	.9616	.9625	.9633
1.8	.9641	.9649	.9656	.9664	.9671	.9678	.9686	.9693	.9699	.9706
1.9	.9713	.9719	.9726	.9732	.9738	.9744	.9750	.9756	.9761	.9767
2.0	.9772	.9778	.9783	.9788	.9793	.9798	.9803	.9808	.9812	.9817
2.1	.9821	.9826	.9830	.9834	.9838	.9842	.9846	.9850	.9854	.9857
2.2	.9861	.9864	.9868	.9871	.9875	.9878	.9881	.9884	.9887	.9890
2.3	.9893	.9896	.9898	.9901	.9904	.9906	.9909	.9911	.9913	.9916
2.4	.9918	.9920	.9920	.9925	.9927	.9929	.9931	.9932	.9934	.9936
2.5	.9938	.9940	.9941	.9943	.9945	.9946	.9948	.9949	.9951	.9952
2.6	.9953	.9955	.9956	.9957	.9959	.9960	.9961	.9962	.9963	.9964
2.7	.9965	.9966	.9967	.9968	.9969	.9970	.9971	.9972	.9973	.9974
2.8	.9974	.9975	.9976	.9977	.9977	.9978	.9979	.9979	.9980	.9981
2.9	.9981	.9982	.9982	.9983	.9984	.9984	.9985	.9985	.9986	.9986
3.0	.9987	.9987	.9987	.9988	.9988	.9989	.9989	.9989	.9990	.9990
3.1	.9990	.9991	.9991	.9991	.9992	.9992	.9992	.9992	.9993	.9993
3.2	.9993	.9993	.9994	.9994	.9994	.9994	.9994	.9995	.9995	.9995
3.3	.9995	.9995	.9996	.9996	.9996	.9996	.9996	.9996	.9996	.9997
3.4	.9997	.9997	.9997	.9997	.9997	.9997	.9997	.9997	.9997	.9998

CHAPTER 1   SECTION 1-1

1. $M = $ Mike's age, $T = $ Tom's age; $M = T + 11$ **3.** $t = x/40$ **5.** $C = 25q + 5n + p$

7. $B = $ Bob's age, $M = $ Mary's age; $B = 2M + 3$ **9.** $t = \dfrac{x}{55} + \dfrac{y}{45}$ **11.** 45

13. **a.** $592 = (N - 6)(C + 10)$, where N is number purchased and C is cost of each   **b.** $R = (N - 6)\left(\dfrac{480}{N} + 10\right)$

SECTION 1-2

1. $R = 10x + 1000$ **3.** $P = 0.5x - 12{,}000$ **5.** $\dfrac{28.6x}{y} + \dfrac{44x}{500} + 35$ **7.** $A = s^2 + s(s + 30) = 2s^2 + 30s$

9. $S = 720 - 4x^2$ **11.** $S = 2s^2 + \dfrac{32}{s}$

SECTION 1-3

1. **a.** $f(0) = -5$ **b.** $f(6) = 7$ **c.** $f(a + b) = 2a + 2b - 5$ **d.** $f(z^3) = 2z^3 - 5$
3. $f(0) = 2, f(4) = 14, f(10) = -96$ **5.** Domain = all real numbers, Range = all nonnegative real numbers
7. Domain = all nonzero real numbers = Range
9. Domain = all real numbers greater than or equal to -4, Range = all nonnegative real numbers
11. Domain = all real numbers except 0 and 2, Range = all nonzero real numbers
13. Taking the first member of each pair as the independent variable and the second member as the dependent variable, only b is not a function since some names may have more than one number. Taking the second member as the independent variable and the first member as the dependent variable, only d and e are functions.
15. $F(9) = 5000(1 + 0.27) = 6350$ dollars   **17.** IQ $= 100a/12$
19. **a.** $C(x) = 90{,}000 + 250x$ **b.** $C(100) = \$115{,}000, C(1000) = \$340{,}000, C(5000) = \$1{,}340{,}000$
c. $f(x) = C(x)/x = (90{,}000/x) + 250$ **d.** $f(100) = \$1{,}150, f(1000) = \$340, f(5000) = \$268$
21. Number of possible responses = Independent variable, Reaction time = Dependent variable, Domain = {2, 3, 4, 5, 6}, Range = {0.35, 0.40, 0.45, 0.50, 0.60}
23. **a.** $C(x) = 1098.92 + 0.58(x - 1000), x \geq 1000$ **b.** $C(1757) = \$1537.98$

SECTION 1-4

1.

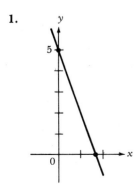

3.

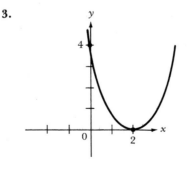

5.

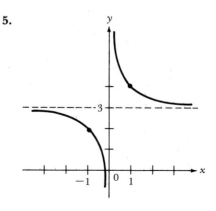

7.

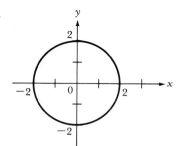

9. b.

Distance	Cost
500	15
1,000	18
1,500	20
3,000	25

11. Let x = number of grams of potatoes and y = number of grams of soybeans; $300 = 0.19x + 0.35y$

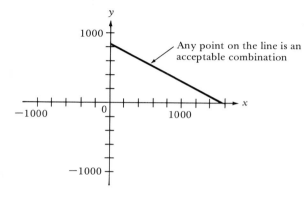

13. a. $t = c/I$ **b.**

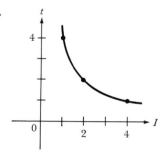

15.

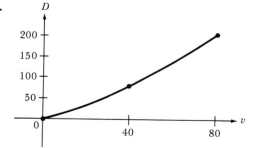

17.

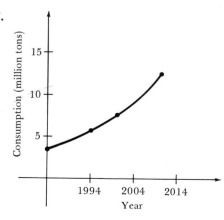

19. **a.–c.**

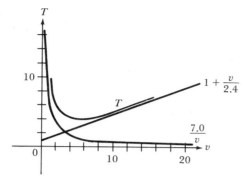

d. Minimum time of about 4.35 seconds at about $v = 4$
21. **a.** Continuous **b.** Not continuous **c.** Not continuous

SECTION 1-5

1. $m = 3$ **3.** $m = \frac{1}{3}$ **5.** $y = 2x + 5$ **7.** $y = \frac{1}{2}x + 3$ **9.** $y = -\frac{2}{3}x + 7$ **11.** $y = -\frac{1}{10}x + \frac{8}{5}$ **13.** $y = 4$
15. $y = 0.09x + 6.75$ **17.** $y = 4.5x - 0.01x^2$
19. $E = \frac{16}{3}c + 68$, where E = energy expenditure (calories per minute) and c = work rate (cartons per minute)
21. $r(v) = 0.03v - 0.12$ **23.** $S(t) = 1200t + 21,600$ **25.** $V(t) = -1500t + 20,000$

REVIEW PROBLEMS (SECTION 1-7)

1. J = Jane's age, B = Bill's age; $J = B - 3$
3. N = number of bicycles purchased, C = cost per bicycle; Revenue = $(N - 2)(C + 35)$
5. $A = lw = (3w + 50)w$
7. **a.** Domain: $0 \le x \le 10$, Range: $0 \le p \le 30$ **b.** $p(0) = 0$, $p(5) = 15$, $p(10) = 30$
c. Percent mutation increases by 3% for each unit increase in dosage.
9. y is a function of x **11.** y is not a function of x **13.** Estimate for 1990 ≈ 606 million cubic feet

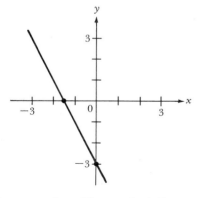

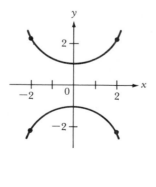

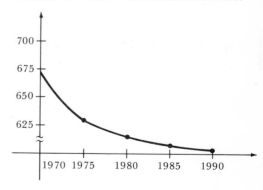

15. $m = -3$ **17.** $y = \frac{1}{6}x + \frac{31}{6}$ **19.** $C(x) = 0.2x + 1.8$ **21.** $x = 10 + y$, $TR(x) = 17,000x - 500x^2$
23. $f(t) = 210t$, $0 \le t \le 25$

CHAPTER 2   SECTION 2·1

1. $\lim_{x\to 4} f(x) = 7$ **3.** $\lim_{x\to 1} f(x) = 2$ **5.** $\lim_{x\to -3} f(x)$ does not exist   **7.** $\lim_{x\to 6} f(x) = 5$ **9.** $\lim_{x\to 4} f(x) = 16$

11. $\lim_{x\to 2} f(x)$ does not exist   **13.** $\lim_{x\to 2} f(x)$ exists   **15.** $\lim_{x\to 2} f(x)$ exists

17. As x approaches 3 from the left, $f(x)$ approaches 9. As x approaches 3 from the right, $f(x)$ approaches 8.

19. $\lim_{y\to 1} g(y) = 2$ **21.** $\lim_{x\to 0} f(x)$ does not exist   **23.** $\lim_{a\to 4} (a^2 - 16) = 0$ **25.** $\lim_{x\to -7} (3 - 7x) = 52$

27. $\lim_{t\to -2} \dfrac{t^2 + 4}{t^3 + 7} = -8$ **29.** $\lim_{x\to -1} 7 = 7$ **31.** $\lim_{x\to 0} \dfrac{\sqrt{1 + x^2}}{x + 1} = 1$ **33.** $\lim_{h\to 0} \dfrac{\sqrt{x^2 + h}}{5} = \dfrac{|x|}{5}$

SECTION 2·2

1. $m = 4; y = 4x$ **3.** $m = -2; y = -2x - 1$ **5.** $m = \frac{2}{3}; 3y = 2x + 10$ **7.** $m = \frac{3}{5}; y = -\frac{5}{3}x + 2$

9. $m = 2; y = -\frac{1}{2}x - 4$ **11.** $m = -\frac{3}{2}; -2x + 3y = 1$

13. A line with positive slope rises from left to right; one with negative slope falls.

15. $y = -3$ **17.** $3x - 7y = 15$ **19.** $4x + 5y = 29$ **21.** $y = -3x$

23. $2x + 3y = 100$; a 1 unit increase in x gives a $\frac{2}{3}$ unit *decrease* in y

SECTION 2·3

1. $\dfrac{f(8) - f(7)}{8 - 7} = 5; \dfrac{f(12) - f(8)}{12 - 8} = \dfrac{7}{4}; \dfrac{f(12) - f(7)}{12 - 7} = \dfrac{12}{5}$

3. $\dfrac{f(0) - f(-3)}{0 - (-3)} = 5; \dfrac{f(4) - f(0)}{4 - 0} = 5$

5. $\dfrac{E(100) - E(50)}{100 - 50} = \dfrac{20 - 50}{50} = -\dfrac{3}{5} = -0.6; \dfrac{E(80) - E(50)}{80 - 50} \approx \dfrac{28 - 50}{30} = -\dfrac{22}{30} \approx -0.73; \dfrac{E(60) - E(50)}{60 - 50} = \dfrac{40 - 50}{10} = -1$

7.

	Average Rate of Change (Thousands per Year)	
	1947 – 1968	*1968 – 1980*
Agriculture, forestry, and fisheries	-178.1	-80.8
Mining	-15.0	-4.2
Transportation and utilities	16.2	41.7
Finance, insurance, and real estate	72.6	113.8
Services	473.8	500.0

9. Percentage increase $= f(t) = \frac{43}{11}t$, where $t =$ number of years past 1956; rate of increase is $\frac{43}{11}\%$ per year; $f(29) \approx 213.36\%$

11. Percentage increase $= p(t) = 6t$, where $t =$ number of years past 1980; rate of change is 6% per year; $p(7) = 42\%$

13. $\dfrac{f(1.1) - f(1)}{1.1 - 1} = 2.1; \dfrac{f(1.01) - f(1)}{1.01 - 1} = 2.01; \dfrac{f(1.001) - f(1)}{1.001 - 1} = 2.001$

SECTION 2-4

1. $f'(-1) = 7$ **3.** $f'(0) = \frac{1}{2}$ **5.** $f'(1) = 6$ **7.** $f'(1) = 4$ **9.** $f'(3) = -\frac{2}{9}$ **11.** $f'(2) = -\frac{3}{4}$ **13.** $y = 6$
15. $y = 2x$ **17.** $y = -4x + 6$ **19.** $y = -6x + 9$ **21.** $y = -x$ **23.** $8y + x = 18$ **25.** $3y = -x + 7$
27. $2y = 4x - 7$

SECTION 2-5

1. $g'(x) = 0$ **3.** $g'(x) = -2$ **5.** $f'(x) = 2x$ **7.** $g'(x) = 6x$ **9.** $f'(x) = -\dfrac{1}{x^2}$ **11.** $f'(x) = 2x + 3$

13. $g'(x) = \dfrac{1}{2\sqrt{x}}$ **15.** $f'(x) = 3x^2$

17. **a.** Three points where sharp "corners" exist   **b.** Point of discontinuity and sharp corner point
19. $S'(r) = 8\pi r$ **21.** $\dfrac{dv}{dt} = 32$ **23.** $f'(t) = \dfrac{5}{2\sqrt{t}}$ **25.** $C'(x) = 60$ **27.** $\dfrac{dR}{dt} = 6 + 0.4t$

SECTION 2-6

1. $y' = 0$ **3.** $f'(x) = 5$ **5.** $y' = 12x^3$ **7.** $f'(x) = 3x^5$ **9.** $f'(x) = 9x^2 + 4$ **11.** $y' = 3x^2 + 12x - 9$
13. $f'(x) = 5x^4 + 12x^3 - 4x$ **15.** $f'(x) = \frac{1}{2}x^{-1/2}$ **17.** $f'(x) = 14x + \dfrac{1}{x^2}$ **19.** $f'(x) = 16x - 2x^{-3}$
21. $f'(x) = \frac{7}{2}x^{-1/2} - 2$ **23.** $g'(x) = -\frac{8}{3}x^{-5/3}$ **25.** $f'(x) = 2x^{-1/3} - \frac{1}{5}x^{-4/5}$ **27.** $f'(1) = 5$ **29.** $y'(2) = 15$
31. $f'(8) = \frac{19}{192}$ **33.** $y = 5x - 3$ **35.** $y = 7x - 7$ **37.** $3y = 4x - 7$

39. $g'(x) = \begin{cases} 1 & \text{if } x < 1 \\ 2x - 2 & \text{if } x > 1 \end{cases}$; $g'(1)$ does not exist

41. $\dfrac{dI}{dr} = -\dfrac{E}{2\pi} r^{-3}$ **43.** $\dfrac{dA}{dP} = -P^{-1.25}$ **45.** $C'(x) = 10x - \dfrac{200}{x^2}$; $C'(10) = 98$

SECTION 2-7

1. Rate of change of revenue with respect to production
3. Rate of change of demand with respect to supply
5. **a.** $C(50) = \$1650$ **b.** $C'(50) = \$18$ **c.** $C(51) - C(50) = \$18.02$ **d.** $C'(60) = \$18.40$
7. **a.** $\dfrac{N(10) - N(0)}{10 - 0} = -700{,}000$; $\dfrac{N(20) - N(0)}{20 - 0} = -600{,}000$ **b.** $N'(10) = -600{,}000$; $N'(20) = -400{,}000$
c. Instantaneous rates of change are *limits* of average rates of change   **d.** $N(40) = 0$; $N'(40) = 0$
9. **a.** $C(25) = \$348$ **b.** $C'(25) = \$11.50$ **c.** $C'(100) = \$10.75$ **d.** Marginal cost seems to decrease
11. **a.** $R(10{,}000) = \$1{,}950{,}000$; $R(20{,}000) = \$7{,}900{,}000$ **b.** $R'(10{,}000) = \$395$; $R'(20{,}000) = \$795$
13. **a.** $S(3) = 102{,}391$; $S(5) = 103{,}975$ **b.** $S'(3) = 794$; $S'(5) = 790$ **c.** Yes, assuming the goal is increased sales
15. $V'(r) = 4\pi r^2$ **17.** $P'(r) = -Tr^{-2}$ **19.** $L'(400) = 0.000025$ inch per degree   **21.** $V'(2) = 36\pi$
23. $TC'(9) = \$1.85$; marginal profit $= TR'(9) - TC'(9) = -\$1.42$ per unit

REVIEW PROBLEMS (SECTION 2-9)

1. $\lim\limits_{x \to -1} f(x) = 4$ **3.** $\lim\limits_{x \to 1} f(x)$ does not exist   **5.** $\lim\limits_{x \to 5} 3x^2 + x - 1 = 79$ **7.** $\lim\limits_{x \to -1} \dfrac{1}{x} = -1$

9. $\lim\limits_{h \to 0} \dfrac{\sqrt{h+5}}{x} = \dfrac{\sqrt{5}}{x}$ **11.** $m = 2$; e.g., $y = 2x + 1$, $y = -\frac{1}{2}x - 4$ **13.** $m = -\frac{3}{5}$; e.g., $y = -\frac{3}{5}x$, $y = \frac{5}{3}x + 1$

15. $y = x$ **17.** $f'(5) = -\frac{4}{25}$ **19.** $f'(-2) = -10$ **21.** $f'(x) = 6x$ **23.** $h'(x) = -5x^{-2}$

25. $\dfrac{dv}{dt} = 32$ ft/sec^2 **27.** $f'(x) = -4x$ **29.** $y' = -4x^5$ **31.** $y' = 8x^3 - 15x^2 + 12x - 3$ **33.** $y' = \frac{1}{2}x^{-3/4}$

35. $g'(x) = 10x - 2x^{-2} + 2x^{-3}$ **37.** $g'(x) = \frac{6}{5}x^{-2/5} - 6x$ **39.** $y = 4x - 5$ **41.** $y - 17x + 60 = 0$

43. $M'(t) = 17 + 22t$; $M'(1) = 39$ **45.** $\dfrac{C(x) - C(0)}{x - 0} = \frac{1}{2}x + 1$; $C'(x) = x + 1$

47. a. $C(50) = \$2310$ **b.** $C'(x) = x + 20$ **c.** $C'(50) = \$70$ per unit   **d.** $C(51) - C(50) = \$70.50$

CHAPTER 2   SUPPLEMENT

1. $\frac{9}{17}$ **3.** 4   **5.** 6   **7.** ∞ **9.** 3   **11.** $\frac{1}{3}$ **13.** -5 **15.** Does not exist   **17.** Assign $f(9) = 3$
19. Continuous at all x **21.** Not continuous at $x = 1$ **23.** Continuous at all x **25.** Continuous at all x
27. Not continuous at $x = 2$ **29.** Continuous at all x **31.** Discontinuous at $x = 6$

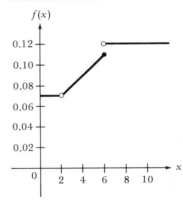

CHAPTER 3   SECTION 3-1

1. Absolute maximum at $x = a$; relative minimum at $x = c$; relative maximum at $x = d$; absolute minimum at $x = b$
3. Absolute minimum at $x = a$; absolute maximum at $x = d$
5. Absolute minimum at $x = c$; absolute maximum at $x = d$; relative minimum at $x = e$
7. $f(x)$ **9.** $f(x)$

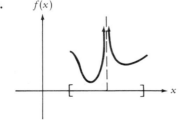

SECTION 3-2

1. $(2, -3)$ **3.** $(\frac{1}{7}, -\frac{83}{7})$ **5.** $(0, 0)$ **7.** $(5, -25)$ **9.** $(-1, 5), (3, -27)$ **11.** $(2, \frac{14}{3}), (3, \frac{9}{2})$
13. $(3, 6), (5, 7)$ **15.** $(0, 0), (2, -4), (4, -1), (3, 0)$ **17.** $(0, 0)$ **19.** Yes
21. Only horizontal lines; every point is a critical point on a horizontal line
23. $(1, 0)$; relative (in fact, absolute) maximum point   **25.** $(0, 0), (8, 4)$

SECTION 3-3

1. Critical points at $x = a$, $x = b$, and $x = c$; increasing on $(-\infty, a)$, decreasing on (a, b), increasing on (b, c), decreasing on (c, d); decreasing on (d, ∞); relative maxima at $x = a$ and $x = c$, relative minimum at $x = b$
3. Relative minimum point: $(2, -3)$ **5.** No optima
7. Relative maximum point: $(-1, 5)$; relative minimum point: $(3, -27)$
9. Relative maximum point: $(2, \frac{14}{3})$; relative minimum point: $(3, \frac{9}{2})$
11. $f''(x) = 4 + 20x^3$ **13.** $f''(x) = 6x + 8$ **15.** $f''(x) = 6x + 12$ **17.** $f''(x) = \dfrac{6}{x^3}$ **19.** $f''(x) = -\frac{2}{3}x^{-4/3}$
21. a. Relative maximum   **b.** Neither   **c.** Relative minimum   **d.** Neither
23. $(\frac{9}{2}, \frac{81}{4})$ is a relative maximum point   **25.** $(3, 0)$ is a relative minimum point
27. $(-2, -4)$ is a relative maximum point; $(2, 4)$ is a relative minimum point
29. $(2, 0)$ is a relative maximum point   **31.** $(\frac{1}{4}, \frac{1}{8})$ is a relative maximum point; $(1, -1)$ is a relative minimum point
33. $f(x) < 0 = f(0)$ for $x \neq 0$; therefore, $(0, 0)$ is an absolute maximum point
35. $f^{(5)}(x) = 0$

SECTION 3-4

1. $f(2) = 2$ is the absolute minimum   **3.** No absolute maximum or minimum
5. $f(0) = 0$ is the absolute minimum   **7.** No absolute maximum or minimum
9. No absolute maximum or minimum   **11.** No absolute maximum or minimum
13. $f(-10) = -58$ is the absolute minimum; $f(1) = 8$ is the absolute maximum
15. $f(0) = 1$ is the absolute minimum; $f(2) = 17$ is the absolute maximum
17. $f(-1) = 0$ is the absolute minimum; $f(2) = 9$ is the absolute maximum
19. $f(0) = 0$ is the absolute maximum; $f(3) = -27$ is the absolute minimum
21. $f(3) = \frac{28}{3}$ is the absolute maximum; $f(-1) = 0$ is the absolute minimum
23. $f(1) = 2$ is the absolute minimum; $f(4) = 4\frac{1}{4}$ is the absolute maximum
25. $f(\frac{1}{2}) = 6$ is the absolute minimum; $f(2) = 16\frac{1}{8}$ is the absolute maximum
27. $f(0) = 7$ is the absolute maximum; $f(3) = -560$ is the absolute minimum
29. $f(1) = 3$ is the absolute maximum; $f(4) = 0$ is the absolute minimum
31. Critical points and endpoints

SECTION 3-5

1. a. $x = 10$ **b.** $f(10) = 800$ **3.** $T(0.5) = 150$ seconds
5. 15, 15   **7.** First number $= 15$, second number $= 5$ **9.** $\frac{1}{2}$ **11.** A 10×10 square
13. Radius $= \sqrt[3]{4} \approx 1.59$ inches; height $= 2r \approx 3.18$ inches   **15.** Length ≈ 15.95 feet; width ≈ 12.54 feet
17. Radius $= \sqrt[3]{100/4\pi} \approx 2.82$ inches; height ≈ 4 inches   **19.** A cube 2 feet on a side
21. Side ≈ 4.64 feet; height $= \dfrac{96}{(\text{Side})^2} \approx 4.46$ feet

SECTION 3-6

1. **a.** 3000   **b.** $C(3) = \$2$ **3.** Minimize $100C(x)/x$; $x \approx 8.77$ hundred pounds
5. Produce and sell 6000 containers   **7.** $4   **9.** About 82 cases   **11. a.** 800,000   **b.** $P(8) = \$440$
13. 42 mph   **15.** $x = 5$ hours   **17.** $350; $350 (in both cases a fee of $350 is optimal)

SECTION 3-7

1. Increasing on $(-\infty, \infty)$ **3.** Decreasing on $(-\infty, \frac{1}{4})$; increasing on $(\frac{1}{4}, \infty)$ **5.** Increasing on $(-\infty, \infty)$
7. Decreasing on $(-\infty, 0)$; increasing on $(0, \infty)$
9. Concave down on $(-\infty, -1)$; concave up on $(-1, \infty)$; $(-1, 5)$ is a point of inflection
11. Concave down on $(-\infty, 1)$; concave up on $(1, \infty)$; $(1, 2)$ is a point of inflection
13. Concave down on $(-\infty, 0)$; concave down on $(0, \infty)$
15. Concave down on $(-\infty, 3)$; concave up on $(3, \infty)$; $(3, 0)$ is a point of inflection
17. y-intercept: $f(0) = -10$; x-intercepts: $x = -2, 5$ **19.** y-intercept: $y = 0$; x-intercepts: $x = 0, -3, -6$
21. y-intercept: $y = 0$; x-intercepts: $x = 0, 2, 4$ **23.** y-intercept: $f(0) = -6$; x-intercepts: $x = 1, -2, -3$

25.

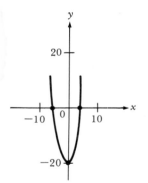

27.

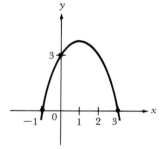

29.

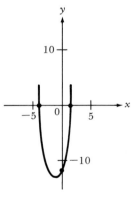

31.

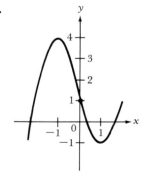

33.

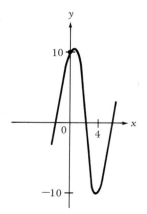

35.

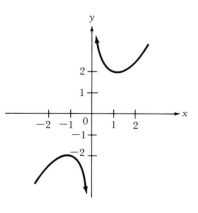

37.

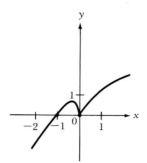

39.

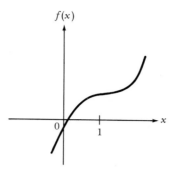

REVIEW PROBLEMS (SECTION 3-9)

1. $(6, 0)$ **3.** $(-5, 575/6)$ and $(6, -126)$ **5.** $(-3, -74)$, $(0, 7)$, and $(3, -74)$
7. $(0, 10)$ is a relative maximum point; $(4, -22)$ is a relative minimum point
9. $(-\sqrt{3}, -2\sqrt{3})$ is a relative maximum point; $(\sqrt{3}, 2\sqrt{3})$ is a relative minimum point
11. No relative maxima or minima   **13.** No absolute maximum; $(-1, -5)$ is the absolute minimum point
15. No absolute maximum or minimum
17. $(-\frac{3}{2}, -\frac{29}{2})$ is the absolute minimum point; $(6, 98)$ is the absolute maximum point
19. $(0, 0)$ and $(3, 0)$ are absolute maximum points; $(-1, -4)$ and $(2, -4)$ are absolute minimum points
21. $(1, 1)$ is the absolute minimum point; $(-1, 5)$ is the absolute maximum point
23. **a.** 6 units   **b.** $P(6) = \$1604$ **25.** Width $= \sqrt{300} \approx 17.3$ feet; length $= 2\sqrt{300} \approx 34.6$
27. **a.** $R(x) = xp(x) = 100x - (x^2/10)$ **b.** $x = 500$ **c.** $\$25,000$
29. Concave down on $(-\infty, 2)$; concave up on $(2, \infty)$; $(2, 0)$ is a point of inflection
31. Concave down on $(-\infty, -4)$; concave up on $(-4, \infty)$; $(-4, 0)$ is a point of inflection
33. y-intercept: $y = -77$; x-intercepts: $x = -11, 7$ **35.** y-intercept: $y = 0$; x-intercepts: $x = 0, 4, -3, 2$
37.

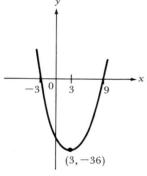

39.

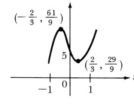

41.

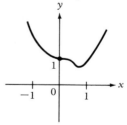

CHAPTER 3 SUPPLEMENT

1. $f(1.1) - f(1) \approx (1.5)(0.1) = 0.15$, % error $= 6.25\%$; $f(1.5) - f(1) \approx (1.5)(0.5) = 0.75$, % error $= 25\%$;
$f(2) - f(1) \approx (1.5)(1.0) = 1.5$, % error $= 40\%$
3. $\Delta y = 0.5$, $dy = 0.5$ **5.** $\Delta y = 0.21$, $dy = 0.2$ **7.** $\Delta y \approx 0.1350$, $dy \approx 0.1414$
9. $x = 100$, $\Delta x = 5$, $\Delta P \approx \$1250$; $x = 100$, $\Delta x = -5$, $\Delta P \approx -\$1250$
11. $f'(0.04) = 2.5$, $\Delta y \approx 0.005$ **13.** About $\$33.30$ **15.** $f(x) = \sqrt{x}$, $x = 25$, $\Delta x = 2$; $\sqrt{27} \approx 5 + 0.2 = 5.2$
17. $f(x) = \sqrt{x}$, $x = 121$, $\Delta x = 3$; $\sqrt{124} \approx 11 + 0.14 = 11.14$ **19.** $f(x) = x^{1/3}$, $x = 27$, $\Delta x = -1$; $\sqrt[3]{26} \approx 3 - \frac{1}{27} \approx 2.963$
21. $r = 6$, $\Delta r = -1$, $\Delta V \approx 4\pi r^2 \, \Delta r \approx -452.4$ cubic centimeters

CHAPTER 4   SECTION 4-1

1. $f'(x) = 5x^4 + 12x^3$ **3.** $f'(x) = 16x - 18$ **5.** $f'(p) = 10p^4 + 12p^3 - 3p^2$ **7.** $y' = 16x^3 - 24x^2 + 6x + 2$

9. $f'(z) = 10z^4 - 28z^3 + 21z^2 + 2z - 9$ **11.** $f'(x) = \frac{21}{2}x^{5/2} + \frac{105}{2}x^{3/2}$ **13.** $f'(x) = \dfrac{-7x^6}{(x^7 + 2)^2}$

15. $g'(t) = \dfrac{t(t - 22)}{(t - 11)^2}$ **17.** $f'(x) = \dfrac{-5x^8 + 18x^7 - 14x^6 - 4x + 6}{(x^7 - 2)^2}$ **19.** $f'(z) = \dfrac{-\frac{9}{2}z^{3/2} - \frac{1}{2}z^{1/2} + 3z^{-1/2} - 30z - 5}{(3z^2 + z + 6)^2}$

21. $g'(t) = \dfrac{6t^2 - 10t - 30}{t^2(t + 6)^2}$ **23.** $f'(x) = \dfrac{-6x^5}{(x^6 - 3)^2}$ **25.** $f'(x) = \dfrac{-6(x + 1)}{x^2(x + 2)^2}$ **27.** $R'(x) = \dfrac{-3(x^2 - 1)}{(x^2 + 1)^2}$

29. $T'(100) = \$147$ **31.** $x = 0$ is a vertical asymptote; $y = 5$ is a horizontal asymptote

33. $x = 2$ is a vertical asymptote; $y = 9$ is a horizontal asymptote

35. $x = 0$ is a vertical asymptote; $y = 8$ is a horizontal asymptote   **37.** Odd asymptote at $x = 3$

39. Odd asymptote at $x = -7$ **41.** Even asymptote at $x = -2$ **43.** No vertical asymptotes

45. $y = 0$ is a horizontal asymptote   **47.** $y = -4$ is a horizontal asymptote   **49.** $y = 0$ is a horizontal asymptote

51.

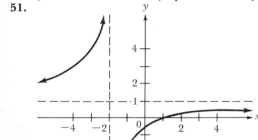

53.

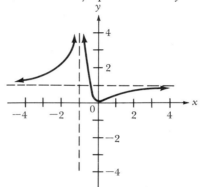

SECTION 4-2

1. $y' = 300(2x + 1)^{149}$ **3.** $f'(t) = 10(6t - 5)(3t^2 - 5t + 6)^9$ **5.** $f'(z) = \frac{5}{3}(5z - 2)^{-2/3}$

7. $f'(x) = (x^2 + 2)^{99}(201x^2 - 800x + 2)$ **9.** $y' = (x^4 + 2)(5x^2 - x)^2(70x^5 - 11x^4 + 60x - 6)$

11. $y' = 600(3x - 2)^{199}$ **13.** $f'(x) = 16(5x + 1)(5x^2 + 2x - 7)^7$ **15.** $f'(p) = \frac{3}{2}(6p - 5)^{-3/4}$

17. $f'(z) = \frac{10}{3}(5z + 4)^{-1/3}$ **19.** $f'(x) = -4(x^2 + 3x + 4)^{-5}(2x + 3)$

21. $C'(a) = (a - 1)(30 - 2a + a^2)^{-1/2}; C'(10) \approx 0.858$ **23.** $C'(5) \approx \$50.91$ **25.** $\dfrac{dy}{dx} = \dfrac{3(2x + 1)}{x^2 + x + 1}$

27. $\dfrac{dy}{dx} = -5.1$ when $x = 2$

SECTION 4-3

1. $y' = -\dfrac{x}{y}$ **3.** $y' = -\dfrac{2y}{x}$ **5.** $y' = \dfrac{-5y^2}{10xy + 4}$ **7.** $y' = \dfrac{-3x - y}{x + y}$ **9.** $y' = \dfrac{3x^2}{10y}$ **11.** $y' = \dfrac{-2xy}{x^2 + 3y^2}$

13. $x + 6y = 13$ **15.** $q' = -20$ when $q = 100$

17. $\dfrac{dA}{dt} = 2\pi r \dfrac{dr}{dt} = 20\pi r \approx 125.6$ square miles per hour when $r = 2$ miles   **19.** $\dfrac{dl}{dt} = -\dfrac{1}{60}$ when $l = 10$

21. $\dfrac{dp}{dt} \approx 1.74$ **23.** $\dfrac{dO}{dt} = 25{,}601{,}600$

REVIEW PROBLEMS (SECTION 4-5)

1. $f'(x) = 12x^2 + 28$ **3.** $y' = 28x^{5/2} + 30x^{3/2} - 3x^{1/2} + 12x^2 + 12x - 1$ **5.** $f'(x) = \dfrac{-3x^2 + 15}{x^6}$

7. $f'(p) = \dfrac{-7p^2 - 48p + 47}{(p^2 - 4p - 7)^2}$ **9.** $f'(x) = \dfrac{6x^2(x^2 + 1)}{(3x^2 + 1)^2}$ **11.** $g'(t) = \dfrac{10t^2 + 110t + 8}{(2t + 11)^2}$ **13.** $R'(x) = \dfrac{-10x^2 + 15}{(2x^2 + 3)^2}$

15. $f'(x) = 840x(7x^2 + 4)^{59}$ **17.** $f'(x) = 205\left(6x + 5 - \dfrac{1}{x^2}\right)\left(3x^2 + 5x + \dfrac{1}{x}\right)^{204}$ **19.** $y' = -350x^4(x^5 - 6)^{-11}$

21. $f'(x) = 60x(2x^2 + 3)^{14}$ **23.** $f'(t) = 26t(3t + 4)(2t^3 + 4t^2 - 11)^{12}$ **25.** $f'(x) = \frac{1}{4}(5x^2 + x + 2)^{-3/4}(10x + 1)$

27. $C'(x) = \frac{1}{2}(x + 2)^{-1/2}$; $R'(x) = \dfrac{20}{(x + 2)^2} - \dfrac{40x}{(x + 2)^3}$; $P'(x) = R'(x) - C'(x)$ **29.** $y' = \dfrac{-3y}{x}$ **31.** $y' = \dfrac{5x - y}{x}$

33. $y' = -\frac{2}{3}y^{1/2}x^{-2/3}$ **35.** $\frac{48}{19}$ feet per second
37. $x = 0$ is an odd vertical asymptote; $y = 7$ is a horizontal asymptote
39. $x = 2$ is an odd vertical asymptote; $y = 3$ is a horizontal asymptote
41. $x = 0$ is an even vertical asymptote; $y = -7$ is a horizontal asymptote
43.

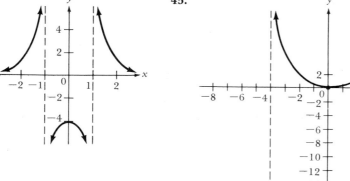

45.

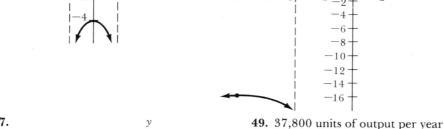

47.

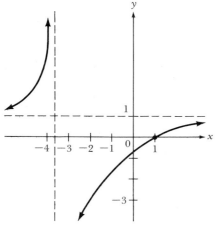

49. 37,800 units of output per year

CHAPTER 5   SECTION 5-1

1. $F(x) = 4x + c$ **3.** $H(x) = \frac{1}{2}x^4 + c$ **5.** $G(x) = -\frac{3}{2}x^{-2} + c$ **7.** $F(x) = x^2 + x + c$ **9.** $F(x) = \frac{10}{3}x^{3/2} + c$
11. $F(x) = \frac{4}{3}x^{3/2} + c$ **13.** Indefinite integral   **15.** $\frac{1}{7}x^7 + c$ **17.** $\frac{1}{21}t^{21} + c$ **19.** $-\frac{1}{3}x^{-3} + c$ **21.** $\frac{3}{2}t^{2/3} + c$

23. $\frac{10}{13}x^{1.3} + c$ **25.** $\frac{1}{3}x^3 + 9x + c$ **27.** $-\frac{4}{5}x^{-5} + c$ **29.** $\frac{1}{3}x^3 + 2x^2 + 3x + c$ **31.** $x^4 - \frac{5}{3}x^3 + \frac{3}{2}x^2 - 2x + c$

33. $x^2 - \frac{2}{3}x^{3/2} + 6x + c$ **35.** $\frac{2}{5}x^{5/2} + \frac{2}{3}x^{3/2} + 2x^{1/2} + c$ **37.** $G(x) = -\frac{2}{3}x^3 + \frac{3}{2}x^2 + x + \frac{5}{6}$ **39.** $a = \frac{5}{2}$

41. $F(t) = t^2 + \frac{2}{3}t^{3/2} + c$ **43.** $c(x) = 70x + 80\sqrt{x} + 50; c(25) = \$2200$

SECTION 5-2

1. Area = total number of miles traveled during a given period of time

3. Area = total number of people that have moved into the suburb over a given period of time

5. Length 1: area $\approx$ 6; length $\frac{1}{2}$: area $\approx$ 7; true area = 8

7. Length 1: area $\approx$ 10; length $\frac{1}{2}$: area $\approx$ 11; true area = 12

9. Area $\approx$ 221   **11.** Area $\approx$ 875

13. 2 subintervals: area $\approx$ 12; 4 subintervals: area $\approx$ 11; 8 subintervals: area $\approx$ 10$\frac{1}{2}$

15. 2 subintervals: area $\approx$ 45.75; 4 subintervals: area $\approx$ 40.125; 8 subintervals: area $\approx$ 37.3125

17. 2 subintervals: area $\approx$ 196; 4 subintervals: area $\approx$ 154; 8 subintervals: area $\approx$ 134.5

SECTION 5-3

1. 9   **3.** 2040   **5.** 74$\frac{2}{3}$ **7.** 22$\frac{1}{2}$ **9.** -20 **11.** $\pi^3 - \pi^2 + 2\pi$ **13.** 10.5   **15.** 36

17. Each definite integral equals 69.   **21.** 36   **23.** 34

SECTION 5-4

1. $\frac{4}{3}$ **3.** 2   **5.** 3$\frac{1}{3}$ **7. a.** 10$\frac{2}{3}$ **b.** 32$\frac{1}{3}$ **c.** 21$\frac{2}{3}$ **9.** 4$\frac{1}{2}$ **11.** 8   **13.** -18 (i.e., a *loss* of $18)

SECTION 5-5

1. Yes   **3.** No   **5.** No   **7.** $\dfrac{(x+5)^3}{3} + c$ **9.** $\frac{2}{9}(3x - 1)^{3/2} + c$ **11.** $-\frac{1}{4}(2x + 7)^{-2} + c$

13. $\frac{1}{6}(x^2 - 3)^3 + c$ **15.** $\frac{1}{3}(x^2 + 8)^{3/2} + c$ **17.** $\frac{4}{21}(x^3 + 5)^{7/4} + c$ **19.** $\frac{1}{22}(x^2 - 2x + 6)^{11} + c$

21. $-\frac{1}{3}(3 - x^2)^{3/2} + c$ **23.** $(x^2 + 3)^{1/2} + c$ **25.** $\frac{1}{3}(x^{3/2} - 2)^2 + c$ **27.** 22$\frac{1}{2}$ **29.** -32 **31.** $\frac{5}{126}$

SECTION 5-6

1. 1152 kilowatts; 989$\frac{1}{3}$ kilowatts   **3.** $\dfrac{-10,000}{\sqrt{T+1}} + 10,000$ students   **5.** 202,520 people

7. 2$\frac{2}{3}$ thousand children   **9. a.** 288 units   **b.** 24 units per hour   **11. a.** $f(x) = \frac{1}{30}, 70 \le x \le 100$ **b.** $\frac{1}{6}$

13. a. 0.84375   **b.** 8437.5 melons

SECTION 5-7

1. a. $15t - \dfrac{t^2}{4}, 0 \le t \le 10; 75 + 5t, 10 < t \le 20; 575 + \dfrac{t^2}{4} - 25t, 20 < t \le 30$ **b.** 50 tickets

3. Total cost $\approx$ \$796; average cost $\approx$ \$79.60   **5. a.** $5000 - 350T^2 - 1400T$ dollars   **b.** $T \approx 2.278$ months

7. a. $x^* = 300$, $p^* = 40$ **b.** \$9000   **c.** \$4500

9. a. The function is nonnegative on $[0, 10]$ and its integral is 1.   **b.** 0.50

REVIEW PROBLEMS (SECTION 5-9)

1. Area = total amount of bonus dollars earned by selling a given number of items above quota   **3.** Area ≈ 48

5. Area ≈ 42 **7.** $\dfrac{x^9}{9} + c$ **9.** $\frac{5}{6}x^{6/5} + c$ **11.** $-2x^{-4} + c$ **13.** $-\frac{3}{16}x^{-4} + c$ **15.** $\dfrac{5x^4}{4} - \dfrac{2x^5}{5} - 2x^2 + 3x + c$

17. $\frac{3}{2}x^2 - \frac{2}{3}x^{3/2} + 2x + c$ **19.** $P(x) = 150x - 0.015x^2$; $P(20) = \$2994$ **21.** 63   **23.** $31\frac{1}{2}$ **25.** 580

27. $74\frac{1}{6}$ **29.** $4\frac{1}{2}$ **31.** $3\frac{1}{5}$ **33.** $21\frac{1}{3}$ **35.** Approximately \$6166.67   **37.** $\frac{1}{3}(2x + 3)^{3/2} + c$

39. $\frac{1}{16}(2x^2 + 6)^4 + c$ **41.** $\frac{3}{20}(x^4 + 5)^{5/3} + c$ **43.** $(x^2 - 2)^{1/2} + c$

45. a. 63 units per month   **b.** 108 units per month

47. a. $h(1) = \frac{5}{11}$, $h(3) = \frac{45}{37}$, $h(12) = \frac{360}{869}$ **b.** $2896\frac{2}{3}$

49. a. The function is nonnegative and its integral is 1.   **b.** 0.75

CHAPTER 5 SUPPLEMENT

1. $\dfrac{x}{3}(x - 2)^3 - \frac{1}{12}(x - 2)^4 + c$ **3.** $\dfrac{x}{15}(3x + 1)^5 - \frac{1}{270}(3x + 1)^6 + c$ **5.** $10x(2 + x)^{1/2} - \frac{20}{3}(2 + x)^{3/2} + c$

7. $\frac{4}{3}x(3x + 4)^{1/2} - \frac{8}{27}(3x + 4)^{3/2} + c$ **9.** $3x^2(4 + x^2)^{1/2} - 2(4 + x^2)^{3/2} + c$ **11.** $7\frac{11}{15}$ **13.** $\frac{21}{3888} \approx 0.00054$

15. $3\frac{1}{3}$ **17.** $\dfrac{x(x - 1)^4}{4} - \dfrac{(x - 1)^5}{20} + c$ **19.** Converges to 1   **21.** Converges to $-\frac{1}{2}$ **23.** Diverges

25. Converges to $\frac{1}{9}$ **27.** Diverges   **29.** Converges to 0   **31.** Converges to 0   **33.** Converges to 62.5

35. a. $\frac{1}{25}$, or 4%   **b.** $\frac{1}{100}$, or 1%   **37.** $\sqrt{5}/48$ **39.** $-\frac{5}{36}$

CHAPTER 6 SECTION 6-1

1. After 3 months: \$5150; after 6 months: \$5304.50; after 12 months: \$5627.54   **3.** $A = a(\frac{1}{2})^{0.77}$

5. $A = 0.20166a$ **7.** $V = 500{,}000\,(1.08)^t$, $0 \le t \le 3$ **9.** 5^x **11.** 5.77 thousand per square mile

13. Model is: Salary = $1800\,(2^x)$, where x is the number of 10 year increments since 1939. 1959 estimated: \$7200, 1969 estimated: \$14,400, and 1979 estimated: \$28,800 all appear to be very good estimates. 1989 predicted: \$57,600

15. 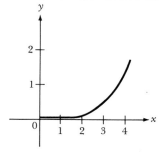   **17.** 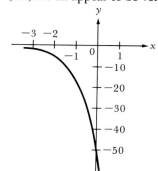   **19.**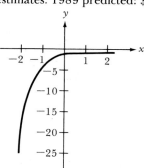

21. A   **23.** B   **25.** C

SECTION 6-2

1. $\log_2 16 = 4$ **3.** $\log_2 \frac{1}{8} = -3$ **5.** $\log_5 0.2 = -1$ **7.** $\log_{1/3} 27 = -3$ **9.** $4^3 = 64$ **11.** $25^2 = 625$
13. $5^{-1} = 0.2$ **15.** $(\frac{1}{2})^2 = \frac{1}{4}$ **17.** $(\sqrt{3})^4 = 9$ **19.** 3   **21.** -1 **23.** 4   **25.** 10   **27.** 45
29. 111   **31.** 90 decibels   **33.** 130 decibels   **35.** 0.001 watt per square meter   **37.** $y = \log_{1/3} x$
39. $y = \log_{1/2} x$ **41.** $y = \log_{1/25} x$ **43.** $y = \log_{1/32} x$
45.

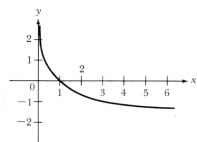

47.

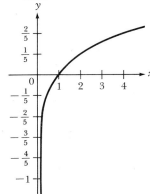

49.

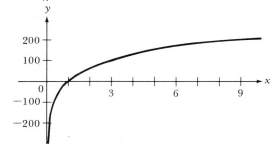

51.

53. $x = \dfrac{\log 25}{\log 2}$ **55.** $x = \dfrac{\log_{10} 3 - \log_{10} 4}{-4}$ **57.** $x = 103$ **59.** $x = 2$

SECTION 6-3

1. $y' = 7e^{7x}$ **3.** $f'(x) = -9e^{-9x}$ **5.** $y' = 40e^{8x}$ **7.** $f'(x) = 2e^{2x+3}$ **9.** $y' = (2x - 4)e^{x^2-4x+3}$
11. $y' = 18x^2 e^{2x^3-4}$ **13.** $f'(x) = 3xe^x(x + 2)$ **15.** $f'(x) = -xe^{-3x}(3x - 2)$ **17.** $y' = \dfrac{e^x(6x - 7)}{(6x - 1)^2}$
19. $y' = \dfrac{2x - x^2 + 1}{e^x}$ **21.** $f'(x) = 3x^2 + 3x^2 e^{-x} - x^3 e^{-x}$ **23.** $y' = 2e^{-0.02t}$ **25.** $N'(t) = 4e^{-0.05t}$
27. $f'(2) \approx \$1071.41$ per year
29. $A' = \dfrac{a}{54.5} \ln\left(\dfrac{1}{2}\right)\left(\dfrac{1}{2}\right)^{t/54.5}$; the derivative represents the rate of change of the amount remaining at 50 days
31. $y' = (\ln 3)3^x$; $y'' = (\ln 3)^2 3^x$ **33.** $y' = 3(\ln \frac{1}{2})(\frac{1}{2})^x$; $y'' = 3(\ln \frac{1}{2})^2(\frac{1}{2})^x$ **35.** $y' = -\frac{1}{2}(\ln 2)(2^x)$; $y'' = -\frac{1}{2}(\ln 2)^2(2^x)$
37. No maximum exists   **39.**

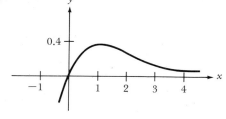

SECTION 6-4

1. $y' = \dfrac{3(2x^2 + 1)}{x(2x^2 + 3)}$ **3.** $f'(x) = \dfrac{3}{x}$ **5.** $f'(x) = \dfrac{18 \ln x}{x}$ **7.** $y' = \dfrac{24(\ln 2x)^2}{x}$ **9.** $f'(x) = \dfrac{2(x - 2)}{(x + 3)(x - 7)}$

11. $y' = \dfrac{2(5x - 18)}{(x - 4)(x - 3)}$ **13.** $f'(x) = \dfrac{1}{x(3x - 1)}$ **15.** $y' = \dfrac{6x}{x^2 + 4}$ **17.** $f'(x) = \dfrac{x^2}{x + 3} + 2x \ln(x + 3)$

19. Relative maximum at $x = -\frac{3}{4}$ **21.** Relative minimum at $x = 1$ **23.** $y' = \dfrac{1}{x \ln 3}$ **25.** $y' = \dfrac{4}{4x \ln 3}$

27. $f'(x) = \dfrac{3x^2}{x^3 \ln 10}$ **29.** $y' = \dfrac{3}{(3x + 7)(\ln 10)}$ **31.** $f'(x) = \dfrac{10(x + 2)}{(x^2 + 4x) \ln 3}$ **33.** $y' = \dfrac{-1}{x(x - 1) \ln 7}$

35. -4342.9 **37.** $f'(x) = (2x + 6)(x^2 - 5x - 7)(2x^3 + 6)\left(\dfrac{2}{2x + 6} + \dfrac{2x - 5}{x^2 - 5x - 7} + \dfrac{6x^2}{2x^3 + 6} \right)$

39. $y' = (x^3 + 2x)(5x^2 - x - 3)(2x^4 - 6x^2 + x)\left(\dfrac{3x^2 + 2}{x^3 + 2x} + \dfrac{10x - 1}{5x^2 - x - 3} + \dfrac{8x^3 - 12x + 1}{2x^4 - 6x^2 + x} \right)$

41. $f'(x) = \dfrac{(2x^2 + 3)^2(e^x - 6x)^3}{(2x^2 + 1)^4}\left(\dfrac{8x}{2x^2 + 3} + \dfrac{3e^x - 18}{e^x - 6x} - \dfrac{16x}{2x^2 + 1} \right)$

43. $f'(x) = \dfrac{(e^{5x} - 7x)(e^x + 3x^2)}{e^{0.5x} - x^2}\left(\dfrac{5e^{5x} - 7}{e^{5x} - 7x} + \dfrac{e^x + 6x}{e^x + 3x^2} - \dfrac{0.5e^{0.5x} - 2x}{e^{0.5x} - x^2} \right)$ **45.** $y' = x^{x^2}(x + 2x \ln x)$ **47.** 1.0   **49.** 3

51. 0.666   **53.**

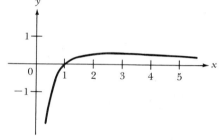

SECTION 6-5

1. $\frac{1}{2}e^{6x} + c$ **3.** $-\frac{1}{4}e^{-4x} + c$ **5.** $\frac{1}{10}e^{5x^2} + c$ **7.** $\frac{3}{8}e^{2x^4} + c$ **9.** $9 \ln|x| + c$ **11.** $2 \ln|x^3| + c$
13. $\ln|4x + 2| + c$ **15.** $\ln|x^2 + 5| + c$ **17.** $\frac{1}{10} \ln|5x^2 - 3| + c$ **19.** $\ln|x^4 - 2| + c$ **21.** $\frac{1}{3} \ln|x^3 - 3x| + c$
23. 53.598   **25.** 22,025.865   **27.** 1.79176   **29.** 23,211.344   **31.** 5725.8   **33.** 50.3%
35. 18.67 years   **37.** $\frac{1}{4}xe^{4x} - \frac{1}{16}e^{4x} + c$ **39.** $e^x(x + 2) - e^x + c$ **41.** $2\sqrt{x} \ln x - 4\sqrt{x} + c$

REVIEW PROBLEMS (SECTION 6-7)

1. After 3 months: \$8200; after 6 months: \$8405; after 12 months: \$8830.50   **3.** 202 million
5. $t \approx 6059$ years   **7.** $\log_3 \frac{1}{27} = -3$ **9.** $\log_{\sqrt{3}} 9 = 4$ **11.** $(\frac{1}{2})^{-5} = 32$ **13.** 6   **15.** 27

17. $y = \log_{1/4} x$ **19.** $y' = 5e^{5x}$ **21.** $y' = -28e^{7x}$ **23.** $f'(x) = \dfrac{2e^{2x}(x^2 + 8 - x)}{(x^2 + 8)^2}$

25. $S'(1) \approx 26.669$; $S'(10) \approx 1.792$ **27.** $y' = \dfrac{2(x + 3)}{x^2 + 6x - 2}$ **29.** $f'(x) = \dfrac{-6}{x(\ln x^3)^3}$ **31.** $y' = 1 + \ln x$

33. Relative maximum at $x = -\frac{11}{5}$ **35.** $y' = \dfrac{4}{(4x - 3) \ln 10}$ **37.** $f'(x) = \dfrac{1}{x(x + 1) \ln 6}$

39. $f'(x) = \dfrac{(x^2 + 2)(e^x - 5x)}{2x + 6}\left(\dfrac{2x}{x^2 + 2} + \dfrac{e^x - 5}{e^x - 5x} - \dfrac{2}{2x + 6} \right)$ **41.** 0.7143   **43.** 1.666   **45.** $\frac{2}{3}e^{2x^3} + c$

47. $\ln|x^2 - 4| + c$ **49.** $\log_8 x + c$ **51.** 266.91 grams   **53.** $t \approx 310.57$ years   **55.** $r \approx 5.8\%$
57. $81,474.33

CHAPTER 7   SECTION 7-1

1. $2t^2(4t) - 4t(2t^2) = 0$ **3.** $3(-4e^{-(2/3)t}) + 2(6e^{-(2/3)t}) = 0$ **5.** $(4e^{2t} - e^{-t}) - (2e^{2t} + e^{-t}) - 2(e^{2t} - e^{-t}) = 0$
9. $y = \frac{1}{2}t^2 + t + 1$ **11.** $y = -e^{6t}$ **13.** $y = \frac{1}{2}e^{2t} + t + \frac{1}{2}$ **15.** $y = \ln(t + 1) + 7$ **17.** $y = \dfrac{(2t - 1)^6}{12} - \dfrac{1}{12}$
19. $y = -\dfrac{1}{t} + e^t - t + 2$

SECTION 7-2

3. $\dfrac{dc}{dt} = -k(C - A)$ **5.** $\dfrac{ds}{dt} = -k$ (a constant)   **7.** $\dfrac{dP}{dt} = aP - bP^2$ **9.** $\dfrac{dq}{dt} = k - aq$ **11.** $ms'' = -ks - as'$
13. $\dfrac{dc}{dt} = 0.002 - \dfrac{c}{50}$, $c(0) = 0$ **15.** $\dfrac{dy}{dt} = ky(c - y)$

SECTION 7-3

1. $y = \frac{2}{3} + ce^{-3t}$ **3.** $y = 4 - 3e^{-t}$ **5.** $y = \left(\dfrac{t^2}{2} - 7\right)e^{-3t}$ **7.** $y = \left(2t + \dfrac{4}{e^2} - 2\right)e^{2t}$
9. a. $Q(t) = Q(0)e^{kt}$ **b.** About 25.36 hours
11. a. $S = ce^{rt} - \dfrac{k}{r}$ **b.** $S(t) = 95,454.545e^{0.11t} - 45,454.545$ **c.** About 3.83 years
13. a. $P(t) = 25,000(1 + e^{0.02t})$ **b.** The model does not take account of crowding
15. a. $\dfrac{dc}{dt} = k(c_0 - c)$ **b.** $c(t) = c_0 + de^{-kt}$
c.

17. $C(t) = 0.2 - 0.5e^{-0.06t} + 0.3e^{-0.1t}$ **19.** About 42,557 years   **21. a.** $mv' = mg - kv$ **b.** $\lim\limits_{t \to \infty} v(t) = \dfrac{gm}{k}$
23. $s(t) = -\dfrac{2}{5}t + 10$; about $9\frac{1}{4}$ minutes

SECTION 7-4

1. $2y^3 = 5t + c$ **3.** $y = \ln|2t + t^3 + c|$ **5.** $y = \dfrac{2}{c - t^2}$ **7.** $y = -t^{-1} + c$ **9.** $P = Ae^{x^2/2}$

11. $\ln|y - 1| = -2t$ **13.** $\ln|y| = \dfrac{-t^2}{2} + t$ **15.** $y = \sqrt{\dfrac{t^4}{2} + 1}$ **17.** $y = \dfrac{2}{1 - 2\ln(t^2 + 1)}$ **19.** $y = \sqrt{\dfrac{2e^{3t} + 46}{3}}$

21. $u^{-3} = -3t^3 + 1$ **23.** $y^3 = \dfrac{3}{2}(\ln x)^2 + 1$ **25.** $y^3 = \dfrac{e^{t^3}}{e^t}$

27. **a.** $\dfrac{dl}{dt} = k(M - l)$ **b.** $l = M(1 - e^{-kt})$

29. **a.** $P = 12.5\,\dfrac{4e^{0.5t}}{4 + e^{0.5t}}$ **b.**

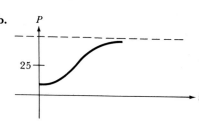

c. At $P = 25$ **31.** $y = e^{e^{-(a/b)t}}$

REVIEW PROBLEMS (SECTION 7-6)

5. $y = \frac{1}{2} + ce^{-2t}$ **7.** $y = \frac{1}{4} + ce^{(4/3)t}$ **9.** $y = e^t$ **11.** $y = (t + 1)e^t$ **13.** About 10%

15. **a.** $\dfrac{dS}{dt} = kS^{1/2}$ **b.** $S = \dfrac{(kt + c)^2}{4}$, where $c \approx 63.25$ and $k \approx 8.38$ **17.** Time of death was about 7:46 A.M.

19. $y = \dfrac{1}{c - \ln t} + 1$ **21.** $y = \sqrt{2\ln t + 1}$ **23.** $y^3 + y = t - \dfrac{1}{t} + 10$

CHAPTER 7   SUPPLEMENT

1. $y' = 4$ when $t = 1$ and $y = 3$ **3.**

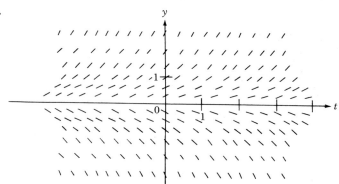

5.

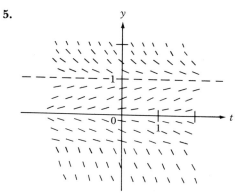

7. $y(1) \approx y_4 \approx 0.4999$

9.

t	y_n	$y(t) = \dfrac{e^t}{3 + e^t}$
0	0.2500	0.2500
0.25	0.2969	0.2997
0.50	0.3491	0.3547
0.75	0.4059	0.4137
1.00	0.4662	0.4754
1.25	0.5284	0.5378
1.50	0.5907	0.5990
1.75	0.6512	0.6573
2.00	0.7079	0.7112
2.25	0.7595	0.7598
2.50	0.8052	0.8024
2.75	0.8444	0.8391
3.00	0.8772	0.8700

CHAPTER 8   SECTION 8-1

1. 56   **3.** 28   **5.** $4a^2 + 3a^4 - 5a^3 + 20$ **7.** -7 **9.** -6 **11.** $2b^2 - 4b^4 + 3b^3 - 15$ **13.** 104

15. 81   **17.** $2c^6 - c^3 + 50$ **19. a.** \$22 million   **b.** \$24 million   **c.** 10   **21.** $\text{Cost} = \dfrac{4.8}{z} + 0.02xz + \dfrac{4.8}{x}$

23.

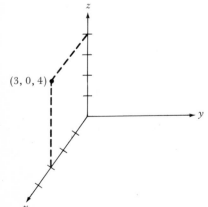

25.

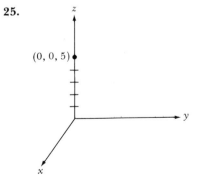

27.

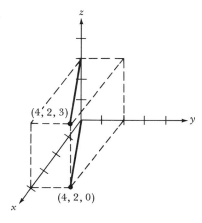

$(4, 2, 3)$
$(4, 2, 0)$

29.

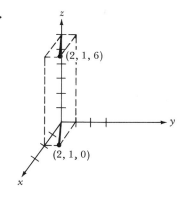

$(2, 1, 6)$
$(2, 1, 0)$

31.

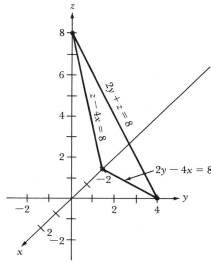

$2y + z = 8$
$z - 4x = 8$
$2y - 4x = 8$

33.

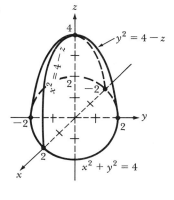

$y^2 = 4 - z$
$x^2 = 4 - z$
$x^2 + y^2 = 4$

35.

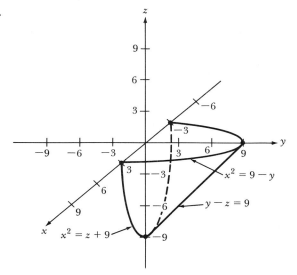

$x^2 = 9 - y$
$y - z = 9$
$x^2 = z + 9$

SECTION 8-2

1. $\partial f/\partial x = 2x + 2y^2 - 2$; $\partial f/\partial y = 4xy$ 3. $\partial f/\partial x = 3x^2y^2 + 2xy^3$; $\partial f/\partial y = 2x^3y + 3x^2y^2$
5. $\partial f/\partial x = 3x^2 - 4xy + 2y^2$; $\partial f/\partial y = 3y^2 - 2x^2 + 4xy$
7. $\partial f/\partial x = 4x^3y^3 - 6x^2y^2 + 10x - 2y$; $\partial f/\partial y = 3x^4y^2 - 4x^3y - 2x + 3$
9. $\partial f/\partial x = 1/2\sqrt{x+y}$; $\partial f/\partial y = 1/2\sqrt{x+y}$ 11. $\dfrac{\partial f}{\partial x} = \dfrac{2xy - 4x}{(y-2)^2}$; $\dfrac{\partial f}{\partial y} = \dfrac{-x^2 - 3}{(y-2)^2}$
13. $\dfrac{\partial f}{\partial x}(2, 1) \approx 5.9366$; $\dfrac{\partial f}{\partial y}(2, 1) \approx 11.8732$ 15. $\dfrac{\partial f}{\partial x}(-1, 1) \approx 5.07949$; $\dfrac{\partial f}{\partial y}(-1, 1) = -1.5$
17. $\dfrac{\partial f}{\partial x}(0, 1) = 4$; $\dfrac{\partial f}{\partial y}(0, 1) = 0$ 19. $\dfrac{\partial f}{\partial x}(-2, -1) = -6468$; $\dfrac{\partial f}{\partial y}(-2, -1) = -3528$
21. $\dfrac{\partial f}{\partial x}(-1, 1) = -28{,}812$; $\dfrac{\partial f}{\partial y}(-1, 1) = 5488$
23. $dQ = 13{,}475.6$ passengers per hour; actual change is approximately 15,721.4 passengers per hour
25. a. $\partial f/\partial x = 48y + 2$ b. $\partial f/\partial y = 48x$
27. a. $\dfrac{\partial U}{\partial x} = (y + 3)^3(2x + 4)$; $\dfrac{\partial U}{\partial y} = 3(x + 2)^2(y + 3)^2$ b. $\dfrac{\partial U}{\partial x}(4, 4) = 4116$
29. Slope of line parallel to xz-plane $= 12$; slope of line parallel to yz-plane $= 10$

SECTION 8-3

1. $\partial^2 f/\partial x^2 = 2$; $\partial^2 f/\partial y^2 = -6$; $\partial^2 f/\partial x\,\partial y = 2$ 3. $\partial^2 f/\partial x^2 = 6x - 4y$; $\partial^2 f/\partial y^2 = 6x$; $\partial^2 f/\partial x\,\partial y = -4x + 6y$
5. $\partial^2 f/\partial x^2 = 6y^2 - y^2e^{xy}$; $\partial^2 f/\partial y^2 = 6x^2 - x^2e^{xy}$; $\partial^2 f/\partial x\,\partial y = 12xy - xye^{xy} - e^{xy}$
7. $\partial^2 f/\partial x^2 = y^2e^{xy}$; $\partial^2 f/\partial y^2 = x^2e^{xy} - (x/y^2)$; $\partial^2 f/\partial x\,\partial y = xye^{xy} + e^{xy} + (1/y)$
9. $\partial^2 f/\partial x^2 = 8y + 2$; $\partial^2 f/\partial y^2 = 2$; $\partial^2 f/\partial x\,\partial y = 8x + 2$
11. $\partial^2 f/\partial x^2 = -1/(x + y)^2$; $\partial^2 f/\partial y^2 = -1/(x + y)^2$; $\partial^2 f/\partial x\,\partial y = -1/(x + y)^2$
13. $\partial^2 f/\partial x^2 = 12x^2 - 2y^2 + 6$; $\partial^2 f/\partial y^2 = -2x^2$; $\partial^2 f/\partial x\,\partial y = -4xy - 1$
15. $\partial^2 f/\partial x^2 = 6xy + 4y^2$; $\partial^2 f/\partial y^2 = 4x^2 + 6xy$; $\partial^2 f/\partial x\,\partial y = 3x^2 + 8xy + 3y^2$

SECTION 8-4

1. $(-2, 0, 9)$ 3. $(\frac{14}{3}, -\frac{13}{3}, -\frac{61}{3})$ 5. $(-1, 3, 0)$
7. $(2, 5, -307.5)$; $(-1, 5, -280.5)$; $(2, -6, 358)$; $(-1, -6, 385)$ 9. $(\frac{3}{2}, 0, -\frac{9}{4})$, relative minimum
11. $(-1, -2, 0)$, saddle point   13. $(-2, 3, -21)$, saddle point
15. $(1, \frac{1}{2}, -\frac{17}{4})$, relative minimum; $(-\frac{1}{3}, \frac{11}{6}, -\frac{331}{108})$, saddle point
17. $(2, 1, 6)$, relative minimum   19. $(0, 0, 0)$, saddle point; $(2, 2, -8)$, relative minimum   21. $x = 24, y = 0$
23. $x = 46, y = 83.5$ 25. Length $=$ width $=$ height $= 4$ inches

SECTION 8-5

1. $(4, 4, 64)$, $\lambda = -16$ 3. $(5, 3, 28)$, $\lambda = -7$ 5. $(50, 50, 2500)$, $\lambda = 50$
7. $(5, -5, 75)$, $\lambda = -15$ 9. $(2, -2, 26)$, $\lambda = -2$
11. Length $= 150$ yards; width $= 150$ yards; area $= 22{,}500$ square yards
13. Plant 1 produces 40 units of x; plant 2 produces 460 units of y; cost $= \$8340$

REVIEW PROBLEMS (SECTION 8-7)

1. 63   **3.** -15 **5.** $3c^4 - 2c^6 - 4c^5 + 15$ **7.** $\text{Cost} = 40xy + \dfrac{120}{y} + \dfrac{120}{x}$

9.

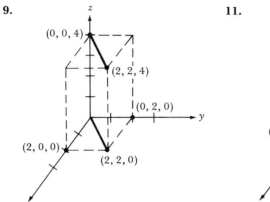

11.

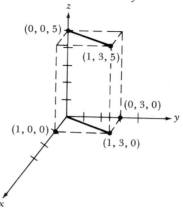

13.

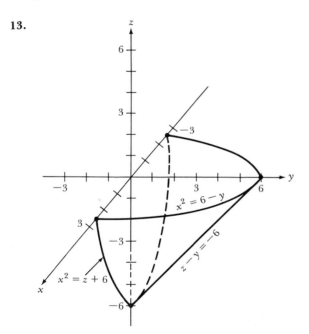

15. \$16,775   **17.** $\partial f/\partial x = 6x + 2xy^2$; $\partial f/\partial y = 2x^2y - 20y^3$ **19.** $\partial f/\partial x = 1/2\sqrt{x - y^2}$; $\partial f/\partial y = -y/\sqrt{x - y^2}$

21. $\partial f/\partial x = (1/x) + ye^{xy}$; $\partial f/\partial y = (1/y) + xe^{xy}$ **23.** $\partial f/\partial x = 12x^2(x^3 + 2y^2)^3$; $\partial f/\partial y = 16y(x^3 + 2y^2)^3$

25. Slope of both tangent lines is 14

27. $\partial^2 f/\partial x^2 = 24x^2 + 12xy + 8y^2$; $\partial^2 f/\partial y^2 = 8x^2 - 30xy - 36y^2$; $\partial^2 f/\partial x\,\partial y = 6x^2 + 16xy - 15y^2$

29. $\partial^2 f/\partial x^2 = y^2 e^{xy}$; $\partial^2 f/\partial y^2 = (-2x/y^2) + x^2 e^{xy}$; $\partial^2 f/\partial x\,\partial y = (2/y) + xye^{xy} + e^{xy}$

31. $\dfrac{\partial^2 f}{\partial x^2} = \dfrac{-2x^2 + 2y}{(x^2 + y)^2}$; $\dfrac{\partial^2 f}{\partial y^2} = \dfrac{-1}{(x^2 + y)^2}$; $\dfrac{\partial^2 f}{\partial x\,\partial y} = \dfrac{-2x}{(x^2 + y)^2}$ **33.** $\dfrac{\partial^2 f}{\partial x^2} = \dfrac{x^3 + y^3}{x^3 y}$; $\dfrac{\partial^2 f}{\partial y^2} = \dfrac{y^3 + x^3}{xy^3}$; $\dfrac{\partial^2 f}{\partial x\,\partial y} = \dfrac{-x^3 - y^3}{x^2 y^2}$

35. $(3, \frac{3}{4}, -\frac{121}{8})$ **37.** $(-2, 1.5, -5.75)$, relative minimum   **39.** $(6, -3, -50)$, relative minimum

41. $(4, 5, -21.5)$, relative minimum   **43.** $x = \frac{8}{3}, y = \frac{5}{3}$ hours   **45.** $(4, -4, 32)$

47. $(0, -1, -1)$ and $\left(-\frac{3}{4}, -\frac{7}{16}, -\frac{155}{128}\right)$

CHAPTER 8 SUPPLEMENT

1. 120   **3.** $\frac{589}{12}$ **5.** $-42\frac{2}{3}$ **7.** $-15\frac{2}{3}$ **9.** $34\frac{1}{6}$
11. $40\frac{4}{9}$ **13.** Approx. -1498.21 **15.** 6319   **17.** Approx. 180.29

CHAPTER 9   SECTION 9-1

1. **a.** $4\pi/45$ **b.** $2\pi/3$ **c.** $5\pi/2$
3. **a.** $-\pi/6$ **b.** $8\pi/9$ **c.** $-13\pi/9$ **5.** **a.** $40°$ **b.** $75°$ **c.** $70°$
7. **a.** $-330°$ **b.** $-315°$ **c.** $-60°$ **9.** $3\pi/2$ **11.** $5\pi/2$ **13.** $-15\pi/4$
15. **a.**   **b.**   **c.**

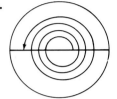

17. **a.**   **b.**   **c.**

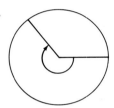

19. 16π

SECTION 9-2

1. $\sin t = 2\sqrt{6}/5$ **3.** $\sin t = \sqrt{35}/6$ **5.** $\sin t = \frac{-3}{5}$ **7.** $13\pi/6,\ 17\pi/6$ **9.** $-5\pi/4,\ -7\pi/4$
11. $\pm 2\pi/3,\ \pm 4\pi/3,\ \pm 8\pi/3,\ \pm 10\pi/3$
13.   **15.**

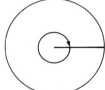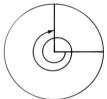

17.   **19.** $t = \pi/4$

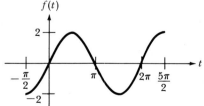

21.   **23.**

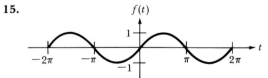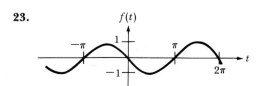

SECTION 9-3

1. $6 \cos 6t$ **3.** $6 \cos 3t$ **5.** $-35 \sin 5t$ **7.** $6t - \pi \sin \pi t$ **9.** $10 \cos(2t + 1)$ **11.** $(3t^2 + 2) \cos(t^3 + 2t)$

13. $4 \sin(1 - t)$ **15.** $-6e^{2t} \sin e^{2t}$ **17.** $(4 \cos 2t)e^{2 \sin 2t}$ **19.** $\dfrac{\cos[\ln(t + 1)]}{t + 1}$ **21.** $2t \cos t^2$

23. $10(t^2 - 2 \cos t)^4(t + \sin t)$ **25.** $e^{2t} \cos t + 2e^{2t} \sin t$ **27.** $12t^2 \sin^3 t^3 \cos t^3$ **29.** $3 \cos 5t \cos 3t - 5 \sin 5t \sin 3t$

31. $\dfrac{\cos t}{\sin t}$ **33.** $\dfrac{(1 + \cos t)(t - \cos t) - (t + \sin t)(1 + \sin t)}{(t - \cos t)^2}$ **35.** $\dfrac{\sin t}{\cos^2 t}$ **37.** $\frac{1}{7} \sin 7t + c$

39. $-\frac{1}{4} \cos(1 + 4t) + c$ **41.** $\frac{1}{4}$ **43.** $\frac{1}{3}$ **45.** $y = 2 \sin t + 1$ **49.** $\theta = \pi/4$

SECTION 9-4

1.

t	$\pi/4$	$2\pi/3$	$5\pi/6$	$5\pi/4$	$11\pi/6$
$\tan t$	1	$-\sqrt{3}$	$-1/\sqrt{3}$	1	$-1/\sqrt{3}$
$\sec t$	$\sqrt{2}$	-2	$-2/\sqrt{3}$	$-\sqrt{2}$	$2/\sqrt{3}$
$\cot t$	1	$-1/\sqrt{3}$	$-\sqrt{3}$	1	$-\sqrt{3}$
$\csc t$	$\sqrt{2}$	$2/\sqrt{3}$	2	$-\sqrt{2}$	-2

5. $5 \sec^2 5t$ **7.** $7 \csc^2(-7t)$ **9.** $6e^{6t} \sec(e^{6t} + 1) \tan(e^{6t} + 1)$ **11.** $4(4t + 3) \sec^2(4t^2 + 6t + 1)$

13. $\dfrac{-5}{(t + 1)^2} \sec^2\left(\dfrac{1}{t + 1}\right)$ **15.** $\dfrac{\sec^2 t}{\tan t}$ **17.** $8te^{-\cot t^2} \csc^2 t^2$ **19.** $\dfrac{2}{t} \csc\left(\dfrac{2}{t}\right) \cot\left(\dfrac{2}{t}\right) + \csc\left(\dfrac{2}{t}\right)$

21. $(te^t \tan e^t + 1) \sec e^t$ **23.** $-2e^t \csc^2(2t) + e^t \cot 2t$ **25.** $\sec t$ **27.** $\dfrac{2t \sec[\ln(t^2 + 1)] \tan[\ln(t^2 + 1)]}{t^2 + 1}$

29. $e^{-2t} \sec t(\tan t - 2)$ **31.** $12t \sec^3(2t^2) \tan(2t^2)$ **33.** $4 \sin 2t \cos 2t$ **35.** $-\dfrac{\tan t + t(t + 1) \sec^2 t}{t^2 \tan^2 t}$

37. $(\sec t)\dfrac{1 + \csc^2 t}{\cot^2 t}$ **39.** $\dfrac{te^t(1 + \cot t) - e^t}{t^2 \csc t}$ **41.** $\dfrac{1}{2} \tan 2t + c$ **43.** $-2\sqrt{3} + 2$

SECTION 9-5

1. $\sin \theta = \frac{3}{5}$, $\cos \theta = \frac{4}{5}$, $\tan \theta = \frac{3}{4}$, $\cot \theta = \frac{4}{3}$, $\sec \theta = \frac{5}{4}$, $\csc \theta = \frac{5}{3}$
3. $\sin \theta = 1/\sqrt{2}$, $\cos \theta = 1/\sqrt{2}$, $\tan \theta = 1$, $\cot \theta = 1$, $\sec \theta = \sqrt{2}$, $\csc \theta = \sqrt{2}$
5. $\sin \theta = \frac{2}{3}$, $\cos \theta = \sqrt{5}/3$, $\tan \theta = 2/\sqrt{5}$, $\cot \theta = \sqrt{5}/2$, $\sec \theta = 3/\sqrt{5}$, $\csc \theta = \frac{3}{2}$
7. $\sin \theta = 2/\sqrt{29}$, $\cos \theta = 5/\sqrt{29}$, $\tan \theta = \frac{2}{5}$, $\cot \theta = \frac{5}{2}$, $\sec \theta = \sqrt{29}/5$, $\csc \theta = \sqrt{29}/2$

9.   **11.**   **13.**   **15.**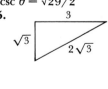

17. $\dfrac{100}{\sqrt{3}} \approx 57.7$ feet   **19.** $60 \tan 70° \approx 164.8$ feet   **21.** $\dfrac{10}{\sin 12°} \approx 48.1$ feet

23. About 0.00169 radian per second   **25.** $\dfrac{\pi}{9} \sec^2(1°) \approx -0.7$ foot per second

REVIEW PROBLEMS (SECTION 9-7)

1. a. $\dfrac{360°}{7}$ **b.** $480°$ **c.** $-150°$ **d.** $-210°$ **3. a.** $-7\pi/2$ **b.** 3π **c.** $-5\pi/2$

5. a. $-2t \sin t^2$ **b.** $12 \cos 6t$ **c.** $\cos^2 t - \sin^2 t - 1$ **d.** $2 \cos t$

7. a. $6(2t + \cos t)^5(2 - \sin t)$ **b.** $-2te^{t^2} \sin e^{t^2}$ **c.** $-2t \sin t^2(e^{\cos t^2})$ **d.** $\dfrac{\sin 2t}{\sin^2 t + 1}$

9. a. $-2 \csc^2 2t$ **b.** $-18t \csc^2 6t + 3 \cot 6t$ **c.** $-3 \csc 3t \cot 3t$ **d.** $-8e^t \csc 4t \cot 4t + 2e^t \csc 4t$

11. About 1 foot **13.** 600 feet **15.** About 63 feet per minute

CHAPTER 10   SECTION 10-1

1. $\frac{1}{4}, \frac{6}{7}, \frac{11}{10}, \frac{16}{13}, \frac{21}{16}$ **3.** $2, 0, 2, 0, 2$ **5.** $1, 0, -\frac{1}{3}, 0, \frac{1}{5}$ **7.** $-\frac{1}{2}, 1, -\frac{9}{8}, 1, -\frac{25}{32}$ **9.** $5, 5, 5, 5, 5$

11. $a_n = 2 + 3n, \; n = 1, 2, 3, \ldots$ **13.** $a_n = \dfrac{3n}{2n + 1}, \; n = 1, 2, 3, \ldots$ **15.** $\lim\limits_{n \to \infty} a_n = 1$ **17.** $\lim\limits_{n \to \infty} a_n = -2$

19. $\lim\limits_{n \to \infty} a_n = 0$ **21.** $\lim\limits_{n \to \infty} a_n = \frac{2}{3}$ **23.** Sequence diverges **25.** $\lim\limits_{n \to \infty} a_n = 0$ **27.** $\lim\limits_{n \to \infty} a_n = \frac{7}{5}$

29. $\lim\limits_{n \to \infty} a_n = 0$ **31.** Sequence diverges **33.** Sequence diverges **35.** Sequence diverges

39. $a_n = 1000 \cdot 8^n, \; n = 1, 2, 3, \ldots$

41. a. $1, 1, 2, 3, 5, 8, 13, 21, 34, 55, 89, 144$ **b.** $\dfrac{a_{n+1}}{a_n} = 1 + \dfrac{a_{n-1}}{a_n}$; therefore, $b_n = 1 + \dfrac{1}{b_{n-1}}$

c. $b = 1 + \dfrac{1}{b}$ implies $b^2 - b - 1 = 0$; thus, $b = \dfrac{1 + \sqrt{5}}{2}$, since b is positive

d. $b \approx 1.618034, \; b_1 = 1, \; b_2 = 2, \; b_3 = 1.5, \; b_4 = \frac{5}{3}, \; b_5 = 1.6, \; b_6 = 1.625$

SECTION 10-2

1. Let $f(x) = e^x - 4x; \; f(0) = 1, f(1) = e - 4 < 0$; thus, $f(x)$ has a root s between 0 and 1; the root s satisfies $0.35 \le s \le 0.36$

3. Let $f(x) = 5x^2 - 2; \; f(0) = -2, f(1) = 3$; thus, $f(x)$ has a root s between 0 and 1; $s \approx 0.63$

5. $x_2 = 2.25, \; x_3 \approx 2.2361111, \; x_4 \approx 2.236068, \; x_5 \approx 2.236068$

7. Let $f(x) = x^3 - 6; \; x_{n+1} = 2\left(\dfrac{x_n}{3} + \dfrac{1}{x_n^2}\right); \; x_1 = 1, \; x_2 \approx 2.666666, \; x_3 \approx 2.0590276, \; x_4 \approx 1.8444283, \; x_5 \approx 1.8175229$

9. Let $f(x) = x^3 - 2x - 5; \; f(2) = -1, f(3) = 16$; thus, $f(x)$ has a root between 2 and 3; $x_{n+1} = x_n - \dfrac{x_n^3 - 2x_n - 5}{3x_n^2 - 2}; \; x_1 = 2,$

$x_2 = 2.1, \; x_3 \approx 2.094568, \; x_4 \approx 2.0945515$

11. $\cos x = x$ for $x \approx 0.7391$

13. $E = 0.5160727; \; \theta - 0.017 \sin \theta - 0.5160727 = 0; \; \theta_{n+1} = \theta_n + \dfrac{0.017 \sin \theta_n + 0.5160727 - \theta_n}{1 - 0.017 \cos \theta_n}$; using $\theta_1 = 0$ gives

$\theta_4 = \theta_5 \approx 0.5245872$ radian, or about $30°$

15. The correct root rounded to three decimal places is 0.749. Using $x_1 = 0.7$, Newton's method produces this in two steps. Successive substitution requires 22 steps.

SECTION 10-3

1. $\sum\limits_{n=0}^{\infty} a_n$ **3.** $\sum\limits_{n=2}^{\infty} \dfrac{1}{n}$ **5.** $\sum\limits_{n=0}^{\infty} (-1)^{n+1} \left(\dfrac{1}{5}\right)^n$ **7.** $\sum\limits_{n=0}^{\infty} (-1)^n \dfrac{2n+1}{3^n}$

9. $\displaystyle\sum_{n=0}^{\infty}\frac{3^{n+1}}{2^n}$ **11.** 1   **13.** $\dfrac{123}{999}$ **15.** $\dfrac{417}{99}$ **17.** Series diverges   **19.** Converges, sum $= 12$
21. Series diverges   **23.** Converges, sum $= \frac{8}{5}$ **25.** Converges, sum $= \frac{1}{4}$ **27.** \$50,500.00   **29.** 180 feet
31. About 52.63 milligrams

SECTION 10-4

1. $1 + 3x + 3x^2 + x^3$ **3.** $x - \dfrac{x^2}{2!} + \dfrac{2x^3}{3!} - \dfrac{6x^4}{4!} + \cdots$ **5.** $1 - x + \dfrac{x^2}{2!} - \dfrac{x^3}{3!} + \dfrac{x^4}{4!} - \cdots$

7. $x + x^2 + \dfrac{x^3}{2!} + \dfrac{x^4}{3!} + \cdots$ **9.** $1 + x + x^2 + x^3 + x^4 + \cdots$ **13.** $x - 2x^3 + \dfrac{2x^5}{3} - \cdots$

15. $x^2 - x^3 + \dfrac{x^4}{2!} - \dfrac{x^5}{3!} + \cdots$ **17.** 0   **19.** $1 - (x-1) + (x-1)^2 - (x-1)^3 + \cdots$

REVIEW PROBLEMS (SECTION 10-6)

1. $\displaystyle\lim_{k\to\infty}\frac{(-1)^k}{k} = 0$ **3.** Sequence diverges   **5.** Sequence diverges   **7.** $\dfrac{220}{6}$

9. 1   **11.** **a.** $\dfrac{111}{99}$ **b.** $\dfrac{2122}{990}$ **13.** \$555,555.56 in total deposits

15. Let $f(x) = e^x - x - 2$; $f(0) = -2$, $f(2) > 0$; the positive root s satisfies $1.140 \le s \le 1.148$; $s \approx 1.14$ **17.** $e^{-3} \approx 1.375$

CHAPTER 11   SECTION 11-1

1. $\{1, 2, 3, 4, 5, 6, 7, 8, 9, 10\}$ **3.** $\{1, 2, 3, 4\}$ **5.** $\{(x, y) | 0 \le x \le 1, 0 \le y \le 1\}$ **7.** $\{2, 4, 6, 8, 10\}$
9. **a.** $\{(1, 1), (1, 2), (1, 3), (1, 4), (1, 5), (1, 6), (2, 1), (2, 2), (2, 3), (2, 4), (2, 5), (2, 6), (3, 1), (3, 2), (3, 3), (3, 4), (3, 5),$
     $(3, 6), (4, 1), (4, 2), (4, 3), (4, 4), (4, 5), (4, 6), (5, 1), (5, 2), (5, 3), (5, 4), (5, 5), (5, 6), (6, 1), (6, 2), (6, 3), (6, 4),$
     $(6, 5), (6, 6)\}$
b. $\{(1, 4), (2, 3), (3, 2), (4, 1)\}$
11. **a.** $\{(x, y, z) |$ each of x, y, z is a nut, bolt, or washer$\}$
b. $\{(n, b, w), (n, w, b), (b, n, w), (b, w, n), (w, n, b), (w, b, n)\}$, where n = nut, b = bolt, w = washer   **13.** $\frac{1}{9}$ **15.** $\frac{8}{9}$
17. **a.** $\frac{1}{13}$ **b.** $\frac{10}{13}$ **c.** $\frac{3}{4}$ **19.** **a.** $\frac{1717}{2000} = 0.859$ **b.** $\frac{1674}{2000} = 0.837$
21. **a.** $\frac{7797}{14,820} \approx 0.526$ **b.** $\frac{2954}{6595} \approx 0.448$ **c.** $\frac{1642}{7797} \approx 0.211$

SECTION 11-2

1. **a.–b.**

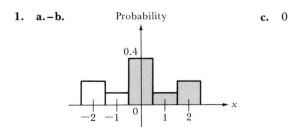

         Probability   **c.** 0

3. \$0.1875   **5.** $17(\frac{10}{156}) + 18(\frac{12}{156}) + \cdots + 23(\frac{15}{156}) \approx 20.2$ miles/gallon
7. $6(\frac{4}{50}) + 7(\frac{8}{50}) + \cdots + 11(\frac{3}{50}) = 8.46$ feet

9. $117(\frac{3}{25}) + 118(\frac{5}{25}) + 119(\frac{4}{25}) + 120(\frac{5}{25}) + 121(\frac{3}{25}) + 122(\frac{4}{25}) + 124(\frac{1}{25}) = 119.64$ pounds
11. $5(0.65)/0.35 \approx \$9.29$
13. Expected revenue without treatment $= 42{,}000(0.3) + 60{,}000(0.7) = \$54{,}600$;
revenue with treatment $= \$54{,}000$; no
15. a.

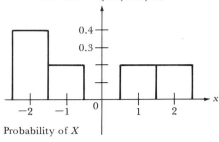

Probability of X   Probability of Y

b. $E(X) = (-2)(0.4) + (-1)(0.2) + 1(0.2) + 2(0.2) = -0.4$,
$\text{Var}(X) = (-2 + 0.4)^2(0.4) + (-1 + 0.4)^2(0.2) + (1.4)^2(0.2) + (2.4)^2(0.2) = 2.64$;
$E(Y) = (-2)(0.1) + (-1)(0.2) + 1(0.2) + 2(0.3) = 0.4$,
$\text{Var}(Y) = (-2 - 0.4)^2(0.1) + (-1 - 0.4)^2(0.2) + (0 - 0.4)^2(0.2) + (1 - 0.4)^2(0.2) + (2 - 0.4)^2(0.3) = 1.84$

SECTION 11-3

1. $P(X = 0) = \frac{1}{2}, P(X = 2) = \frac{1}{8}, P(X = 5) = \frac{1}{64}$ **3.** $\frac{2}{3}$ **5.** $P(Y \le 3) = 0.9919, P(Y \ge 4) = 1 - P(Y \le 3) = 0.0081$
7. $a = \frac{3}{2}$ **9.** $P(k = 1) = e^{-12}0.12 \approx 0.0000737, P(k \le 3) = e^{-12}\left(1 + 12 + \dfrac{12^2}{2} + \dfrac{12^3}{6}\right) \approx 0.00229$
11. $1 - e^{-0.8} \approx 0.55$ **13.** $e^{-3.5} + e^{-3.5}(3.5) \approx 0.14$ **15.** $1 - e^{-5}(1 + 5 + \frac{25}{2}) \approx 0.88$

SECTION 11-4

1. $0 \le d \le \sqrt{2}$ **3.** $f(x) \ge 0, \displaystyle\int_0^2 \frac{1}{2}x \, dx = 1$ **5.** $f(x) \ge 0, \displaystyle\int_{\pi/2}^{\pi} \sin x \, dx = 1$ **7.** $a = \frac{1}{20}$ **9.** $a = 12$
11. $F(x) = 3x^2 - 2x^3, 0 \le x \le 1$ **13.** $F(x) = e^x - 2, \ln 2 \le x \le \ln 3$ **15.** $f(x) = \frac{1}{6}, 2 \le x \le 8$
17. $f(x) = 2e^{2(x-1)}, 0 \le x \le 1 + \frac{1}{2}\ln(1 + e^{-2})$ **19. a.–b.**

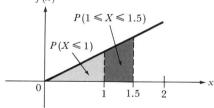

21. $P(X \le 2) = \displaystyle\int_{1/2}^2 \frac{1}{2x} \, dx = \frac{1}{2}(\ln 4 - \ln 1) \approx 0.69$ **23.** $P(X \le 0.5) = 0.5$ **25.** $F(5) = 0.5$
27. a. $F(2) = \frac{1}{8}$ **b.** $P(X > 4) = 1 - F(4) = \frac{5}{8}$ **c.** $F(5) - F(2) = \frac{3}{8}$ **29.** $k = 2.8$ **31.** $k = 6.6$
33. $E(X) = \frac{4}{3}, \text{Var}(X) = \frac{2}{9}$ **35.** $E(X) = -1 + 3\ln 3 - 2\ln 2 \approx 0.91, \text{Var}(X) \approx 0.013$

REVIEW PROBLEMS (SECTION 11-6)

1. {0, 1, 2, 3, 4, 5, 6, 7, 8, 9, 10, 11, 12}   **3.** $\int_0^1 x^2\, dx = \frac{1}{3}$

5. $P(2 \le X \le 4) = 0.4$ Probability

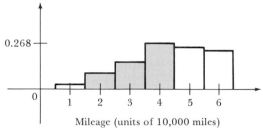

Mileage (units of 10,000 miles)

7. $E(X) = 1.05$, $\text{Var}(X) \approx 2.59$ **9.** $E(X) = \frac{1}{2}$
11. a. $P(3 \le X \le 4) = F(4) - F(3) = \frac{37}{117}$ **b.** $f(x) = \frac{3}{117}x^2$ **13.** $P(X \ge 30) \approx 0.68$; about 68%
15. $P(k > 3) \approx 0.14$

CHAPTER 11   SUPPLEMENT

1. $E(X) = \dfrac{a+b}{2}$, $\text{Var}(X) = \dfrac{(b-a)^2}{12}$ **3.** $\frac{1}{2}$ **5.** About 25   **7.** $F(x) = 1 - e^{-\lambda x}$ **9.** $P(2 \le X \le 4) \approx 0.018$

11. $P(X > 1000) \approx 0.29$ **13.** $P(X < 10{,}000) \approx 0.28347$; $0.28347 \times 10{,}000 \approx 2835$; no, do not issue recall
15. $\mu = 0$, $\sigma = 4$ **17.** $\mu = 1$, $\sigma = 3$ **19. a.** 0.0139   **b.** 0.84   **c.** 0.93   **21.** $P(X > 75{,}000) \approx 0.0475$
23. $P(X > 8.5) \approx 0.0228$ **25.** Score > 602 **27.** $P(X < 200) \approx 0.3446$ **29.** $\sigma = 3$

CHAPTER 12   SECTION 12-1

1.

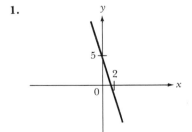

3.

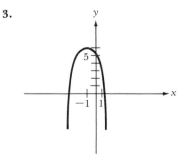

5.

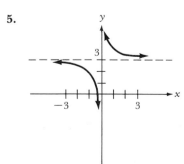

7.
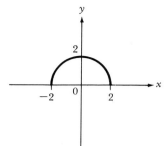

9.
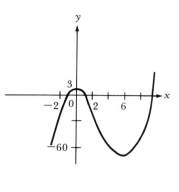

11.
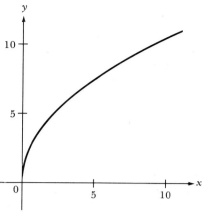

13.
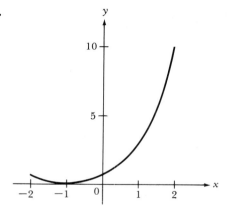

15.
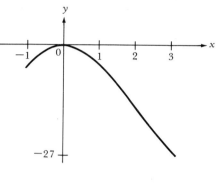

17.
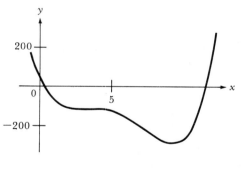

19.
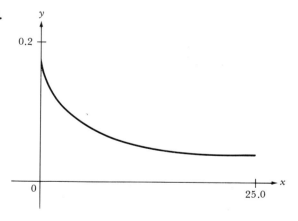

SECTION 12-2

1. 7   **3.** 8   **5.** ∞   **7.** −10   **9.** $\frac{3}{4}$ **11.** $-\frac{2}{9}$ **13.** 12   **15.** Undefined   **17.** 0   **19.** Undefined

SECTION 12-3

1. 2.094552   **3.** -0.6299609 **5.** From $x_1 = 1$: ≈ 0; from $x_1 = -5$: -6; from $x_1 = -10$: -6 **7.** 2.236068
9. 1.817121   **11.** Minimum: $x \approx -0.9992106$, $f(x) \approx 0$; maximum: $x \approx 1.996159$, $f(x) \approx 8.976971$
13. Minimum: $x \approx 2.99616$, $f(x) \approx -26.96539$; maximum: $x \approx 0.5269064$, $f(x) \approx -1.519497$
15. Minimum: $x \approx 5.99616$, $f(x) \approx 0.1715446$; maximum: $x \approx 2.003841$, $f(x) \approx 0.6645407$

SECTION 12-4

1. INTEGRAT 10: 10.395, 100: 9.135446; TRAPEZD 10: 9.045001, 100: 9.00045
3. INTEGRAT 10: 21.68729, 100: 22.41314; TRAPEZD 10: 22.56229, 100: 22.50064
5. INTEGRAT 10: 14.595, 100: 14.95545; TRAPEZD 10: 15.045, 100: 15.00045
7. INTEGRAT 10: 42.96408, 100: 36.6805; TRAPEZD 10: 36.16408, 100: 36.0005
9. INTEGRAT 10: -31.78771, 100: -31.99787; TRAPEZD 10: -31.78771, 100: -31.99787
11. Area ≈ 35.99996 **13.** Area ≈ 21.33332 **15.** Area ≈ 14.6666

SECTION 12-5

1.
```
ENTER INITIAL VALUES FOR X,Y  1,.01
ENTER THE STEP SIZE  1
ENTER THE MAXIMUM VALUE OF X  10

X                Y
2                .61
3                37.21
4                2269.81
5                138458.4
6                8445962
7                5.152037E+08
8                3.142742E+10
9                1.917073E+12
10               1.169414E+14


ENTER INITIAL VALUES FOR X,Y  1,.01
ENTER THE STEP SIZE  .5
ENTER THE MAXIMUM VALUE OF X  10

X                Y
1.5              .185
2                3.4225
2.5              63.31626
3                1171.351
3.5              21669.99
4                400894.8
4.5              7416554
5                1.372062E+08
5.5              2.538316E+09
6                4.695884E+10
6.5              8.687386E+11
7                1.607166E+13
7.5              2.973258E+14
8                5.500527E+15
8.5              1.017598E+17
9                1.882555E+18
9.5              3.482728E+19
10               6.443045E+20
```

3.
```
ENTER INITIAL VALUES FOR X,Y  0,1
ENTER THE STEP SIZE  .2
ENTER THE MAXIMUM VALUE OF X  1

X                Y
.2               .836
.4               .7269856
.6               .6577766
.8               .6193624
1                .6069751


ENTER INITIAL VALUES FOR X,Y  0,1
ENTER THE STEP SIZE  .1
ENTER THE MAXIMUM VALUE OF X  1

X                Y
.1               .9095
.2               .8354667
.3               .775146
.4               .7263894
.5               .6875276
.6               .6572763
.7000001         .634666
.8000001         .6189898
.9000001         .6097668
1                .606718
```

5.
```
ENTER INITIAL VALUES FOR X,Y   0,.25
ENTER THE STEP SIZE   .5
ENTER THE MAXIMUM VALUE OF X   3

X                Y
.5               .3532715
1                .4726252
1.5              .5950735
2                .7061979
2.5              .7965529
3                .8639251


ENTER INITIAL VALUES FOR X,Y   0,.25
ENTER THE STEP SIZE   .25
ENTER THE MAXIMUM VALUE OF X   3

X                Y
.25              .29953
.5               .3542679
.75              .4131332
1                .4746038
1.25             .5368526
1.5              .5979574
1.75             .6561351
2                .7099411
2.25             .7583891
2.5              .8009764
2.75             .8376325
3                .868619
```

7.
```
ENTER INITIAL VALUES FOR X,Y   1,2
ENTER THE STEP SIZE   .2
ENTER THE MAXIMUM VALUE OF X   2

X                Y
1.2              3.1
1.4              4.471428
1.6              6.159438
1.8              8.209203
2                10.66591
```

9.
```
ENTER INITIAL VALUES FOR X,Y   1,0
ENTER THE STEP SIZE   .1
ENTER THE MAXIMUM VALUE OF X   2

X                Y
1.1              .1045455
1.2              .2182163
1.3              .3402471
1.4              .4699914
1.5              .6068956
1.6              .7504803
1.7              .9003265
1.8              1.056065
1.9              1.217367
2                1.383938
```

SECTION 12-6

1. $x \approx 1.50, y \approx -1.52, f(x, y) \approx -2.25$ **3.** $x \approx 5.00, y \approx -2.00, f(x, y) \approx -9.00$ **5.** $x \approx 2.00, y \approx 1.00, f(x, y) \approx 6.00$

SECTION 12-7

1.

VALUE OF RANDOM VARIABLE	NO. OF OBSERVATIONS	PROBABILITY
17	10	6.410256E-02
18	12	7.692308E-02
19	30	.1923077
20	41	.2628205
21	26	.1666667
22	22	.1410256
23	15	9.615385E-02

EXPECTED VALUE = 20.19872

VARIANCE = 2.633588

3.

VALUE OF RANDOM VARIABLE	NO. OF OBSERVATIONS	PROBABILITY
6	4	.08
7	8	.16
8	11	.22
9	18	.36
10	6	.12
11	3	.06

EXPECTED VALUE = 8.46

VARIANCE = 1.6484

5.
```
VALUE OF RANDOM VARIABLE    NO. OF OBSERVATIONS          PROBABILITY
-------------------------------------------------------------------------
     117                         3                           .12
     118                         5                           .2
     119                         4                           .16
     120                         5                           .2
     121                         3                           .12
     122                         4                           .16
     124                         1                           .04

EXPECTED VALUE =  119.68

VARIANCE =  3.3376
```

7.
```
ENTER THE VALUE FOR A   .25
ENTER THE VALUE FOR B   .5
ENTER THE NUMBER OF RECTANGLES  1000

SIZE OF SUBINTERVAL IS         .00025

THE VALUE OF THE INTEGRAL IS            .3437969

ENTER THE VALUE FOR A   .5
ENTER THE VALUE FOR B   1
ENTER THE NUMBER OF RECTANGLES  1000

SIZE OF SUBINTERVAL IS         .0005

THE VALUE OF THE INTEGRAL IS            .4996248
```

9.
```
ENTER THE VALUE FOR A   3
ENTER THE VALUE FOR B   8
ENTER THE NUMBER OF TRAPEZOIDS  1000

SIZE OF SUBINTERVAL IS         .005

THE VALUE OF THE INTEGRAL IS            4.945174E-02
```

11.
```
ENTER THE VALUE FOR A   -.3
ENTER THE VALUE FOR B   .3
ENTER THE NUMBER OF TRAPEZOIDS  1000

SIZE OF SUBINTERVAL IS         6.000001E-04

THE VALUE OF THE INTEGRAL IS            .2358231
```

13.
```
ENTER THE VALUE FOR A   -3
ENTER THE VALUE FOR B   0
ENTER THE NUMBER OF TRAPEZOIDS  1000

SIZE OF SUBINTERVAL IS         .003

THE VALUE OF THE INTEGRAL IS            .3413448
```

15.
```
ENTER THE VALUE FOR A   -9
ENTER THE VALUE FOR B   4
ENTER THE NUMBER OF TRAPEZOIDS  1000

SIZE OF SUBINTERVAL IS         .013

THE VALUE OF THE INTEGRAL IS            1
```

SECTION 12-8

1.

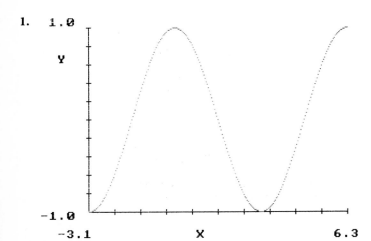

3.

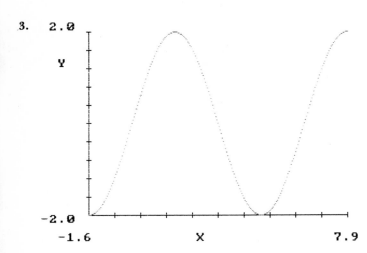

5.

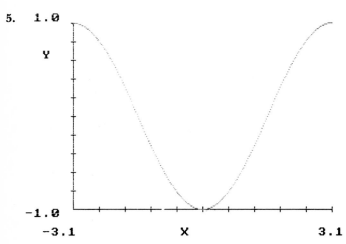

7.
```
TYPE 1 IF THIS IS A MINIMIZATION PROBLEM
TYPE 2 IF THIS IS A MAXIMIZATION PROBLEM
? 1
ENTER THE VALUE OF A  0
ENTER THE VALUE OF B  6.28318
ENTER THE SIZE OF THE FINAL INTERVAL  .01
ESTIMATE OF OPTIMAL POINT

X =          3.140312
F(X) =       -.9999992
```

9.
```
ENTER THE VALUE FOR A  0
ENTER THE VALUE FOR B  3.14159
ENTER THE NUMBER OF TRAPEZOIDS  1000

SIZE OF SUBINTERVAL IS       3.14159E-03

THE VALUE OF THE INTEGRAL IS              1.999999
```

ANSWERS TO ALGEBRA REVIEW EXERCISES

1: VARIABLES AND CONSTANTS
a. Variable: travel time; constants: 300 miles, m miles per hour
b. Variables: man's age, daughter's age; constants: age multiple, 10; sum of ages, 33
c. Variable: gallons of gasoline; constants: miles per gallon, 22; monthly driving distance, 500 miles
d. Variables: x square miles, value of lost timber; constant: value per square mile, $3000
e. Variables: length of one side, cost of box; constant: cost per square foot, $0.30

2: ALGEBRAIC EXPRESSIONS
a. $10x^2 + 4x$ **b.** $6x\sqrt{x} + 24\sqrt{xy}$ **c.** $14x^2 - 12x$ **d.** $7x^2 + 3x$ **e.** $4x^2 + 6xy - 2y^2$

3: MULTIPLYING ALGEBRAIC EXPRESSIONS
a. $24x^2 + y^2 + 10xy$ **b.** $\dfrac{3y^4}{x} + 6y^3 + \dfrac{y\sqrt{x}}{x} + 2\sqrt{x}$ **c.** $-xy^2 + 2xy - x$ **d.** $x^2 - y^2$ **e.** $x - 2\sqrt{x} - 3$

4: RECTANGULAR COORDINATES
a.

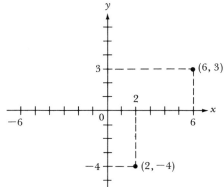

b.

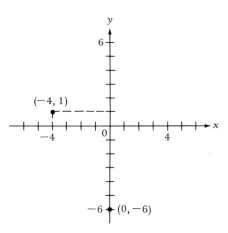

c.

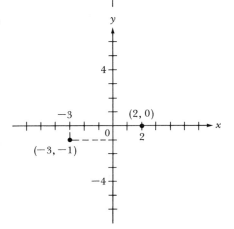

5: SIMPLIFYING ALGEBRAIC EXPRESSIONS

a. $\dfrac{x^3}{7}$ **b.** $\dfrac{16y^2}{y+8}$ **c.** $\dfrac{y^2}{3}$ **d.** $\dfrac{x+y}{y^3+4y}$ **e.** $\dfrac{2x+h}{h+2}$

6: ADDING AND DIVIDING FRACTIONAL TERMS

a. $\dfrac{xy+2x+1}{y^2}$ **b.** $\dfrac{17x^2-x+2}{3x+3x^2}$ **c.** $\dfrac{3x^2+2y^2+9xy}{x^2y+xy^2}$ **d.** $\dfrac{6x^2}{4x^2+21x+5}$ **e.** $\dfrac{-3}{5(x+5)}$

7: RATIONALIZING EXPRESSIONS INVOLVING RADICALS

a. $\dfrac{2-x}{\sqrt{2x}-x}$ **b.** $\dfrac{1}{\sqrt{x^2+3}+2}$ **c.** $\dfrac{-2x+1}{1+\sqrt{x+2x^2}}$ **d.** $\dfrac{-1}{3-\sqrt{x}}$ **e.** $\dfrac{x}{x+\sqrt{x}}$

8: EXPONENTS AND ROOTS

a. $\frac{3}{2}x^{-1/2}$ **b.** $8x^{-2/3}$ **c.** x^2 **d.** $\sqrt[3]{x^2}$ **e.** $\dfrac{1}{\sqrt{(x+2)^3}}$

9: THE QUADRATIC FORMULA

a. 2 or $-\frac{1}{2}$ **b.** There are no real roots. **c.** $\dfrac{-1\pm\sqrt{10}}{3}$ **d.** 4 **e.** 0 or -1

10: DIFFERENCE OF TWO SQUARES

a. $36x^2-25$ **b.** $(3+7x)(3-7x)$ **c.** $(2x^2-3y)(2x^2+3y)$

11: INTERVALS

a. $\{x|2<x<7\}, \{x|3\le x\le 4\}, \{x|0<x\le 1\}$ **b.** $[-3,5], (2,\infty), (0,1]$ **c.** $[-1,18), \{x|-1\le x<18\}$

12: SOLVING ONE-VARIABLE EQUATIONS

a. $x=1$ **b.** $x=0$ or $x=\frac{1}{2}$ **c.** $x=\pm 3$ **d.** $x=\frac{3}{2}$ **e.** $x=-\frac{15}{2}$

13: SOLVING QUADRATIC EQUATIONS BY FACTORING

a. $x=7, x=3$ **b.** $x=\frac{5}{3}, x=\frac{1}{4}$ **c.** $x=-4, x=-\frac{1}{2}$ **d.** $x=-2$ **e.** $x=\frac{3}{2}, x=-\frac{5}{3}$

14: FACTORING CUBIC EXPRESSIONS

a. $(5-x)(25+5x+x^2)$ **b.** $(x-1)(x-2)(x-3)$ **c.** $(x+1)(x-1)(x-2)$ **d.** $(x-1)(x-1)(x-1)$
e. $(x+2)^2(x-1)$

15: COMPOSITION OF FUNCTIONS

a. $f(g(x))=2/x^2; g(g(x))=1/2x^2$ **b.** $f(g(x))=x^2; g(f(x))=x^2$ **c.** $f(x)=1/x; g(x)=x+2$
d. $h(x)=x^3; g(x)=x^2+2x-5$ **e.** $f(x)=\sqrt[3]{x}; g(x)=2x+1$

16: SUMMATION NOTATION

a. $3(2)+3(3)+3(4)+3(5)=42$ **b.** $f(x_1)+f(x_2)+f(x_3)$ **c.** $(-3)^2+(-2)^2+(-1)^2+(0)^2+(1)^2=15$
d. $2+4+1+6-1=12$ **e.** $-1+1+2=2$

17: LAWS OF EXPONENTS

a. y^2 **b.** 6^{y-3y} **c.** 4^{2x^2} **d.** $b^{3x}b^{-y}=\dfrac{b^{3x}}{b^y}$ **e.** $b^{-x}b^2=\dfrac{b^2}{b^x}$

18: SOLVING SYSTEMS OF EQUATIONS

a. $x=3, y=-1$ **b.** $x=1, y=1$ **c.** $(1,-2), (-1,2)$ **d.** $w=0, x=2, y=\frac{1}{2}$ **e.** $x=2, y=4, z=3$

DIFFERENTIATION RULES

Power rule

$$\frac{d}{dx}\,x^p = px^{p-1}$$

Constant rule

$$\frac{d}{dx}\,[cf(x)] = cf'(x)$$

Sum/difference rule

$$\frac{d}{dx}\,[f(x) \pm g(x)] = f'(x) \pm g'(x)$$

Product rule

$$\frac{d}{dx}\,[f(x)g(x)] = f(x)g'(x) + g(x)f'(x)$$

Quotient rule

$$\frac{d}{dx}\left[\frac{f(x)}{g(x)}\right] = \frac{g(x)f'(x) - f(x)g'(x)}{[g(x)]^2}$$

Reciprocal rule

$$\frac{d}{dx}\left[\frac{1}{g(x)}\right] = \frac{-g'(x)}{[g(x)]^2}$$

General power rule

$$\frac{d}{dx}\,[g(x)]^p = p[g(x)]^{p-1}g'(x)$$

Chain rule for $y = f(g(x))$

$$\frac{dy}{dx} = \frac{df}{du} \cdot \frac{du}{dx}$$

where $u = g(x)$, $y = f(u)$

Rules for exponential functions

$$\frac{d}{dx}\,e^x = e^x$$

$$\frac{d}{dx}\,e^{g(x)} = g'(x)e^{g(x)}$$

$$\frac{d}{dx}\,b^x = (\ln b)b^x$$

Rules for logarithmic functions

$$\frac{d}{dx}\,\ln x = \frac{1}{x}$$

$$\frac{d}{dx}\,\ln g(x) = \frac{g'(x)}{g(x)}$$

$$\frac{d}{dx}\,\log_b x = \frac{1}{x \ln b}$$

$$\frac{d}{dx}\,\log_b g(x) = \frac{g'(x)}{g(x) \ln b}$$